GREENLAND

GREENLAND

EDITED BY NIELS ELERS KOCH

ROWMAN & LITTLEFIELD
Lanham • Boulder • New York • London

Editor-in-Chief:
Niels Elers Koch, Professor, dr.agro., dr.h.c.

Published by Rowman & Littlefield
An imprint of The Rowman & Littlefield Publishing Group, Inc.
4501 Forbes Boulevard, Suite 200, Lanham, Maryland 20706
www.rowman.com

86-90 Paul Street, London EC2A 4NE, United Kingdom

Design: Imperiet.dk
Graphic design and repro: Narayana Press
Printed in USA 2023

Front-page photo: Ilulissat in May
© Aningaaq Rosing Carlsen/Visit Greenland, 2021

The editing of this book took place in the period 2020–2023 and was completed in March 2023

British Library Cataloguing in Publication Information Available

Library of Congress Cataloging-in-Publication Data

ISBN: 978-1-5381-8124-9 (cloth)
ISBN: 978-1-5381-8125-6 (epub)
ISBN: 978-87-7181-100-1 (entire work)
ISSN: 2445-6020

Printed in India

TABLE OF CONTENTS

EDITORIAL PREFACE

This volume describes Greenland. The volume constitutes part of the description of Denmark in *Trap Denmark. Trap Denmark* is based on a solid foundation and long-standing and unique tradition not seen elsewhere in the world. In 1860, the Cabinet Secretary of King Frederik VII of Denmark, J.P. Trap, completed the publication of the first edition of the Statistical-Topographical Description of the Kingdom of Denmark. The work briefly described Greenland in two pages under the heading 'Dependencies' together with the Faroe Islands, Iceland and the West Indies. Since then, *Trap Denmark*, as the work was later called, has been the authoritative source of topographical knowledge about the Kingdom of Denmark. *Trap Denmark* has subsequently been published four times: The 2nd edition in the period 1872-1879, the 3rd edition in the years 1898-1906, the 4th edition in the years 1920-1932 and the 5th edition in the period 1953-1972. The 2nd-4th editions had no descriptions of Greenland, but as part of the 5th edition, an independent 689-page book with maps of Greenland was published in 1970.

Since the publication of the 5th edition, the Danish and Greenlandic societies have changed radically. Therefore, a new edition of *Trap Denmark* was needed, which collects topographical and statistical information about the Kingdom of Denmark in a manageable way and disseminates it in a modern language both in print and digitally. By 2014, enough funds had been raised for the task of describing Denmark to begin. In mid-2016, it was also decided to try to raise funds for a description of Greenland and the Faroe Islands. In this connection, two groups were set up to prepare the description of Greenland: A working group consisting of Inge Høst Seiding, Karl Brix Zinglersen, Andreas Westergaard-Nielsen and Stefanie Høy Brink and a steering group consisting of Daniel Thorleifsen, Morten Meldgaard, Bo Elberling and Niels Elers Koch.

By spring 2020, enough funds had been raised for a description of Greenland and the Faroe Islands. The volume on Greenland is called *Trap Greenland* and is published both in Greenlandic, Danish and English in three independent books and digitally in the three languages at trap.lex.dk. The funds for this come in particular from the Danish foundations behind *Trap Denmark*: Queen Margrethe's and Prince Henrik's Foundation, The A.P. Møller and Chastine Mc-Kinney Møller Foundation, The Augustinus Foundation, the Aage and Johanne Louis-Hansen Foundation and the G.E.C. Gad Foundation. However, as a precondition, these foundations wanted Greenland to be described primarily by Greenlanders to Greenlanders, and for some co-funding to come from Greenland. The Greenlandic institutions, foundations and other sponsors that have contributed to *Trap Greenland* are shown in the overview at the bottom of page 4.

We are extremely grateful for the support and for the great help we have received from all our collaborators in Greenland and Denmark.

The work is a description of the Kingdom of Denmark. In consultation with national and local researchers and experts, the editors have selected the key information – from the most common to the most special facts. Since it must be possible to read the individual sections separately, there will be repetitions of important contexts. The content is created by a large team of researchers and experts and has been quality assured and edited by professional, text and image editors. In addition, *Trap Denmark's* Professional Committee has been quality assured and established the overall editorial principles and criteria that can be seen at trap.lex.dk. This is also the place to find definitions and explanations of the technical terms used.

In addition, this 6th edition consists of an extensive digital universe with additional topographic information from collaborators, see trap.lex.dk. The digital universe makes it possible to get relevant information about any particular place in Greenland. It is neither a copy of the book nor a digital book, but a dynamic, searchable and more personal tool that will inspire and guide users on their digital journey exploring the great material about Greenland in Greenlandic, Danish and English. *Trap Greenland's* digital universe is a national gift from the foundations and the other sponsors behind the project – available for free to everyone.

Despite all efforts, mistakes cannot be avoided. Trap Greenland's editorial board will therefore appreciate any information about errors and mistakes. Please submit such information to info@trap.dk.

Discover Greenland – with Trap.

Trap Greenland's editors

PATRON'S INTRODUCTION

STINE HEILMANN

It is with great pleasure that I have agreed to serve as patron of the new edition of *Trap Greenland*, which is published as a volume in the work *Trap Denmark*. The first edition of the work was published in 1858-1860 and has subsequently been revised four times, most recently in 1953-1972. This new and 6th edition has been rewritten and is supplemented by a digital version.

Both the printed work and the digital version provide a comprehensive description of Greenland with a myriad of information about the country and every town in the country. With the digital version, we get a practical and live reference tool that can be continuously updated and which is available no matter where you are in Greenland, provided there is mobile coverage.

In a globalized world, it is important to know one's country and hence its preconditions. This way, it is easier to put into words what the Unity of the Realm has to offer, and how and why Greenlandic, Danish and Nordic culture serve as inspiration for many other countries.

Trap Greenland has been created in an interaction between Greenlandic and Danish experts, so professionals across the Unity of the Realm have helped to create a contemporary and detailed picture of Greenland available to everyone thanks to generous donations from several foundations in Denmark and foundations and other sponsors in Greenland.

Enjoy the book.

Crown Prince Frederik of Denmark

MORLIGHEM M.FL. 2021: ICEBRIDGE BEDMACHINE GREENLAND

GREENLAND

Greenland is an autonomous country in the Unity of the Realm – a national community with Denmark and the Faroe Islands. It is the world's largest island with an area of 2,486,000 km^2, a distance from north to south of about 2,670 km and a total coastline of more than 44,000 km. About 80 % of the country is covered by the ice sheet. At the same time, it is the world's most sparsely populated area, with a total population of only about 56,000 people. Along the coasts, an area of well over 410,000 km^2 is ice-free, and this is where the population lives in towns and settlements. Today, more than 70 % live in the seven largest towns: the capital Nuuk and the towns of Sisimiut, Ilulissat, Qaqortoq, Aasiaat, Maniitsoq and Tasiilaq, but originally, Inuit lived in small hunting communities. Greenland's neighbours are Canada to the west and northwest, Russia on the other side of the North Pole, and to the southeast is Iceland and then the Faroe Islands, Norway, Denmark and the rest of Europe.

Greenland has a long formation history, which includes virtually all types of geological processes throughout Earth's geological history. From cracks in the contiguous supercontinent of Pangea 175 million years ago to giant layers of lava from Earth's interior ten million years ago, which have formed an up to 7 km thick basalt layer. But first and foremost, it is ice that characterises Greenland. The ice sheet has covered the island for about seven million years, but its prevalence has varied due to changing glacials and warmer interglacials. Glaciers and ice streams from the ice sheet reach out towards the sea, forming large icebergs such as at Qeqertarsuup Tunua (in Disko Bay). The ice sheet has long been the subject of interest by international researchers, and the link between increased melting and global warming has only strengthened interest. Greenland has mineral deposits of coal, lead and zinc, and attempts have also been made to extract platinum, gold and uranium. There are polar bears and muskoxen in parts of Greenland, but it is especially life in the sea that has given the Inuit the gifts

Photo pages 8-9. On an August day in Kulusuk, southeast Greenland, a small group of children jumped into the water. The settlement of roughly 240 inhabitants is located on the rocky island of the same name. Kulusuk Airport is the main gateway to the Tasiilaq district.

JONATHAN NACKSTRAND/AFP PHOTO/RITZAU SCANPIX, 2019

GREENLAND AT A GLANCE
Total area 2.486.000 km^2
Covered with ice 81 %
Highest point approx. 3.700 m (Gunnbjørn Mountain)
Coastline 44.087 km

Population
1950 23.642 persons
1980 49.773 persons
2021 56.421 persons

Population density ice-free areas 0,14 persons per km^2
Population density throughout Greenland (2021) 0,025 persons per km^2

Average age (2021) 36 years

Average income before tax per person over 14 years of age, 2020 263.429 DKK

Taxable income, average per person 217.295 DKK

Municipalities 5
Dioceses 1
Parishes 17

Kalaallit Nunaat
Grønland
17,00
Glædelig Jul
Juullimi pilluaritsi
Maria Bach Kreutzmann pinx.
2021

THE MEANING OF THE PLACE NAME

The name is known from 1075 when Adam of Bremen is the first to mention a *Gronland*. In Norse sources from the 12th century (*Islendingabok and Landnamabok*) and the 13th century (*Grœnlendinga Saga and Erik the Red Saga*), and in preserved later transcripts thereof, the name appears in the forms *Grœnland* and *Grønland*. In a Danish language context, the name appears as *Grønland* in the 16th century. The name is a compound of the Norse adjective *grǿnn* »green« and the noun *land* »land area, large island«, with the overall meaning »the green land area« or »the large green island«. The choice of name is explained explicitly in *Islendingabok*, which states that when Erik the Red, exiled from Iceland, discovered and settled in Greenland in the 980s, he decided to give the place a good and alluring name, so that others would also want to move there. In addition to the name's real estate value, the somewhat milder climate of the Viking Age is also thought to have helped justify the name, as the southern tip of Greenland was lush and grassy and allowed growing grain in the summer months. In the clerical administration of the Middle Ages, Greenland constituted its own diocese with the name *Garðar*, which is a plural form of the West Norse noun *garð* »farm«, named after the place in the Eastern Settlement where the episcopal residence was located (near what is now Igaliku).

In modern Greenlandic, Greenland is called *Kalaallit Nunaat* »Land of the Kalaallit people«, named after the largest Inuit ethnic group in the country. Older sources also use the name *Inuit Nunaat*, Land of the Inuit.

The 2021 Christmas stamp is designed by the young artist Maria Bach Kreutzmann. It combines tradition and renewal; the women's hand and face tattoos, beaded collar, earrings and hairtop are drawn in a sharp graphic line underpinned by a contrasting and rich colour palette.

TUSASS, 2021

of nature and thus their livelihood: whales, seals, fish and shrimp.

The first people immigrated to the country from Siberia, Alaska and Arctic Canada about 4,500 years ago. Then Icelandic Vikings arrived, who settled as peasants in the southwest and adopted the name Greenland. In the late 12th century, Inuit immigrated from Alaska to Greenland, laying a foundation for today's population. In the Nordic view, Greenland was part of the kingdom, and therefore in 1721 the missionary Hans Egede arrived, supported by King Frederik IV and by Norwegian traders, and now began a concrete colonisation of Greenland with a Christian mission and the construction of trading posts along the west coast of Greenland.

The key to the colony project was blubber from Inuit whale and seal hunting. Originally, the blubber was used by the family and the settlement, and the catch was an important part of Inuit culture. Now the blubber was made into a commodity and part of international trade. By the late 19th century, the market lost interest in blubber; fishing industry was now to replace seal hunting, and the country faced a massive wave of modernisation according to the Danish role model, focusing on urban development. Lots of Danish labour was now to be used in construction, education and administration in particular.

The Danisation, as it was called, brought improvements in some areas, but also triggered frustrations, resistance and desire for secession. A Greenlandisation sprouted with greater interest in traditional Inuit culture. The rock band Sume saw the light of day and gained great popularity with songs in Greenlandic. In 1979, a referendum led to the introduction of the Home Rule, and in 2009, a self-government arrangement for Greenland replaced the home rule arrangement. The self-government covers most internal affairs and can be extended to, for example, independent police, legal system and border control, while foreign, security and defence policy are not covered, as that still falls under the Unity of the Realm.

COAT OF ARMS OF GREENLAND

A polar bear is the obvious choice for a coat of arms for Greenland, and in 1666 the polar bear was added to the royal coat of arms of Denmark as the arms of Greenland. Since 1819, the arms have been permanently depicted on the royal coat of arms of Denmark.

The coat of arms is used by the Danish authorities in Greenland with the royal crown on the escutcheon. The Government of Greenland and the Greenlandic authorities use the coat of arms without the royal crown and with the polar bear's left forepaw raised, in contrast to the Danish authorities, who use a version with the bear's right forepaw raised. The two versions of the weapon were adopted in 1987, at which point a Greenlandic wish to have the polar bear raise its left forepaw was allowed, since a polar bear always uses its left forepaw to defend itself.

Blazoning (description): Against a blue background, an upright seated silver bear with a red tongue.

Greenland is administratively divided into five regional municipalities, containing a total of 17 towns and 54 towns and settlements spread over an enormous geographical area. There are no road connections between the towns and settlements, so transport is by sailing and flying and in some places by snowmobile. In North and East Greenland, dog sledding is also used. The long distances and sparsely populated town and settlements mean that Greenland is characterised by island communities, where each town and settlement has to be self-sufficient in terms of all societal functions, such as electricity, school, health care, shopping, etc. Politically, priority is given to a welfare society combined with continued dispersed towns and settlements outside the towns. Usually there is no market platform for private companies in the small towns and settlements, which is why many services are provided by the public sector.

Under the ice at Tasiilaq in southeast Greenland, a red jellyfish (*Cyanea capillata*) moves silently through the cold water, trailing its fishing line like a long, burning veil. The red jellyfish is just one of a handful of jellyfish found in Greenland waters and the only one whose nettle cells are so powerful that they can »burn« in a painful way. They are found along the east and west coasts up to the northern part of Baffin Bay.

ALEXANDER BENEDIK/CATERS/RITZAU SCANPIX, 2017

Greenland has always been the subject of outside attention. The Danes have left their clear mark, but the Americans, too, have had a big impact. During and after World War II, in which Denmark was occupied by Germany, the United States established a number of military facilities and had thousands of soldiers in Greenland. Today, the global climate debate focuses on the melting of the Greenland ice sheet, and the location and resources of Greenland are key to the geopolitical ambitions of several nations for the Arctic. In future, global interest in Greenland will only be reinforced.

Globalisation is happening at the same time as a renewed interest in traditional culture. This is manifested, for example, in Greenlandic art and culture. The traditional drum dance and song have thus regained great popularity and is on the UNESCO World Heritage List for Intangible Cultural Heritage, while in the more recent visual arts, movements away from the national narrative can be seen. The combination is seen, for example, in Greenlandic fashion and design. Where previously, mainly products that referenced Greenlandic culture were created, the gaze is now turned to more global trends – while using the traditional symbols of the Inuit.

Greenland is important to Inuit. And to the world.

ERFALASORPUT – THE FLAG OF GREENLAND

The flag is designed by the Greenlandic painter and visual artist Thue Christiansen. He became Greenland's first Minister for Culture and Education in 1979. He has described how the flag symbolises the ice and the sun, and the red and white colours show Greenland's ties to Denmark and the Danish flag, Dannebrog.

Greenland was given home rule in 1979, at which point the question of a Greenlandic flag became relevant. Several proposals were submitted in the 1970s, and in 1980 the Greenland Government arranged a flag competition. More than 600 proposals were submitted. However, a flag committee headed by Greenland's first prime minister, Jonathan Motzfeldt (1938-2010), failed to come to a decision.

In 1984, it was decided to choose between two different designs: Thue Christiansen's red-and-white flag with the sun disk and a Nordic cross flag, green with a white cross, proposed by the Danish editor and heraldist Sven Tito Achen.

There were three main positions in the debate on the Greenlandic flag: One group wanted a flag inspired by the Dannebrog, which would be reminiscent of the Danish Realm and the Nordic collaboration. Another group wanted a flag without a cross, which would emphasise the uniqueness of Greenlanders as Inuit. Finally, some wanted the Dannebrog to remain the only flag in Greenland.

The idea of a referendum was abandoned. Instead, in 1985, the Greenland Parliament made a decision by ballot. Christiansen's flag won with 14 votes, Achen's flag received 11 votes. The new Greenlandic flag officially flew for the first time on 21 June 1985.

Today, the Erfalasorput is a national symbol that unites all Greenlanders and is seen everywhere in towns and settlements and on vessels registered in Greenland. In Denmark and the Faroe Islands, anyone can fly the Greenlandic flag without special permission.

The appearance and use of the flag is regulated by an act of parliament from 1985. An executive order from 2008 sets out a number of official flag days on which flags are flown from public buildings. The national authorities in Greenland, including the national ombudsman, the courts, the defence and the police use the Dannebrog as a swallow-tailed flag.

Grønlands flag er vandret delt af hvidt over rødt og har en cirkelrund skive, som er forskudt mod flagstangen. Skivens øverste halvdel er rød, den nederste halvdel er hvid. Flagets navn er Erfalasorput, som betyder »vores flag« på grønlandsk. Flaget blev indført i 1985.

NATURE AND LANDSCAPE

BO ELBERLING, 2016

Cape Morris Jesup, Greenland's northernmost point (740 km south of the North Pole). This is where the Lincoln Sea in the west meets the Wandel Sea in the east. The small island, Inuit Qeqertaat, is located immediately north of Cape Morris Jessup, where other small islets are sometimes also visible

Greenland has an area of just under 2.5 million km^2. Lying between the 59th and 84th latitudes and the 11th and 74th longitudes, the country is bounded on the west by Davis Strait and Baffin Bay, on the northwest by Smith Strait and Nares Strait, on the north by the Lincoln Sea and the Wandel Sea, collectively known as the Arctic Ocean. To the east, the country is bounded by the Greenland Sea and the Denmark Strait, and finally on the south by the Atlantic Ocean.

Layer upon layer of snow and ice forms the ice sheet. With a volume of about 3 million km^3, it covers about 81% of Greenland. This makes the ice sheet the world's second largest freshwater reservoir – second only to Antarctica at the South Pole.

Most of the year, temperatures are below freezing, but should the ice sheet melt, it will equate to a total global sea mirror rise of seven-eight metres. The central part of the ice sheet is about 3,200 m above sea level, and from here excess ice moves towards the coast, where the ice calves off in or near the sea to form impressive icebergs.

A stable high pressure is often formed over the ice sheet, which

Photo pages 16-17. During October, the midnight sun sets over icebergs near Ilulissat.

STEFAN EISEND/LOOK/RITZAU SCANPIX, 2008

means that the ice sheet not only dominates the island's climate, as low pressure from the west is pushed south, but also affects the further course of the low pressure passages and thus affects the weather in Europe.

Greenland has been covered by the ice sheet for millions of years, but its distribution has varied considerably in step with changing glacials and warm interglacials in the last two million years. During the last glacial period, the ice sheet covered all of the country, and in most places, it extended into the ocean. As the ice retreated to its current position, it exposed the Greenland glacial landscapes, which are similar to those known from many places in Iceland and Denmark. Since then, the landscape has been characterised by a number of minor changes, and the cold winters have caused the permafrost to spread to more than 80 % of the present ice-free part of Greenland.

The ice-free part of the country (equivalent to 410,449 km^2) extends 2,670 km from Greenland's southern tip, Uummannarsuaq (Cape Farewell), to the northern tip, Cape Morris Jesup. Uummannarsuaq is located at the same latitude as Oslo. Greenland is the northernmost contiguous land area on Earth.

The ice-free part of the country is similar in size to Sweden and constitutes an up to 250 km wide zone along the coast. Here the geology is exposed and dominated by a shield of a Precambrian bedrock, which, for example, in the Isua area near Nuuk in West Greenland consists of crystalline rocks, which are considered among the world's oldest rocks. On several occasions, the bedrock shield has in many places been overlaid by sediments to later be deformed, subjected to metamorphosis and folded and raised like mountains. In North and East Greenland, sediments continued to be deposited in the Palaeozoic geological era (from 541 to 252 million years ago) and were then subjected to the mountain range formation that formed the Caledonian fold belt in East Greenland and the Ellesmerian fold belt in North Greenland. Later, large sedimentary basins in West Greenland were filled with deposits from the Cretaceous more than 60 million years ago, and in connection with the Tertiary opening of the North Atlantic, large parts of South Greenland were exposed to extensive volcanic activity.

The geology of Greenland contains a number of important mineral deposits of e.g. coal, lead, zinc and cryolite, which have previously been exploited commercially. The earliest mining is known from the Stone Age, when the Saqqaq culture utilised a special deposit of killiaq (a very hard slate) as their preferred material. However, most mineral deposits are difficult to access, making mining unprofitable in most cases, but in recent years, repeated attempts have been made to extract e.g. gold, platinum, uranium. Maarmorilik and Mestersvig are new mines used for large-scale extraction, and they have had major local impact and in both cases caused considerable pollution.

The continental shelf around Greenland is a geological continuation of the land areas. The subsoil is dominated by a crystalline bedrock, which is covered by younger sediments and basalts. On deeper water, these deposits are replaced by actual ocean floor rocks consisting of material formed in connection with the seabed dispersal.

Most of the ice-free part of the country is below the height of 1,500 m. The highest point, however,

Uummannarsuaq (Cape Farewell), Greenland's southernmost point (centre of the photo). A large part of the Arctic sea ice drifts along Greenland's east coast and circles Uummannarsuaq, making it difficult to navigate. The Greenland Ice Service circulate daily ice charts.

NASA, 2006

is Gunnbjørn Fjeld (about 3,700 m), the highest mountain north of the Arctic Circle. It is located on the east coast near Tasiilaq and is a nunatak, which is a mountain top or rock that reaches through glaciers or the ice sheet. In several places, glaciers from the ice sheet reach all the way to the ocean, meaning that the ice-free part of the country consists of at least three non-contiguous land areas separated by glacial ice. This has had an impact on the spread of e.g., caribou and muskoxen. Likewise, nunataks may have served as refuge for animals and plants during the glacial periods and thus had a decisive impact on the spread of biological diversity since the last glacial period.

The topography of Greenland includes high mountains along the coasts, while large parts of the ice-covered area are below sea level. The northernmost part of the country, Peary Land, is not covered by an ice cap because the air is too dry to produce enough snow.

From north to south, Greenland can be divided into several climate zones, vegetation types, etc. The differences are decisive for the living conditions – not least the length of summer, which is relevant for the presence of permafrost, and the mean temperature in summer, which is decisive for agriculture and tree growth. Add to that the water balance, which to the south is dominated by a precipitation surplus and thus leaching of nutrients and salts. To the north, precipitation is so low that it is in fact an Arctic desert, where salts precipitate on top of the earth. Everywhere, plants and animals can be found that have adapted to the natural conditions and made it possible for humans to follow.

In many places, the coast is characterised by deep fjords, high mountains and an archipelago filled with islets and reefs. The coastline is more than 44,000 km measured on maps in the scale of 1:250,000. However, the length of the coastline depends on which scale is used. In satellite images with a resolution of 0.5 m, the Danish Agency for Data Supply and Infrastructure has found a coastline of approximately 105,000 km. All towns and settlements are located along the ice-free coast (see p. 10), and the population is concentrated along the west coast. The

With its area of 57.000 km^2, Peary Land is the northernmost part of Greenland, named after Robert E. Peary's expedition (1891-1892). The area stretches from Victoria Fjord in the west to Independence Fjord in the south and southeast and from the ice sheet in the south to the Arctic Ocean in the north. Peary Land is almost ice-free and part of the National Park in North and East Greenland. Many signs of early settlements have been found in the area.

BO ELBERLING 2016

The photos to the left: In August 1867, H. Rink photographed the territory around Kobbefjord southeast of Nuuk and then coloured the photographs by hand. To the right are Sebastian Marker Westh's photographs of the same locations 151 years later. In the top photograph from 2018, the head of the Kobbefjord can be seen to the right. The lower ones are taken facing towards Qaqarssuaq. Note, in for instance, the retreat of glaciers over the period and the trend towards increased vegetation.

H. RINK, 1867, SEBASTIAN MARKER WESTH, 2018

northeastern part of the country constitutes the world's largest national park, the National Park in North and East Greenland, covering an area of 1 million km^2.

CLIMATE

A world invisible to humans appears early in Greenland culture – a world populated by *inuat* – creatures or states that control the conditions of the physical world. The weather *inua* (see p. 156) was known as an erratic creature that made hunting difficult by producing storms and changeable weather. With the mountainous coastlines, deep fjords and migratory low-pressure systems, changeable weather is part of Greenland. This section describes regional differences in the country's climate, focusing on the period from approx. 1873 onwards, from which we have systematic climate measurements.

Since the late 18th century, a number of institutions have collected high quality climate data, and the Danish Meteorological Institute (DMI) today publishes their observations annually with time series dating back to 1784

ANDREAS WESTERGAARD-NIELSEN, 2022

Low-pressure system on the way between Iceland and Greenland. The spiral shape of clouds is shaped by the Earth's rotation counterclockwise due to being in the northern hemisphere. Low pressure creates a vacuum that sucks air from high-pressure areas; clouds follow. The photo was taken by the Aqua MODIS satellite in September 2003.

for Nuuk. After 1870, the measurements became more reliable thanks to the scientist H. Rink, among others (see Figure 1).

Several national and international organisations are currently collaborating on climate and ecosystem monitoring in Greenland, of which the Danish Meteorological Institute (DMI) and Asiaq (Greenland Survey) are responsible for climate measurements in the coastal regions. In total, the country has about 100 automatic climate stations, covering all the climatic zones. Since the 1970s, satellite measurements have become increasingly important for measuring the climate and physical environment. In particular, changes to the surface, such as the height of glacial ice, vegetation cover and sea ice cover, are being monitored closely.

Atmospheric circulation and wind systems

Above Greenland, air masses generally move from west to east in the troposphere. The reason is the global energy balance that supplies more solar energy at low latitudes than in the polar regions. It creates a pressure difference which, in interaction with the rotation of Earth, drives the general air flow. The air in the northern hemisphere is divided between the cold polar air mass from the north and the warm tropical air mass from the south. The two air masses meet at the polar front. This develops polar lows which are important for the distribution of precipitation. The polar front moves with the seasons and is furthest south in winter, where it is decisive for the weather in northwestern Europe.

Front systems

With a maximum elevation of approx. 3.700 m.a.s.l., Greenland acts as a barrier in the troposphere. Lows formed at the polar front and the related front systems that form west of the country are typically pushed up by this barrier. This results in relatively large amounts of precipitation in the coastal mountain areas and reduced precipitation from the coast towards inland. Depending on how far north the front systems meet the coast, they move either north parallel to the coast, or south around the southern tip of the country. Sometimes the front systems also move across the ice sheet. In addition, the Icelandic low is formed west of Iceland, which transports the relatively warm air masses north past Iceland. The warmer air masses can cause extremely unstable weather conditions in South Greenland and bring with them large amounts of precipitation (see map p. 25).

Cooling the air masses

Above the ice sheet, most of the solar radiation is reflected back into the atmosphere. It causes a low temperature and a year-round snow cover in the areas where the ice sheet maintains or increases its mass. The cooling over the ice sheet makes the air masses heavier, thereby moving them downwards from the sloping sides of the ice sheet towards the coast. Such an air flow, called catabatic wind, is frequent on the ice sheet and can cause wind speeds between 6 and 8 m/s. Along the often mountainous coast, a number of wind systems driven by temperature and elevation differences are developing at the same time. For example, shelter effects can occur when wind meets a mountain range, whereby strong, dry foehn winds can develop. A particularly powerful variant is called Piteraq, which means 'that which attacks you'. The phenomenon is known around Tasiilaq and has led to a tradition of building houses parallel to the wind direction, and with the doorway in the sheltered side.

Fjord winds

Greenland's complex fjord topography also causes local differences. Here, fjord winds can occur depending on a fjord's orientation, width and depth. They arise through a combination of pressure differences due to unequal warming of land and sea, as well as a jet effect due to the channelising shape of fjords. The winds

Air temperature 2 m above ground for the normal period 1990-2019 based on the regional climate model MAR (Modèle Atmosphérique Règional). The model is based on the Danish Meteorological Institute's measuring stations in the coastal areas. Note the difference in temperature scale between the two maps.

PARTLY BASED ON DATA FROM DMI

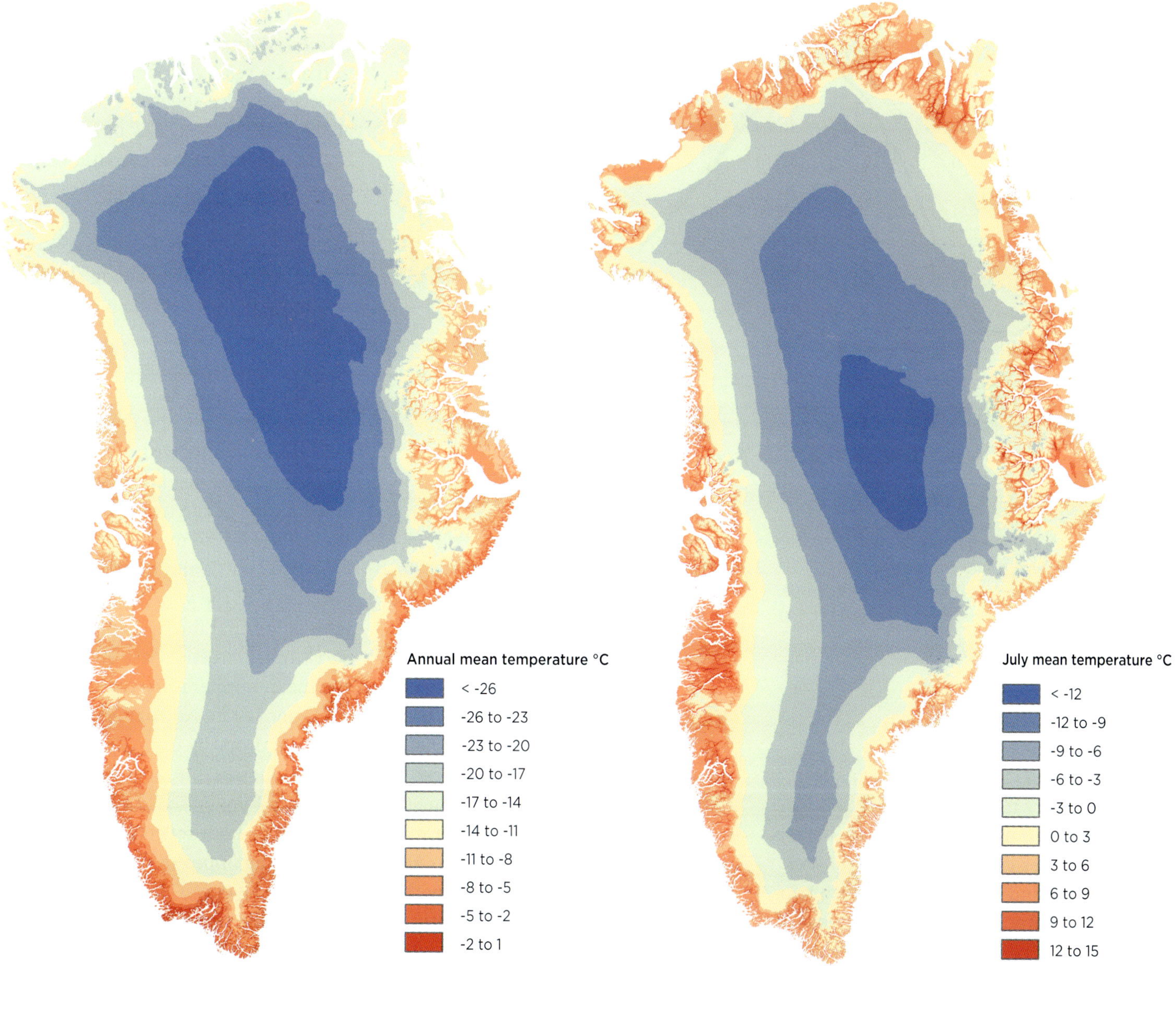

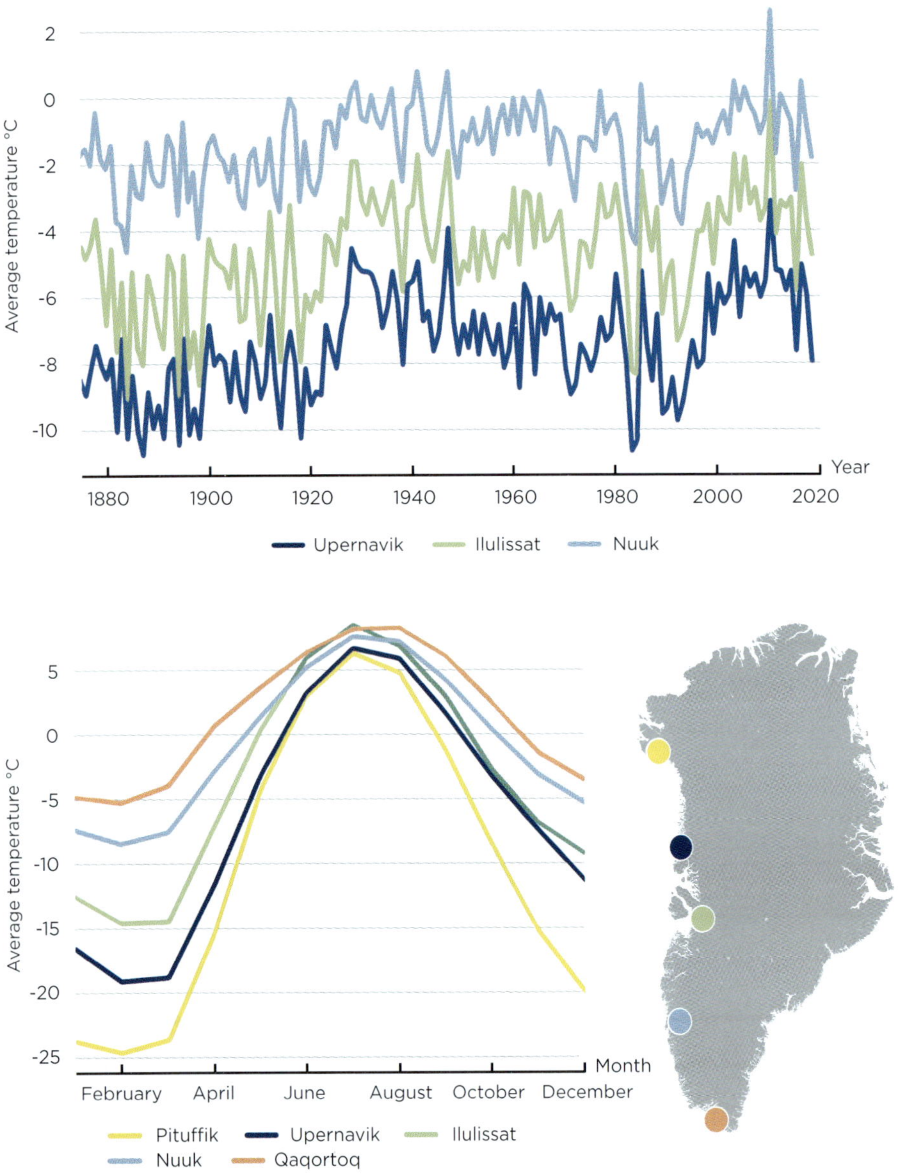

Figure 1. Selected time series of air temperature for West Greenland. Top: annual average temperature trend in the period 1875-2019. Bottom: seasonal variation in mean temperature per month in the period 1990-2019.

BASED ON DATA FROM DMI

can occur relatively quickly and locally, and if you sail in the fjords, it is necessary to assess whether you will be able to make it back home before the waves come, or whether to seek shelter and postpone the return trip.

In winter, most fjords freeze over and sea ice forms around the coasts, except for the stretch between Disko Island and Narsarsuaq on the west coast. The sea ice particularly affects the climate in winter, where, for example, North Greenland at times almost has a continental climate with very low humidity and precipitation and severe cold. During spring and the first part of summer, the ice begins to melt, and the encounter of the cold sea surface with the warmer spring air can form widespread fog along the coasts where there is open water.

Temperature and precipitation

According to Köppen's climate zone classification, Greenland has a polar climate where all months of the year have an average temperature below 10°C. The ice-free areas have a polar tundra climate. Areas covered by glaciers have a polar ice climate, where the monthly average surface temperature remains below freezing.

Due to large differences in elevation and a relatively short distance between the ice sheet, ice-free land areas and the coast, it is possible to experience incredible climatic variations over distances of just a few kilometres. Overall, both temperature and precipitation increase from north to south, where the growing season is also longer.

Temperature variations over the ice sheet

Over the ice sheet, the air temperature varies mainly with latitude and elevation above sea level. At Summit Station, one of the highest points on the ice sheet, the mean temperature is approx. -31°C (and minimum temperatures all the way down to -70°C), with an annual fluctuation of approx. 30°C. To the west or east of Summit, the mean temperature rises approx. 0.8°C per 10 km. Normally, the temperature drops by 0.5-1°C per 100 elevation metres, but over the ice sheet there are long periods of low radiation and/or high reflection, which leads to the opposite phenomenon: inversion layers.

The stable temperature conditions result in a widespread and stable high pressure over the ice sheet, which is only interrupted by changed radiation conditions or front systems that sometimes pass (see p. 22). Weather changes in the ice-free areas are often controlled by migratory low-pressure systems that arise from fluctuations in the polar front. When a low pressure passes, it causes a relatively rapid change of weather with slightly rising temperatures, precipitation and sometimes storms.

Temperature variations from north to south

The temperature variation from north to south is most significant in the winter. North of 66°33 north latitude (the Arctic Circle), the sun does not rise above the horizon during parts of the winter period, and thus the earth's surface does not heat up significantly. In January, for example, the average temperature in Pituffik (76°32N) will be 20°C colder than in southern Qaqortoq (60°43N). As summer approaches and the days gradually become longer (in the northern regions the day length increases for a short period by more than 40 minutes per day), the temperature difference from north to south will decrease (see map p. 23).

During the summer period, the area north of the Arctic Circle has midnight sun, and the significant change in radiation means that in July there is only a few degrees difference between Qaqortoq and Pituffik.

Temperature variations from coast to land

You can experience another interesting temperature gradient when you move from the coast over land and towards the ice sheet. In summer, the coastal areas will be cooled by the cold sea, whereas the deep inland valleys at the same latitude may well be 5°C warmer on average. In addition to the cooling from the sea, the difference is also due to a more stable inland climate in summer, with less fog and cloud formation to block for the sunlight. In winter, the coastal areas will be warmer as the sea water is often warmer than the atmosphere, thereby warming land. In areas with widespread sea ice, the climatic difference between coast and inland will be smaller, as the ice isolates the seawater from the atmosphere.

Just like over the ice sheet, inversion layers can form over ice-free land areas. In winter, inversions occur as a result of cooling of the lower air layers due to snow cover, while in summer, they are more likely to form near the coast where the lower air layers are cooled by cold sea water. In some fjords, you see the lushest vegetation in a belt at 100-200 m above sea level, which is precisely an effect of warmer air masses in summer at this elevation.

Temperature measurements from the end of the 1980s onwards show that Greenland's ice-free areas have become warmer during that period. Measurements from Qeqertarsuaq on Disko Island show, for example, an increase of approx. 2°C during the period, especially due to warmer winters. At the same time, the temperature change in several parts of the country seems to have stagnated since 2000 – in some places slight cooling is seen in spring and autumn. The measurements show the complexity of temperature changes and that the geographical area and measurement period must be taken into account.

Precipitation and precipitation distribution

The distribution of precipitation is linked to the temperature in the atmosphere and the access to surfaces with water that can evaporate, but is also determined by wind conditions and topography. Warm air masses moving over open water from the

Spatial precipitation maps (rain plus snow, as water equivalents) for the climate normal 1990-2019, based on the regional climate model MAR (Modèle Atmosphérique Règional). The model is based mainly on DMI measuring stations in the coastal areas. It is difficult to measure and model precipitation in areas such as Greenland with snowfall and rich topography, and the map is subject to uncertainties. However, the general north-south and east-west differences can be seen.

PARTLY BASED ON DATA FROM DMI

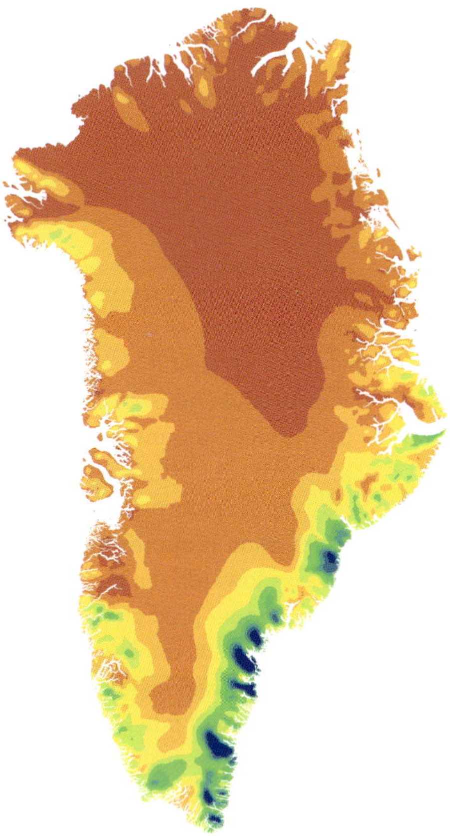

Average rainfall mm/year

40 to 200	1,200 to 1,500
200 to 400	1,500 to 1,800
400 to 600	1,800 to 2,100
600 to 900	2,100 to 2,400
900 to 1,200	2,400 to 3,000

LARS H. OSTENFELDT, 2020

Ice sheet monitoring includes snow drilling, which measures how much snow has fallen. The more snow falls, the better the ice is protected from melting. Masashi Niwano, PhD, from the Meteorological Research Institute in Japan, and Jason E. Box, Professor of Glaciology at the Geological Survey of Denmark and Greenland (GEUS).

south towards Greenland contain large amounts of water vapor, which are released as precipitation when the air meets land and is pushed upwards by the steep mountain sides at the southern tip of the country. In the southwest at Qaqortoq, the annual precipitation is approx. 990 mm and decreases towards the north. In Pituffik, the annual precipitation is thus only approx. 130 mm.

In the northern regions, most of the precipitation will fall as snow, while most of the precipitation in Qaqortoq is rain. This means that to the north, the precipitation accumulates as snow and then melts and flows off to rivers and lakes in spring. This makes a small part of the precipitation available to the terrestrial ecosystems. Where the precipitation falls as snow, it will have an insulating effect on the soil temperature in relation to the air temperature, and thereby it will have an impact on, for example, the presence of permafrost.

Precipitation is also affected by coast-inland dynamics. The colder winter air inland may contain less water vapour. As there are also fewer water surfaces from which the water can evaporate, inland areas will receive less precipitation than coastal areas. Kangerlussuaq on the west coast (67°0N) gets less than 1961-90 mm of precipitation per year (1961-1990), whereas Sisimiut (66°56N) at approximately the same northern latitude but 130 km further towards the coast gets approx. 340 mm (1961-1990). The stable weather in Kangerlussuaq was, moreover, a decisive reason for locating a strategically important airport there.

Projections for the future climate of Greenland

Towards 2100, climate change is expected to increase globally. The Arctic region as a whole is expected to see an intensified temperature rise, mainly due to a meltback of sea ice and its effects. Over both ice-free and glacier-covered areas, temperatures are expected to rise, although the southeastern part may experience a cooling due to weakened cold ocean currents. As a warmer atmosphere may contain more water vapour, warming in Greenland is expected to result in increased autumn and winter precipitation, and more frequent extreme and unpredictable weather events.

THE ICE SHEET

The ice sheet covers 81% of Greenland and glaciers and ice streams extend to the coast, where icebergs calve and meltwater forms rivers and fills lakes. The ice sheet is the second largest mass of ice on Earth, surpassed only by Antarctica. The volume of the ice is about 3 million km^3 and with a mean thickness of 1.8 km. The centre of the ice sheet reaches an elevation of about 3,232 m above the current sea level. If the ice were distributed over Denmark, it would be about 70 km thick. The ice is fresh and is thus also the Earth's second largest reservoir of frozen fresh water. The volume of the ice corresponds to a global sea water rise of about 7-8 m. Studies of the bottom material under the ice as well as studies of sediments in the sea around Greenland indicate that there has been ice in Greenland for at least 7 million years. The size of the ice sheet has changed in line with the warm and cold periods of the past, and the current ice cap has covered the country for at least 1 million years. The top of the ice sheet's surface has been mapped using satellite measurements and the bottom using radar measurements. Such a heavy mass of ice has over time pushed the rock surface by up to a kilometre down in the central parts. Today, the bottom in the central parts is more than 200 m below the current sea level.

The ice sheet means a lot to Greenland and the Earth's climate. Snow falls on the ice cap and in the central highlands, it is cold all year round, meaning that the snow does not melt in summer. The white areas reflect about 80 % of incoming sunlight and thus have a very high albedo – an expression of the proportion of sunlight reflected from a given surface. This means that the large ice cap helps to keep the Earth cold.

Melting of the ice sheet

In the lower areas along the edge of the ice sheet, snow and ice melt in summer. In many areas this forms blue meltwater lakes and rivers that drain the meltwater via well-like shafts (moulins) down to the bottom of the ice, where it flows under the ice out to the edge. In the lower areas, the winter snow melts away and the darker ice surface is exposed and melts. It is the natural cycle that ice from the centre of the ice cap flows out toward the edge to melt. The melting surface is much darker than the white snow surface seen in the central parts of the ice cap and this intensifies the melting along the edge.

As the temperature rises, the melting area will reach higher up on the ice sheet in summer and more ice will melt. At the very edge you will also find shafts that occur when the ice floats over the unevenness of the rock surface or increases the speed and is pulled apart. The

Large and small cracks and caves in the ice sheet are common and result from the ice moving and cracking. Some of the cracks are deepened by meltwater, leading to very long and deep cracks, which are dangerous and have always presented a challenge for expeditions on the ice sheet. More recently, the ice is melting faster, which has led to more cracks, especially along the edges of the ice sheet. Here, Professor of Glaciology Alun Hubbard is seen carrying out fieldwork 175 metres below the surface of the ice sheet.

LARS H. OSTENFELDT, 2021

LARS H. OSTENFELDT, 2020

The colour of the surface is crucial for the ice albedo and thus how much solar energy it takes to melt the ice. Meltwater from the ice sheet disappears into so-called moulins (meltwater cracks), which can be several hundred metres deep.

shafts can typically be 20 m deep and can be difficult to see in winter and spring when covered in fresh snow.

The ice sheet as a resource

The ice sheet is of great importance both regionally and globally; regionally in terms of water supply, fishing and transport, globally in terms of sea level rise.

Meltwater from the ice sheet, which accumulates in lakes, is an important part of the water supply to towns and settlements, and in the big towns, water from rivers and lakes is also used for hydropower. In places where the ice flows into the sea and calves as icebergs, there is a very large supply to the sea of very nutritious fresh water. This means that the neighbouring area is often rich in living resources. The calving icebergs are also important for transport and the possibility of oil extraction from the seabed.

Exploration of the ice sheet

The great ice sheet has been alluring for expeditions all the way back from Fridtjof Nansen, who crossed the ice at the southern end in 1888, Robert Peary in the north in 1892 and Lauge Koch in the middle in 1913.

The first major international scientific expeditions across the ice were the EGIG expedition in the years 1956-1972. The expeditions were organised by the International Commission on Snow and Ice and had participation from Denmark, Germany, Austria, France and Switzerland. EGIG measured a profile over the ice and made meteorological and glaciological observations. Since then, numerous expeditions have been on the ice sheet to measure the ice, sample ice cores for palaeoclimatic studies and to understand the dynamics and melting of the ice. Satellite observations of changes in elevation, surface temperature and ice flow have since 1979 revolutionised the knowledge we have about the ice sheet.

Recent climate research on the ice sheet

The ice sheet contains important information about the climate of the past. Every year, snow falls on the ice. The annual layers of snow are slowly compressed into ice and sink into the ice cap, thus the ice consists of horizontal annual layers. The ice sheet is in constant motion, and it takes an average of 10,000 years from a snowflake lands on the central parts of the ice sheet until the snow has turned to ice and has flowed with the ice out to the edge to melt or been carried further out to the sea as an iceberg. At the bottom, the ice is up to 1 million years old, but the lower ice is so compressed and mixed that it has not so far been found in stratigraphic order older than 130,000 years back in time.

The ice sheet is a unique source of information about the climate of the past, not least due to the long climate series and the high temporal resolution.

Ice core drilling

By drilling a 3 km long ice core in the central part of Greenland, the climate can be mapped with annual precision. One can i.a. see that the climate has changed between glacials and interglacials.

During the previous interglacial, the Eemian interglacial 115,000 to 130,000 years ago, it was 5°C warmer in Greenland than now, and there was still a lot of ice. Based on measurements of the old ice and the fact

SCIENTIFIC EXPLORATION IN A HISTORICAL PERSPECTIVE

With the colonisation of Greenland in the early 18th century, knowledge about the country's people and nature started being collected. The missionary Hans Egede (see p. 113) became the first storyteller with the publication of *The New Perlustration of Greenland or Natural History* in 1741, which is considered to be the first real physical geography and ethnographic documentation of living conditions in Greenland. The pioneer in the field of fauna was the missionary Otto Fabricius, who in 1780 published *Fauna Groenlandica*.

The first to carry out a systematic study of the minerals was the German geologist Karl Ludwig Giesecke, whose work was continued by the geologist Hinrich Johannes Rink (see p. 154). The geology has been supplemented with zoological and botanical studies, and the mapping of the country has continued using modern aids. Aerial photography was used for the first time during Lauge Koch's expeditions from the 1920s to the 1940s. Since 1946, the geological surveys have been organised by the Geological Survey of Greenland (which later became part of the Geological Survey of Denmark and Greenland).

Studies of the sea and marine wildlife have been carried out since 1879 and have been crucial to fishing becoming a main industry. Since 1946, this research has been organised in the Greenland Fisheries Survey, which in 1995 became part of the Greenland Institute of Natural Resources.

To support scientific research in Greenland, the Danish government set up the Commission for the Management of the Geological and Geographical Surveys in Greenland in 1878, which was replaced in 1931 by the Commission for Scientific Investigations in Greenland. Since 1879, these organisations have published the monumental work *Monographs on Greenland*, in which results of research in all disciplines have been published.

In collaboration with the Home Rule Government at that time, the Danish Polar Centre, established in 1989, served as secretariat and co-coordinator for the scientific studies in Greenland, until the centre was closed in 1988.

Since the end of the 19th century, systematic meteorological observations have been made in Greenland, which are of great importance for climate research and for European weather forecasts. The drill cores from the ice sheet, which have been retrieved since 1980, have been groundbreaking for the international understanding of past climate fluctuations.

UDDANNELSES- OG FORSKNINGSMINISTERIET, 2018

Water, air bubbles and dust in the ice sheet represent a climate history going back more than 100,000 years – from when the ice fell as snow. Professor Dorthe Dahl-Jensen of the Niels Bohr Institute shows an ice core to Minister of Education and Research Tommy Ahlers.

HANNIBAL HANSCHKE/REUTERS/RITZAU SCANPIX, 2021

The melting Sermeq glacier, located around 80 km south of Nuuk. Greenland has many glaciers, where the glacier Sermeq Kujalleq approx. 50 km east of Ilulissat is the most productive. Sermeq Kujalleq moves with approx. 40 m a day and calves around 46 km³ of ice annually.

that all the ice cores drilled in both South and North Greenland contained ice from the Eemian interglacial, it is estimated that the ice cap was 20 % smaller during the warm Eemian period than now.

During the glacial period, there were 25 events of 1,000 to 3,000 years' duration where the temperature over Greenland, over a period of 100 years, suddenly rose by 7°C to 15°C and then slowly fell again. Climate reconstructions show that there has been a changing climate back in time. However, measurements of greenhouse gases in the entrapped air bubbles in the ice show that there have been no periods where the concentration of carbon dioxide has exceeded 300 ppm and thus not even close to the concentrations of over 400 ppm seen in recent times.

Global warming and sea level rise

There is significant focus on the impact of the ice sheet and the surrounding smaller ice caps and glaciers on the global sea level rise. The warming over the ice sheet is 2-3 times the global warming, which gives rise to massive melting. The melting from the ice sheet comes both from melting along the edge of the ice and from calving icebergs from the many fast-flowing ice streams around Greenland (see photo on p. 59). Since the year 2000, the mass loss from the ice sheet has increased, and in the period 2010 to 2020, the ice sheet has lost an average of 240±60 km³ of ice annually, corresponding to a global sea mirror rise of 0.7 mm per year.

Forecasts for the mass loss from the ice sheet are very uncertain, particularly because it is difficult to predict the development of ice streams. The ice streams flow faster than the surrounding ice and calve icebergs into the fjords. Sermeq Kujalleq (formerly known as Jakobshavn Glacier) is the best known with speeds up to 14 km per year. In recent times, the speed of the ice stream has increased from 7 to 14 km per year, but most recently dropped to around 10 km per year, although it continues to get warmer over the ice sheet.

The many hundreds of ice streams around Greenland have almost all increased their speed, thereby giving off more ice to the sea. Researchers have a large network of observation stations for monitoring the melting from Greenland, which is seen as one of the largest sources of increased global sea level rise in the future.

GEOLOGY

Greenland is an island with a distinctive shape, which virtually everyone in the world can recognise just by looking at the outline of the country. When describing the geology of country, it is important to distinguish between when it is about landscapes, when it is about what materials the rocks, the mountains are made of, and when it is about where the material the rocks are made of came from. All this is also related to when and how Greenland became an island.

The deep structures

Roughly speaking, the country looks like a giant ice cream cone with the tip facing south. The east and west coasts form the cone's acute angle, while the north coast forms an obtuse angle with the tip facing the Arctic Ocean.

If you let your eye run up along the two 'cone coasts' to the east and west, you quickly notice that the vast majority of fjords, valleys and glaciers along the west coast are oriented NE-SW, roughly parallel to the country's northwest coast and Nares Strait towards Canada. Similarly, the fjords are oriented NW-SE along the east coast, roughly parallel to the eastern north coast.

The reason is the continental plate collisions that happened between 1,900 and 1,700 million years ago: An ocean plate from the south thrust under what was to become Greenland, and old existing continental parts merged into one whole. The part of country situated south of Arsuk was created of new magma formed by the melting of the Earth's mantle and sediments and ocean floor which was pulled with the ocean plate down into the depths. The rest of the country is dominated by recycling of older continental fragments.

It could be seen as a geological collision. A number of smaller continental blocks were hit from the south and the collisions propagated through the system as if a large truck had hit a queue of smaller cars. Some were simply badly deformed and others were stacked on top of each other. In the geological version, smaller sea basins – the roadway between the cars – were folded into the continents as they collided. The stacked 'cars', the continental plates, rose like mountain ranges of a scale like the Himalayas today. With this, the deep bedrock under Greenland was welded together, and the country has never divided since.

Experiments have shown that if you squeeze a rock block from the south and north, it will start to crack in two orientations, NE-SW and NW-SE. The fine details in the coastline, the fjords, are a result of the fact that the ice has had an easier time eroding the land along the cracks that propagated through the country when the continental collision took place some 1,800 million years ago.

The country finds its form

Greenland has a very long formation history, which includes virtually all types of geological processes throughout the Earth's geological history. The land we call Greenland was not formed as an island; it is a fragment of a continent that was broken up into smaller pieces by the continental drift across the face of the Earth.

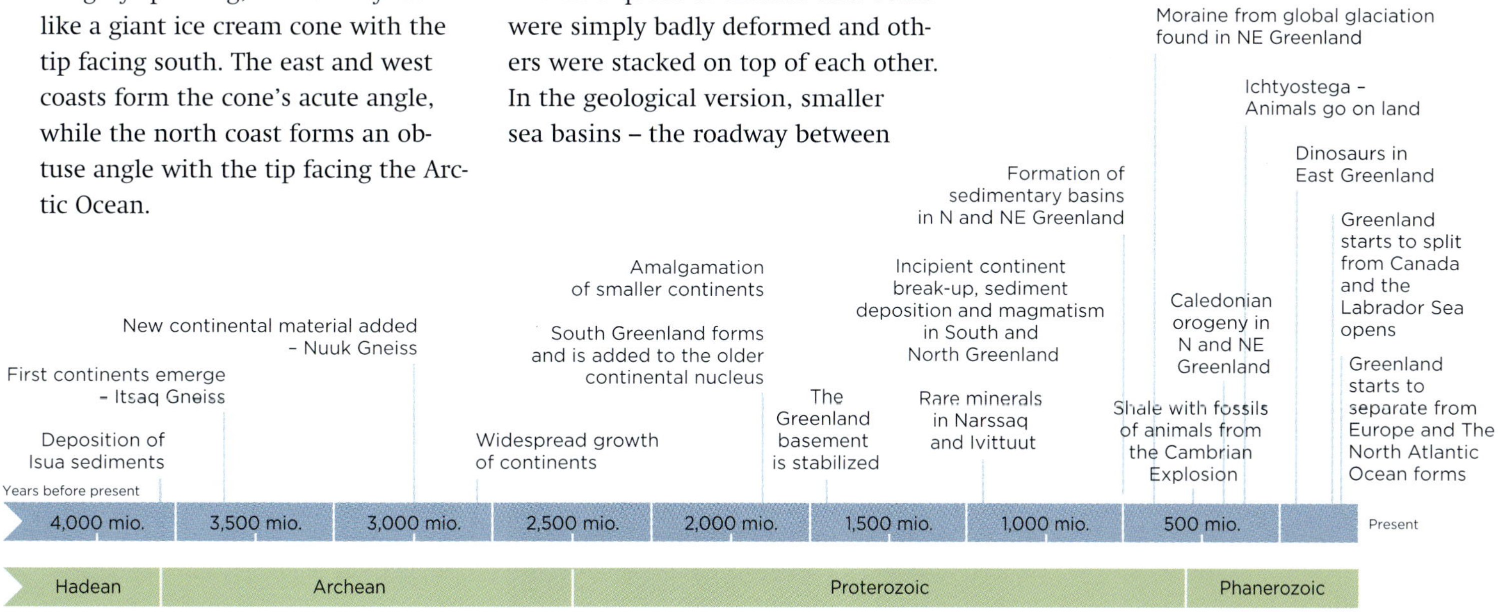

The geology of Greenland must therefore be understood in the context of its surrounding areas; the country is a piece of a puzzle which includes Scandinavia, parts of Russia, Canada, the Faroe Islands and the British Isles.

The supercontinent is broken up

175 million years ago, all the continents of the Earth were joined in one so-called supercontinent, stretching like a belt from the North Pole to the South Pole. The idea that all the Earth's continents were once joined together in one continent, was fostered by Alfred Wegener during the Denmark Expedition to Northeast Greenland in 1906-1908. He called this whole primeval continent Pangea. Wegener suggested that the land masses are in constant motion, drifting around the Earth's surface. 330 million years ago, all continents had collided with each other and had begun to move as a total mass, Pangea.

175 million years ago, the supercontinent began to crack and different parts drifted in different directions. Africa started to drift away from South America 150 million years ago. The cracks between the two parts were filled with basaltic magma which rose from the hot interior of the Earth. The cracks propagated northward, like a zipper that quietly opened from south to north. As the basalt-filled crack landscape evolved, it was flooded by the sea, forming what we know today as the Atlantic Ocean.

90 million years ago, the crack system had reached the level of South Greenland. Greenland, along with Europe, was drawn east, while Canada and the United States drifted west, and the Labrador Sea opened up between West Greenland and Labrador.

Sea basins are formed

When a continent breaks up, it happens due to forces pulling in different directions. First, the land yields as much as it can, by the crust being pulled thinner until cracks appear and finally a final fracture. This development is important for the geology along the coast of West Greenland, where the deformed crust sank in some places and was flooded by the sea. This led to the formation of a number of basins in the seabed off Greenland, including in Qeqertarsuup Tunua (Disko Bay) and around Nuussuaq Peninsula.

The basins were gradually filled with sediments and dead plants, animals and planktonic organisms which over time were transformed into coal, oil and gas. It is that part of the creation history which forms the basis for the hopes for oil and gas deposits off Greenland, and which created the coal layers that were mined during the last century's mining in Qullissat (see p. 117).

Lava builds up basalt layers

Greenland could very well have been part of Europe today had Iceland not entered history. Before Greenland and Canada were completely separated, the continent passed over a stream of hot material that rose from several hundred kilometres depth in the Earth's mantle – known as a 'hotspot'. The hot material began to melt, and large amounts of magma rose upwards towards the surface. It flowed out through the many cracks and zones of weakness that had emerged during the progressive split between Greenland and Canada and through zones of weakness along what is now the east coast of Greenland.

The coincidence between the fracture and the hotspot created the giant layers of lava that cover Qeqertarsuaq (Disko), Nuussuaq and Svartenhuk in West Greenland, Baffin Island

BO ELBERLING, 2021

Highly deformed rocks dominate the Greenlandic bedrock formed 20-30 km down into the earth's crust beneath now eroded mountain ranges. Most of the bedrock was formed more than 2,800 million years ago, including here at Danmarkshavn, where bands of gneiss can be seen. The land area south of Arsuk is an extension formed about 1,800 million years ago.

BO ELBERLING, 2021

in Canada and the Blosseville Coast in East Greenland.

Over the course of 10 million years, millions of cubic kilometres of lava built up to 7 km thick basalt layers which today dominate the landscape in the western part of Qeqertarsuup Tunua (Disko Bay) and north to Upernavik, and the Blosseville Coast in East Greenland.

The continent cracks

The continental plate continued its drift across the warm current, while volcanism continued. The constant flow of hot material from the Earth's deep mantle created a huge bulge under the country. This led to land uplift and to the crust being stretched. What is now Greenland's east coast had formed the coast of a now vanished sea, the Iapetus Ocean, 600 million years ago. This ancient fracture zone proved to be a zone of weakness in the entire Europe-Greenland region of Pangea. The continent cracked and the opening of the Atlantic Ocean, which had otherwise tried to wedge itself between Greenland and Canada, jumped to between Greenland and Europe. Northeast Greenland and Norway became geological mirror images of each other, and many features of South Greenland's geology can be followed through Scotland, northern Scandinavia and into the Kola region of Russia.

Pieces of the thick layers of basalt from the east coast of Greenland followed Europe to the east, where they now form the Faroe Islands, the Hebrides and parts of Ireland. We can still see that the Faroe Islands were connected to the Blosseville Coast, because layers of lava from a few specific volcanic eruptions 55 million years ago have one half in East Greenland and the other in the Faroese mountains. The hot stream from the mantle still produces large amounts of lava and is called the Iceland Hotspot because the volcanic island of Iceland is where the volcanism can be seen on the surface.

Columnar structures in basaltic lava at Kuanit near Qeqertarsuaq. Basaltic lava flowed to the surface through deep cracks formed during North America's and Europe's separation about 60 million years ago. The columnar structures are formed in connection with the solidification and cooling of the lava layers. As newly formed basalt cools, it shrinks and the characteristic hexagonal columnar structures appear.

The continent's old coast – North and Northeast Greenland

North Greenland and Northeast Greenland are geologically very fascinating areas where it is still possible to see the coastline of the continent, which was created by the tectonic plate collision 1,800 million years ago. It is possible to stand on the interface where sediments have settled on

BO ELBERLING, 2016

The island of Tartupaluk ('newly reformed'), called Hans Ø, lies in the middle of the Kennedy Channel in the Nares Strait between Canada and North Greenland. In this northern part of Greenland, the bedrock is covered by kilometre-thick layers of sediments deposited between 1,200 and 450 million years ago. The island's flaky structure is due to almost horizontal sediment layers. A dispute over ownership was resolved in 2022, when Greenland, Denmark and Canada reached an agreement to share the island. The new boundary follows the natural geological uplift seen in the photo.

top of the gneiss from the old bedrock. It is reminiscent of the sabkha environments found in Arabia today. A marsh-like coast with dunes and lakes, which have occasionally dried out, leaving drought cracks and salt crystals, which can still be seen as imprints in the sandstone today, almost two billion years later.

If you walk across the huge sediment platform from the edge of the ice sheet towards the coast, you encounter younger and younger deposits. You can stroll through a billion years of geological history, and see the results of dramatic upheavals in the Earth's climate that have changed the conditions of life on Earth. After the deposition of the sandstone layers, thick layers of basalt are found, formed by a new fracture 1,250 million years ago. This is followed by 400 million years with no preserved deposits, most likely because they have eroded away.

Supercontinent Rodinia

Greenland enters history again about 850 million years ago. The entire North American-Greenland continent, Laurentia, was now part of a coherent supercontinent, Rodinia, which in many ways corresponded to the later Pangea. North Greenland was still on the edge of the continent, with the coast at the current ice edge line. From here, a shallow shelf stretched about 100 km north, and was replaced by a deeper ocean basin towards what is now the Arctic Ocean.

At the bottom of the deposits, and thus first, coarse sand is deposited, close to the shore. Higher up in the deposits, and thus later in time, finer sand and mud are deposited, which typically represent deeper and calmer water, and then limestone and then sandstone again. Together, the deposits are about 1,000 m thick, and they are all deposited at a depth of a few metres in the sea.

But how does that add up? A sea basin that can hold 1,000 m thick layers of sediment must be at least 1,000 m deep? This actually illustrates the dynamics involved in the movements of the continental plates and the opening and closing of oceans. The thing is that stretching the continental crust has led to the sinking of a basin on the shelf. As the basin has sunk, it has become filled with sediment. The sediment has weighed the crust and pressed it down, and the stretch has created space for the basin to sink at a rate making it shallow throughout the period, or at times dry.

From microbes to the earliest fauna

Many of the limestones in North and Northeast Greenland are made up of millimetre-thin layers, which form irregular undulating layers and cauliflower-like structures. It shows that part of the limestone is formed by microorganisms, which have accumulated in layers on the seabed, and protected themselves from the external environment with layers of mucus. Small limescale particles have been continuously trapped in the mucus layer, and have blocked the sunlight. The microorganisms have therefore grown up through the lime layer, and have formed a new mucus layer. The process has been repeated over and over again and has created the thick flaky deposits – stromatolites. They represent the more than three billion years of Earth's history, in which animals and plants had not yet come into being, and all life consisted of microbes.

In the sandy layers on top of the stromatolites in North and Northeast Greenland, however, some small tubular structures begin to appear in the sandstone. These are burrows, made by some kind of early animal, which is probably reminiscent of sandworms. With this, we enter a whole new era of Earth's history. A world of animals and plants – perhaps not a modern world, but something that is beginning to resemble it.

The total glaciation

The theory called *Snowball Earth* puts forward the thesis that, for yet unknown reasons, the Earth's climate became significantly colder about 700 million years ago – a theory that builds on findings of moraines from an early glacial period.

From the plate tectonic evolution of the Rodinia continent, a prior tropical environment, represented by sabkha deposits and limestone, can be detected. At the time when they were deposited, Greenland was right on the equator, the hottest zone on Earth, and the country was still on the equator when the moraine layers were formed. This shows that the climate at the equator 717 million years ago was like at the South Pole today.

Ella Island, or Ella Ø, in King Oscar Fjord in Northeast Greenland is one of the finest examples of the transition between tropical and glacial climate, but there are similar deposits from this period on all continents. Earth had probably been glaciated twice or more over a period of 90 million years, and the climate crisis was apparently a unique event in Earth's long history. If so, life somehow – open cracks in the global ice cover, a floating ocean beneath the ice – managed to get through the unimaginably

The Earth was supposedly impacted by a global ice age about 650 million years ago. At that time, the continents were united in a supercontinent, and the part that is now Greenland lay on the equator. Nevertheless, thick glacial deposits are found in Northeast Greenland. These layers indicate that the Earth was covered by ice caps from the poles to the equator for about 20 million years.

WALTER MYERS/SCIENCE PHOTO LIBRARY/RITZAU SCANPIX, 2010

long winter, and traces of the earliest animals on Earth occur shortly after the end of the global glacial period.

Evolution picks up speed – Animals emerge

Just north of J.P. Koch Fjord in Peary Land is the Sirius Pass. It intersects limestone and black shales from the earliest part of the Cambrian, which marks the transition to the Phanerozoic, the geological Aeon, where animals and plants occupy the Earth. In 1984, in the black shale at the mountain of Buen, an expedition from the Greenland Geological Survey, now part of GEUS, found some fossils that did not look like anything ever seen before. However, they did resemble some fossils found in the Burgess Shale in Canada. Fossils are mostly made up of skeletons, but these were preserved traces of animals without skeletons or with preserved soft parts that had never been studied before.

The Sirius Pass suddenly provided insight into some of the first animals that emerged in the ocean and follow how today's animal groups evolved in an astonishingly short time. When a shale block from the Buen Formation is split, it shows a picture of the Cambrian explosion frozen in time. It gives a snapshot of a brief hectic evolutionary event in which all the genera of animals that populate the Earth today emerged. As far as is known, the age of the layers of Sirius Pas can be determined at around 513 million years.

A supercontinent is breaking up

By this time, the Rodinia supercontinent was breaking up, and a sea, the Iapetus Ocean, had started to wedge itself down between Greenland and Norway. The Baltic continent was separated from Laurentia and along the coasts of the two new continents, kilometres of sediment layers were deposited until the ocean closed again and Baltica and Laurentia collided again 425 million years ago, as part of the construction of the Pangea supercontinent.

Formation of sedimentary basins

However, this did not mean that this was a period of geological calm in the region. The formation of the Caledonian mountain range was, like most mountain ranges, followed by long-term adjustments of the continental crust. The joint where the continents were 'glued' together, and where the Iapetus Ocean had been, continued to form a zone of weakness in the crust. Over the following many millions of years, a number of basins emerged along what are now the coasts of Greenland and Norway. Some were connected to the sea and were filled with limestone, shale, and other marine sediments. Others were continental basins filled with red desert sand, and others still represent shallow coastal swamps. These basins can now be found consecutively through the mountain landscapes from Kangertittivaq (Scoresby Sound) and up through Crown Prince Christian Land which forms the northeast corner of Greenland.

These sedimentary basins are, individually, small geological treasure chests, illustrating different parts of the Earth's recent geological history. During the Danish East Greenland Expedition in 1931, remains of Ichthyostega, the 'four-legged fish', were found at Celsius Bjerg in the fjord complex north of Mestersvig. Here, we find deposits formed in a shallow swamp area during the Devonian period about 370 million years ago. Ichthyostega had features reminiscent of fish, but also limbs like the Tetrapods – the four-legged vertebrates on land, and they were able to both walk and swim through the swamps. They are considered the earliest form of land animal.

Climate catastrophe and mass extinction

Further south, on the Jameson Land peninsula, just north of Kangertittivaq, there are thick deposits from the geological time periods Triassic, Jurassic and Cretaceous and, where the peninsula is connected to the mainland, also layers from the older time periods Carboniferous and Permian. Again, a climatic drama can be followed through the geological layers.

The Permian period ended rather abruptly 252 million years ago, with the most violent mass extinction the Earth has ever experienced. More than 90 % of all animal species living in the sea and 70 % of the many species of vertebrates that populated the terrestrial soil died within a very short period of time due to a severe climate crisis. It seems that the Earth's average temperature rose around 10°C, due to a sharp increase in the CO_2 content of the atmosphere. There are indications that a combination of volcanism and fires in the thick deposits of coal that had formed in the preceding Carboniferous triggered the climate catastrophe.

Recent times – Mesozoic

The catastrophe has left its mark on the northern and eastern Jameson Land. Here you will find the most incredible red mountain landscapes made of desert sand from the scorched earth.

The violent wave of extinction of animal species, as always, led to a huge leap in animal evolution. New species of animals and plants emerged. One of the groups of animals that rose from the ashes after the Devonian extinction were the

dinosaurs. In the mountains around Carlsberg Fjord, you will find dinosaur skeletons, and the traces of dinosaurs that walked around more than 200 million years ago are still clearly visible in the layers that formed the Earth's surface at that time.

Red layers of sand and mud at Carlsberg Fjord in East Greenland. About 200 million years ago, large herbivorous dinosaurs left their footprints in a wet mud layer. The fossilised footprints and skeletal remains of the great dinosaurs are now embedded in the rocks and tell of a very different Greenland in the distant past.

In central Jameson Land, there are also thick deposits from the Jurassic and Cretaceous, where the sedimentary basins again sank and were flooded by the sea, as a result of Pangea's incipient fragmentation. Greenland was still connected with Norway, the Faroe Islands and Scotland, and Iceland did not exist yet. During that period, a number of sedimentary basins were formed in the zone between Greenland, the British Isles and Scandinavia, just as similar basins arose along the west coast of Greenland. These basins are the source of the Norwegian oil wealth and the source of the Danish and British oil fields in the North Sea. This also makes them the source of Greenland's hope to find oil within its own territory.

MINIK ROSING, 2010

Iron ore in rock face in isua. The rock was formed in the ocean more than 3,700 million years ago and consisted of alternating layers of the minerals quartz and magnetite. Magnetite is a magnetic mineral composed entirely of iron and oxygen. Since iron ore was discovered in the 1960s, several large projects have sought to exploit the iron deposit. The mountain ranges of isua are among the oldest, providing the most comprehensive picture of the surface environment in early earth history.

Sedimentary layers deposited on the seabed more than 3,700 million years ago. The grey layers are formed by mud avalanches that have rolled down underwater slopes. The black layers are slates that have been formed by the deposition of sediment during quiet periods between mudslides over thousands of years. The shale's black colour is due to myriads of microscopic carbon particles formed by dead organisms that have sunk to the bottom of the ocean.

CHRISTIAN KNUDSEN, 2005

Greenland chooses side

At the end of the Cretaceous, the Labrador Sea first opened between Greenland and Canada. Since then, the old quarry from the Iapetus Ocean sprang up again and Europe began drifting eastward, while Greenland continued along with North America to the west. The thick layers of basalt, from the lava that flowed out through the cracks, are now seen in the steep carbon black rock walls along the Blosseville Coast, Svartenhuk, Qeqertarsuaq (Disko) and Nuussuaq, and the highest mountain in the Arctic, Gunnbjørn Fjeld, consists of lava that flowed out back when the North Atlantic was born.

When life started to take shape

Like everywhere else, the bedrock in Greenland is rather uneventful, some would say boring. However, a tiny piece of bedrock from Isua 150 km north-east of Nuuk calls for special attention. In the late 1960s, it was discovered that the grey gneiss in Nuup Kangerlua (Godthåb Fjord) was more than 3,600 million years old, and thus significantly older than anyone had imagined that rocks could be.

At the same time, the Danish cryolite company Kryolitselskabet Øresund discovered an occurrence of iron ore of a special striped or banded type, which was deposited as a sediment in the sea in the Earth's early childhood. It turned out that there are very small pockets of well-preserved sedimentary rocks and lava rocks in Isua, and that the layers are more than 3.700 million years old.

Among the old sediments was a very small group of rocks, the size of a desk, from which layers of the finest, well-preserved, black shale protruded. Upon closer study, it turned out that the black colour of the shale comes from millions of tiny inclusions of graphite, carbon. It was further uncovered that the carbon has a characteristic ratio of the two stable isotopes, carbon-12 and carbon-13. A ratio known only from living organisms. The small group of rocks in Isua thus became a birth certificate for life on Earth.

Greenland's geological resources

Humans have exploited geological resources in Greenland throughout the country's 4,500-year cultural history. Rock crystal, diabase, quartzite and soapstone are found in all parts of Greenland and have been used as tool materials by all Inuit cultures. Soapstone was also mined by the Norse, who left runic inscriptions in soapstone quarries along the west coast.

Killiaq is a fine-grained, slate-like, volcanic rock found on Nuussuaq and on a small island in Qeqertarsuup Tunua (Disko Bay). Tools made of Killiaq from both the Greenlandic Stone Age cultures, Saqqaq and Dorset (see p. 92), are known from virtually the entire West Coast. Killiaq has clearly been traded as a semi-finished product and is an example of organised extraction of special raw materials.

The people of the precolonial Inuit culture extracted Telluric (terrestrial) iron from some, globally, extremely rare natural deposits in basalt on Qeqertarsuaq, and at Cape York, meteoric iron was extracted and cutting edges were made for knives and harpoon tips by cold forging.

Since colonisation in 1721, focus has been on exploiting Greenland's natural resources. Initially, it was mainly oil from seals and whales, but over the centuries, the mineral resources increasing became the object of interest. Shortly after his arrival in, Hans Egede (see p. 113) reported deposits of the then very expensive mineral graphite. In 1852, mining of copper ore started at Alanngorsuaq Fjord (Coppermine Bay), south of Arsuk, but mining was quickly abandoned. In 1905, they tried again, now with smelting works, but gave up again in 1914.

In recent times, there have been several more or less successful mining projects. At Maamorilik in the Uummannaq Fjord, 11 million tonnes of lead and zinc ore with a grade of 14.3 % zinc, 4 % lead and 29 g silver per tonne were extracted during the period from 1973 to 1990.

An iron deposit at Isua near Nuuk, a large zinc deposit at Citronen Fjord in North Greenland and a number of deposits of molybdenum, platinum metals and gold have been investigated on the east coast, but have not yet been profitable.

Cryolite
– Greenland's biggest mining success

In the area near Arsuk, a so far unknown chalk-white mineral was found in the 1770s, which in 1779 was described and named cryolite – 'ice stone'. It became the mission of

IKAITE COLUMNS AT IVITTUUT

In many places in Greenland, unusual geological formations can be found that have been exploited locally as raw materials. One example is the now closed mining settlement of Ivittuut where cryolite, together with a number of other rare minerals is embedded in granite formed more than 1 billion years ago.

In a nearby fjord, it has caused the leaking groundwater to form giant submarine columns made of the rare calcium carbonate mineral ikaite. Within a very small area, hundreds of columns up to 20 meters high »grow« from the seabed. Since 1995, a number of interdisciplinary studies have been conducted to shed light on this unique natural phenomenon.

RICHARD MARTIN, 1995

Greenland is home to a wide range of special mineral deposits. In the Ikka Fjord in southwest Greenland, columns of the mineral ikaite rise from springs in the seabed. The mineral is only formed in cold water, and it degrades quickly if the columns are taken out of the sea. The first columns found were boxed and sent for analysis in Denmark, but they disappeared before the boxes arrived.

J. LAUTRUP, GEUS, 2021

HARTMANN'S, 2021

To the left are five cabochon-cut gemstones from the Greenlandic underground: nuummit with yellow reflections, greenlandite (dark green), tugtupite (pink), amazonite (green), lapis lazuli (blue). To the right: Rings in white gold with faceted rubies surrounded by diamonds.

GREENLANDIC GEMSTONES

The most interesting Greenlandic gemstones are *ruby* and pink sapphire. The deep red to pink colour of the stone is due to a 0,5-1 % chromium content. Rubies were first found in 1966, but it was not until 2017 that the Aappaluttoq ruby mine opened 150 km south of Nuuk. The raw stones are exported to Asia, where they are cut. The Government of Greenland receives a royalty of 5,4 % of the sales price. The rubies formed deep in the Earth's crust 2.7 billion years ago.

As the first place in the world, the pink *tugtupite* was found in 1957 in the 1.1 billion-year-old Ilimassaq intrusion near Narsaq. It is often referred to as Greenland's national gemstone.

Greenlandite is a variety of quartz with the gemmological name aventurine quartz. The green colour is created by microscopic grains of the chromium-containing mineral fuchsite. The stones can be found in the 3.7 billion-year-old Isua area, and Greenlandite can thus be described as the world's oldest gemstone.

Nuummite is an inconspicuous, black, 3 billion-year-old lava rock with two intergrown amphiboles. Only when polished do the stone's coloured reflections appear in golden and bluish nuances (iridescence).

Lapis Lazuli is known only from the Tupertalik site near Maniitsoq in the 1.8 billion-year-old volcanic rock carbonatite. The mineral has been a coveted gemstone since ancient times. The Greenlandic stone has a less intense blue colour than the classic azure stones from Afghanistan.

Other Greenlandic gemstones include *amazonite, Greenlandic moonstone, sodalite, agate* and other *quartz varieties.* Ruby and to some extent tugtupite are sold internationally, the others are used for local production of gemstones.

Greenland's biggest mining success, and one of Denmark's biggest industrial adventures. The industrial magnate C.F. Tietgen and the chemist Julius Thomsen started the exploitation of the cryolite deposit at Ivittuut in 1854 – with a view to producing soda and enamel. It was soon discovered that cryolite could be used as an excipient in the production of aluminium, which until then had been more expensive than platinum and gold. By using cryolite in the process, aluminium became cheaper than copper, and thus it became available as a utility metal. Greenland dominated the cryolite market until the deposit had been depleted in 1987, by which time more than 3.5 million tonnes of cryolite had been mined.

Access to the strategically important cryolite was a contributing factor to the American presence during World War II, when the income from cryolite paid for Greenland's supply of goods.

Extraction of uranium, thorium and rare earth elements

The area around Narsaq in South Greenland has had almost mythological status for almost 150 years. The area has some very unusual rocks, which contain very rare or unique minerals and have high content of a number of elements, such as uranium, thorium and the rare earth elements (REE), which are otherwise largely absent in ordinary rocks.

In 1956, Niels Bohr proposed that uranium from Kuannersuit (Kvanefjeld) could support Danish nuclear power, and a major exploration effort was launched. Towards the end of the century, REEs became important metals because they are used in many electronic components and in components for green energy production. This meant renewed interest in the area with two potential projects; one for the extraction of REE and zirconium, and one for REE, uranium and zinc. The uranium deposit has a relatively low grade but is a very large overall resource. Throughout the project's almost 70-year lifespan, the issue of uranium extraction, based on both environmental and political considerations, has given rise to a, at times, heated debate in Greenland (see p. 178).

The possibility of oil and gas extraction

Greenland is considered to have a potential for oil and gas roughly equivalent to the amount extracted in the Danish part of the North Sea. Deposits to that extent would have a significant effect on Greenland's economy, but so far no actual discoveries have been made. With an increasing shift in the energy market from oil, gas and coal to renewable energy sources, and a time horizon of decades from a possible discovery to production, it is doubtful whether there will ever be oil or gas production in the country.

Extraction of mineral resources

Extraction of mineral resources in Greenland is regulated through extensive legislation. This will reduce the negative societal or environmental impact of the exploitation of geological resources. Therefore, there is reason for being optimistic that the nascent mining industry can make a positive contribution to the development of the Greenland society.

THE ICE-FREE LANDSCAPES

The landscapes are determined by their underlying parent material and shaped by the erosion of repeated glacial periods and the redistribution of sediments. During the last glacial period maximum about 20,000 years ago, ice covered virtually all of the land and continued well beyond the current coastline. During the glacial period, large amounts of water were bound as ice, which weighed down the land, and at the maximum extent of the glacial period, the sea level was approx. 150 m lower than now.

Continuous permafrost means that an area has continuous permafrost in the soil. Discontinuous permafrost means that 30-80 % of the landscape has permafrost in the soil. Here, changes in the permafrost can have a significant impact on the stability of infrastructure such as airports, etc. In South Greenland, permafrost is largely absent

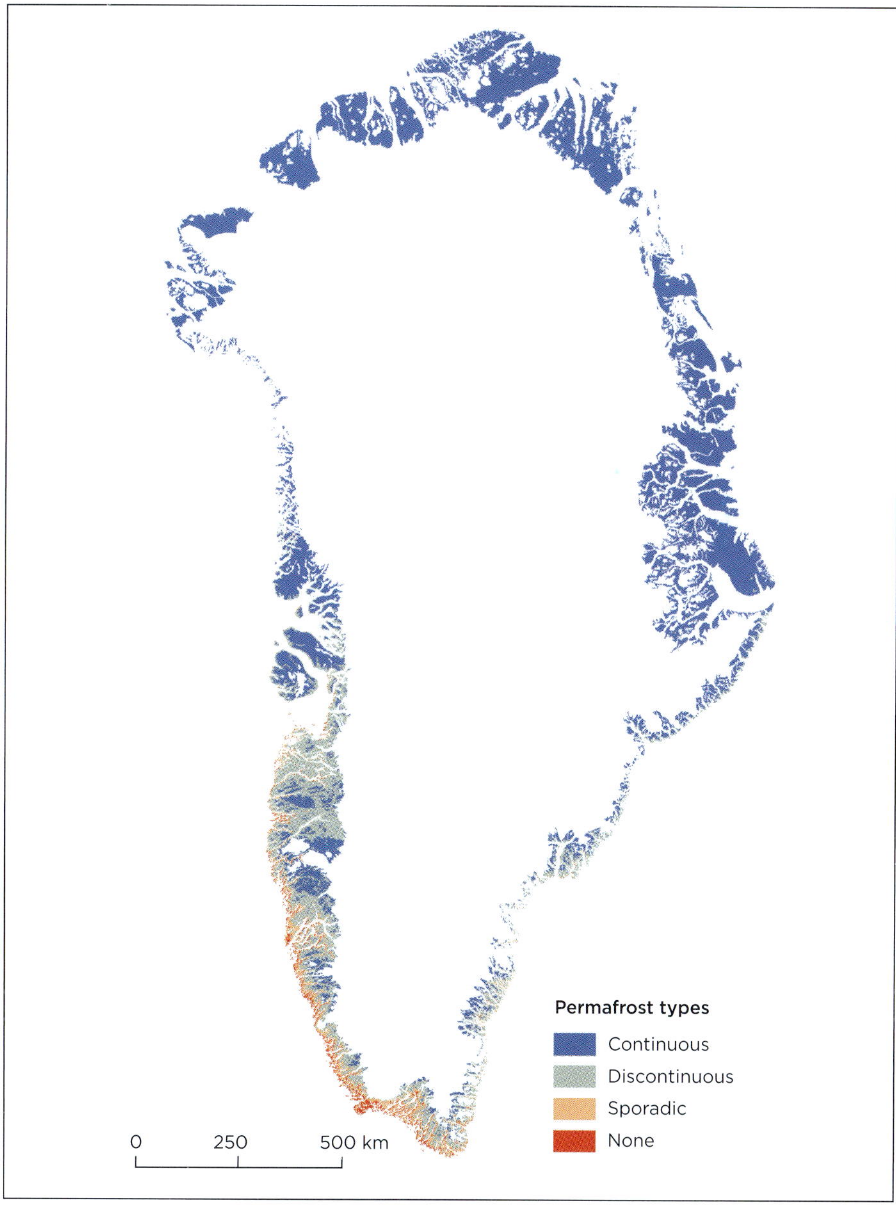

DET ISFRIE GRØNLAND, BO ELBERLING/CENTER FOR PERMAFROST, GYLDENDAL 2021

Polygon soils are widespread in areas with continuous permafrost. Polygon soils near Brønlundhus in Peary Land.

The landscape can change rapidly when the permafrost thaws. This is a so-called reverse thaw slide, where the surface of an entire landscape slides down a slope due to gravity. This exposes an ever-increasing part of the permafrost, and the process can thus be self-reinforcing.

During and immediately after the glacial period, ice and meltwater brought large amounts of sediment from the central parts of country and out to the areas that today make up the relatively shallow banks with water depths of less than 200 m off the coast. Most Greenland landscapes with mountain peaks, deep valleys, round rock landscapes, moraine landscapes, dead ice landscapes and meltwater plains, fjords and archipelagos today are a result of these processes. Since the retreat of the ice, coastal areas have slowly been uplifted. In many places along the coast, the remnants of former coastlines are seen in the form of beach ridges and coastal cliffs at an altitude of up to 140 metres.

Permafrost

The ice sheet insulates so well against the winter cold that the Earth's own heat production ensures that there is no permafrost under the ice sheet. Permafrost – soil and sediments that have been frozen for at least two years in a row – is thus only found in the ice-free areas where it was formed when the ice was fully retreated after the last glacial period about 10,000 years ago. Since then, most of the permafrost in Greenland has increased in thickness. At Qeqertarsuaq (Disko) in West Greenland, the depth of the permafrost has been measured at 350 m and at Qaanaaq in North Greenland at 450 m below the Earth's surface. In South Greenland, permafrost is only found in depressions or high up in the mountains.

The landscape formed until the end of the last glacial period has subsequently been characterised by repeated freeze-thaw processes and snow – an interplay which has created very special periglacial landscape forms. Solifluction occurs, for example, where the upper soil layers move on top of the underlying frozen soil, while ice wedges form when meltwater seeps into frost cracks in the soil and subsequently freezes and expands repeatedly, winter after winter. This forms wedge-shaped ice bodies, seen from above as polygon-shaped patterns.

Changes in landscapes are also linked to climate change. When the permafrost thaws, the meltwater runs off, parts of the landscape collapse and erosion increases. Such physical degradation of the landscape can accelerate the consequences of climate change and has already had significant consequences for infrastructure in some places in Greenland, when, for example, buildings, pipelines, roads and debris sites are destabilised or destroyed.

Release of greenhouse gases

Today, the ice sheet is retreating, and in the new exposed landscapes new permafrost is forming. Pioneer plants immigrate rapidly and spark soil development in a matter of years. The immigration of plants and the formation of the soil environment are closely linked. In the soil, carbon accumulates from the plants, and at the same time, there is a constant decomposition of both the soil's pool of organic matter and a weathering of the soil's other particles. The degradation of plant material is slower than plant production during the growing season due to the generally cold conditions. This is why Arctic soils are generally characterised by an accumulation of carbon. However, as the landscapes in Greenland are relatively young, the amounts of accumulated organic matter are limited with peat formations of typically up to half a metre compared to much older landscapes such as in e.g. Siberia, where peat formations can be more than 10 metres thick.

The Arctic has long been in focus as a source of greenhouse gases due to the accumulation of organic matter in the soil environment and a warming that exceeds the average global warming. However, there are many indications that the Greenland nature on land does not make a significant contribution to the atmosphere's increasing content of greenhouse gases. The natural carbon dioxide account virtually balances (i.e. the plants absorb the same amount as is released from the soil), the uptake of methane in the widespread dry soils exceeds the release of methane from wetlands and the amounts of nitrogen in the soil environment are so limited that the release of nitrous oxide is very small.

BO ELBERLING, 2012

Permafrost drilling at Zackenberg in Northeast Greenland is a prerequisite for characterising the permafrost's content of frozen layers. The release of meltwater followed by erosion with a future thaw is a cause of concern – not the least in relation to infrastructure. Moreover, the decomposition of organic matter can contribute to the release of greenhouse gases.

Weathering of the landscapes

The landscapes are also slowly undergoing physical and chemical degradation. Due to large variation in the geological parent material, there is a major difference in how quickly the landscapes weather and thus decompose. In many places, the Earth's surface consists of bedrock, and in those places the weathering is slow. At Qeqertarsuaq (Disko) in West Greenland, on the other hand, the parent material is basalts, volcanic rocks, which weather faster and release large amounts of iron, which colour the landscapes and rivers reddish. In North Greenland, the source material is carbonate rocks, e.g. dolomite which weathers much faster than bedrock of, for example, gneiss.

The landscapes are also being transformed today. Earthquakes, landslides and calving icebergs can cause tsunamis and subsequent floods. In summer 2017, a landslide caused a 90-metre-high tsunami wave, which was 10 metres high when it hit the small fishing town of Nuugaatsiaq in West Greenland and unfortunately washed houses and four people into the sea more 30 kilometres away from the landslide.

LARS HOLST HANSEN, 2011

Arctic foxes and muskox carcasses at Zackenberg in Northeast Greenland. The arctic fox, like other terrestrial mammals, is well adapted to the cold climate with a thick and heat-insulating coat. Furthermore, the body is more compact, and extremities such as ears, snout and tail are shorter than in its more southern relatives. All to minimise heat loss.

Flora and fauna on land

Life on land largely disappeared in Greenland during the last glacial period, except perhaps from a few lowland areas and mountain peaks surrounded by ice (nunataks, see p. 20). Animal and plant life on land is therefore today determined by the species' ability to migrate and adapt to the physical conditions, including the extreme climatic conditions with low winter temperatures, snow, ice and the dark polar winter.

The diversity of species on land in Greenland is generally lower than southern rural areas and it generally

Grazers such as caribou and arctic hare depend on access to food all year round, even in the cold, dark winter months. Icing events, where a layer of ice forms on vegetation or snow, can prevent animals from reaching food and lead to increased mortality levels. Icing events are believed to be the cause of the extinction of caribou in Northeast Greenland around 1900.

LARS K MIKKELSEN/RITZAU SCANPIX, 1996

LARS GEJL/BIOFOTO/RITZAU SCANPIX, U.Å.

decreases from south to north. Local differences in climate, soil and nutrient availability give rise to different habitat types (e.g. between coast and inland, and between mountains and valleys). The Greenland flora and fauna is gradually changing from a lush birch forest scrub in sub-Arctic South Greenland to the polar desert in High Arctic North Greenland.

As the climate warms, the distribution patterns of flora and fauna are expected to change. Bushes, shrubs and trees grow larger and increase their distribution when there is enough water and nutrients. With higher temperatures, southern species are expected to move north and shift the distribution of more cold-adapted species, which as far as possible must move further north or up mountain sides to follow the temperature and precipitation conditions to which they are adapted.

The breeding season is short, so it is important to be ready when conditions are right. The arrival of migratory birds, calving of caribou and muskoxen, and the emergence of insects are therefore carefully attuned to the changing seasons and the growing season of the plants. A balance that is challenged by climate change and a premature and more varied melting of the snow cover.

Terrestrial mammals

Greenland only has seven naturally occurring species of terrestrial mammals: muskox, caribou, polar wolf,

RELEASE OF MUSK OXEN

Muskoxen were originally only found in North and Northeast Greenland. In 1962-63, a total of 27 muskoxen were moved from Northeast Greenland and released in the Angujaartorfiup Nunaa area south of Kangerlussuaq in West Greenland. This area is, among other things, characterised by only a limited amount of precipitation. In winter, it offers the muskoxen easier access to food, which has resulted in lower mortality and higher reproduction compared to Northeast Greenland. The population therefore grew rapidly, and today the animals can also be found north of the Kangerlussuaq fjord. The most recent count (in 2018) of muskoxen in the areas north and south of Kangerlussuaq estimated that the population consists of approx. 22,000 animals.

Since the release of muskoxen at Kangerlussuaq, muskoxen have been moved from here to seven other locations in West Greenland from 1986 to 2014. The main purpose of the releases has been to boost local hunting opportunities. Signs of overgrazing can be seen in some release areas, however, the effect of releases on vegetation composition in the new areas has yet to be sufficiently investigated.

LEMMINGS – AN IMPORTANT LINK IN THE ARCTIC FOOD CHAIN

Lemmings and their predators are known for their cyclic population fluctuations. Multiplication of the lemming population approx. every four years provide predators with optimum breeding conditions and thus result in larger populations of predators. This leads to increased predation and thus a pronounced reduction in the lemming population. Fewer lemmings mean fewer food sources, resulting in increased mortality among predators. The cycle can thus begin again.

The northern collared lemming lives in North and Northeast Greenland. It is an important herbivore and a vital prey for predators such as snowy owls, long-tailed skua, stoats and arctic foxes. In years with many lemmings, predation on the breeding birds of the tundra is lower. The lemming is thus a central element in the food chain in Greenland as well as in the rest of the Arctic.

The climate is changing in the Arctic, and this is changing the living conditions of lemmings and the species they interact with in the food chain. In some places in the Arctic, the lemming's well-known cyclic population fluctuations seem to have stopped and been replaced by much more unpredictable fluctuations with great variation from one year to the next. The role of climate change in this phenomenon and how it will affect food chains in the Arctic remains uncertain.

ERIK THOMSEN/BIOFOTO/RITZAU SCANPIX, 2008

In summer, the northern collared lemming is greyish brown, camouflaging it against the rocks and bare earth. With the arrival of autumn, however, the fur gradually lightens, and when the snow has settled over the land, the lemming's coat is mostly white. At the same time it develops an outgrowth under two of the claws on the front paw. These 'double claws' counteract the wear from the frozen ground and provide a little extra insulation.

10 COMMON BIRD SPECIES IN GREENLAND

Whether a bird is common depends on where you are, when you are there, the size of the population and the bird's appearance and behaviour. The ten bird species shown here have been chosen based on their distribution, numbers, protection status and how often they can be observed, but they have also been chosen to represent different groups of species and what is common in different parts of the country.

JULIUS NIELSEN, 2021

Miteq siorartooq (GRL), *Somateria mollissima* (LAT)

The common eider is found in large parts of Greenland and a sought-after hunting prey. Protection of the species during the breeding season (implemented in 2001) turned decline into increase, aided by large flocks of four or six chicks.

JULIUS NIELSEN, 2022

Nattoralik (GRL), *Haliaeetus albicilla* (LAT)

The white-tailed eagle is often seen near towns and settlements in West Greenland and has spread north in recent years. Unlike other birds of prey, such as the snowy owl, gyrfalcon and peregrine falcon, the white-tailed eagle hunts mainly at sea.

JULIUS NIELSEN, 2019

Qupaloraarsuk (GRL), *Plectrophenax nivalis* (LAT)

The snow bunting is the most common passerine in the country. It breeds throughout Greenland and is often seen in small flocks on spring and autumn migrations through towns and settlements, where it is a frequent visitor at the feeding table.

LARS WITTING, 2006

Appaliarsuk (GRL), *Alle alle* (LAT)

With an estimated 60-80 million individuals in Greenland, more than 80 % of the global population, the little auk is a national protected species of Greenland. The species breeds in large colonies mainly at Qaanaaq and Scoresby Sound.

LARS WITTING, 2006

Appa (GRL), *Uria lomvia* (LAT)

The thick-billed murre is a traditionally important hunting species in Greenland. The species is particularly sensitive to hunting, as a breeding pair produces only one young per year. Despite management measures, the population has been in decline for decades.

LARS WITTING, 2007

Qaqulluk (GRL), *Fulmarus glacialis* (LAT)

The northern fulmar is a characteristic bird of the open sea, where it soars low over the water in search of food. It is common in coastal and open seas, and is often seen following larger ships.

JULIUS NIELSEN, 2019

Aqisseq (GRL), *Lagopus mutus* (LAT)

The rock ptarmigan is an abundant breeding bird in most of Greenland. The plumage changes to white in winter, and despite adapting to a snow-covered landscape, it is a frequently killed prey.

JULIUS NIELSEN, 2019

Tulugaq (GRL), *Corvus corax* (LAT)

The raven is a prominent figure in the Inuit creation story and is included in many Greenlandic legends and myths. The species is common in Greenland and can be seen all year round in towns and settlements throughout Greenland.

LARS WITTING, 2005

Naajaannaq (GRL), *Larus glaucoides* (LAT)

The Iceland gull is one of the most common gull species in Greenland. It breeds throughout West Greenland and more sporadically in Southeast Greenland. The winter months are spent on the open sea in Greenland or to the south.

CARSTEN EGEVANG, 2013

Nerleq (GRL), *Anser albifrons* (LAT)

The Greenlandic white-fronted goose is an endemic subspecies of the greater white-fronted goose. It breeds in West Greenland and is one of six goose species found in Greenland.

LOUISE MURRAY/SHUTTERSTOCK/RITZAU SCANPIX, 2011

SEABIRDS

Seabirds live most of the year at sea, but breed on mountains and islands along the coasts. In winter, they seek out the ice-free areas along the continental shelf in the North Atlantic (see p. 59). Greenland has large seabird populations with the little auk colonies in the North Water area by Qaanaaq with more 30 million pairs being one of the world's largest seabird populations. *The little auk* is a small auk that feeds on zooplankton and small fish. It is estimated that 80 % of the world's population breeds on the sides of the pebble mountains.

The thick-billed murre has 19 breeding colonies of very varying size along West Greenland and two smaller colonies in East Greenland. The largest thick-billed murre colonies are found in northern Upernavik and in the area around Qaanaaq. The populations along West Greenland have been in decline for decades, primarily due to hunting and disturbance, however, climate change may also be a contributing factor. The thick-billed murre feeds on smaller fish and lays its solitary egg on rock ledges on steep cliffs.

Other typical seabirds are the *Arctic fulmar*, which breeds on steep cliffs from Qeqertarsuup Tunua (Disko Bay) and northwards to Qaanaaq and along East Greenland, and the *kittiwake*, which breeds everywhere along the coasts where there are steep cliffs and access to ice-free areas in summer.

Among ducks, the *common eider* is the most widespread seabird in Greenland. It is associated with coastal areas where the pairs build nests on small islands out of reach of foxes. Most eiders breed in West Greenland, but there are also small colonies in East Greenland. In winter, about half a million eiders from Canada and Greenland congregate in the shallow areas along Southwest Greenland where they feed on the plentiful mussel beds.

Unlike the common eider, the king eider does not breed in colonies but in breeding pairs scattered on the tundra near lakes. In winter, large concentrations of over one million king eiders from North Canada and North Greenland congregate on Store Hellefiskebanke.

Black-legged kittiwake (Latin: *Rissa tridactyla*, Greenlandic: Taateraaq) is adapted to the marine environment and breeds all along the west coast and more sporadically in the east. Breeding colonies are on steep cliffs facing the water, in fjords and along outer coastlines. The decline in the population means that the black-legged kittiwake is listed as 'Vulnerable' on the Greenlandic red list.

A special seabird is the *Arctic tern*, which migrates the longest of any animal from Greenland to Antarctica. In Greenland, Arctic terns breed in colonies on islands and islets. The world's largest colony is located on Grønne Ejland in Qeqertarsuup Tunua, but the population has dropped considerably; presumably due to the growing fox population.

Other small populations of seabirds in Greenland are black guillemot, long-tailed duck, harlequin duck, red-breasted merganser, great cormorant and the gulls glaucous gull, Iceland gull and great black-backed gull.

arctic fox, stoat, arctic hare and collared lemming (polar bear is considered a marine mammal). This is due to the country's isolated location with limited immigration opportunities via the passage offered by the sea ice to other land areas.

Polar wolves are few in number and occur sporadically in Northwest, North and Northeast Greenland. Muskox, stoat and collared lemming are native only to North and Northeast Greenland, caribou only to West Greenland (after they went extinct in Northeast Greenland around 1900), while arctic fox and arctic hare are widespread in most of the country. Caribou and muskoxen have been released in several places in West Greenland, including at Kangerlussuaq, where the muskox population has grown to several thousands of animals since the release of 27 animals in 1962-1963.

There are no reptiles and amphibians in Greenland.

Bird species

Approx. 235 bird species have been registered, of which about 60 breed regularly in the country. Most migrate south in winter, the arctic tern as far away as Antarctica. More than half of the breeding birds are associated with either land or freshwater. There are more bird species in Northeast Greenland than in West Greenland, which is due to more stable and predictable weather conditions. Lesser black-backed gull and Canada goose are examples of birds that have immigrated in recent times and have established large populations over the past 30 years.

There are no endemic bird species in the country due to the short amount of time since the last glacial period, but four species have developed distinct Greenlandic subspecies, namely the Greenland greater white-fronted goose, mallard, dunlin and rock ptarmigan.

Insects and other small animals

The number of species among insects and other small animals is uncertain. Currently, Greenland has more than 1,200 registered species of insects, spiders, mites and millipedes. Inland areas with many hours of sun and overall high summer temperatures have a much higher diversity of insects and other small animals than the coastal zone. Insect species and small animals have immigrated to Greenland from both North America and Europe, but since Greenland is an isolated island, the diversity is quite low.

Vegetation

The vegetation is Arctic, meaning that it consists primarily of perennial herbs and dwarf shrubs. The tallest species are willow and birch trees, which in areas offering a high level of protection can grow more than 2 metres tall. Forests of low birch trees are only found in the inner fjord ar-

To the left: The four spot orb weaver (Latin: *Araneus quadratus*, Greenlandic: Aasiak sisamanik toornilik) lives only in South Greenland, where it often makes wheel spindles in willow thickets or between stones and boulders. Here it sits in hiding, waiting for the vibration of its web to reveal that an insect has fallen into its trap. To the right: Arctic colias (Latin: *Colias hecla*, Greenlandic: Pakkaluaq sungaartoq) is widely seen all over Greenland, where it lives in dry grassland areas. They tend to sit with their wings folded, so the colourful exterior side is usually only seen in flight.

KIRSTEN HJØRNE, 2015

10 COMMON PLANT SPECIES IN GREENLAND

About 500 native plant species can be found in Greenland. The plants can be found in plant belts, which cover the High Arctic, Low Arctic and Subarctic. The number of species decreases the further north you get. The following species have been chosen based on their distribution and how frequently they are observed. Moreover, they represent different forms of growth, e.g. dwarf shrubs, herbs and sedges and are easily recognisable and often characteristic species in the areas in which they grow.

KATRINE RAUNDRUP 2009

Tikiusaaq (GRL), *Campanula gieseckiana* (LAT)

Greenland bluebell is found in large parts of the country. The 2-3 cm edible flower is bluish purple, and the plant, which grows 10-40 cm tall, cannot be confused with any other.

IDA BOMHOLT DYRHOLM JACOBSEN, 2021

Ivissuaq nakangasulik (GRL), *Luzula spicata* (LAT)

Spiked woodrush is a grass-like plant that grows 15-30 cm tall. It has drooping inflorescences that resemble ears of corn. It is common on dry heaths, in moist substrate and on gravelly rocky slopes.

KATRINE RAUNDRUP 2016

Orpigaq (GRL), *Salix glauca* (LAT)

Grey willow is found in many areas of Greenland. The species is described as a dwarf shrub, but in sheltered areas it can reach a height of up to 250 cm. It flowers with male and female flowers on separate plants.

KATRINE RAUNDRUP 2012

Paarnaqutit (GRL), *Empetrum nigrum* (LAT)

Black crowberry is a dwarf shrub that grows in large areas of Greenland. The flowers are quite small, but the plant produces edible slightly bitter black berries, which are widely used in Greenlandic cooking.

KATRINE RAUNDRUP 2009

Ukaliusaq (GRL), *Eriophorum scheuchzeri* (LAT)

The flowers of the Scheuchzer's cotton grass are quite small. Only once they actually finished flowering do you see the characteristic white top, which is formed by long hairs on the mature seeds. The plant is found on moist substrate throughout the country.

IDA BOMHOLT DYRHOLM JACOBSEN, 2021

Avaalaqiaq (GRL), *Betula nana* (LAT)

Dwarf birch is common in large parts of Greenland, where it often thrives on heaths, in fell fields, bogs and marshlands. The species is described as a dwarf shrub, but can grow up to 1m tall.

IDA BOMHOLT DYRHOLM JACOBSEN, 2019

Seernaq (GRL), *Oxyria digyna* (LAT)

Mountain sorrel grows in damp and gravelly rocky areas, as well as herb slopes and snowfields throughout the country. The edible green leaves have a high vitamin C content and are eaten either raw or stewed.

KATRINE RAUNDRUP 2006

Umerluusaq (GRL), *Dryas integrifolia* (LAT)

Entireleaf mountain-avens forms cushions of 2-3cm in size, of single white flowers. The plant grows on dry and gravelly heaths throughout the country. In East Greenland the closely related white dryas is found.

KATRINE RAUNDRUP 2006

Niviarsiaq (GRL), *Chamaenerion latifolium* (LAT)

Dwarf fireweed grows 15-30cm tall and has large purple flowers. The edible plant is Greenland's national flower and grows on gravelly soil throughout the country.

ANDERS TVEVAD/BIOFOTO/RITZAU SCANPIX, 1996

Kuanneq (GRL), *Angelica archangelica* (LAT)

The garden angelica grows 70-200cm tall and with its 10-20cm flower heads cannot be mistaken for any other species. The edible plant grows in moist areas and is common in the southern part of Greenland.

ANDERS MØLLER, 2008

The flora in the South Greenland valleys, such as Blomsterdalen (Flower Valley) at Narsarssuaq, is among the lushest in the country. A rich profusion of flowers and metre-high willow trees are a common sight in the most protected valleys of the inland areas.

eas of South Greenland which are climatically referred to as sub-Arctic. The large expanse of the country creates vastly different living conditions for the plants, and diversity decreases considerably from south to north.

The distribution of species differ biogeographically, since South Greenland is the only region with more eastern species than western species. The distribution of species is determined by differences in plant dispersal possibilities, preferences and physical environment. Some plant species are widespread across different climate zones (e.g. koenigia), while others are associated with a specific habitat type and are only known from a few locations (e.g. elephant's head). About 520 plant species have been registered, of which about 30 are endemic.

The spread of fungi, lichens and mosses is conditional on the interaction between light, humidity, temperature, substrate and, in the case of fungi, the vegetation. It is estimated that there are approx. 1,600 species of fungi, almost 1.000 lichens and about 600 moss species. The spores of the fungi spread easily with the wind and fungi can therefore be found almost everywhere. Lichens grow on rocks, soil or other bare surfaces and they can withstand low temperatures and drying out for extended periods. Mosses are widespread and depend more on the substrate than on the climate.

The cultural landscapes

From north to south, the country's biotopes range from the High Arctic desert in Peary Land, where musk-oxen made for the primary game, to Southwest Greenland's subarctic environment that allows agriculture. Human immigration and dispersal and the subsequent traffic and trade have primarily taken place along the coast, meaning that especially the communities on the east coast experienced isolation from the more populous areas of west coast for extended periods of time. Early concentrations of towns and settlements occurred in archipelago environments at the mouths of the large fjords and near areas with open water in the ice-covered sea areas with rich marine resources.

Raw materials were also obtained in the form of driftwood, soapstone and iron, and the Saqqaq and Dorset cultures exploited deposits of killiaq and other suitable rock types for making stone tools (see p. 92).

Man has for millennia used and interpreted the landscapes in different ways, and traces can be seen everywhere in the form of towns and settlements with remnants of early houses made of peat and stone, stone built caches, shooting blinds and cairns that show the way or were used for battue. On the surface, traces of the earliest people Saqqaq, Independence I and Greenlandic Dorset are usually seen only in the northernmost regions where hearths and stone circles made of the stones used to keep the tent skin close to the ground are visible (see p. 92) where they were left perhaps millennia ago.

In West Greenland, it is the Inuit's and the Norse's peat and stone-built dwellings and, for the Norse, also churches and storage houses that stand out.

Terrestrial game played a major role in West Greenland between Qeqertarsuup Tunua (Disko Bay) and Nuup Kangerlua (Godthåb Fjord), where the ice-free land areas are home to the largest and most stable caribou populations. In this part of the country, coastal marine resources together with the inland caribou populations form the basis for the characteristic coast-inland economy where life in coastal winter settlements in summer was replaced by living in tents, fishing at estuaries and inland caribou hunting. This lifestyle and the resulting visible remains are preserved and disseminated in the World Heritage site, Aasivissuit – Nipisat (see p. 371). In contrast to the Inuit, the Norse had a pattern of towns and settlements that was strongly oriented towards the inner fjord and the more productive grazing areas. In southern Greenland the cultural landscape of the Norse as well as contemporary sheep farming is protected in the World Heritage site Kujataa.

The towns mostly evolved around the 18th and 19th-century colonies. The trade at the colonists depended largely on the hunters who were able to provide blubber, skin and fish. Later, these locations have presented challenges in terms of establishing airports or providing drinking water or areas for urban development.

THE FRESH WATERS

Greenland has vast areas of fresh water. Water cascades from rock walls, accumulates in streams, ponds and large lakes and eventually flows into the sea. All year round, estuaries with red, brown and clear water supply enormous amounts of fresh water to the sea. Melting from the many glaciers in summer and precipitation throughout the year are the direct cause. A network of streams has formed over time in valleys and on plains where precipitation and meltwater accumulate in small streams which in turn join into large rivers, which flow into the sea.

Many rivers and streams run through more stagnant waters and form a number of freshwater ecosystems along the way. These include mires, ponds, lakes and river basins, each of which is characterised by physicochemical conditions that in turn provide the basis for a special composition of plants and animals.

Due to widespread permafrost (see p. 42) and bedrock, water seepage is limited in many places, which increases the near-surface runoff and means that groundwater is largely absent. Accordingly, water accumulates on the surface and constitutes a significant element in the ecosystem.

The fresh waters are affected by the climatically determined runoff of meltwater from snow and ice and its content of suspended mineral particles and dissolved inorganic salts. Organic material from plants and

Communal houses were multi-family winter homes developed in the latter half of the 17th century and used until the beginning of the 20th century. To the left are peat walls from a winter dwelling at the Innap Nuua site near Sisimiut. The house measured 7x3 m and probably housed four to five families or 20 to 30 persons. The roof, covered with discarded boat skins and peat, was removed when leaving the winter house in spring. In this way of living, the living space was aired out and cleaned before it was used again. The drawn reconstruction to the right, illustrates the Norse settlement Sandnæs/Kilarsarfik in Amealkfjorden near Nuuk with cultivated fields and cattle that are driven to the mountains.

JENS FOG JENSEN, 2016

JENS ROSING, U.A.

microorganisms can also be added in areas with vegetation that are adjacent to lakes and streams. Overall, however, freshwater systems are typically nutrient deficient. This generally means that freshwater systems are frequent and constantly affected by the weather, the shape of the landscape as well as the processes that occur in the landscape.

Homeothermic springs

Some very unique ecosystems can be found in and around homeothermic springs, which are known from several places in Greenland such as Uunartoq and at Qeqertarsuaq (Disko). Green vegetation is present all year round and it is possible to find rare plants like orchids. Some springs are only a few degrees warm, while others reach higher temperatures and are used by locals and visitors as 'hot baths'.

Salt lakes

Another unique type of freshwater is lakes with a high saline content, which have formed because the land has risen after the last glacial period that former old lacunae have been cut off from the sea. Other salt lakes have formed because meltwater with a high saline content is continuously being led to the lake and because water evaporates from the surface at the same time.

Marshland

The rivers that flow into fjords and along coasts often form large delta areas (marshland) that serve as a transition zone to the marine ecosystem. The marsh is an extremely dynamic environment with repeated floods and deposits from a large catchment area. It provides a food base for waders that migrate to Greenland from e.g. Europe, and is a pantry for seabirds and fish. Even land animals, such as foxes, are attracted by the availability of food. Seen from a larger perspective, freshwater ecosystems make up a considerable part of Greenland's nature and therefore play a major role in the ecological cycle.

Freshwater flora and fauna

The biology of Arctic lakes and streams differs from lakes in many other climatic zones by having

a short growing season of a few months. The growing season is considered to be the period when the water is warm enough for animals and plants to grow and multiply. Winter is a long period with very little light because there is a thick layer of ice and snow on top of the water. Towards the bottom, however, the temperature rises and the temperature may be 2-4°C, while it is 0°C just below the ice. North of the Arctic Circle there is no light at all for a few months, since the sun is constantly below the horizon. The ice thickness typically reaches 1 to 2 m, meaning that very shallow lakes and ponds freeze solid during winter. Lakes with greater water depths (more than 3 m) will never freeze solid and therefore allow animals and plants to survive without going into hibernation. The thickness of the snow layer varies depending on the snowfall and the wind conditions.

The concentration of nutrients is often quite low, since the incoming

BO ELBERLING, 2016

Typical coastal landscape with many lakes and rivers that lead the water into a fjord. Here an area near Kangerlussuaq in West Greenland.

KIRSTEN S. CHRISTOFFERSEN, 2021

Moraine landscape with many small lakes and streams. Here from Zackenbergdalen (Zackenberg Valley) in Northeast Greenland.

water mainly comes from melted snow and ice. The meltwater can also transport silt (glacial flour) and soil from the surrounding area, making the water turbid.

Despite low temperatures and poor light conditions, plants and animals are active long before the ice melts. The intensity of sunlight increases significantly during spring (March to May), meaning that sufficient light can penetrate the snow and ice layer and thus kick-start photosynthesis. The amount of phytoplankton increases with the intensity of the light during these months despite low water temperatures. When the ice has melted and there is full sunlight, the growth of phytoplankton often stagnates because the nutrients are depleted.

Food chain in lakes

The food chain in the Greenland lakes is short and consists of a grazer food chain where phytoplankton is eaten by zooplankton (daphnia and copepods), which are themselves the prey of invertebrates (e.g. beetle larvae) and fish. There is also a degradation chain where the organic substances produced by algae and from animals that eat algae or each other are consumed by bacteria. The decomposers themselves are eaten by single-celled animals (protozoa), which in turn end up as food for daphnia and copepods. This closes the cycle and only a small part of the production ends at the bottom. Nevertheless, an active food chain can also be found at the bottom, including bacteria, unicellular organisms, chironomid larvae, flies and beetles as well as tadpole shrimp, an ancient crustacean order, which feed on the material that sinks to the bottom.

The key component in the lake bed, however, is the 'algal mats', which are millimetre-thick complex layers of algae and bacteria. The algal mats can cover large areas of the lake bed. An important component of the mats are filamentous blue-green algae that are able to bind nitrogen, which is of great importance in an otherwise nitrogen-deficient environment.

Another type of blue-green algae group, which belongs to the genus, the cynaobacteria *Nostoc*, forms spherical colonies that can grow to a considerable size. They are popularly referred to as sea tomatoes and can reach a considerable age because they grow slowly and have no enemies. On the other hand, the algal mats are a source of food for the benthic animals, but actually also for the zooplankton species from water bodies, which get a valuable food supplement on the lake bed by 'grazing' on the mat surface.

The tadpole shrimp (*Lepidurus arcticus*) is an interesting animal because it is so well adapted to Arctic life. It can be found both in the water body and on the bed. It is omnivorous and does not mind eating members of its own species.

Organisms in streams

The streams generally do not offer living ground for many types of organisms because of the considerable differences in the rate of flow throughout the season. In spring, the water flows rapidly, making it difficult for animals and plants to find a stable habitat. Nevertheless, chironomid larvae and flies can be found in the bottom layer and on the underside of rocks. The benthic organisms in lakes, ponds and streams are important prey for fish, and when aquatic stages of chironomid larvae and other insects hatch into flying adult stages, they provide food for many bird species.

Aquatic plants

Mosses are frequently seen at the bottom of lakes, ponds and streams with clear water, but Greenland does not have many species of actual aquatic plants, and often only a few species dominate a lake. Most hydrophytes, apart from the mosses, have floating leaves or other types of aerial leaves. This increases the uptake of carbon from the air, as fresh water in Greenland is very low on bicarbonate, which is the carbon source of plants. The most common species are pond weed (*Potamogenton* sp.), mare's tail (*Hippurus vulgaris*), water-starwort (*Callictriche sp.*) and bur-reed (*Sparganium sp.*). In the northernmost part of Greenland, you usually only find mosses.

Freshwater fish

The Arctic char (*Salvelinus alpinus*) is Greenland's most common freshwater fish. Its habitat is lakes with a water depth of more than 4-5 m, since the ice layer otherwise does not provide enough space and especially not enough oxygen for the fish to survive. It is not uncommon to find several variations of Arctic char in the same lake. There can be both a migratory population if the lake has access to the sea via a river, and a stationary population, which in turn can consist of two forms. The migratory form can grow up to 60-70 cm and spends several months in the sea where it eats a lot of food so that it can swim back to the lake where it originally came from to spawn.

The two non-migratory forms are shorter and more slender, which is most clearly seen in their dwarf form, being only about 15-17 cm when sexually mature. The different forms are the result of the scarcity of food in the lakes.

Although the occurrence of Arctic char is often sparse, they have a great impact on the volume and composition of zooplankton and tadpole shrimp in the lakes. It is generally an 'either-or' situation; when Arctic char are present, there are no large daphnia and only very few and small tadpole shrimp.

Salmon (*Salmo salar*) is common in the sea around Greenland but only breeds in the river at Kapisillit – which means salmon in Greenlandic – in the innermost part of Nuup Kangerlua (Godthåb Fjord). The salmon here has probably been isolated for many thousands of years and is genetically very unique. Unlike Arctic char, salmon roe requires relatively high temperatures to develop. In the river system at Kapisillit, a number of small shallow lakes can be found which are relatively easily warmed by the sun in summer, and this probably explains the occurrence of salmon in this area.

Birds of freshwater systems

Birds associated with the freshwater systems especially include geese, which use lakes and delta areas during the period they spend in Greenland to breed and moult. In some areas there can be so many geese that it reduces the vegetation cover and at the same time, nutrients are supplied via goose droppings to the lakes. Other common bird species at lakes are common loon, common goldeneye and red phalarope, most often with just one couple per lake.

Floods and emptying of glacial lakes

The water level in the rivers is at its highest when the snow melts in spring, during heavy rain showers and when high air temperatures cause massive and rapid melting from glaciers and ice sheet. It is called an outburst flood when the water levels and runoff are extraordinarily high and the water can flood land areas and cause great damage to roads, bridges and buildings.

Kangerlussuaq and Qeqertarsuaq are places where heavy rain and rapid melting of snow and ice have created large outburst floods. In July 2012, the

KIRSTEN S. CHRISTOFFERSEN, 2010

Dwarf form of mountain trout from upper Red River, Qeqertarsuaq (Disko). The mature and spawning fish are less than 20 cm long.

road between the two bridges crossing the Watson River in Kangerlussuaq was washed away by the water masses, and in 2014, the Røde Elv (Red River) in Qeqertarsuaq ran on both sides of its bridge after heavy rainfall.

In Narsarsuaq and at the research station in Zackenberg, outburst floods are seen in connection with a sudden emptying of lakes dammed by glaciers. Along the edge of the ice sheet and large glaciers lie lakes that are dammed by ice. When these lakes become too large, the water pushes its way out under the ice and empties into the river in an outburst flood. The same process is repeated at yearly intervals.

OLE GEERTZ-HANSEN, 2007

The emptied glacial lake at Kangerlussuaq in September 2007. The line on the glacier shows the water level a few days before emptying. There is about 30 metres difference.

Drinking water supply

Most towns and settlements use water from lakes and rivers for drinking water after it has been filtered and UV-purified or chlorinated. In the towns, water can be used directly from the tap, but in many towns and settlements, the citizens have to fetch water for the household from a tapping house.

North Greenland is very dry and cold in winter, and the surface water freezes solid, making it impossible to get water from lakes and rivers. In areas with insufficient surface water to supply citizens and industry, drinking water is made by desalinating seawater and melting growlers. In a place like Qaanaaq, this can lead to problems maintaining a good water supply during the winter period.

All lakes and rivers used for drinking water are protected by a water barrier zone, an area where activities are regulated.

Hydropower

In areas where water flows down from great heights in the terrain, it can be used for hydropower. Five hydropower plants generate electricity for six of the largest towns in Greenland. In 2018, they covered 67 % of the electricity consumed by citizens and industry.

The largest and oldest hydropower plant is Buksefjorden, which supplies electricity to Nuuk. The supply cable that carries electricity from the power plant to the town stretches over 58 kilometres and crosses, among other things, the huge Ameralik Fjord south of Nuuk. Here, the cable runs from a mountain top on one side of the fjord to a mountain top on the other, and at 5,376 metres, it is the longest free-hanging high-voltage line in the world.

Monitoring the hydropower potential

Hydropower is only used to a limited extent today, and it has great potential. However, hydropower has a great impact on nature, which must be held up against the benefits of security of supply and affordable green energy. The hydropower potential has been monitored since the 1970s, and provides valuable information about how much meltwater is formed from the ice sheet from year to year. The fresh water bodies are also monitored in terms of volume and content of various nutrients and thus also the nutrient supply to the many fjords.

Water volumes are monitored in many places in order to gain insight into how much water is flowing past the potential hydropower plants. Asiaq – Greenland Survey undertakes these surveys on behalf of the Naalakkersuisut (Government of Greenland). Asiaq has 21 active survey stations on the west coast from Qasigiannguit in the north to Paamiut in the south where precipitation and melting glaciers produce vast amounts of water in the rivers. In addition, several survey stations are operated by research

OLE GEERTZ-HANSEN, 2017

The ruby mine at Qeqertarsuatsiaat is located at the edge of a lake. Lowered water levels and turbid water are the immediate impact on the lake and river. The water level will rise again when the mine is closed.

projects, including the Zackenberg River in Northeast Greenland and the Watson River in Kangerlussuaq.

Pollution and wastewater

For people living along the coast, wastewater is led directly to the sea by way of pipelines or ditches. The settlements and parts of the towns have no sewage system, so here toilet waste is collected and emptied into the sea, while the grey water (kitchen and washing) is discharged into ditches and drains or to the ground. In some places, this can result in eutrophication (nutrient load) of suburban lakes.

Nalunnguarfiup Tasia lake in Sisimiut is an example of a place where seepage from the surrounding buildings and keeping of dogs have given rise to large amounts of filamentous algae.

Considerable amounts of fertiliser are used for agriculture in South Greenland, and leaching of nutrients to the streams causes local eutrophication phenomena such as the presence of filamentous algae in the streams.

Mining

The only industry in Greenland that makes discharges to freshwater systems is the mining industry. Most of the debris from mines consists of waste rock, i.e. rock that must be removed to get to the ore, and 'tailings', i.e. remnants of the crushed ore once the saleable minerals have been extracted from the ore.

Waste rock and tailings are often deposited at the bottom of freshwater lakes since this reduces the release of mainly heavy metals compared to deposit on land. The actual discharge of tailings and waste rock in lakes causes reduced visibility due to fine-grained suspended material. Once the material has sedimented, further degradation and release of unwanted substances occurs slowly. In addition to waste rock and tailings, there may be discharges of process chemicals from the processing of the ore and of wastewater from the mining town.

One example is the gold mine at Nalunaq, closed in 2013, where gold was mined using e.g. cyanide. During mining operations, measurements showed slightly elevated values of a number of substances such as cadmium in water and trout in the nearby river. However, four years after the closure, levels were back to normal.

The ruby mine near Qeqertarsuatsiat is another example of a mine located on the shores of a lake where the water level has been lowered 10 metres and where waste rock and tailings are being deposited in the lake causing slightly elevated levels of some metals and reduced clarity in the lake.

The mining industry has its own set of laws and executive orders, which set out recommended thresholds for a number of substances in the discharge to freshwater systems. Modern mining industry is relatively new in Greenland, and environmental regulation is still being developed.

THE SEA AND THE FJORDS

Greenland is surrounded by vast bodies of water with different temperatures and production conditions, which affect all life in the coastal areas – i.e. areas less than 3 nautical miles from land. Cold polar water that flows south on both sides of the country meets with warm Atlantic water and creates favourable production conditions on the continental shelf.

The location high up north results in considerable seasonal fluctuations in ice conditions and the biological production in the seas. Most of the seas and fjords around Greenland are covered by ice in winter. The exception is Southwest Greenland and the so-called polynyas, which do not freeze even though they are surrounded by sea ice.

The seas, sea currents, ice and the living resources

The stretches of sea around Greenland have experienced large fluctuations in both temperature and salinity, and both are also expected to change in future. From 1920 to 1930, the sea temperature along West Greenland rose by 1.5°C, and remained at this high level until the 1960s. During the period 1970 to 1990, the temperature fell, but then rose again, peaking in 2016. Since then, sea temperatures have decreased, but based on the current climate projection, it is expected that the surface temperature in the seas around Greenland will increase by an average of 0.3°C per decade until year 2100. This means that, by that time, large areas will be without ice during summer.

The ice around Greenland has also undergone major changes in the past 40 years since the amount of open water has increased dramatically. In Baffin Bay, the sea ice area has fallen by 4 % per decade, and the area currently accounts for only a third of the area in 1979. In the Greenland Sea, the decline is 9 % per decade.

Life in Greenland and the exploration and exploitation of Greenland have always been closely tied to the sea with its living resources, transport options and ice-filled limitations. Hunting of marine mammals and especially whales have driven several people across the ice from Canada, and commercial whalers from Europe have sailed to Greenland. Today, the fishery resources are crucial to society, whereas income from other resources in the sea, e.g. oil and gas, have yet to gain any traction.

Greenland is also located at the gateway to the Northwest Passage, which

Figure 2. Map of ocean currents. The cold ocean currents from the Arctic Ocean are shown in light blue and the warm North Atlantic Current is shown in fading red.

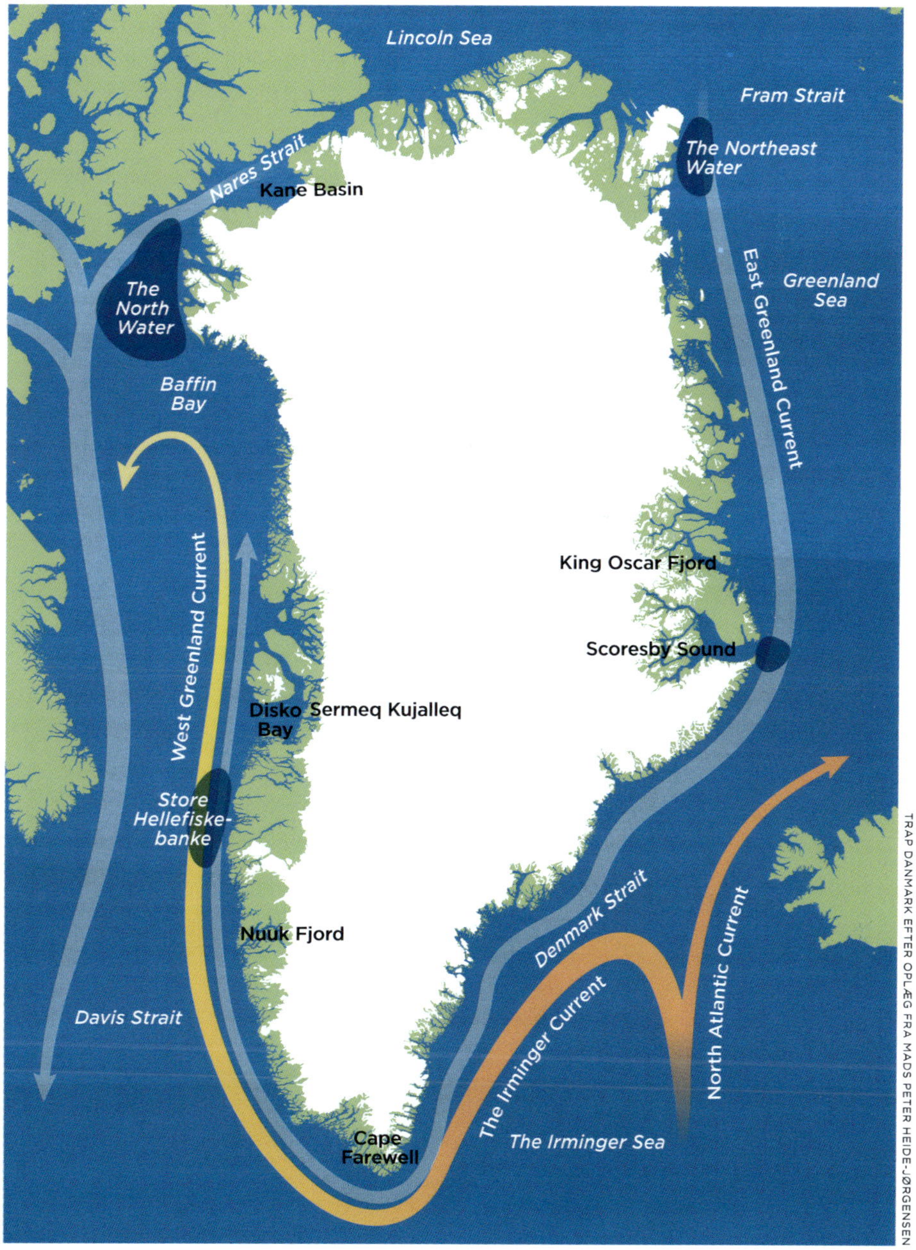

TRAP DANMARK EFTER OPLÆG FRA MADS PETER HEIDE-JØRGENSEN

ill-fated explorers have tried to map for centuries. With reduced ice formation in North Canada, the Northwest Passage may in the future gain new importance as a transport route between the Atlantic and Pacific Oceans.

Baffin Bay and Davis Strait

Baffin Bay lies between northern Canada and Greenland and is a large (690,000 km^2) and up to 2,000 m deep basin, which is covered with ice in winter and ice-free in summer. The eastern part of Baffin Bay is shallow with a water depth of less than 200 m. A continental shelf intersected by deep canals connects the deep water outside the continental shelf with the fjords and glaciers along the coast.

To the south, Baffin Bay joins Davis Strait, which stretches with a 300-600 m deep threshold from Greenland to Canada and forms the link to the North Atlantic. Sea depths in the southern part of Davis Strait are greater than 2,000 m. The eastern part of Davis Strait, unlike Baffin Bay, has a rather narrow continental shelf (less than 50 km).

The narrow and relatively low straits in North Canada constitute the primary supply of cold polar water to Baffin Bay. On the east side along Greenland's west coast, the West Greenland Current transports warm and salt water of Atlantic origin to the north.

Lincoln Sea, Arctic Basin and Greenland Sea

North of Greenland lies the Lincoln Sea. For most of the year, it is covered with perennial sea ice and which sends cold polar water down through Nares Strait to Baffin Bay. Water from the Pacific Ocean, which flows along North Canada, meets Atlantic water in the Lincoln Sea, from where the mixed water passes down the Nares Strait.

The Greenland Sea of about 1,2 million km^2 off the northern part of East Greenland is adjacent to the Norwegian Sea and the Denmark Strait and constitutes, via Fram Strait, the link between the Arctic Basin and the Atlantic Ocean. The western part is dominated by the cold East Greenland Current which moves south on the Greenland shelf. At Uummannarsuaq (Cape Farewell), the current bends to the north and continues in decreasing strength up to Qeqertarsuup Tunua (Disko Bay). To the east, a tributary of the Gulf Stream moves north with warm Atlantic water, which is cooled in the Greenland Sea and sinks to the bottom.

LIFE IN THE OPEN WATERS

When the ice disappears during spring, and the light can once again penetrate the water, a cascade of production begins in the water column. The melting of the sea ice has left a thin layer of fresh water on the surface, and here the phytoplankton exploits the rays of the sun for an intensive production. Nutrients are supplied from the deeper saltier water, but during summer the surface water becomes so mixed up that the primary production decreases in intensity and moves to greater depths. The periodicity of the ice and light means that production in the seas around Greenland has a distinct seasonal rhythm, which differs from lower latitudes. This rhythm is of great importance for zooplankton, which constitutes the next link in the food chain. The daphnia, which overwinter in deep water, migrate up through the water column in spring to graze on the intense production of phytoplankton. The fatty daphnia are eaten by fish larvae and certain birds and whales and are thus a vital link in the transport of energy, in the form of, e.g. omega-3 fatty acids, up through the food chain and thus also for the recruitment of new fish for the fishing industry.

To the south of East and West Greenland, in slightly warmer waters without ice cover in winter, the North Atlantic species of krill can be found, which primarily graze on the phytoplankton. Krill, in turn, is prey for a wide variety of boreal fish, birds and marine mammals, which head to the northern regions to forage on the concentrations of krill.

Rising ocean temperatures will affect the composition of zooplankton communities, and this will have consequences for the entire food chain in the ocean.

Krill (*Meganyctiphanes norvegica*) is an important link in the food chain in the southern marine areas east and west of Greenland.

In March, sea ice typically reaches maximum size, covering an area almost seven times the size of Greenland. As the salt is released when the ice forms, it floats on the seawater.

PATRICK KELLEY, USGS/US GEOLOGICAL SURVEY, CREATIVE COMMONS, 2009

The Greenland Sea is one of the most important places where the sinking of the warm water from the Gulf Stream takes place. This process is crucial for the circulation in the oceans. With a future increase in the surface temperature in the Greenland Sea, it is believed that this heat exchange, and thus the pump behind the Gulf Stream, could be affected.

Irminger Sea/Denmark Strait

Off Southeast Greenland lies a deep (2,000 m) basin, which is supplied with cold water from the East Greenland Current and warm water from a tributary of the North Atlantic Current. The area, which is called the Irminger Sea, is important for the formation of the Irminger Current, which brings warm water up along West Greenland and thus keeps the sea ice-free in winter and stimulates rich marine production.

Denmark Strait lies in the northern part of the Irminger Sea where cold and heavy polar water sinks beneath the light water from the North Atlantic Current.

Qeqertarsuup Tunua (Disko Bay)

The southwestern part of Qeqertarsuup Tunua has a 700 m deep channel, which supplies the bay with warm water from the West Greenland Current. The bay is usually covered by ice from January to April, but open water all year round has become more common in recent decades.

The Sermeq Kujalleq glacier in the eastern part of Qeqertarsuup Tunua is the largest supplier of icebergs in Greenland. It brings nutrients from land into the sea, and the melting of the light fresh water creates turbulence. The phytoplankton exploits the nutrients which, together with the light, set in motion the food chain.

The fjords

There are more than 150 named fjords in Greenland, and on top of this comes a large number of side fjords, e.g. King Oscars Fjord in East Greenland alone has 14 side fjords. The marine production in these fjords varies considerably, depending on the masses of water and active glaciers that supply the fjords with water and sediments.

In the largest fjord in West Greenland, Nuuk Fjord, lies the capital Nuuk. The fjord has an average depth of 250 m, but also has depths over 600 m. It is 190 km long and covers an area of 2,000 km^2. Fresh water from three glaciers, which lie on land, and melting of icebergs from another three glaciers, which run into the sea, mix in the fjord with more saline water from the West Greenland Current, allowing for a biological production that is atypical for Arctic areas. Only the inner parts of the fjord are covered with ice in winter.

The fjord system in Kangertittivaq (Scoresby Sound) is about 350 km long (about 10,000 km^2), and it is the world's largest fjord. Eleven side fjords with depths of up to 1,000 m branch off around the large Milne Land island. The fjord, which is cold and covered with sea ice from December to June, is supplied with meltwater from 12 glaciers and from the East Greenland Current.

Ice conditions along the coasts

Sea ice can take many different forms – land-fast ice, pack ice, perennial ice, fjord ice, drift ice, just to name a few. West of Greenland has a pulsating annual ice formation in Davis Strait and Baffin Bay. The ice first forms in the northern and western parts of Baffin Bay in December and spreads southwards during winter. Eventually, the ice reaches the coast of Greenland, and the maximum extent is seen in

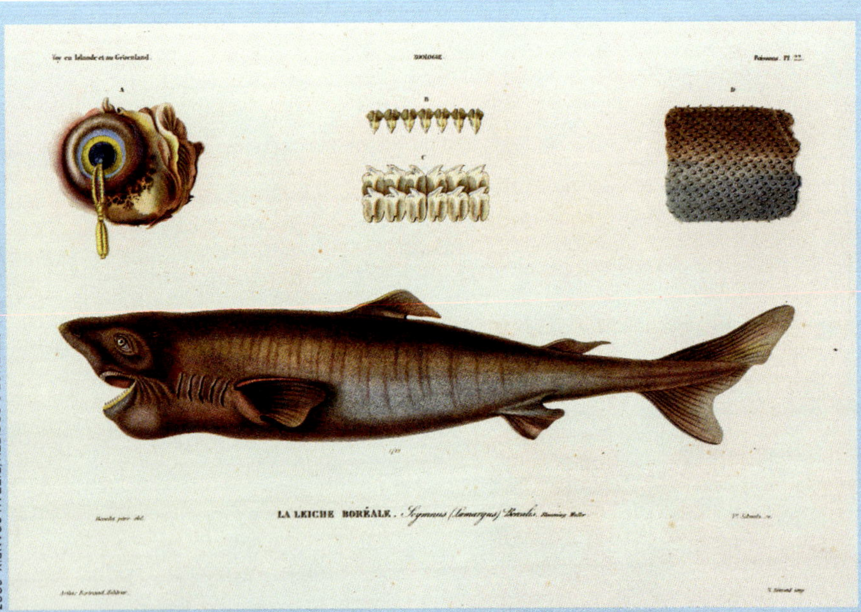

SSPL/SCIENCE & SOCIETY/RITZAU SCANPIX, 2003

GREENLAND SHARK

The Greenland shark can be found round all of Greenland and generally lives in the cold waters of the North Atlantic where it can be found from the surface down to a depth of 3 km. Until World War II, the Greenland shark was subjected to commercial exploitation because of its valuable liver oil, which was used as e.g. an excellent machine oil. Today, the sharks are mostly a nuisance to the fishery of Greenland halibut, and its meat is only used as dog food.

Although the Greenland shark reaches sexual maturity late and only gives birth to a few young, the population in the North Atlantic has apparently survived earlier commercial exploitation. Despite the historical exploitation, very little is known about the biology of sharks and their role in the ecosystem. They are apparently omnivorous, as remains from both snails, varies species of fish, seals and even large whales have been found in the sharks' stomachs. It is not known whether the sharks only go for carcasses, or whether they are fast enough to catch live prey such as seals, which are often found in their stomachs.

What makes the Greenland shark particularly special is its potential lifespan of up to 200-300 years, making it one of the longest-living animals on earth. This has been demonstrated through Cc14 analyses of the eye lens and is supported by observations showing that the sharks grow extremely slowly.

Lithograph of Greenland shark. This shows details of the teeth, skin and eye, as well as the characteristic parasite the Greenland shark often carries. The lithograph is by the French artist Auguste Mayer. Together with mainly naturalists Paul Gaimard and Eugene Robert, Mayer participated in a scientific expedition to Iceland and Greenland in 1835-36.

March, when the ice can reach as far south as to Nuuk.

In East Greenland, the ice is always in strong motion since the East Greenland Current carries more than 800,000 km^2 of ice per year from Fram Strait and further south at 10 km/h. The ice from Fram Strait consists of both perennial and one-year pack ice from the Arctic Basin, the multiyear sea ice, also called storis (great ice). The storis that drifts with the current down along East Greenland continues for some years around Cape Farewell where it is caught by the West Greenland Current and can drift all the way up to Nuuk Fjord. In recent years, ice exports from Fram Strait have increasingly consisted of one-year ice due to climate change, and storis is now rarely seen along the southern part of West Greenland.

Land-fast ice forms long the coast and inside the fjords in both West and East Greenland and is stable and immobile all the way to the shore. This is the ice that is used for dog sledding in the northern areas. In the northernmost parts of Greenland, the land-fast ice can remain throughout summer, but most fjords are ice-free from June.

Polynyas

All over the Arctic, you find coastal areas that are ice-free throughout winter. These areas are termed using the Russian word 'polynyas', and they are found on both the east and west sides of Greenland. The polynyas are formed in the same places every winter and occur either when warm water flow into an area inside the ice throughout winter, or as the result of the wind and current removing the newly formed ice. Polynas can also be formed as a combination of both mechanisms.

The constant formation of new ice in the polynyas leads to brine rejection, which contributes to the nutrient turnover in the water column and thus to the primary production. The polynyas are also characterised by open water in early spring when the sun can stimulate primary production. This means that seabirds can exploit the food chain in the upper part of the water column and thus start producing young in early spring. The polynyas are important for marine mammals because they allow seals and whales to get up for air throughout winter.

One of the most productive polynyas in the Arctic is the North Water Polynya located in Smith Sound between Greenland and Canada. It is kept open by strong northern winds that blow in Smith Sound throughout winter. With the formation of an ice bridge across Smith Sound, the wind gets a blow-off edge where new ice is constantly being pushed south. The North Water Polynya is recognised as a critical habitat for a large number of animal species. The future of the polynya is uncertain since the ice bridge in Smith Sound has become more unstable in recent years.

Another large but rather unproductive polynya is the Northeast Water, located in the northeasternmost corner of Greenland. The third large polynya is located off the Scoresby Sound fjord Kangertittivaq and, like the North Water Polynya, is an important area for a large number of birds and marine mammals.

Marine resources

The primary production in the seas around Greenland shows great seasonal variation, which is due to the six-month 'winter night' with no or little sunlight combined with heavy ice cover – and the corresponding intensive period in summer with open water, considerable nutrient supply and sunlight around the clock. The ice dampens the effect of the wind on the mixing and thus the circulation of nutrients in the water column.

In the free water masses, a number of organisms and bacteria live off the spring and summer primary production of planktonic algae. The most important organisms in the grazing in the upper water layers are daphnia and krill, which also constitute the food source for fish larvae, fish, seabirds and marine mammals. In shallow water (less than 200 m), a large part of the algae production sinks to the bottom, where a number of animals survive by filtering particles, e.g. excrement and residues from the production in the water surface. The benthic animals form part of a number of food chains, which include fish, eider, bearded seal and walrus. In deep water, only a small part of the surface algae production reaches the bottom.

Fish and other marine resources

There are about 250 fish species in the seas around Greenland. Mainly the populations that are exploited have been studied, while for most species, it has only been noted that this species has been observed. Fishing for a limited number of species, primarily shrimp, cod and Greenland halibut today constitutes the country's most important export and accounts for a quarter of the gross domestic product. Fishing is regulated through quotas which distinguish between coastal fishing with small boats and seagoing fishing with large factory trawlers. To support coastal fishing, factory facilities have been established in 11 towns and smaller facilities in several settlements.

The Atlantic cod is perhaps the fishing resource that has had the greatest impact on the country's development. Kitchen middens have revealed very large fish bones from cod, which shows that the earliest cultures in the country fished for cod. The first reports of cod fishing off Greenland date from the mid-19th century, but from the 1920s, fishing on the banks in West Greenland grew in scale. Fishing peaked in the late 1960s with record catches of half a million tonnes a year. The period from 1920 to 1970 was warm, and the influx of cold water after 1970, meant that cod disappeared from the banks and has not returned in the same numbers since.

MADS PETER HEIDE-JØRGENSEN, 2013

Fishing for cod with pound nets at Maniitsoq. In 2016, the catch of the inland cod stock increased to 34,000 tonnes, but it has since decreased to half that amount, and it is uncertain whether the catch is sustainable in the long term.

ALEXANDER BENEDIK/CATERS NEWS/RITZAU SCANPIX, 2017

On the seabed off Sisimiut, among red-brown algae and small purple sponges, a polychaete worm hunts for prey. The small, striped animal belongs to a group of bristle worms with relatively hard plate sections on their backs, which has given them the name scalebacks.

THE HIGHLY DIVERSE BOTTOM FAUNA

Bottom-dwelling animals constitute some of the most diverse species of marine fauna. The most common invertebrates stem from groups of mussels, worms, sea urchins, ascidians and crustaceans. Fish from the group of flatfish and lumpfish species are also part of the bottom fauna. Greenland's marine areas of about 2.2 million km2 have an immensely diverse bottom fauna, however, systematic studies have only been carried out in a few areas, and it is believed that the approximately 2,100 different species described so far only represent a fraction of the bottom fauna.

The deep-water shrimp is found in deep water (over 50 m) and soft bottom from Uummannarsuaq (Cape Farewell) to Qimusseriarsuaq (Melville Bay) and along most of the East Greenland coast. The shrimp has a peculiar life form as it changes sex from male to female when fully grown. It feeds on dead organic matter, algae and small worms, and often embarks on nocturnal expeditions up through the water column in search of plankton. Shrimp fishery began in the 1930s by Sisimiut, but it was not until the 1950s that the inshore fishery in Qeqertarsuup Tunua gained commercial importance. With the decline in cod in the 1970s, offshore shrimp fishery also increased in both West and East Greenland, and since the 1980s, fishing for the deep-water shrimp has been the most economically important type of fishery in Greenland. Of the two other bottom-dwelling species caught in small numbers in West Greenland, the snow crab is the most important, and the scallop is currently of no importance. Other common species in the bottom fauna include mussels, starfish, sea urchins, brittle stars and a number of other crustaceans.

The Davis Strait has large populations soft corals and sea pens; off Nuuk, at a depth of 300 to 600 metres, there is an area of about 500 km^2 with a high concentration of soft corals, and in East Greenland there are populations of large 2-3 m high sea fans, which are also a type of coral. These areas densely populated with corals are also habitats for a number of species of starfish, shrimps, snails, sea anemones, sponges, crabs, bottom-dwelling fish, sea lilies and basket stars. Since these areas are of particular biological importance, restrictions on trawling have been introduced in some of the areas with high concentrations of corals.

Several small stocks of cod that are not particularly affected by temperature fluctuations can be found in the fjords, but these stocks are negligible in comparison with the stock on the banks during the warm period.

Greenland halibut can be found in the cold water masses in both East and West Greenland. Spawning takes place in Davis Strait, and larvae and fry are carried north with the water masses into the banks and into Qeqertarsuup Tunua (Disko Bay). Greenland halibut is also found offshore (more than 3 nautical miles from land) at a depth of 1,000-1,500 m along the slope of the deep pit of Baffin Bay. Today, Greenland halibut is the most important export fish in Greenland.

The deep-sea shrimp is found mainly on the edge of the banks at depths from 50 m to 600 m where it feeds on worms, dead organic material, algae and various small animals. The shrimp is also a food source for cod, Atlantic halibut, Greenland halibut and salmon. Fishing for deep-sea shrimp started in the 1950s in Qeqertarsuup Tunua, and as fishing for cod declined during the 1960s, shrimp fishing grew more important. Since the 1980s, deep-sea shrimp has been the most economically important fishing resource in Greenland.

Scallops and crabs are fished to a lesser extent and have some economic significance locally.

Other important species for fishing include red perch, which is caught in deep water in Southeast Greenland, lumpfish, which is caught near the coast in West Greenland to exploit their roe, and capelin, which is caught in East Greenland. Fishing for mackerel, herring and blue whiting, which are pelagic species, developed surprisingly rapidly in East Greenland after 2011, but is now in decline. Some bluefin tuna have also been caught in East Greenland. Finally, we have the North Atlantic salmon, which is caught to a lesser extent, and which has a small spawning stock in Nuuk Fjord. Several other species are not part of the fishing, but are important for the ecosystem: polar cod, Arctic cod, sand lance and Greenland shark.

COASTS

Greenland is dominated by an about 44.000 km long coastline between the ice sheet and the sea (see p. 20). Most human activity takes place near the coasts. Most towns and settlements are located directly on the water where important industries such as hunting and fishing have easy access to open water. However, most of the coastal areas are uninhabited, which has created space for natural coastal development where coastal processes have been able to run their natural course.

The coasts are dominated by deep fjords where sedimentary deposits such as beaches, alluvial cones and deltas are common, and which primarily originate from a sediment supply from rivers and streams. In the catchment area, weathering has broken down rocks into small particles that can be transported to the shores by wind and water. In addition, the rivers erode and transport sediments on the way to the shores. Deltas form near the mouth where the valley system flattens, the flow of water decreases and sediments therefore deposit.

SEALS IN GREENLANDIC WATERS

Five seal species can be found year round in Greenlandic waters. They include the harp seal, the ringed seal and the spotted seal.

The three populations of harp seal feed their young on breeding grounds in the drift ice at Newfoundland, in the Greenland Sea and in the White Sea. From here, they embark on extensive migrations to the far north on both the east and west coast of Greenland. The largest population of harp seal (approx. 7 million) can be found between Canada and Greenland. There are approx. 400,000 harp seals in the Greenland Sea, and together with the ringed seal, it is the most abundant seal species.

Ringed seals can be found everywhere in Greenlandic waters, both in fjords close to the glaciers during winter and far out in the pack ice. The number of ringed seals is unknown, but the species is widespread throughout the Arctic where it can be both largely stationary but also make long migrations.

The hooded seal can also be found throughout Greenlandic waters, but is to a greater extent associated with deeper waters than other seals where it can exploit its ability to make long, deep dives lasting more than 50 minutes and to depths of more than 1,000 metres. It gives birth to its young on the ice on breeding grounds in the Greenland Sea, at Newfoundland and in Davis Strait, and it prefers to stay on the sea ice during the shedding season in summer. The population that breeds in the Greenland Sea is, for unknown reasons, in decline and currently stands at less than 100,000 hooded seals, while the population that breeds in Canada is growing and is thought to be in the order of 600,000.

Bearded seals rest on the ice floes as they drift across the banks where they forage on the seabed, but they also seek out sandy beaches and rocky islands to rest on. Sandbanks in the fjords in the southern part of Greenland are home to the harbour seal, which is the rarest seal species in Greenland.

FLIP NICKLIN/MINDEN/RITZAU SCANPIX, U.A

The ringed seal is the most widespread and probably most numerous seal in Greenland. It has always been important game and is still an important part of hunting life in Greenland.

WHALE SPECIES AND WHALING

CARSTEN EGEVANG, 2022

The bowhead whale (*Balaena mysticetus*), showing its distinctive two peaks on its back, occurs in West Greenland in spring in Qeqertarsuup Tunua (Disko Bay), where it feeds on copepods. It is also found year round in north-east Greenland

Three whale species can be found all year in the waters surrounding Greenland: the bowhead whale, the beluga whale and the narwhal. At 18 metres and weighing up to 50 tonnes, the bowhead whale is one of the largest whales and has a very thick layer of blubber of up to 40 cm. The beluga whale and narwhal, both medium-sized toothed whales, are also well insulated with a blubber layer of 10 cm, and they reach a weight of 1.5 and 1 tonnes, respectively. Both the bowhead whale, narwhal and beluga whale reach sexual maturity relatively late, have a high life expectancy, but only get one calf every three years. Populations therefore grow slowly and both whale hunting and disturbances can have a major impact on populations.

Currently, four species of baleen whales (fin, minke, humpback and bowhead whales) and two species of small whales (beluga whales and narwhals) are subject to scientifically based quotas. In addition, there are four species of small whales (long-finned pilot whale, dolphin, orca and harbour porpoise) that are not subject to quotas.

In summer, ten whale species seek out the productive sea areas in West and Southeast Greenland. These are all species that migrate to warmer areas in winter.

BOWHEAD WHALE

From the mid-1600s until 1900, the bowhead whale was the target of whale hunting, both east and west of Greenland. After this period, populations were so low that it was difficult to find the whales. Until about year 2000, the species remained extremely rare, but the population has increased in both West and East Greenland in recent years.

The bowhead whale can only be found in West Greenland in winter where it is estimated that about 1,400 whales, mainly adult females, stay in Qeqertarsuup Tunua (Disko Bay). In summer, this population stays in northeastern Canada where the calves are born, and the population is estimated at 6,000 animals.

The bowhead whale is mainly found in cold polar waters, and its migrations seem to follow the currents of the polar waters. There is a quota of two whales per year for the West Greenland population, but they have not been caught since 2015. The second population stays in the northernmost corner of East Greenland, but they also migrate up into the Arctic Basin, to northern Svalbard and across to Franz Josef Land. This population is now thought to consist of a few hundred whales and is protected. Bowhead whales feed on copepods, which are filtered through the whales' baleen.

BELUGA WHALE

The beluga whale is only found as vagrant in East Greenland, but two populations can be seen in winter in West Greenland. One population stays in the coastal open waters between Qeqertarsuup Tunua and Maniitsoq and numbers about 8,000 animals. The second population stays in the polynya of the North Water area and numbers about 2,000 animals. Both populations stay in river deltas in northern Canada in summer and make annual migrations to and from wintering grounds, and both populations are caught subject to quota regulation and are believed to be stable or increasing. They feed on polar cod, squid, shrimp and capelin, but also on Atlantic cod when they are available in West Greenland.

NARWHAL

The narwhal consists of several populations that stay in Greenland during summer. The best known population can be found in Inglefield Gulf where the old method of harpooning from a kayak is still practised. Other populations can be found in Qimusseriarsuaq (Melville Bay) in West Greenland and in Kangerlussuaq, Kangertittivaq (Scoresby Sound), Young Sound and Dove Bay in East Greenland. In winter, narwhals migrate from the coast to deep waters where they can dive to depths of 1,800 metres for Greenland halibut. The narwhal is a typical cold-water species that does not thrive in the warming seas and the largest populations are concentrated in cold waters. Their choice of food is limited to a few species and consists mainly of squid, polar cod and Greenland halibut. Catches are subject to quotas and several populations are in decline. Populations in Southeast Greenland are thought to be threatened with extinction due to whale hunting and climate change.

HARBOUR PORPOISE

The most important toothed whale that visits Greenland in summer is the small harbour porpoise. More than 2,000 harbour porpoises are caught each year and they are found in a genetically and behaviourally isolated ecotype in West Greenland that is distinct from the other North Atlantic populations. The harbour porpoise avoids the ice, and although very stationary in terms of summer grounds in Greenland, they make extensive migrations down into the Atlantic in winter. The harbour porpoise is a high-energy whale that quickly reaches sexual maturity but only lives for a very short time (less than 15 years) and has high energy needs, so it must constantly forage for food. In Greenland, harbour porpoises feed on small schooling fish and squid, and in the Atlantic where they spend the winter, they are thought to eat small fish from the mesopelagic layer at depths of 200-300 m. Quotas have recently been introduced on the whale hunting of harbour porpoise.

ORCA

The orca can be found sporadically in Greenland but has been observed and caught regularly in East Greenland during the past decade. The orca mainly feed on fish in the North Atlantic, but in Greenland they also feed on marine mammals such as minke whale, narwhal and seals. It is not known how many orcas there are in Greenland, and the species is not protected.

MODERN WHALING

Of the 12 species of whale that migrate to Greenland in summer, the fin whale, the minke whale and the humpback whale are the most important for modern whale hunting. All three species are subject to quotas issued by the International Whaling Commission. They are relatively fast whales that make long migrations between southern breeding and wintering grounds and northern foraging grounds. All three species have large populations in the Atlantic, and it is the sub-populations that can be found in Greenland during the open-water period where they consume a large proportion of their annual food requirements by feeding on krill, capelin, sandeel, blue whiting and cod.

A small pod of male narwhals lie on the water surface of Melville Bay. Males can be identified by the 1,5-3 m long, twisted tusk, which is the left canine tooth that extends from the upper jaw. Females may also have a tusk, but it is considerably shorter.

CARSTEN EGEVANG, 2019

KRISTIN LAIDRE/UNIVERSITY OF WASHINGTON/NYT/RITZAU SCANPIX, 2015

On an early spring day, with snow still covering the landscape, a female polar bear and her two cubs cross a glacier in southeast Greenland. After hibernating in the mountains over the winter, they head to shore to catch seals and replenish depleted reserves.

THE POLAR BEAR

The polar bear is found in five different sub-populations in Greenland: Kane Basin, Baffin Bay, Davis Strait, Arctic Basin and East Greenland, and can be found almost anywhere in the country, although they are rare in inhabited areas. It is believed that 300 polar bears live in Kane Basin, 3,000 in Baffin Bay and 2,000 in Davis Strait.

The Greenlandic catch of the charismatic polar bear comes from populations in Kane Basin, Baffin Bay, Davis Strait and East Greenland. Overall, the quota in 2022 was 155 bears. In addition to the quota, there is a small number of 'problem bears' that are shot in self-defence. The three populations in West Greenland are shared with Canada, and constitute about 5,500 bears. It is not known how many bears there are in East Greenland.

The polar bear's main food is the ringed seal, but other seal species and other mammals are also polar bear prey. The hunt for seals takes the bear across vast ice-covered seas, and bears can be found far from land in both Baffin Bay and the Greenland Sea. In summer, when the ice retreats, most bears must remain on land and feed on the berries they can find, carcasses drifting ashore or land mammals. Hungry bears can also attack humans, depots and remains of human activities.

In November, female polar bear head for mountain slopes where they can dig a lair in the snow. There are many lair areas in East Greenland, but few in West Greenland, which is likely due to human activity. The female gives birth to one or two cubs at the end of the year, but it is not until February-March that the cubs are big enough to come out of the lair with their mother. After four months of fasting, the mother is busy catching seals to keep the suckling going, so lair areas lie close to areas with access to seals. Ice formation on the sea and in fjords is therefore crucial for the polar bear's survival, and it is unquestionable that the reduced ice cover resulting from global warming will result in the polar bear losing large parts of its Arctic habitat. In Greenland, the loss of habitat will be greatest in southern areas, particularly in Southeast Greenland, but areas in West Greenland where bears come from Canada will also be affected as ice cover in Baffin Bay diminishes.

Most of the coasts are rocky coasts where coastal processes such as waves and tides have a relatively small impact in terms of changing the shape of the coast. Waves, however, affect rocky shores through breaking and erosion of the rocks, but it is a slow process, and even over decades, no major changes can be observed in coastal areas with rocky shores. A small part of the coasts are sedimentary coasts with beaches where loose sediment can be eroded, transported and rearranged by coastal processes associated with waves, tides and wind. This chapter mainly focuses on this type of coast.

Greenland's coasts are unique. There are not many places in the world where the sediment supply from the glaciers to the coastal areas takes place over such short distances as between 10 and 50 km. A number of conditions specific to the Arctic affect the coast. The two primary conditions are sea ice which dampens coastal activity and permafrost which reduces erosion. Both are affected by the ongoing climate change and may lead to increased sediment supply and coastal activity.

Sedimentary shores and coastal processes

Sedimentary shores are found where sedimentary deposits are eroded, creating new landscapes such as deltas, barriers or beaches. New deltas are formed along coastlines with large sediment supply, and a distinction is made between: deltas exposed to wind and waves from different directions and deltas at the head of fjord systems with limited exposure.

In the case of exposed deltas, the coastline is affected by wave processes. Deltas with limited exposure to wind and waves are dominated by river runoff, while the wave impact along the delta front is very limited. Barriers and beaches are being created under the influence of waves where the sediment sources originate from eroding cliffs, sediment-bearing rivers and sediment transport from coastal, low-water areas.

Waves, tides and ice

The processes that affect the sedimentary shores are primarily controlled by waves, tides and ice. The wave impact in the fjord systems is local. The free stretch, wind speed and wind duration determine the wave height, which is a measure of how much energy is available for erosion, transport and sediment deposition.

The free stretch is often limited to a few kilometres, meaning that the wave height in a fjord rarely exceeds 1 m – even during a storm. The wind in the fjord systems is often controlled by the local topography where the wind accelerates when it is forced into the fjord. This affects the wave activity, and the highest waves are created in continuation of the fjord orientation and beaches with considerable rebedding of sediments are formed at the head of the fjord.

On exposed coastlines with open sea, swells originating from wind fields outside the coastline in question play a significant role. When the waves hit the shore at an angle, a transport of sediment is generated along the shore, which can lead to the formation of recurved spits or barriers. Along delta fronts, waves often create bars and redistribute sediments transported via delta flows. When a delta channel changes, it often means that no sediment is added to the old delta flow and it is being eroded, while the active delta flow receives sediments that accumulate. To this should be added the impact of

ARAT KROON, 2013

The beach with small dunes in Tuapaat on the south coast of Qeqertarsuaq (Disko). Waves and currents along the coast carry the sand to the beach, and the wind blows from the beach to the dunes. In the background, drift ice and icebergs in Qeqertarsuup Tunua (Disko Bay).

tides, which means that the wave energy is distributed differently during low or high tide.

The ice has both a direct and an indirect impact on the coastal areas. The ice dampens waves and thus reduces the intensity of the processes along the coasts. When the ice melts, wind and waves can transport drift ice toward the shores, affecting the shores directly by pushing ice and sediments onto the beach.

Short-term coastal changes

Extreme events often have a major local impact on the coasts. Storms in coastal areas have two direct impacts: the wave height, and thus the wave energy, increases and the water level rises.

The rising water level is mainly a consequence of the wind. High water levels and increased wave energy create storm deposits on the beach, which can be seen as storm ridges as well as erosion along the cliffs. The sediments in these storm ridges are often gravel and sand. Large rocks on beaches have often been eroded directly from a cliff, while the smaller particles have been carried away. Some large stones may be ice-rafted debris, which have come with icebergs and left behind when the ice melts.

Extreme events on the coasts are seen in connection with landslides, calving icebergs or earthquakes. This can create a tsunami wave, which can grow several metres high along shallow coastlines and flood coastal areas.

In relatively small fjord areas, this type of wave often reflects on the opposite side of the fjord and resonates in the basin.

Long-term coastal changes

Long-term processes that affect sedimentary shores play a role on time scales from decades to millennia. Relative sea level changes are defined as the result of vertical changes in the water level (eustatic changes) and vertical changes in the land crust (isostatic changes).

Water level rises are directly controlled by melting from ice caps and heat expansion of the seas by heating. Land uplift occurs when the ice melts (glacioisostasy), relieving the pressure on the land. In general, there is a north-south gradient in Greenland where the southern areas have generally experienced a relative fall in sea levels during the first part of the Holocene epoch, followed by a relative rise in sea level in the past period. In the northern part of the country, a relative fall in sea levels has been observed throughout the Holocene. In addition, the gradient is often strong in the relative sea level from the inner fjord areas close to the ice caps to outer fjord areas close to the open sea, where the ice caps have melted away over a long period of time.

Relative sea level changes have been explored and studied on many Greenland coastlines. In areas with a relative fall in sea levels, coastal deposits can be seen to lift up slowly. This means that old beach ridges, an indicator of old beaches, are located high up in the landscape in relation to the current beach and the current sea level. Dating of beach ridge systems at different levels above sea level has been used to reconstruct relative sea level changes at different coastal locations, and these sea level curves are used to reconstruct climate change back in time.

Beach ridge systems have been a typical home for the early cultures (see p. 90). Often they consist of gravel and sand that drain the water efficiently and are less sensitive to permafrost. Archaeological finds and settlement structures are often associated with beach ridge systems where the oldest culture lies highest in the terrain.

BO ELBERLING, 2017

Archaeologists at work at a Norse settlement in Nuuk Fjord. Some Norse settlements are difficult to find as they are overgrown by an increased growth of mainly willow.

AGRICULTURE

Agriculture and livestock farming in Greenland root back to the Norse. More than 1,000 years ago, Erik the Red and his successors came from Iceland and brought along a Nordic peasant culture. The livestock provided the Norse with wool for their clothes and milk for the production of butter and cheese. A combination of livestock farming and hunting became the basis of Norse lifestyle for 500 years. In South Greenland and in the inner fjord areas in the Nuuk area, often right up to the edge of the ice sheet, the Norse found a mild climate and good grazing areas. You can still find many signs of early livestock farming in the form of stables, irrigation canals and animal folds. The Norse disappeared completely during the 15th century (see p. 102).

JETTE BANG PHOTO/ARKTISK INSTITUT, 1939

Sheep on their way to a slaughterhouse in 1939 near Qaqortoq. Greenlandic sheep are imported from the Faroe Islands and Iceland, and, to a very limited extent, from Scotland, England and Norway. Greenlandic sheep belong to the Northern European short-tailed sheep breed.

Farming after the Norse

Present farming have had a slow start since the Norse disappeared. Hans Egede (see p. 113) was the first to introduce livestock, including sheep, near Nuuk. In 1783, the Norwegian Anders Olsen and his Greenlandic wife Tuperna founded an agricultural culture in Igaliku based on cattle farming and hunting (see p. 118). This initiative turned into a lifestyle where people with an Inuit background became livestock farmers.

Livestock farming

In the early 1900s, agriculture experienced a renaissance due to declining hunting options. In 1906, the Greenlandic priest Jens Chemnitz introduced sheep from the Faroe Islands to the former Moravian mission station Narsarmijit (Frederiksdal). The experiment got off to a promising start. The Greenland colonial administration saw the potential in the experiment, and in 1915, sheep husbandry manager Lindemand Walsøe introduced 180 sheep from Iceland to Qaqortoq (Julianehåb). In the same year, the Greenland colonial administration set up a sheep farming station in Qaqortoq. This kick-started a development.

The purpose of the sheep farming station (today: Upernaviarsuk Research Station) has since been to educate and support sheep farmers with new knowledge, financing the start-up of new farms and offering ongoing consultancy. In the same area as Erik the Red in the 980s, Greenlandic sheep farmer pioneer Otto Frederiksen founded the first sheep farm in Qassiarssuk (formerly Brattahlíð) in 1924 on the shores of Tunulliarfik Fjord just 12 km from the ice sheet.

In summer, the grasslands closest to the towns and settlements (the so-called home fields) were kept free from grazing, while sheep, goats and cows were led to the natural vegetation in the mountains. In late summer, the farmers would then be able to harvest the home fields for winter fodder.

Keeping livestock in Southwest Greenland was and still is limited by the length of the winter and how much winter feed can be obtained during the summer season. Both during the Middle Ages and up to recent

ANINGAAQ R. CARLSEN/VISIT GREENLAND, 2019

Sheep in Inneruulalik. In summer, the sheep are in the mountains and feed on the natural vegetation. It is a traditional form of sheep farming that can be traced back to Norse settlements in Greenland more than 1,000 years ago.

times, only livestock manure was used as fertilizer. Today, chemical fertilizers are also used for the home fields and concentrates are used for supplementary winter feed.

Farming today

The climate has been crucial for the development of agriculture since the time of the Norse. In the period from 1927 to 1960, the number of sheep in South Greenland increased from about 5,000 to about 50,000 ewes. A remarkable increase – not least because the climate in the same period was characterised by both warming and cooling. Accordingly, there were some very harsh winters in this period, during which the stock was greatly reduced.

It is a legal requirement that all animals be housed and fed in winter, and since the 1980s, the number of ewes has remained stable at around 20,000. Today, there are about 50,000 animals on summer pastures.

Of the more than about 1,000 ha of cultivated land (2014), 99 % is used for the production of perennial grass and annual green fodder crops for winter feed. The remaining 1 % of the cultivated area is mainly used for potatoes. More than 242,000 ha are used as natural mountain pasture during the summer.

Organisation of and consulting services for farmers

Originally, the Greenlandic sheep farmers were organised into smaller associations, but in 1951, Savaatillit Peqatigiiffiit Suleqatigiissut (SPS, The Cooperative Sheep Farmers' Associations) headquartered in Qaqortoq was formed and has since represented the farmers' interests. SPS handles important tasks related to the annual price negotiations regarding meat prices, feed purchases, communication with public authorities and the annual general meeting, which brings together virtually everyone in the industry. Although sheep husbandry accounts for most of its activities, SPS handles all areas within agriculture.

Today, the Greenlandic Agricultural Council is an advisory body to the Naalakkersuisut (Government of Greenland). They issue opinions on new measures for regulating agriculture and advises the agricultural industry on financial support and development schemes.

The Agricultural Consulting Services in Qaqortoq is the public consulting service for agriculture and undertakes research work in Upernaviarsuk. The research station and the agricultural school in Upernaviarsuk are organised under Food College Greenland, Inuili, in Narsaq.

The future of agriculture

Climate change may mean that new fields and perhaps completely new regions in Greenland may become interesting for farming purposes. There are natural uncertainties associated with assessing the possibilities for a future agriculture sector. This is particularly true when it comes to precipitation. For several years, lack of precipitation during summer has led to desiccation and poor grass harvest. In addition, in many areas the soil is coarse-grained and thus poor at holding on to water, and for the same reason many farms use irrigation during the summer. Furthermore, in most places, the soil is poor in nutrients, so liming, fertilization with both livestock manure and chemical fertiliser as well as irrigation are prerequisites for increased agricultural production. I addition, new weed species and pests also pose a challenge.

Sheep farming is not the only type of livestock farming. Up until now, cattle and caribou have been introduced as a niche production of meat and perhaps commercial milk production in the long run. This is supplemented by cultivation of potatoes and turnips, especially in South

Field near Qassiarsuk. Despite the close proximity to the ice sheet, the climate in summer is favourable for cultivation. This year's harvest is pressed and baled and will be used as fodder in the coming winter.

MADS PIHL, VISIT GREENLAND, 2015

BO ELBERLING, 2020

Urban garden from Qeqertarsuaq in West Greenland. Gardening in Greenland has slowly gone from being an essential household supplement to being a beautiful ornamental garden and a place to relax.

HORTICULTURE

Greenland's cold climate and rocky landscapes with soils poor in organic matter are not the obvious place for horticulture. Nevertheless, the history of horticulture goes back to the Danish-Norwegian missionary Hans Egede, who in front of his residence in Nuuk established the first garden in 1728 (see p. 354). During the 1700s, kitchen gardens were built in front of the homes of the Europeans, also at the Moravian mission stations in the South Greenland fjords, where the climate was more favourable than in the coastal colonies. The yield was modest, but an important contribution in the fight against scurvy caused by the vitamin C-deficient diet at the time.Well-fertilised soil from old camp sites and house plots as well as soil from bird colonies was added to the thin topsoil. Manure and latrine were also used as fertiliser. The garden plots were fenced in by soil ridges and picket fences to provide shelter and keep free-range livestock out.

The gardeners did many experiments with European herbs such as lettuce, spinach, chervil and carrots. Beets and kale that could be stored for the winter fared the best. The introduction of cold frames and greenhouses during the 19th century extended the growing season and resulted in increased interest in horticulture. Gradually, room was also made for lawns, flower beds, gravel paths and gazebos, and the gardens started being used as outdoor living spaces.

Greenlanders had no tradition of horticulture, except in Igaliku where animal husbandry and horticulture had been common since the late 18th century (see p. 118). This changed after an active effort by the sheep farming station established in Qaqortoq in 1915. Sheep farmers grew vegetables, primarily turnips and potatoes. In 1930, the first bilingual guide on plant farming was published based on many years of experience in planting gardens, fertilising, sowing and treating a number of garden plants that thrived well in Greenland.

In 1971, a new bilingual guide made another attempt at promoting horticulture, and since the 1990s garden associations have been established in Qaqortoq and Nuuk. Today's urban gardens are mainly flower gardens supplemented by a couple of cold frames for vegetables. The research station has provided robust perennials, including wolfsbane (Aconitum), leopard's bane (Doronicum), nootka lupine and aven (Dryas), which are now supplemented with imported perennials from Iceland. In addition, there is a wealth of annual summer flowers as well as rose root, dwarf fireweed and angelica retrieved from nature.

As something new, berry bushes, small trees and woody shrubs are also seen in the gardens, e.g. Alaska willow, balsam poplar, northern mountain ash, green alder, and conifers such as Engelmann spruce, juniper and Siberian larch.

Greenland. Future prospects for developing the agricultural industry in South Greenland are good, with new products and ideas being developed and formulated.

Right of use of the agricultural land

Land in Greenland is common property. There is no such thing as private property on land, but the individual farms have an individual right to use associated agricultural land. This means that the individual farms and the Government of Greenland's sector management are very closely linked. During the past 20 years, the cultivated area has increased by almost 30 %, while the number of farms has decreased by 20 %. This trend is seen in many places: The profitable farms grow in size and are supported by various ancillary incomes, e.g. tourism, construction, fishing and hunting, which they combine with farming.

Challenges of cultivating new land

Lack of good agricultural land has always been and remains a significant challenge. The cost of clearing scrubs and removing rocks to establish fields is significant. In addition, farms often need both drainage and irrigation in combination with fertilisation and liming. Today, the agricultural industry receives significant support in the form of support for the establishment of new fields, purchase of new stables and irrigation systems as well as a subsidy per livestock just as is the case with much agricultural production in other parts of the world. However, meat, wool and plant products are supplied to the Greenland society, thereby strengthening the level of self-sufficiency. The future will have to show whether sheep husbandry will move north as a result of a warmer climate.

ANDERS MØLLER, 2009

Hot springs on Qeqertarsuaq are protected as a unique habitat type. They support a rich flora and fauna and often provide the northernmost distribution for many otherwise southern species.

BIODIVERSITY AND NATURE MANAGEMENT

Biodiversity includes all living organisms on land, in freshwater, in the sea, on glaciers and in the air. The species that live on Greenland have adapted to the extreme living conditions of the Arctic climate with low temperatures, prolonged snow and ice cover and periods with no or permanent sunlight. However, living conditions vary significantly from north to south, from the deep sea to the coast and from the coast to the ice sheet. Within these gradients, life conditions are also controlled by ocean currents, salinity, nutrient conditions, geological conditions, permafrost, glacier dynamics, snow and ice conditions, temperature, etc. These conditions also determine primary production, and subsequently the food availability in the Arctic ecosystems.

Greenland is the contiguous land area of the Arctic, which has the longest north-south stretch (more than 2,600 km). Greenland's land area spans across several climate zones and, by using climatic criteria, it is divided into High Arctic, Low Arctic and Sub-Arctic ecosystems, which in turn can be divided into floristic provinces based on differences in vegetation. Individual adaptation and dispersal possibilities mean that species are distributed differently. Some species are widespread throughout the country, some are associated with certain habitat types, while others are only known from a few localities.

The marine biodiversity

Greenland's marine biodiversity is contingent on the water circulation and ocean currents that characterise the entire North Atlantic and the Arctic sea areas. The North Atlantic Ocean Current is an extension of the Gulf Stream and is relatively warm and saline. This current mixes with the cold and relatively fresh water from the Arctic Ocean and, closer to the coast, with the fresh water supply from rivers, glaciers and the ice sheet. Sea ice, nutrient availability

TREE PLANTING

The cold sea around Greenland induces low summer temperatures which limit tree growth. Summers in the inner South Greenland fjord complexes are warmer, and in some places willow scrub can grow up to 3-5 metres, and birch scrub can grow up to 5-8 metres in sheltered areas with optimum conditions. Compared to similar areas in North America, however, the number of tree species is modest, the trees are small and forest-forming conifers are not present.

Alpine juniper is the only natural conifer in Greenland, and it is likely that the Labrador Sea has prevented the spread of the Northeast American conifer species after the last Ice Age. However, for over 100 years, there has been interest in introducing and planting conifers, especially in South Greenland where, for many years, firewood was harvested from the birch scrub forest in the inner fjords.

As early as 1898, the need for forestry was discussed in order to be able to provide future generations with firewood.

The earliest attempt at introducing new tree species was made by Moravian missionaries, who in 1846 sowed Norway spruce seeds in the inner parts of the Aluitsup Kangerlua fjord (Lichtenau Fjord). The plants are mentioned in 1923 but have since disappeared. Other seedings have also disappeared. One exception is the 'Rosenvinge Trees' from 1893 at Qingua in the fjord Tunulliarfik a little north of Narsarsuaq. Here, three old Scotch pines were measured at 8 metres in 2013.

Thousands of trees were planted under the management of Poul Bjerge from 1953 onwards. A result of this work is the small coniferous forest at Qanasiassat in the Tunulliarfik fjord where some lodgepole pine and Siberian larch trees measured at over 12 metres in 2016.

Since the 1970s, afforestation experiments have also been made further north near Nuuk, in Qooqqut Valley and Ameralla Fjord as well as near Kangerlussuaq. Here, the largest trees measured at 5-6 metres in 2019.

Today, the largest collection of non-indigenous tree species is found in the Arboretum Groenlandicum/Kalaallit Nunaata Orpiuteqarfia (Greenlandic Arboretum) located at Narsarsuaq. The arboretum was opened in 2004, but the majority of the plantings date from the 1980s and 1990s. During that period under the management of Søren Ødum and Kenneth Høegh, over 100,000 trees were planted spread across more than 100 species. Far from all the species have survived in the local climate, but an astonishing number of species survive, and some grow well and are able to reproduce naturally.

The afforestations at Qanasiassat and Narsarsuaq show that you can grow trees in South Greenland as long as you plant the right species. In recent decades, the trees have grown particularly well, which is partly due to the ongoing climate change. Forecasts show that with the future warmer conditions, the area where boreal forest can grow will increase significantly in size. The considerations from the 1890s regarding forestry in South Greenland have therefore become relevant again.

KENNETH HØEGH, 2021

Siberian larch trees in Narsarsuaq, planted in 1994. Larch is adapted to the drying foehn winds, which are particularly frequent in South Greenland.

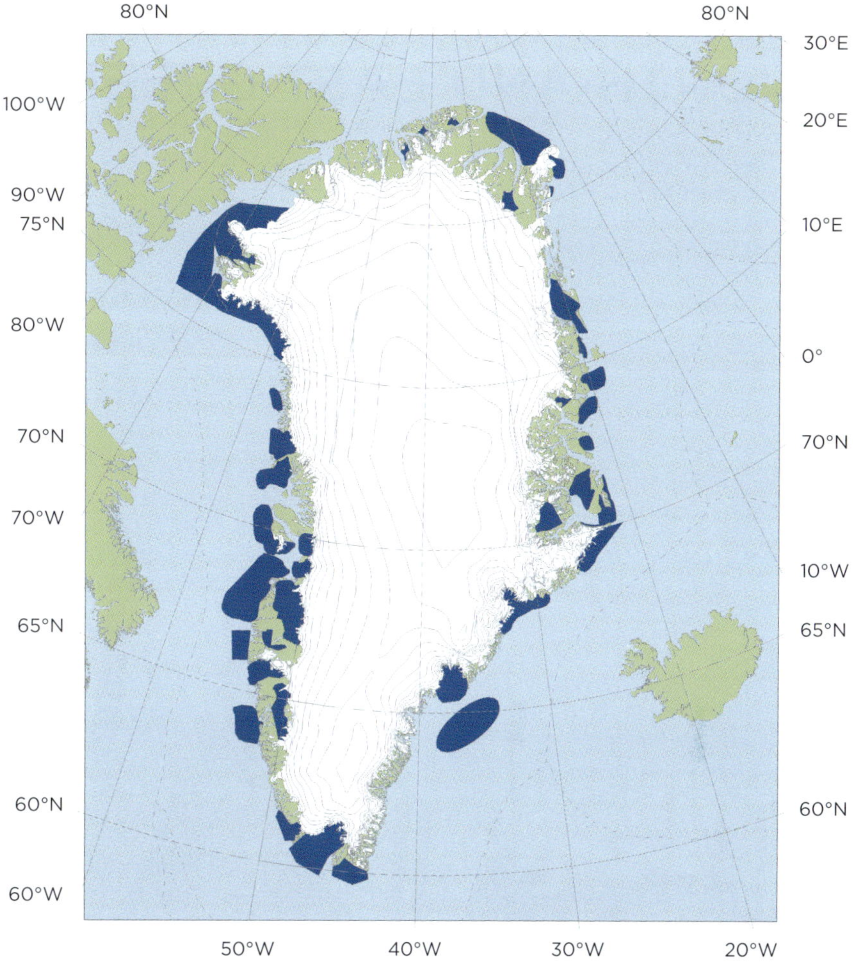

Areas of importance for Greenland's biodiversity. The number and location of areas may vary as knowledge of species increases, remaining areas are mapped, or species distributions and numbers change over time.

NATUREMAP.EAMRA.GL

and seasonal variations in physical conditions play a major role in the productivity of marine ecosystems.

On the basis of these complex physical and biological conditions, Greenland's marine areas can be divided into larger and smaller marine ecosystems. In particular, the coastal areas can be divided into small area types based on more detailed knowledge of physical conditions (e.g. banks, polynyas, fjords and the coast), productive conditions (e.g. primary production of algae and zooplankton) and important areas for sustaining life in the sea (e.g. foraging areas, spawning grounds and rearing areas).

Arctic species

The biodiversity of the Arctic, including Greenland, generally includes fewer species compared to temperate and tropical areas, but many of the Arctic species are unique in how they adapt to the extreme climate conditions that prevail in the Arctic.

Many of the species that occur in Greenland are also widespread in other regions of the Arctic. However, there are also species found only in Greenland (endemic species), as well as species where much of the world's total population lives in Greenland (national responsibility species). Naalakkersuisut (Government of Greenland) regularly updates a list of endangered species in Greenland, a so-called Red List, where the main focus is on endemic species and national responsibility species.

Example of the diversity of animals that can be found on the seabed in Greenland. The photo depicts cauliflower corals, feather stars, soft corals, sponges, sea anemones, brittle stars, hydroids and bryozoans with branched limestone skeletons in a coral garden on the continental slope in West Greenland. The laser dots (green) are placed 20 cm apart; the dot to the left is partially hidden.

MARTIN BLICHER, 2020

THE NATIONAL PARK IN NORTH AND EAST GREENLAND

EINAR GADE-JØRGENSEN / ARKTISK INSTITUT, 1963

The world's largest national park of 970,579 km^2 covers the north-eastern part of Greenland. The national park stretches from Hall Land in the northwest to Jameson Land in the southeast. The national park covers a coastline of more than 18,000 km, large parts of the ice sheet and the world's northernmost land area where different peoples have been able to sustain life for thousands of years based on the High Arctic animal species.

Besides early unit cultures and expedition huts, there are still several hundred huts from the Danish/Norwegian hunting period. Many of these are now maintained by Nanok – an association that restores selected huts each summer.

Outside the National Park, to the south in Jameson Land, lies Nerlerit Inaat Airport (Constable Point). The airport serves as the main gateway to the national park as well as to the town of Ittoqqortoormiit.

The area was designated a national park by the Danish Nature Protection Agency in 1974. Access to the park is extremely limited. Apart from hunters from Ittoqqortoormiit and very few tourists, only employees from the weather service, the military and researchers come to the national park. Less than 50 permanent residents undertake monitoring and research at the six stations in the national park.

MESTERSVIG

Mestersvig is an abandoned mining town that since 1988 has served as a small logistic platform for the Sirius Dog Sled Patrol, which has kept a runway open and supervised expeditions.

DANEBORG

Daneborg to the south is the headquarters of the Sirius Dog Sled Patrol, whose task is to enforce Denmark's sovereignty in North and Northeast Greenland and to keep track of expeditions and enforce conservation regulations in the National Park.

ZACKENBERG RESEARCH STATION

Established in 1995, the Zackenberg Research Station is a scientific station 22 km from Daneborg. The station, which officially opened in 1997, was established in collaboration between the then Danish Polar Centre and the University of Copenhagen. Located in a High Arctic climate zone that reacts earlier than other zones to changes in global weather, the research station's main task is to study and measure climate fluctuations and their impact on ecosystems.

Since its establishment, Zackenberg has served as a field station from May to October. The station is owned by the Self-Government and operated by the Aarhus University.

DANMARKSHAVN

Further north lies Danmarkshavn, now a weather station, formerly a base station for the Denmark Expedition (1906-08). As part of a global network of radiosonde stations, the staff launches a weather balloon with measuring instruments twice a day.

STATION NORTH/ VILLUM RESEARCH STATION

Station Nord is Danish Defence's northernmost military station where five men are tasked with keeping a runway open all year round. The station was established in the period 1952-56 as a weather and telecommunications station with a runway. In 2014, Villum Research Station was established at Station North, and the station serves as an interdisciplinary research platform and a springboard for research expeditions into North Greenland. The station is open all year round, owned by the Self-Government and operated by the Aarhus University.

BRØNLUNDHUS

Brønlundhus is the northernmost research station in Greenland. It was established by the Danish Pearyland Expedition in 1948 and served as a base for the expedition. Today, the Centre for Permafrost at University of Copenhagen maintains a monitoring program focusing on climate and permafrost at the station.

A Nansen sled on the ice sheet, 1963. The photo was taken by Einar Gade-Jørgensen, who participated in several Greenland expeditions, including with archaeologist Eigil Knuth's second Danish Pearyland Expedition, from which the photograph originates. Eigil Knuth was on several expeditions in this rugged northeastern part of Greenland, where the country's northernmost research station, Brøndlundhus, is located.

SYSTEMATIC ENVIRONMENTAL MONITORING

There are several monitoring programmes in Greenland that follow changes in nature:

From 25 measuring stations and using data from aircraft and satellites, Promice monitors the ice sheet mass balance, focusing on the ice edge. The observations include measurements of melting, climate and ice movement and are communicated to the public through polarportal.dk. PROMICE is managed by the Geological Survey of Denmark and Greenland (GEUS) in collaboration with DTU, Technical University of Denmark, and Asiaq - Greenland Survey.

GEM (Greenland Ecosystem Monitoring) is a monitoring programme established to document climate change in Greenland and its impact on ecosystems, including biodiversity and greenhouse gases. GEM, which dates back to 1996, started at the Zackenberg Research Station (see p. 79). Since then, the program has been extended to include Kobbefjord Research Station and Kangerluarsunnguaq (start-up in 2005) and at the Arctic Station in Qeqertarsuaq (start-up in 2016). The programme is coordinated by Aarhus University in collaboration with the Greenland Institute of Natural Resources, Asiaq, University of Copenhagen, GEUS and DTU. The rationale is to continuously measure and document climate change in Greenland and its impact on the overall ecosystem, including biodiversity and greenhouse gases.

Current research and environmental monitoring build on traditions that can be traced back to botanist Morten Porsild, who was the initiator of the first research station in Greenland, the Arctic Station, in 1906. Here, they focused on the importance of carrying out research close to a local community and all year round.

In 1906, botanist Morten P. Porsild established a research station on the island of Qeqertarsuaq (Disko) near the town of the same name. He served as stationmaster for 40 years. In 1953, the station was transferred to the University of Copenhagen under the name 'Arctic Station'. Since then, it has been the centre of a wide range of research, mainly in the natural sciences, and it is one of the few field stations north of the Arctic Circle that is staffed all year round.

BO ELBERLING, 2018

Through the Greenlandic Nature Protection Act and international treaties, Greenland has a special obligation to protect the country's biodiversity, which is why there is a special management focus on these red list species.

Nature has always been the foundation for the survival of humans in the Arctic climate, and it still plays a major role for the economy, culture, health and well-being. Our use of nature can have an adverse effect on species and ecosystems. This may be through unsustainable fishing and hunting, or through changes in habitats (e.g. bottom trawl, agriculture, mining and the establishment of infrastructure), pollution (from local sources and long-range transboundary pollution), introduction of invasive species, noise from transport or other forms of disturbance.

Climate change and biodiversity

Biodiversity is also affected by climate change. Already today, there are signs of changing dispersion patterns, lusher vegetation and 'ecological mismatch' – a temporal shift in the growth of plants so that it does not match caribou calving and the emergence of insects, for example.

With higher temperatures, southern species can spread to and establish themselves in Greenland, either naturally or with the help of humans, and potentially pose a threat to the species that already exist here today. The High Arctic ecosystem to the north is particularly vulnerable to climate change. Some species are expected to be able to adapt to minor changes in the climate, while others will be challenged by limited dispersal options to the north. In the long run, climate change might therefore pose a threat to the High Arctic ecosystem.

The mapping of Greenland's biodiversity is being challenged by the size of Greenland and the financial resources available for such studies. In this context and what is especially important is the knowledge about the distribution of species (e.g. habitats, important breeding, foraging, rearing, shedding/moulting and hibernation areas), species' vulnerability to disturbances and climate change (e.g. endangered species) and knowledge of particularly unique habitat types (e.g. areas with high primary production, hot springs and salt lakes).

Whereas relatively good data is available about most exploited animal species, knowledge of other groups of species is more sporadic. Researchers have mapped the areas that are considered valuable in relation to protecting Greenland's biodiversity (see map p. 77).

ESTHER HORVATH/ALFRED-WEGENER-INSTITUT, 2019

SEASONAL POLLUTION IN NORTHEAST GREENLAND

Unexpected results from airborne surveys in the 1970s showed that the air over Greenland was polluted and that the pollution of the Arctic atmosphere required permanent measuring stations in Greenland. The SAGA (Studies of Aerosols in Greenland Arctic) project began in 1979 and included four stations, including Station Nord in Northeast Greenland (see p. 79).

Surprisingly, the results showed that northern Greenland receives substantial seasonal pollution every year. The concentrations of sulphur and heavy metals in March-April had a pronounced maximum, which at Station Nord far exceeds the levels in Denmark.

It became clear that the pollution in Northeast Greenland originated from atmospheric long-range transboundary transport from urban and industrial areas in Russia and Europe. The monitoring of atmospheric content and atmospheric precipitation and its significance in the Arctic food chains remain a central focus of current research at Villum Research Station/Station Nord.

At the Villum Research Station, a helium-filled balloon is launched by researchers from the Leibniz-Institut für Troposphärenforschung, Universität Leipzig. In collaboration with Aarhus University, the institute measures wind speeds, turbulence, energy radiation and black carbon, which are soot particles that are released, for example, when coal is burned. Soot particles are an important component of the atmosphere because they absorb sunlight.

Nature management

Nature in Greenland is affected by climate change and various forms of human disturbance. To protect Greenland's biodiversity, the Naalakkersuisut has therefore implemented laws and executive orders that regulate the exploitation and protection of Greenland's nature.

Nature protection in Greenland is generally managed according to Greenland Parliament Act no. 29 of 18 December 2003 on nature protection (the Greenlandic Nature Protection Act), but a number of other laws also relate directly to nature management in Greenland. Today, the management of Greenland's biodiversity is divided between several ministries that attend to nature and environmental protection in relation to conservation efforts and in connection with the exploitation of living resources, mineral resources and genetic resources.

Protection of natural areas

Nature in Greenland is managed through the protection of natural areas and species, regulation of the exploitation of living resources as well as regulation of activities that have a negative impact on nature. The protection of natural areas includes; protected areas designated by conservation orders, preservation of certain sites important for certain species (bird protection areas, seabird colonies, walrus landing sites, polar bear hibernation areas and areas closed to fishing or use of fishing methods that damage the seabed), protection of certain habitats and species no matter where they occur (hot springs, salt lakes, protected species), areas with restrictions related to mineral resource activities as well as protection lines along coasts, rivers, hot springs and salt lakes.

Exploitation of natural resources

The exploitation of the living resources (e.g. shellfish, fish, birds and mammals) is regulated by Greenland Parliament Act no. 18 of 31 October 1996 on fishing (as amended) and Greenland Parliament Act no. 12 of 29 October 1999 on catching and hunting (as amended), while the exploitation of genetic resources is regulated by Greenland Parliament Act no. 3 of 3 June 2016 on the exploitation of genetic resources and related activities. Underlying executive orders enable authorities to stipulate limitations on and conditions for exploitation, e.g. in the form of quotas and hunting seasons.

Foreign activities create pollution

Greenland's nature is also directly or indirectly affected by activities in other countries. Global emissions of greenhouse gases and other pollution affect Greenland's ecosystems and species. Pollution can reach Greenland by air or by sea currents and can be concentrated via the food chain. This means that some long-range atmospherically transported heavy metals end up in some of the species eaten in Greenland, including substances that are known to harm human health and learning ability (including mercury and cadmium).

Sustainable exploitation

Greenland also shares many exploited populations with other countries, such as some species of fish, marine mammals and birds. Sustainable exploitation of these populations therefore requires that the overall exploitation is sustainable across the countries that exploit the individual populations. This is one of the reasons why Greenland engages actively in international co-operation in the field of nature and the environment, including in the Convention on Biological Diversity, in the Arctic Council's work groups as well as in international, multilateral and bilateral agreements and co-operation within fisheries and hunting, e.g. the International whale hunting Commission and fisheries organisations.

Preserved and protected areas

The Greenlandic Nature Protection Act provides the framework for the designation of conservation areas in Greenland by giving authority to prepare conservation executive orders. Up to 2021, 12 national conservation areas have been implemented. Furthermore, in pursuance of the Greenlandic Nature Protection Act, 12 wetlands of international importance (Ramsar Areas) have been designated, some of which overlap with the 12 national conservation areas. These areas together represent 1,002,514 km^2

Protected areas in Greenland. Conservation orders and Ramsar areas are included.
aturemap.eamra.gl

NATUREMAP.EAMRA.GL

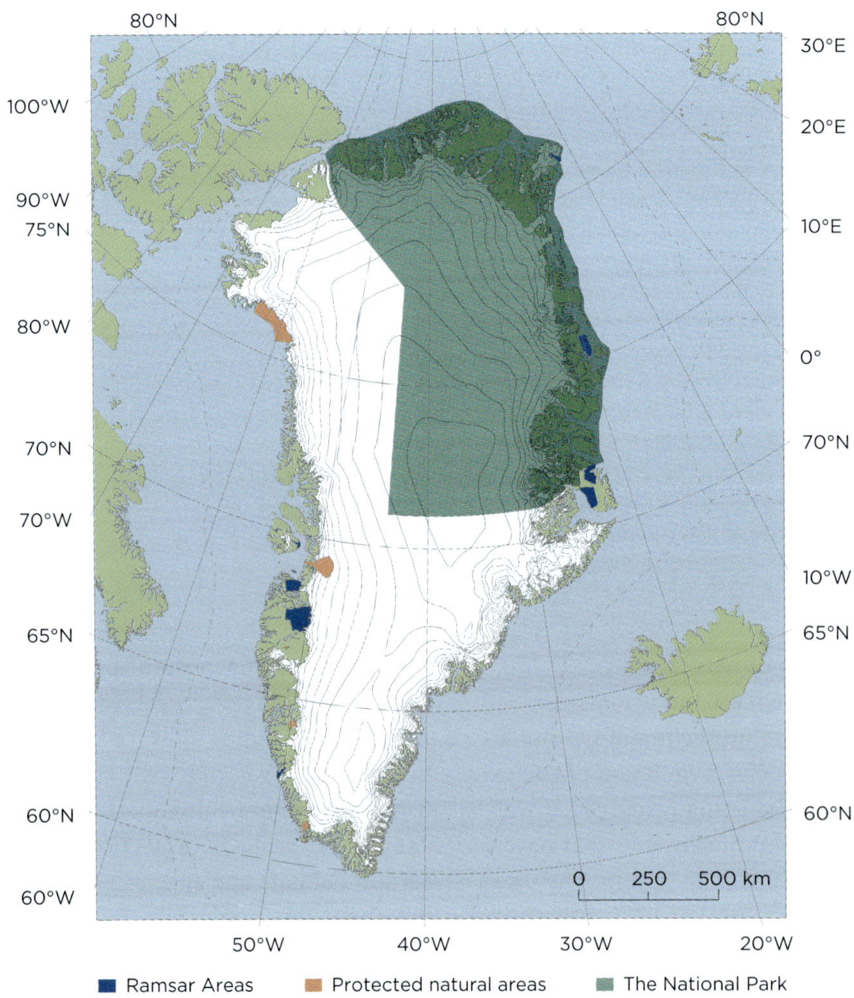

PAUL ZIZKA PHOTOGRAPHY/CATERS/RITZAU SCANPIX, 2016

Northern Lights over Kangia (Ilulissat Icefjord), home to the fastest flowing stream of ice in the Northern Hemisphere, the Sermeq Kujalleq Glacier. The area was designated a UNESCO World Heritage Site in 2004 to protect its scenic beauty and natural and cultural heritage qualities.

(about 41.6 % of Greenland's land area incl. the ice sheet and 4.2 % of the sea area). The National Park in North and East Greenland alone represents 970.579 km^2 (including both land and sea areas).

Based on criteria defined by international organisations and conventions, some of the protected areas also have a special international status. Kangia (Ilulissat Icefjord) has been designated as a UNESCO World Heritage Site due to the area's outstanding character with the fast-moving glacial current from the ice sheet with calving glaciers in the fjord and the stranding of colossal icebergs at the mouth of the fjord. Under the auspices of the Ramsar Convention, 12 Ramsar areas were designated in 2016. The National Park in North and East Greenland (established in 1974) was designated as a UNESCO Biosphere Reserve from 1977-2018, after which its international status was deregistered due to changes in designation criteria, which Greenland did not believe the area could live up to.

Subject to other legislation, areas have been designated that also aim to protect nature, biodiversity or certain species from exploitation or destruction of habitats. These include areas with bans on certain types of fishing, 40 bird protection areas (2019), as well as protection of bird colonies, walrus landing sites and polar bear hibernation areas. In addition, there is the protection of important areas for vulnerable species in relation to the exploitation of mineral resources as well as protection zones to protect the coast, rivers, salt lakes and hot springs against development, overexploitation and other destruction.

Through municipal statutes, the Greenland municipalities can impose restrictions on traffic, resource exploitation, development, etc. in specified areas. In addition, informal agreements can be made locally between fishermen/hunters to regulate exploitation in certain areas.

HISTORY

The Greenlandic name for Greenland, Kalaallit Nunaat, means 'land of the Greenlanders'. The first word, *Kalaallit*, is in all likelihood derived from a European term for Inuit in Greenland.

In Greenlandic parlance, *Kalaallit* was widely used in Southwest Greenland during the 19th century, whereas in Northwest Greenland *Inuit* was used for longer. Later, *Kalaallit* became the common national self-designation. Thus, the country's name has references to the Inuit culture, the encounter with Europeans, colonisation, and the formation of the nation in recent times.

The country's earliest history begins well over 4,500 years ago with a migration from the areas we know today as Siberia, Alaska and Arctic Canada. The first people in the country are seen to be directly connected to people in Arctic Canada, and it appears they have inhabited large parts of the Greenland coast.

Since then, the last migrating Vikings came from Iceland and settled as farmers in Southwest Greenland. This is when the country was given the name that is still used in Danish – Greenland (see p. 13).

At the end of the 12th century, when people came to the country again, they came from the west, once again as a result of a migration. Over the course of just a few generations, the migration led Inuit from Alaska across Canada and eventually into a country where the Norse were about to give up their existence as farmers. The newly arrived Inuit laid the foundation for today's Inuit population, the Kalaallit. They encountered Europeans regularly who hunted whales along the coast, and from 1721 the European presence became permanent.

The Danish-Norwegian kingdom began to take an interest in the country the Norse had left, but which in the Nordic consciousness was still part of the kingdom. Therefore, the Danish king supported the project which the missionary Hans Egede wanted to establish with Norwegian traders, and a few years later, the king also showed goodwill towards the Moravian Brethren Congregation, which was to support the work of Hans Egede.

This began an actual colonisation with Christian missions and trade, and alongside the construction of trading posts along the west coast from Nanortalik in the south to Upernavik in the north, the life and culture of the Inuit were drastically changed. The colony project depended on the Inuit hunting and experience as the blubber, the primary trade product, came from Kalaallit seal and coastal whale hunting.

In *Oqaluttuaq* (*Tales*), Greenlandic artist Nuka K. Godtfredsen tells the story of the Inuit in four volumes. The portraits shown on pages 84-85 are the supporting characters in the second volume of the series, *The Ermine* (2012), whose action takes place in the 12th century in the field of tension between Tunit, Inuit and Norse. The cartoon has been made in close co-operation with archaeologists from the National Museum in Copenhagen, and the physical details, such as the costumes shown here, have been reproduced as accurately as possible.

KONRAD NUKA GODTFREDSEN, 2012

With the establishment of the state-owned Royal Greenland Trading Department (KGH) in 1774, the Danish state took over the colonisation entirely; the KGH was the actual colonial administration until the early 1900s, when trade and administration were separated.

Inuit hunting had been free before – it had to feed the family and the settlement. Now it became an integral part of a trading economy, which had to be borne by the selling of whale oil and seal skins on the world market. In the late 19th century, when blubber lost its importance as a commodity, the focus turned on reforms of industry and societal structure designed to make the economy of the colony self-sustaining again.

Many of the changes were of great importance both culturally and economically. In the early 1900s, colonial rule decided that fishing should bear the economic development, and so sealers had to change professions. This brought about a drastic cultural and economic change that was reinforced with the modernisation reforms after World War II.

The hunting profession remained central to the Inuit culture, but the modernisation that followed the post-war new arrangement had, for the first time in colonial times, been entirely based on a Danish role model, both culturally, linguistically, administratively and politically. At the same time, modernisation brought about the largest ever addition of Danish labour – in construction, education and administration, increasing the impact Denmark had on social development.

Marine mammals were the focal point of Inuit life, culture, and European colonisation. While skins and especially blubber were sought-after commodities by Europeans, all parts of the animal were vital to the Inuit before and after colonisation. Marine mammals were not just food. They were used for clothing, tents, kayaks and boats, building materials, heating and lighting. The photo to the left shows a man in the Thule district skinning a seal (1936), while the photo to the right shows a girl and a woman cooking over an open fire (ca. 1960).

Everywhere in daily life at workplaces, schools, and in the social and health care system, the population saw this development as a new colonisation. It brought improvements in living conditions in a number of areas, but it also brought frustrations which, over time led to a mobilisation of political resistance, which became fundamental to the demand for increased secession from Denmark.

Although the colonial era formally ended with the new arrangement in 1953, the country did not achieve the political equality with Denmark that was expected. The home rule implemented after the after the referendum in 1979 was the first step on the road to a more independent country and the building of a society on the population's own terms.

In 2009, that development continued when self-government was introduced. This happened after a referendum in which there was a large majority in favour of Greenland as an autonomous country in the Unity of the Realm with Denmark and the Faroe Islands.

PALEO-INUIT

About 5,000 years ago, Paleo-Inuit migrated from northeastern Siberia across Alaska and Arctic Canada to Greenland.

The ancient times of Greenland are described on the basis of interpretations of archaeological remains. The first people left no written sources about who they were, what their religion was, how their societal structure had been built up, or what prerequisites they had for their existence. The interpretation of the material remains has been done in part by drawing parallels to the Inuit and their accounts in the form of legends and myths, among others.

The assumption that a Stone Age culture existed emerged in the 1830s, and over time archaeologists began to realise that the country's ancient times were more complex than first thought.

However, the existence of an earlier and older culture than the Inuit culture was not definitively recognised until around the 1950s. The archaeologists gave the Arctic Stone Age peoples the term 'palaeo-eskimos' (palaeo = old), which is no longer used, as the word 'Eskimo' is generally perceived by the Inuit as derogatory. The pan-Inuit organisation ICC (see p. 142),

GREENLAND'S HISTORY IN BRIEF

Approx. 2500-1800 BC: Paleo-Inuit. The Independence I culture immigrates from Canada via Ellesmere Island and settles in Northeast Greenland and further down the East Coast.

Approx. 2400-700 BC: Paleo-Inuit. The Saqqaq culture immigrates from Canada via the Nares Strait and settles along the west, south and east coasts.

Approx. 800 BC-1 AD: Paleo-Inuit. Greenlandic Dorset culture immigrates from Canada via Baffin Island and settles along the west coast.

Approx. 1-700: Presumably no settlements in the country.

Approx. 700-1350: Paleo-Inuit. The Late Dorset culture immigrates from Canada and settles in the north-westernmost part of the country. Late Dorset marks the final stage of the Paleo-Inuit.

985/86: Erik the Red and his family leave Iceland and settle in Brattahlíð (Qassiarsuk) in South Greenland.

Shortly before the year 1000-ca. 1450: The Norse settle in southwestern Greenland.

1124: Garðar (Igaliku) in South Greenland becomes a diocese.

From approx. 1150: The Inuit immigrate from Arctic Canada.

12501300: The Norse builds seven stone churches, including Hvalsey Church (Qaqortukulooq) in South Greenland.

1261: The Norse voluntarily submit to the King of Norway.

1378: The last bishop in Garðar (Qassiarsuk) dies. However, bishops are entrusted the Greenland diocese until the Reformation in 1536.

1475: The dating of the Inuit mummies found at Qilakitsoq near Uummanaq in North Greenland.

16th century: The dating of the Inuit mummies found at Uunartoq in South Greenland.

17th century: The dating of the Inuit mummies found at Pisissarfik near Nuuk in Southwest Greenland.

17th and 18th centuries: European whale hunting and expeditions in Greenland lead to the meeting of the Inuit and Europeans.

1721: With the Danish king's assent, missionary Hans Egede travels to Greenland and founds the Håbets Koloni (Hope Colony) together with Det Bergen Grønlandske Compagnie (The Bergen Greenland Company). Colonisation is initiated.

1728: The missionary Hans Egede founds the Godthåb Colony.

1733: With the Danish king's assent, missionaries from the Moravian Brethren Mission travel to Greenland to strengthen the missionary work.

1734: The Danish king grants merchant Jacob Severin the right to all trade and seafaring in Greenland.

1750: Severin returns his trading rights in Greenland. Det Almindelige Handelskompagni (The General Trade Company), a company of landowners, merchants and civil servants, takes over the trading rights.

1776: The Danish government dissolves Det Almindelige Handelskompagni. The Danish State takes over the company's shares and thus its trading rights, and then establishes the state-operated trading company, the Royal Greenland Trading Department.

1814: Denmark retains supremacy over Greenland despite defeat in the England Wars.

1861: The first Greenlandic newspaper, *Atuagagdliutit*, is published.

1863: After a reform of the colonial policy designed to improve the economy of commerce, governorships are established in all colonial towns.

1900: The Brethren Mission leaves Greenland.

1905: The Act on Church and School Services in Greenland is adopted.

1908: The Act on the Governance of the Colonies in Greenland of 27 May repeals the governorships. 62 municipalities are created with municipal councils, but this system does not come into effect until 1911.

1910: Knud Rasmussen founds a privately owned trading station at Thule.

1911: New local government act for Greenland separates trade and administration.

1931-33: Norway lays claim to Greenland, but the International Court of Justice in The Hague affirms Denmark's sovereignty over Greenland.

1940—45: Since lines of communication between Greenland and German-occupied Denmark is severed during World War II, the country becomes part of the United States' sphere of interest and several military facilities are established.

1948: In the post-war years, colonial policy reforms are prepared and the Greenland Commission is established.

1952: After World War II, a defence accord is concluded with the United States, and Thule Air Base is established near the Inughuit settlement of Uummannaq. The population of the settlement is forcibly relocated by the Danish state to a newly built settlement, Qaanaaq.

1953: Following a number of acts of the Danish Parliament and a change in the Danish constitution, the colonial period formally ends as Greenland is annexed to Denmark as a part of the country.

1964: The birthplace criterion is introduced.

1979: Home rule is introduced and Greenland gets its own parliament (Inatsisartut) and government (Naalakkersuisut).

1985: Greenland secedes from the EC.

2004-2008: The Self-Government Act is adopted by referendum.

Map of settlements from 2500 B.C. to 1500 A.D. The signatures cover the entire ice-free area, as a more precise determination is not possible.

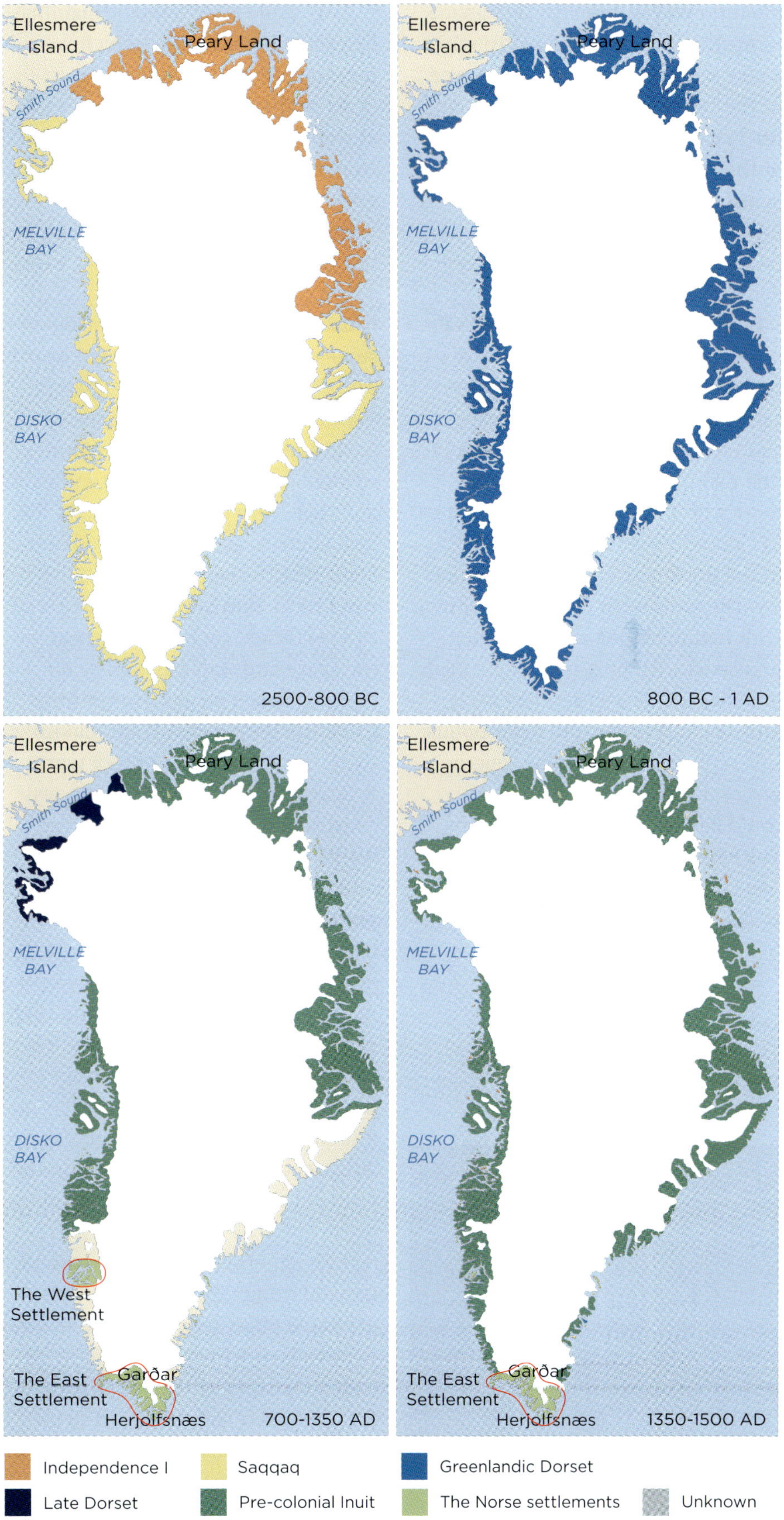

EFTER DEN STORE DANSKE ENCYKLOPÆDI

adopted by resolution 2010-2011, to replace the word Paleo-Eskimo with Paleo-Inuit because the word Eskimo is not part of the Inuit languages.

Archaeologists have named the Paleo-Inuit cultures according to their found sites in Independence Fjord, Saqqaq and Kinngait (Cape Dorset).

The Paleo-Inuit Culture

About 4,500 years ago, the first people from Arctic Canada immigrated across the Nares Strait to Greenland. They left traces in the landscape testifying to the fact that the first people have lived in all regions of the country.

Findings from Paleo-Inuit cultures have been dated from about 2500 BC to about 1350 AD. The remains are primarily stone tools and debris from making tools, as well as the remains of dwellings, because preservation of biodegradable materials such as bone, tusks and wood is far poorer.

The stone tools can most often be related to hunting, and the women, children and elderly of early hunter-gatherer societies are thus not as well represented as the men, which may give the impression of a cultural history predominantly characterised by masculinity.

Independence I
(approx. 2500-1800 BC)

The earliest Paleo-Inuit in Greenland are named Independence I by archaeologist Eigil Knuth, who first found remains from the culture in Independence Fjord in Northeast Greenland. The Independence I people immigrated from Arctic Canada to northwestern Greenland roughly 4,500 years ago, and they dispersed to the north, across Peary Land and further down the northeastern part of the country. Relative to subsequent cultures, the geographical distribution of the Independence I people was limited and population density was low.

Despite several variations in tool types, housing and game, some distinguishing features of this people can be highlighted. They are especially known for their exceptionally fine making of stone tools, made mostly of flint-like raw materials such as chalcedony and quartz, available in the northeastern part of the country. The tools were made by knapping with hammerstone and were then processed with pressure flakers made of bone, tusks or antler. The stone tools include arrow and lance points, side scrapers, end scrapers, microblades, burins and others, which could have had different functions. The chipped edges are serrated.

The sewing needles with round needle eyes from the Independence I culture were essential tools for sewing or repairing leather and footwear. The photo shows thin and long needles made of bird bones. The longest is 4.7 cm long.

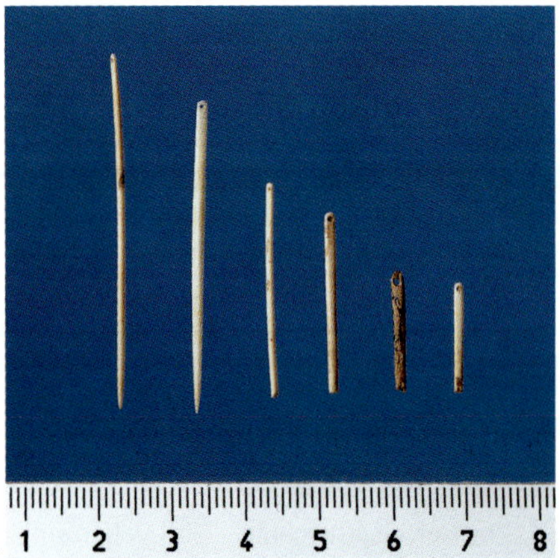

JOHN LEE/NATIONALMUSEET I KØBENHAVN, U.A.

Tools made from bone or mammalian tusks include awls, which were presumably used to poke holes in skins, and side spurs for fish spears, as well as thin sewing needles made from bird bones with round drilled eyes of the needles.

Independence I lived throughout the year in round or oval tents with a skin covering. The dwelling often had an indoor mid-passage made of stones with a centrally located hearth (see p. 190). In the warmer seasons, the dwelling was characterised by having an outdoor hearth, typically square boxes built of thin stone slabs, and they were used as both light and heat sources, as well as for cooking. Some hearths were filled with fist-sized rocks that have been construed as fire-cracked rocks used to heat the space and water as well as for cooking. Based on analyses of some dwellings with debris from making tools, the distribution of tool types is interpreted as if the dwelling was constructed with gender preferred workspaces.

In many midden deposits there is a preponderance of bones from muskoxen, which is why the archaeologists believe that Independence I primarily hunted muskoxen in the High Arctic, and that it was the hunt for these as well as the then warmer climate conditions that guided their migration to northernmost Greenland. Driftwood, which is deposited on the coasts of the Northeast Greenland by ocean currents in the Arctic Ocean, was also another important resource base. Although Independence I also exploited other terrestrial and marine mammals, it was the interaction between the warmer climate, the hunt for muskoxen and driftwood to use as fuel, which allowed them to live in the northernmost settlement in the world for a few hundred years.

Although the towns and settlements were concentrated in Northeast Greenland and thus geographically separated from the contemporary Saqqaq culture, which settled in the more southerly part of the country, there is great similarity between their tools and types of dwelling. Therefore, the Independence I and Saqqaq people are considered today to be one and the same people, characterised by differences in their resource base.

The Independence I people disappeared at the same time as a colder climate period began, but it may also have been societal dynamics that underpinned the end of the culture.

Saqqaq (approx. 2400-700 BC)

Following archaeological findings at the Saqqaq settlement in Qeqertarsuup Tunua (Disko Bay), the first people who settled on the west coast were named the Saqqaq culture by archaeologist Jørgen Meldgaard. Towns and settlements from the culture can be found along the west, south and east coasts (see p. 89). This suggests that they spread southwards from the north-west of the country and onwards up to Northeast Greenland.

The Saqqaq people immigrated to the country largely at the same time as the Independence I culture, but a meeting between the two cultural groups has yet to be proven.

Despite Saqqaq's long time span and wide geographic distribution, the culture is characterised by being homogenous and conservative. Most finds have been made in and around Qeqertarsuup Tunua as well as Sisimiut, where larger and extensive concentrations of towns and settlements can be found. Of these finds, most are dated back to the first 800 years of the culture's presence (approx. 2400-1600 BC), and then gradual changes are traced followed

To the left: The Saqqaq settlement Qeqertasussuk was located on a headland on the island of the same name. The permafrost in the middens preserved knots of hair and human bones for thousands of years before they were found by archaeologists. To the right: The settlement of Qeqertasussuk was inhabited all year round during the first period of the Saqqaq culture. Still, in the later phase of the culture, it was only occupied during the summer. The photo shows a drawing of what life on the summer settlement might may have looked like.

by extensive changes in the choice of raw materials and the construction of hunting and household tools, as well as in the selection of game and hunting methods.

Tools from the culture have been found in the towns and settlements that are made from driftwood, bone, tusks and antler. Among these are harpoon heads, shafts, sewing needles, spoons, bowls, and needle cases, some of which are simply decorated with lines and kayak-like vessels (see p. 103). Particularly well-preserved finds come from the Qeqertasussuk and Qajaa settlements in Qeqertarsuup Tunua and Nipisat south of Sisimiut. The finds have been in permafrost soil layers and provide a unique insight into a culture that existed several thousand years ago. For example, it has been possible to sequence a nearly complete genome from one DNA molecule. The genome, which turned out to originate from a man, was extracted from a knot of hair found in the compost heap in Qeqertasussuk. Finds of thin wooden strips have been interpreted as drum frames, which resemble Inuit skin drums, and finds of dog bones testify that the Saqqaq people kept dogs.

The killiaq (silicate slate) type of stone was the preferred raw material for stone tools. Only two sources are known, on the Nuussuaq Peninsula and in the southern part of Qeqertarsuup Tunua. The distribution of killiaq across vast distances attests to a vast network along the vast stretches of coastline. Raw materials for stone tools also included chalcedony and quartz, especially during the early period of the culture.

The stone tools were manufactured in the same way as by the Independence I culture; however, some tool types were given a final processing with polishing, especially in the later phase of the culture. The distinctive stone tools include two-sided weapon points, burins, end and side scrapers, lances, knife blades and axe blades (see p. 92). Hunting gear was nicely and precisely worked for the hunting of a wide variety of marine and terrestrial mammals.

The dwellings were round or oval tents with a central mid-passage and hearth for use in the cooler seasons (see p. 192). Tent rings with flagstone platforms and with a central hearth have been interpreted as winter dwellings. Dwellings from the warmer seasons often had hearths outside the tents. Soapstone vessels and lamps for lighting, heating and cooking have not been found from the early Saqqaq culture. Variations of stone hearths and fire-cracked rocks, on the other hand, characterise the early period of the culture. Around the year 1500 BC, soapstone vessels and small round soapstone lamps started to gain ground.

The Saqqaq people disappeared after 1,800 years of presence. It happened in parallel with climate change, which may have had significant implications for the presence of humans and animals. Saqqaq is the longest-lasting culture in Greenland.

Greenlandic Dorset
(approx. 800 BC to 1 AD)

A new immigration took place around 800 BC. The first finds from the culture known as Greenlandic Dorset were made in the 1920s in Cape Dorset (Kinngait) on Baffin Island in Arctic Canada. Over time, the culture was divided into three phases. The first two stages of culture were formerly called Independence II and Dorset I/Early Dorset, but are now assumed to be from one and the same cultural phase now known as Greenlandic Dorset.

Greenlandic Dorset spread rapidly along the west coast, often settling in the same places that the Saqqaq culture inhabited before. Like the Independence I peoples some 1,700 years earlier, Greenlandic Dorset spread northwards in the country and remained in Northeast Greenland, but moved further south to places where there were polynyas, open water areas surrounded by sea ice during the winter months (see p. 62). Polynyas attract marine mammals and birds and, therefore, they provided good hunting opportunities, especially at the end of the critical winter period.

Greenlandic Dorset were generally more marine oriented in their choice of game, but also exploited terrestrial mammals. For example, a specialised focus on caribou hunting has been demonstrated at some towns and settlements in the inland fjord systems of the west coast.

There are regional variations in the finds of tent dwellings, which are adapted to colder and warmer seasons. Dwellings for the cold season are partly built with peat on either side of the space. As with the previous cultural groups, stone tent rings with a central mid-passage or a central hearth were also used. Finds of parts of snow knives and sled shoes testify that in Greenlandic Dorset there was a more specialised use of snow and ice for dwellings than in the past, just as they have had sledges for winter use.

Dwellings for the warmer seasons are characterised by round or oval tent rings with outdoor box hearths. Double tent rings testify to a new form of dwelling in which two tents were built together (see p. 192).

New distinctive structures are triangular stone tile surfacing in dwellings that are interpreted as summer residence dwellings, as well as the fan-shaped ends of the central mid-passage and the use of both multi-coloured and black stone tiles.

It is presumed that, for periods, Greenlandic Dorset in Northeast Greenland also lived in dwellings built of snow.

In addition to sled shoes and snow knives, there are some tool finds made from bone, tusks and antler, including harpoon heads, lances and thin needles that are pointed at both ends and equipped with a carved ob-

To the left: Harpoon tips were important fishing tools. Here they are seen in various shapes and sizes adapted to the specialised fishing methods used by the Greenlandic Dorset. The longest is 10.4 cm long (Photo: John Lee, The Danish National Museum) To the right: The indispensable knife blades (top row) and end scrapers (bottom row), without handles, are made of various stone materials and in different shapes and sizes. The largest blade is 4.8 cm long.

JOHN LEE/NATIONALMUSEET I KØBENHAVN, U.A.

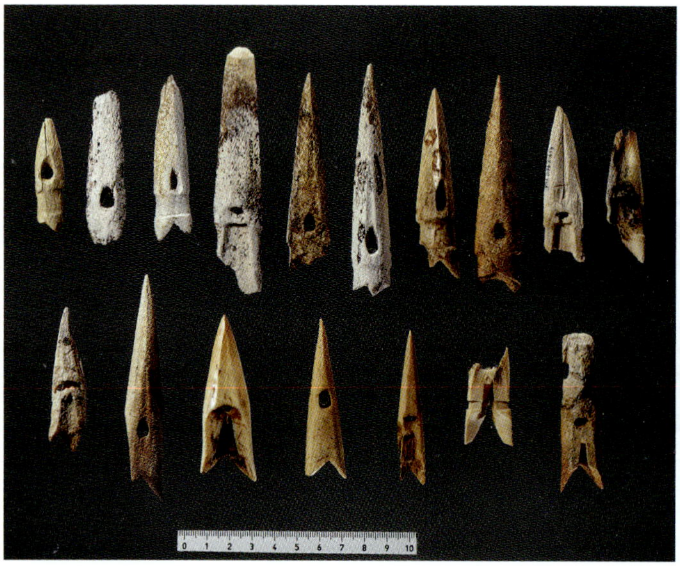

JOHN LEE/NATIONALMUSEET I KØBENHAVN, U.A.

long needle eye. Finds of round and oval soapstone lamps or parts of vessels testify that these were used as sources of light or as containers.

A distinctive feature of Greenlandic Dorset is the use of colourful stone tools made of agate and chalcedony and rock crystal. Heat treatment before chipping stone tools is another characteristic, just as some stone tools have been polished. The stone tools were made with lashing notches by attaching them to a shaft or weapon, and end scrapers from the culture are characterised by bulging sides.

The possibility of a meeting between the Saqqaq people and Greenlandic Dorset is temporally not impossible; however, so far there is no evidence that they met. After Greenlandic Dorset, the country was depopulated for roughly 700 years until a new immigration from the west repopulated the country. While the country lay empty of human life, the Dorset culture however continued to exist in Arctic Canada, but the settlement moved to the more southern regions of Labrador and Newfoundland. This period of the Dorset culture is called Middle Dorset by archaeologists and no remains have yet been found from this period.

Late Dorset (approx. 700-1350 AD.)

The last Dorset period is characterised by the repopulation of the previously abandoned High Arctic areas. The Dorset culture, which is an offshoot of the Canadian culture, the Late Dorset, immigrated to Greenland and stayed in the north-westernmost regions of the country for about 600 years (see p. 89).

The late Dorset culture is the first archaeologically recognised and generally best-known of the Paleo-Inuit cultures. Although the evolution and relationship to the Canadian Middle Dorset and the Late Dorset is not entirely understood, it is assumed that the culture of Late Dorset developed locally from the Middle Dorset culture of Arctic Canada and Newfoundland. The archaeologists do not know the reason why the Late Dorset people spread to the High Arctic areas, but climate change during the medieval warm period is a generally accepted reason for their expansion into the High Arctic. However, an interplay is often seen between several conditions when cultures change.

The Late Dorset people developed a distinctive and conspicuous artistic tradition, which was usually small carved figurines made of bone, tusks, driftwood and soapstone, among others. They depict tools, people and animals in both naturalistic and abstract versions. At times, the figurines are seen decorated with carved symbols and signs. The tradition of making small figures is believed to be promoted by the Late Dorset culture's use of the metals meteorite iron and copper for cutting tools. The figurines had different functions; some were related to spiritual practices, personal amulets, toys, ritual objects, while others illustrate the relationship between humans and animals as well as their surroundings.

The Late Dorset people also stand out in that they constructed buildings, which the archaeologists call longhouses or megaliths. The buildings are made of large stone blocks set in 8-43 m-long parallel rows with internal width dimensions of 4-6 m (see p. 193). It is uncertain whether the structures were originally covered with a skin roof, or if skin tents were erected inside the structures. On the whole, the purpose of the longhouses is not known. However, there is general agreement that these large buildings were constructed to be used for larger social gatherings, rather than for winter or summer dwellings.

The winter dwellings were depressed square dwellings which could measure 4 × 5 m and were constructed with peat walls. The finding of snow knives and the use of soapstone lamps with fuel from oil squeezed out or melted from the thick fat layers of marine mammals indicate that for periods they were able to live in snow huts.

CANADIAN MUSEUM OF HISTORY, NIHF-4:93, CD1995-0273-028, U.A.

A polar bear figurine from the Late Dorset culture of arctic Canada carved out of walrus tusk. It has an extended neck and head and trailing/hanging fore and hind limbs. The figurine is interpreted as an abstract representation of a human (the shaman) who is transformed into a polar bear, or a naturalistic representation of a swimming polar bear hunting or simply going on its way. The length of the figurine is 5 cm.

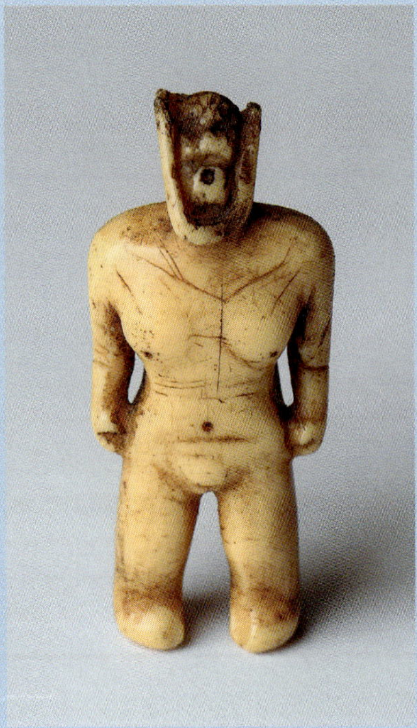

HANS LANGE/NUNATTA KATERSUGAASIVIA ALLAGAATEQARFIALU, 2022

In the late Dorset period, human figures were sometimes depicted with details such as clothing, hair and gender-specific aspects such as genitals. This carving out of a walrus tusk depicts a man with his navel and nipples clearly visible. The high collar around the neck is seen on several other figurines and is characteristic of the Dorset people's outerwear. The length of the figurine is 6 cm.

THE STRONG TUNIIT PEOPLE

Inuit legend in both Arctic Canada and Greenland tells of the meeting with a people called *Tuniit* or *Tunersuit*. Although legends differ, it is clear that they portray the same people.

The Inuit recount the Tuniit as big and strong and who spoke a different language and had a different way of life. They built square slightly buried dwellings, used neither bow nor kayak, and for seal hunting on the ice they used very small soapstone lamps.

The Inuit legends of the Tuniit say something about the depth of time in the hand down of knowledge (see p. 108), and assumptions of who these people were have changed over time. It has been proposed that they were Norse, that they were indigenous people in North America, and it has been proposed that they were early Inuit. Today, general consensus is that the Tuniit lived in the Late Dorset culture.

It is possible that the Inuit and Dorset people knew of or met each other. The Inuit immigrated at the Pikialasorsuaq polynya (North Water) in approx. year 1150, and the latest dating from the Dorset culture in the area is from the early 14th century.

Due to optimum preservation conditions and early ethnographic descriptions, the study of that part of cultural history has some advantages. However, more extensive research into the Inuit storytelling tradition is needed to be able to use it critically in the interpretation of ancient times.

The summer dwellings are very similar to the central mid-passage dwellings of the earlier cultures; however, flat stones with a recess have often been found at one of the sides of or in the central hearth. These are interpreted as pot holders.

One of the most notable things in all the Dorset peoples is that the use of the bow drill, bow and arrow disappears and no finds have been made which may suggest the use of ocean-going vessels as with the Inuit.

For Late Dorset, the population remained in the area at the Pikialasorsuaq polynya (the North Water), situated between Ellesmere Island and Northwest Greenland. Their hunting implements were custom designed for blowhole hunting on sea ice and hunting from the ice edge. The design of weapon tips and harpoon heads shows that they were specially made for hunting marine mammals.

Soapstone vessels and lamps are usually small round or oval vessels. The use of colourful types of stone is still prevalent; most especially agate, quartzite, rock crystal, basalt, shale and chalcedony were used.

Meteoric iron carved from meteor deposits at Savissivik in the northern part of Qimusseriarsuaq (Melville Bay) has been used for weapon tips and cutting tools, which were hammered into shape. Meteoric iron from Savissivik has also been found on Baffin Island and more westerly islands in Arctic Canada. Together with finds of natural copper from Kogloktok (Coppermine River) in the northwestern part of Arctic Canada, this testifies to a wide-ranging exchange network.

Finds of items from the Norse both in the northwest of Greenland, Ellesmere Island and Baffin Island also indicate that the network could include the Norse. If the situation was such, then the meeting between Late Dorset and the Norse closed the circle around the globe for human migration to the east and west more than 60,000 after populations emigrated from Africa.

THE NORSE

Nordic farmer-hunters – the Norse – settled in Greenland in the decades shortly before 1000 AD and thus initiated some 450 years of European history in the Arctic. It was one of the last migrations of the Viking Age where, over a few centuries, settlers from especially Norway had settled on the islands of the North Atlantic. The settlers in Greenland came from Iceland. This North Atlantic *landnam*, a Norse word for taking new land, is described in Nordic sources and especially in the Icelandic saga literature. In the sagas, political and familial disputes in the homelands are highlighted as the main reason why, during the Viking Age, thousands of people dared to set out across the sea in open ships with everything they owned. The same theme is found in the saga about the namer of Greenland, Erik the Red, who after three years of exile from Iceland and accompanied by a small fleet of kinsmen settled in Greenland in 985/986 AD.

More recent research has nuanced the view of the North Atlantic landnam. The development of ocean-going ships enabled sea voyages and, at the same time, settlers were well helped by a relatively mild and stable climate regime, known as *the Medieval Warm Period*. The mild climate also contributed to the Nordics experiencing large population growth, with land shortages as a result. But the tale of free farmers seeking the land of opportunity in the North Atlantic is romanticised to some extent. Instead, the landnam presumably needs to be viewed more closely as the strategy of ambitious families to gain wealth and social status by acquiring new lands and resources.

New research has suggested that the towns and settlements were founded as bases for hunting exotic Arctic luxury goods, primarily walrus and narwhal tusk, as well as polar bear furs. Trade in these goods also shows that the towns and settlements were part of a growing European cultural, commercial and religious community, a connection the Norse maintained until the discontinuation of settlement around 1450 AD.

Our knowledge of the Norse

Apart from owners' marks, short Christian prayers and pagan spells written in runes on pieces of wood and soapstone, and a single letter drawn up at the diocese of Garðar in 1409, the Norse have left nothing in writing. Everything written about them is authored by strangers. This

Aerial photo of the only approx. 3 km wide isthmus between Igaliku Fjord and Tunulliarfik Fjord. The isthmus connects the most densely populated parts of the Norse Eastern Settlement. This strategic location was one of the reasons why the Norse episcopal residence of Garðar was established precisely here, where the settlement of Igaliku is located today with surrounding farms and fields.

MORTEN RASMUSSEN/B OFCTO/RITZAU SCANPIX, 2008

also applies to the Icelandic sagas, where the stories about Greenland were largely written only from the 13th century, after a few centuries of oral tradition. In particular, *the Vinland sagas – Erik the Red's Saga* and *Graenlendinga Saga* – tell about the 'landnam' in Greenland and the travels to Vinland on the American continent. From the 13th century, some more detailed sources exist, including notes in the Icelandic annals, the Norwegian prose work *Kongespejlet (The King's Mirror)* and Ívar Barðarson's *Beskrivelse af Grønland (Description of Greenland)*.

Despite the fragmentary state and problematic origin of written sources, they have been the defining factor for the research into the Norse for many years. In the 1830s, the Royal Nordic Society of Antiquaries initiated a systematic collection of written sources about the Norse Greenland, which since then provided the understanding framework for the topographical and archaeological studies in the Norse settlements. The studies became progressively more systematic and professional, starting in the late 19th century under the Commission for the Management of the Geological and Geographical Surveys, from 1921 under the National Museum of Denmark and from 1982 with the Greenland National Museum and Archives as an active research party and responsible authority. However, it was only with the introduction of new scientific approaches, methods and international interdisciplinary collaborations in the 1970s, that the Greenlandic Norse seriously gained a voice of their own through archaeology.

The Norse towns and settlements

The Norse settled in two parts of Southwest Greenland. The largest settlement was the Eastern Settlement, where more than 560 archaeological sites have been recorded. The Eastern Settlement was spread over an ice-free area of about 24,000 km^2, stretching more than 300 km from the southern tip of the country and up to present-day Arsuk. About 350 km further north lay the smaller Western Settlement with only about 110 registered sites and an area of about 10,000 km^2 around the Nuuk and Ameralik fjord systems. In the Eastern Settlement, the settlement existed from the edge of the ice sheet and out to the far reaches of the fjords, while Western Settlement farms were located almost exclusively in the inner parts of the fjords. This difference in the settlement reflects the slightly

To the left is a gown found during the excavation. Few were buried in coffins. Most often, the dead were laid on the ground wrapped in discarded clothing, which indicates that the Norse dressed as they did in Europe. The high front and back pleats of this homemade robe have given it an elegant drape, and fibre analysis of the wool shows that it was originally mottled brown and white, while the patch on the chest was white. The gown is carbon 14 dated to the first half of the 15th century. To the right: Excavation of coffins in the cemetery at Herjolfsnæs/Ikigaat in the southernmost part of the Norse Eastern Settlement, 1921. This was the first of a series of extensive archaeological investigations of Norse ruins, which until around World War II were carried out by the National Museum of Denmark.

PETER DANSTRØM/NATIONALMUSEET I KØBENHAVN, 2003

POUL NØRLUND/NATIONALMUSEET I KØBENHAVN, 1921

more favourable environmental conditions and higher population figures in the Eastern Settlement.

In both Norse settlements, the settlement was concentrated and most extensive in the inner fjords, where lush valleys, mountains and plains along the edge of the ice sheet form the climatic and environmental niche best suited for farming with livestock. Subarctic and more continental areas are found here characterised by colder winters and warmer summers, as well as a more arid and stable environment than in the Low Arctic and coastal zone along the Davis Strait. Although the distance between the two zones is only 60-100 km, there is a striking difference between the sparsely wooded fjord and archipelagic environment and the regular 4-6 m high scrub forests of willow and birch in the inland's most lush and protected valleys. Even in the inland fjords, however, the Norse's farming was on the borderline of the possible, and almost all farms were located on sunny mountain slopes right down to the fjord or up through low-lying valleys with rich vegetation.

Because the limited vegetation resources had to maintain almost self-sufficient Norse farms, they were almost all individually situated and relatively dispersed. However, far from all farms were independent. Many were in, or over time became part of, a dependency relationship with the richest chieftains' farms as a kind of tenant or serf farm. However, the lack of written information from Norse Greenland makes it impossible to determine whether a just as complicated inheritance and property relationship developed as in the rest of the North Atlantic. Much suggests, however, that the towns and settlements experienced increasing socio-economic centralisation around a small community elite residing on chieftains' farms that were almost comparable to estates or manors with large adjoining lands.

MORTEN RASMUSSEN/BIOFOTO/RITZAU SCANPIX, 2008

The photo shows the preserved door between the stable and the barn of the Norse episcopal residence at Garðar/Igaliku. Ruins at the episcopal residence are characterised by being the largest in Greenland, a clear indication of the farm's significant status. Consequently, the bishop was able to overwinter about 100 cows in two large cowsheds, a very considerable achievement even in comparison with other contemporary North Atlantic chieftains farms.

The chieftains' farms can be recognised by their agriculturally favourable location, as well as the size, number and combination of adjacent buildings, including banquet hall and church. The chieftains' farms stand out so clearly that with great probability it was possible to identify several with named places in the medieval writings, among them Erik the Red's Brattahlíð (Qassiarsuk), Hvalsey (Qaqortukulooq) and Herjolfsnæs (Ikigaat) in the Eastern Settlement and Sandnes (Kilaarsarfik) and Anavik (Ujarassuit) in the Western Settlement. The Diocese of Garðar (Igaliku) occupies a special position as the absolute largest farm in Greenland and is particularly distinguished by its huge barn and storage buildings and a cathedral. The legal institutions of the Norse, the thingsteads, also appear to have been linked to the chieftains' farms, as well as the trade and hence the redistribution of imported goods, not least the precious iron which was not produced in the country. The chieftains' farms were the religious, socio-economic and legal hubs of the Norse society. It was also only the chieftains who owned ships of the necessary size for hunting voyage to the country's remote regions.

Outside of the towns and settlements, stretched the great Arctic wilderness, which the Norse called the unsettled land. Here walrus, narwhal, polar bear and other Arctic animals thrived – the source of tusk and skin

ARNOLD MIKKELSEN/NATIONALMUSEET I KØBENHAVN, 2017

Small slate rune stone – length 11.2 cm – found in 1821 on the island of Kingittorsuaq 15 km north of Upernavik at latitude 72. The interpretable part of the rune inscription is as follows: »Erlingr Sigvatsson and Bjarne Thordarson and Endridi Oddson on Saturday before Rogation Day, built cairn(s) and ...«. The discovery of the rune stone so far north is indisputable evidence of the Norse's far-reaching hunting and discovery journeys to the High Arctic.

The youngest medieval church is located at the settlement of Qassiarsuk, designated as the place where Erik the Red settled in Greenland and is called Brattahlíð. The fact that this farm had a significant status can be seen from the churches found and examined on the site. Several flat tombstones, some with rune inscriptions, were also uncovered during the archaeological investigations and have since been returned and can be seen today.

WERNER FORMAN ARCHIVE/SHUTTERSTOCK/RITZAU SCANPIX, 1970

products that were rare and therefore very valuable outside of Greenland. These commodities formed the basis for the Norse's continued trade interaction and contact with the Nordic countries and Europe. Written and archaeological sources also show that the hunt for Arctic luxury goods in the remote areas was frequent, well organised and of great value. A particularly important area was Qeqertarsuup Tunua (Disko Bay) some 500 km north of the Western Settlement. The area has been identified by the written sources as the northern hunting grounds. But the hunting led the Norse even further north, and one of the most reliable and important finds from the remote hunting grounds is the rune stone from the island of Kingittorsuaq near Upernavik at 72.5° north. The rune stone briefly reports about three Norsemen who erected cairns on the island in April sometime in the 13th century.

Arctic hunters, Scandinavian farmers

Animal bones found in debris layers by the farms show that marine mammals, primarily seal, also played a vital role in the Norse's subsistence economy, especially on the smallest farms. Already in the earliest years of the settlement, seals represent 40-60 % of the total number of animal bones, and the proportion grows to up to 80 % on some farms towards the 1400s. A completely consistent picture of the large and growing importance of marine mammals has also been illustrated by isotope analyses directly on the skeletons of the Norse. These show that already in the earliest Norse, protein from marine mammals could account for 15-50 % of the diet, while in the last period the proportion was 50-80 %. Overall, isotope analyses show that at the be-

Different, well-preserved household utensils made of wood and soapstone from the Norse farms in Greenland. Owing to the excellent preservation conditions of the past in the Arctic, archaeological investigations of the farms have revealed a wealth of organic material, including, perhaps surprisingly, a lot of wood. New analyses indicate that the vast majority was driftwood and, therefore, among the resources, the Norse had to fetch from the remote areas.

JOHN LEE/NATIONALMUSEET I KØBENHAVN, 2010

ginning there was a greater spread in who had access to a diet from agriculture and the ocean, but that everyone – regardless of status in society – ultimately ate more marine mammals.

Bones of whole seals are found in quantities in the debris layers of the Norse farms close to the fjords and inland. The bones are most often of young seal species which, in the Middle Ages, were found along the Greenland coasts in their millions every season. This suggests that seal hunting was a coordinated activity in which hunters cooperated in periodic mass hunting of seal in the mouths of the fjords. Once the hunt was over, the prey was distributed, perhaps according to status in the community or the individual's share in the hunt. The Norse also hunted stationary seal species and virtually all of the wild land animals and birds found seasonally in the environment. Apart from the hunting of caribou, which was a major source of meat especially in the Western Settlement, the exploitation of other wild species appears to have represented only a minor supplement to the diet.

Despite the importance of hunting, the Norse were entrenched in an agriculture culture based on farms and livestock. As elsewhere in the North Atlantic, their agriculture can be described as pastoral farming, that is, a use aimed at meat and dairy farming by grazing and feeding cattle, pigs, goats and sheep, with the latter also providing wool for textiles. Horses were primarily transport and draught animals, dogs were used for hunting and for guarding, while cats could keep dwellings and pantries free from mice, which were also introduced to Greenland with the Norse's agriculture package. The cattle were the most valuable livestock, not only as meat and dairy animals, but also as a currency and therefore also as a mark of status. This may explain why, throughout the settlement, the Norse kept relatively many cows, even though they were difficult to keep over winter in the Arctic environment. On smaller farms, sheep and, as something particularly Greenlandic, goats, became by far the most important meat and dairy animals over time. Some farms in the peripheral regions seem to have almost specialised in this production. On the other hand, over time pigs were almost completely phased out and eventually only existed on the richest farms.

For the sedentary Norse, the farm with its household, animals, land and buildings was the socio-economic core unit. The centre and largest building of the farm was the main house that often, and especially in the later period, housed not only an extended family but also livestock. These so-called centralised farms were adapted to the Arctic conditions which, at the same time, were heat-efficient, material saving and allowed animals to be attended to without having to go outside. Scattered around the main house were other buildings such as stables, barns, drying and storehouses, saunas, forges and paddocks for the animals. The buildings on the farm, all constructed in turf, stone and wood, lay on or around a home or infield, a fertilised and therefore relatively lush grass or meadow area where hay was grown for winter feed for the livestock. Outside the infield lay the outfield, the uncultivated land that stretched out into the Arctic expanses of the unsettled areas. The outfield was an important resource area for the Norse. Here, there were pastures and fodder was collected for the livestock, fuel for the long winter, soapstone for lamps, dishes and vessels, bone and antlers for tools, driftwood for buildings, furniture and utensils and, even further out in the outfield, the necessary marine mammals, birds and other game.

Recent research has shown that the majority, perhaps up to two-thirds of known Norse localities were shielings. Shielings were buildings in the outfield where only a smaller portion of the household took up seasonal residence, often to allow the livestock to summer graze while continuing dairy farming. If these shielings are omitted from the count, there were only 200-300 regular farms, representing a total population figure of maximum about 3,000 people at the height of the settlement. This low population figure is consistent with the estimated number of people buried in the Norse's cemeteries.

Christianity and royal power at 'the ends of the earth'

No pagan tombs have been found in the Norse towns and settlements, and archaeological dating of the earliest churches show that they were built just after the 'landnam'. They were erected on the chieftain's farms near the house, thus signalling the cohesion between the community elite and the religion. The first churches were small, averaging only about 10 m^2, and constructed in peat, stone and driftwood. The surrounding churchyards were circular and had a diameter of 14-20 m. Everyone was buried in the churchyard, while the small church building was only for use by the owner of the church and his household. Larger churches with a Romanesque floor plan – with nave and smaller chancel in the east – appeared in Greenland from the 12th century at the same time as four-sided churchyards were introduced.

The chancels of the Romanesque churches with the most sacred, the altar, were reserved for the priesthood, and perhaps the new floor plan should be seen as a sign of a growing clerical influence, following on from the establishment of the Greenland diocese at Garðar in 1124. The bishop consecrated and supervised the priests, and presumably the owners of the church required the bishop to fulfil the liturgical requirements and collect church taxes. According to Icelandic accounts, it was the Greenlanders themselves who took the initiative and asked the Norwegian King Sigurd Jorsalfar to ap-

LENNART LARSEN/NATIONALMUSEET KØBENHAVN, U.A.

Head of a crozier found in a tomb in Igaliku. Only the 14 cm high head carved in the walrus tusk was preserved. The bishop's right hand has held the crozier, which lay across the grave. On his ring finger the ring shown was placed; the valuable gemstone in the socket had been removed before the burial. Only a few of the Norwegian bishops sent to Greenland died there. One of them was Ólafur (bishop during the period 1246-1280), which is consistent with the fact that the skeleton in the bishop's tomb is carbon 14 dated to 1223-1290.

point a bishop. The king appointed the Norwegian priest Arnaldur, and given the bishop's great economic and spiritual power, one could wonder why all subsequent Greenlandic bishops were Norwegian. In Iceland, it was the richest lineages that occupied the offices of bishop right up to the 1300s. Perhaps none of the chieftains were wealthy enough to seize the office of bishop, or they saw greater advantage in supporting Norwegian appointed bishops.

In 1261, seafarers returned home to Norway after four winters in Greenland. They conveyed the news that the Greenlanders agreed to pay taxes and levies to the Norwegian king. This was the result of several years of active Norwegian foreign policy in the North Atlantic. At the coronation of King Håkon IV in Bergen in 1247, the papal envoy had declared that no people should be without a king, and that same summer Bishop Ólafur was sent to Greenland to get the Norse to submit. It is presumed that the Greenlandic towns and settlements – which is historically known about as the abandonment of the Icelandic Free State in 1262 – voluntarily submitted, with a demand that the Norwegian king should secure the navigation and thus trade and supplies. The king was not heavily represented in the country, and only a few royal farms are mentioned fleetingly in the written sources. The true interest of the royal power lay in the revenue from the customs duties that the merchant ships from Greenland had to pay when they arrived in the commercial town of Bergen.

The ruins of Hvalsey Church, the largest and most well-preserved Norse ruin in Greenland. The last written document about the Norse in Greenland refers to a wedding held here in 1408 A.D. The church was built later than 1300 A.D., probably under the direction of British builders, and may be an expression of the increased influence of the Norwegian king and church in Greenland. However, excavations have revealed that there must also have been an earlier church on the site.

With limited administration, the royal power presumably depended on the clergy, who had more than 100 years of experience in the country. The construction of seven large one-room stone churches after 1250-1300 is perhaps precisely the

ARNOLD MIKKELSEN/NATIONALMUSEET I KØBENHAVN, 2005

An angakkoq under a trance. Figure carved in driftwood and stained with soot and whale oil from the blubber lamp. The Angakkoq's hands are tied behind his back, and he is sitting in front of a skin that hangs in front of him. At his feet is the Greenlandic drum and two helping spirits. All the remedies will aid him on his soul journey to the other world - silap alla. The figure was collected by Therkel Mathiassen in 1932 in Ammassalik, East Greenland.

expression of the close alliances between the Norwegian church and royal power and powerful Greenlandic large farmers. The grandest representative of the one-room churches is the very well-preserved Hvalsey Fjord church in Qaqortukuluup Imaa. The church is built entirely in stone, and the walls have been sealed with mortar made from local beach shells. The construction technique deviates from what we otherwise know, and presumably the unique stone churches were constructed under the direction of British builders who had been sent to Greenland by the Norwegian king.

The depopulation of the Norse towns and settlements

The depopulation of the Norse towns and settlements began as early as the mid-13th century, but seriously accelerated during the 1300s. Around 1400, the Western town and settlement was deserted, and 50 years later the Eastern Settlement was also abandoned. A lot of explanations have been presented through time as to why this happened. Today, there is broad agreement that the depopulation was due to an unfortunate coincidence of several conditions. The town and settlement was, at its starting point, vulnerable. Geographically, the Norse towns and settlements were situated far from the world on which the Norse depended. The worsening of the climate in the 13th century – the beginning of the Little Ice Age – put the already fragile agriculture under severe pressure and led to increased reliance on seal hunting. Initially, the transition was successful, but perhaps also masked the fact that the community order upon which the power and authority of the leading relatives rested began to falter.

Foreign trade, which was also under the control of chieftains, also failed, and there was very little they could do, since contacts to the Nordics and Europe were entirely dependent on visiting merchants. The Norse themselves did not own oceangoing ships. The growing geographical, economic and cultural isolation created great external pressure on the Norse community, which they could not counteract themselves. In addition, they also experienced an internal pressure. The total population probably never exceeded 3,000 people, and the population declined sharply in the 1300s. Regardless of whether parts of the population migrated or they perished due to increasing poor living conditions, accidents travelling to and from hunting grounds or more frequent encounters with a growing population of Inuit, with the calculated number of inhabitants, it did not take more than an average annual loss of approximately 10 young fertile people before the phasing out of the Norse community was a natural demographic consequence. Lack of labour in society, including for the long hunting trips, may have resulted in reduced production of the very commodities for which foreign merchants travelled to trade, making it less attractive for them to risk the dangerous journeys to Greenland. A vicious spiral that was amplified by falling prices on, in particular, walrus tusks in European markets.

The attempt to secure trade and contact with Europe was precisely one reason why in 1261 the Norse voluntarily submitted themselves to the Norwegian king. But the countermove had limited effect. Severe plague outbreaks in Norway in 1347-52 weakened the country, and made it impossible for the king to fulfil his obligations. Things did not improve when in 1380, Norway entered into a dynastic connection with Denmark, and the administration of the Twin Kingdom moved to Copenhagen, where interest in the poor remote settlement in Greenland was probably not great.

There is no knowledge of how life proceeded for the last of the Norse. The church pulled out in 1378 when the last bishop at Garðar died. A legal sequel reports about a wedding at Hvalsey in 1408, while the last ship we know of left Greenland in 1420. Among the passengers were guests from the wedding in 1408.

THE INUIT CULTURE, PRECOLONIAL PERIOD (APPROX. 1150-1721)

The Inuit culture is a unified term that embraces several groups of people who still inhabit the Arctic region. The area covers Nunatsiavut, Nunavik, Nunavut, Inuvialuit in Canada, Inupiaq in Alaska and Yupit in Alaska and Siberia, and Kalaallit in Greenland.

The Inuit immigrated and settled in Northwest Greenland in approx. 1150. A precise dating has not yet been possible, but preliminary dating and typological developments indicate an Inuit presence as early as about 1150.

The first Inuit had travelled over 4,000 miles in the space of a few generations from the island of St. Lawrence between Siberia and Alaska to Greenland. The Inuit culture made use of large open skin boats for walrus hunting and whale hunting, as well as kayaking for seal hunting. In winter, dog sleds were used, which allowed the Inuit to hunt over long stretches and quickly occupy new areas.

The earliest archaeological finds in Northwest Greenland have distinct stylistic features in common with the Punuk culture, which specialised in whale hunting in the Bering Sea. In particular, ceramic lamps, excavated in Northwest Greenland, made of clay from Alaska, testify to the rapid movement across Arctic Canada.

Inuit immigration and settlement

When the Inuit arrived in Northwest Greenland, the fjords were already inhabited by people from the Late Dorset culture (see p. 93). The cultural encounter between these two groups can be traced archaeologically and in the oral tradition. Tuniit or tunersuit (see p. 94) is described in several of the myths as inland fighters, incredibly strong and shy people. They spoke kutattut, which to Inuit sounded like a kind of child's language.

PRE-COLONIAL INUIT CULTURE

The period of Inuit culture that runs from the immigration in around year 1150 to the colonisation in 1721 was named the Thule culture by Danish archaeologist Therkel Mathiassen. Since the 1920s, the term has been used to denote Inuit material culture until either the year 1600, where frequent contact with European whalers has been documented, or until 1721, where Denmark formally colonised Greenland. Mathiassen, who documented the culture through finds and excavations, named it after Knud Rasmussen's 5th Thule expedition (see p. 121), in which he himself took part, and the term is still partly used.

Since an archaeological designation of a still living culture emphasises the ethnocentric view that has often been on Greenland, the term is not used here. Instead, the term *Inuit culture* is used both for the prehistoric and the contemporary culture, and for the period approx. 1150 to 1721 specifically, the term »pre-colonial Inuit culture« is used.

The abandoned settlement of Nuulliit in the Qaanaaq area offers remains from several archaeological periods, among them winter dwellings constructed by the Inuit in the 1300s. The winter dwellings were partly dug into the surrounding terrain, and the

Nuulliit whale hunting scene carved into walrus tusk. At the bottom is an umiaq with the harpooner standing in the stem, arms raised and holding a harpoon, ready to harpoon a whale. The oarsmen are placed at regular intervals behind the harpooner. At the top are a series of symbols that look like upside-down Ys. These are interpreted as stylised whale tails and testify to the success of the hunters.

JOHN LEE/NATIONALMUSEET I KØBENHAVN, U.A.

walls constructed in stone, peat and sometimes whale bones. Plank beds were erected along the walls, and the dwelling was presumably lit by small ceramic lamps brought from Alaska, and later by small soapstone lamps. The up to 3 m long entrance tunnel was partially dug down and thus functioned as a cold trap. What sets these winter dwellings apart from later winter dwellings is a special kitchen extension in the front wall next to the entrance tunnel and with access from the space in the dwelling. Analyses of soil layers show that blubber and fatty bones have served as fuel for the kitchen fire.

Yet another element that characterises the early period of Inuit in Greenland are the large 'qassi'. These structures are erected in the same

THE WOMAN
- DUALITY AND BALANCE

Because of the female cycle, and because she can give birth, the woman was seen as the bearer of a special duality. Like her husband, she had to observe a number of taboo rules (see p. 158), including when she was preparing the game. Both before, during and after the death of an animal, it was necessary to show it respect by observing taboos and thus to thank it for allowing itself to be killed. If the rules were not observed, it would lead to a failed catch.

The taboo rules that the woman had to observe varied from place to place, but might include not using the same ulu (a special female knife) for fish and caribou, for example, or keeping different types of meat separate during cooking.

By observing these rules, the woman helped to ensure the survival of the community, just as everything from the tattoos on her skin, the patterns of her clothing and the way she handled her needlework helped to create a balance between this world, *Sila*, and the invisible world, *Silap aappaa*. The taboo rules were thus incorporated into the woman's daily routine.

KONRAD NUKA GODTFREDSEN, 2012

Nuka K. Godtfredsen's illustration from *The Ermine* (see p. 84-85) depicts summer activities: Fish spears are used to catch trout, which are cooked over a fire and also hung up to dry so that the fish can be used for supplies later in the year.

Map of Greenland with the five major summer camps (aasiviit) and trade routes. Sermermiut traded mainly in baleen, Taseralik mainly in caribou skins. Nuuk Fjord's rich deposits of soapstone were traded at Nepisat, while along the southwest coast there were deposits of driftwood. In southernmost Greenland, caribou skins and fox skins were traded in particular.

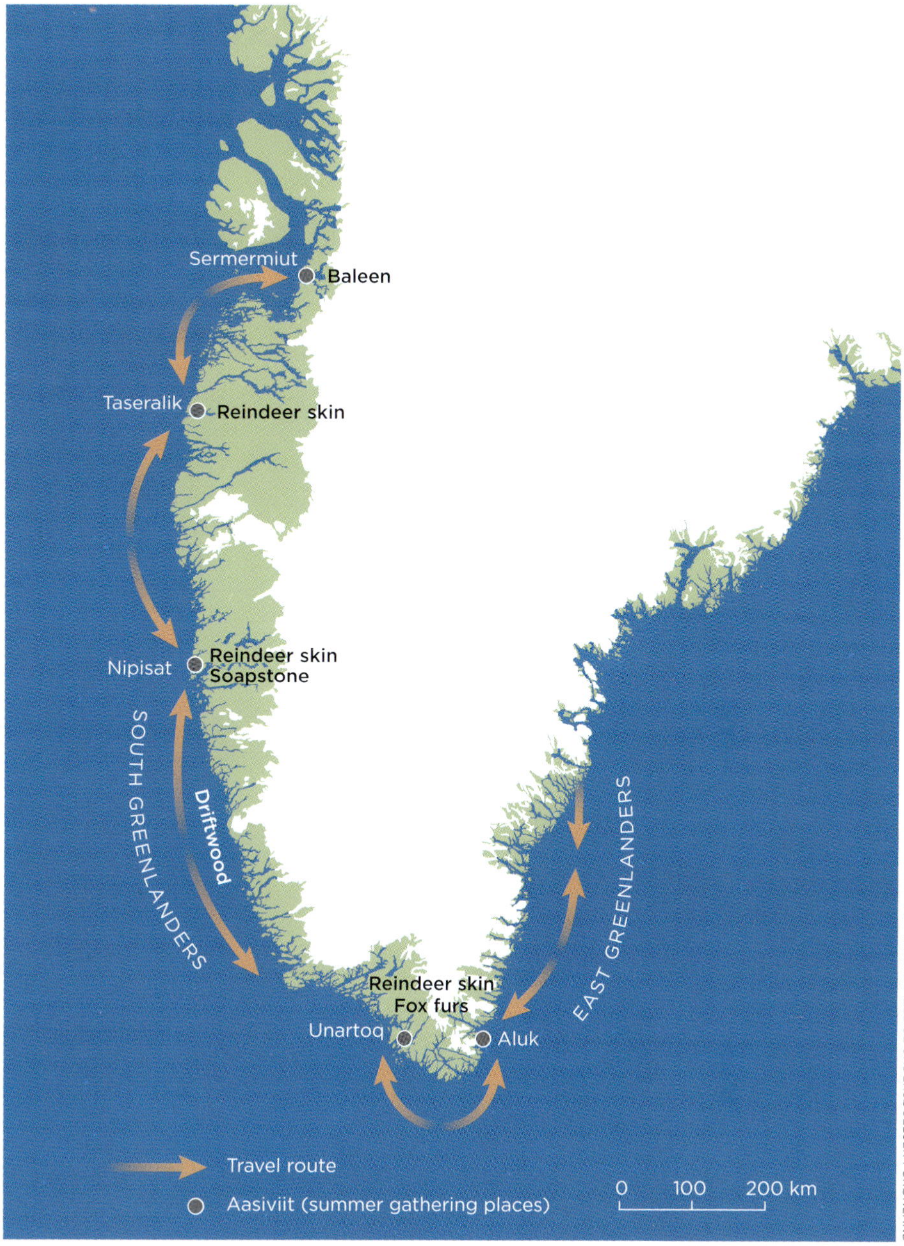

EFTER GULLØV, H.C. (1997). *FROM MIDDLE AGES TO COLONIAL TIMES: ARCHAEOLOGICAL AND ETHNOHISTORICAL STUDIES OF THE THULE CULTURE IN SOUTH WEST GREENLAND 1300-1800 AD.* KOMMISSIONEN FOR VIDENSKABELIGE UNDERSØGELSER I GRØNLAND

materials as the winter dwelling, but yet on a larger scale, have narrower plank beds and no traces of cooking. A 'qassi' served as a workshop, meeting place where communal hunting was planned, and possibly also as a party house. Traces of tool-making and the large quantities of whale baleen emphasise the orientation of the Inuit towards marine resources, the exploitation of which was a collective task. The organisation of the Inuit whalers was necessary to ensure good and safe whale hunting. A whale hunting team was led by an umialik, i.e., someone who owns an umiaq (a large open skin boat). It took a resourceful family and a good reputation to be able to acquire the title of umialik and to attracting a handful of hunters who would work together as a team and set off in an umiaq in search of the bowhead whale. If the hunting season was successful, the caught whale would ensure food for more families through the winter, while the umialik would rise in reputation and influence.

One such whale hunting scene is engraved in a piece of walrus tusk; here the harpoon with lifted lance is shown in the bow of the umiaq.

The harpooner wore a waterproof one-piece suit with a rope of whale baleen attached to it. Having sailed completely quietly up to a sleeping bowhead whale, the harpooner's job was to jump onto the whale and impale it by the blowhole. When the whale then tried to escape, the harpooner would fall into the water and the team, using the rope, could pull him up, and then they could continue hunting the whale.

Pre-colonial Inuit hunting culture along the west and east coast

Over the course of quite a few generations, the Inuit had migrated both north and south, along the west and east coasts of Greenland. The Inuit followed the seasons and the availability of animals. In the spring, they moved from their winter settlement and began their nomadic season to gather supplies for the next winter. They travelled in large open skin boats and kayaks, wandered inland and slept either in hunter's beds, a recess quickly built with stones, in rock caves, under open skies or when they stayed in an area for extended periods of time, in tents made of skins.

THE MUMMIES FROM QILAKITSOQ, PISISSARFIK AND UUNARTOQ

WERNER FORMAN ARCHIVE/SHUTTERSTOCK/RITZAU SCANPIX, 1970

This naturally mummified infant was buried with its mother in the single rock tomb at Qilakitsoq around 1475.

Three mummy finds in Qilakitsoq, Pisissarfik and Uunartoq offer unique insight into the early skin products of the Inuit. The making and patterns of the well-preserved garments tell both about regional affiliations and the gender of the wearer, and in some cases, amulets hidden in the garments give insight into the Inuit worldview.

The mummies, consisting mainly of women and children, were buried close to where they lived. The mummification happened because of the low humidity and cold of the Greenlandic climate.

THE QILAKITSOQ MUMMIES

The Qilakitsoq mummies from the Uummannaq area in Northwest Greenland have been dated to 1475. Two stone-covered graves with mummified bodies were found by hunters in 1972, and in 1978 the Greenland National Museum excavated eight buried people – six women and two children aged around six months and four years. The mummified women and children were all buried with extra clothes, extra skins and grave goods. In total, there are 78 pieces of clothing, and the find gives a rare insight into what life was like in that place and at the time they lived.

Typical clothing consisted of an inner anorak of bird skin where the feathers were turned towards the body and the edges were sewn with caribou skin. The outer anorak could be made of either sealskin or caribou skin, in which case the fur would face outwards. Both the inner and outer anorak had hoods. The trousers were either thigh-length or went down to the knee and were made of sealskin or caribou skin.

The kamiks were made of hairless sealskin, and under these they wore stockings made of sealskin or caribou skin. The lower part of the kamik was lined with grass for comfort and warmth.

The garment was aesthetic with symmetrical patterns created using dark and light leather.

CHILDREN'S GARMENT

The four mummies found at Pisissarfik in Nuuk Fjord are estimated to have been buried in the 17th century. Among these four were two infants. The Uunartoq mummies from South Greenland have been dated to the 16th century. Among an unknown number of mummies there were three infants. Together with the two children from Qilakitsoq, they provide a rare insight into the children's world, which can be difficult to interpret archaeologically.

The infant from Qilakitsoq was dressed in a sealskin anorak with a close-fitting hood and trousers with integrated feet. The four-year-old boy was dressed in an inner and outer anorak, trousers, socks and kamiks – all made of sealskin. The trousers were sewn with large stitches and many knots, and the kamiks, although well made, had many patches under the soles and scratches on the legs. The wear marks testify to a lively childhood, the high quality of the stockings and the ornamentation on the outer anorak are testament to a mother's love.

The total of six mummified infants, separated in time and space along the west coast of Greenland, tell us about both the high infant mortality rate and the care and attention the infants received during their short lives. Their garments show that the differences in men's and women's garments only become significant later in life.

Although the women were seasoned seamstresses, the work of sewing sinew through hides has been hard, and especially in winter the poor light from a blubber lamp has required concentration and precision. The ornamentation on the clothes is therefore attributed a greater function than pure aesthetics – it probably had a symbolic meaning.

WOMEN'S FACE TATTOOS

The five oldest women from the grave at Qilakitsoq had almost identical line tattoos on their foreheads, and it is not yet known whether the women from Pisissarfik and Uunartoq are tattooed since the mummies have not yet been examined for tattoos. However, it is very likely that these women also had facial tattoos, since historical sources tell of the tradition from the beginning of the 17th century until colonisation in 1721 on the west coast and until 1895 on the east coast.

The tattoos, together with the patterns on the garments, can be interpreted as a visual language with a protective function. Many of the same patterns recur on hunting tools and the hunting scene engraved on the walrus tooth from Nuulliit (see p. 103).

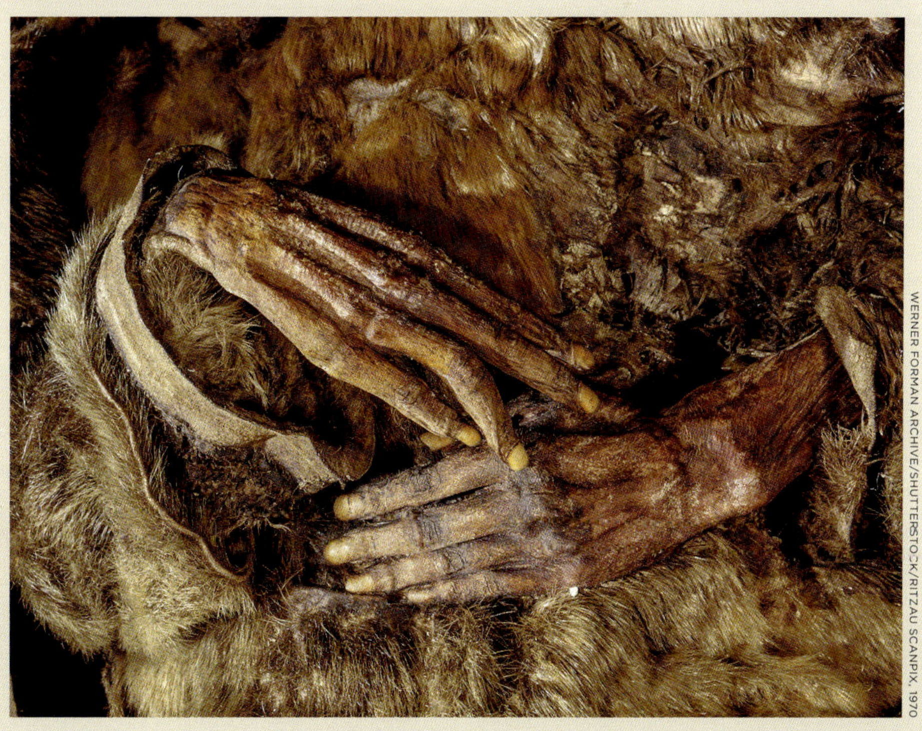

WERNER FORMAN ARCHIVE/SHUTTERSTOCK/RITZAU SCANPIX, 1970

The naturally mummified human remains are in very good condition. This woman was buried with five other women and two children in Qilakitsoq.

LEIFF JOSEFSEN/SERMITSIAQ, 2012

Mother of the Sea, sculpture by Christian Rosing, Nuuk. The Mother of the Sea, Sassumap Arnaa, is one of the most important figures in Inuit mythology. She is determined and, at the same time, »maternal« in her upbringing of the people, and the story of her emergence and influence on people's lives is a particular Inuit trait.

The harp seal was an especially important resource during the spring months. The seals, which migrated from Canada and southwards, were caught at the mouth of the fjords. At the same time, fishing started for the small capelin, ammassak, which played an especially important role in the planning of the winter supplies. The shoals of ammassat were shovelled up along the shore and laid to dry on the bare rocks or heather to add some natural aroma. After the drying was complete, the ammassat were laid side by side and rolled into a large bundle, which was kept dry and cool until winter.

On land, caribou and hares were hunted with bow and arrow and trout was caught in the rivers. The caribou hunt was an organised, collective enterprise. Shooting hides were set up along the caribou's migratory routes, and cairns and humans were used to scare the caribou in the right direction, where the hunters were standing by with bow and arrow. After the killing of a caribou, the belly was quickly opened, and the women and men then roughly dismembered the prey before the heavy trek back to the tent camp.

On water, seals were hunted from a kayak by using harpoons and hunting float which would ensure that the seal did not sink before the sealer could reach it. At Sermermiut in Qeqertarsuup Tunua (Disko Bay), fishing lines and nets made of whale baleen have been excavated, so cod, Atlantic wolffish and Greenland halibut have probably also been part of the menu.

Multi-family homes

In the mid-1600s, the rectangular communal house made its appearance. Before, when up to three families lived together under one roof, now four to six families could do the same, or about 20-30 individuals. They shared the catch, supplies and fuels for both light and cooking. The communal house was conceived and erected by the Inuit whale hunting teams and their families who constructed them along the coastal stretches, and only rarely inside the fjord systems.

The archaeologically recorded communal houses usually measure 4-5 m in width and 8-10 m in length, while some have measured to be over 20 m. The house had up to three windows that consisted of gut skins, which had been gutted, dried and stretched before being stitched together and firmly fixed on a wooden frame. The peat- and stone-built walls were covered internally with skins that kept the draught out and at the same time kept the dried ground in place. The plank bed ran the entire length of the main room, which was divided into family units separated by curtains made of skin.

In the limited space in front of each unit, there was a wooden lamp stand for the soapstone lamp, over which a soapstone pot could be hung, while wet skin clothing hung from a drying rack under the ceiling. On the edge of the plank beds, the women sat and sewed skin garments with great care.

New and old tales

The Inuit did not use written language, which is why oral tradition played a crucial role. A distinction was made between newer and old tales. The newer ones dealt with events the storyteller had heard of first-hand, or from a second or third source, and the old ones were passed down from the ancestors over several generations.

The old tales were to be passed down as originally told as possible. They had entertaining elements, but at the same time they had to convey the norms and values of society to the next generations. Another important element of the old tales was that by telling from an 'us' point of view, they created a sense of community.

In these tales it is not only the Inuit who are depicted as being compared with Tuniit (see p. 94), Europeans or other cultures. Tales are also told about mythical and supernatural beings, e.g., Sassuma Arnaa (Mother of the Sea), and how the Sun (Maliina) and Moon (Aningaaq) came into being (see p. 156).

The tales put into words how the world had come into being and what taboo rules the Inuit had to adhere to in order to maintain balance, e.g. by warning of the dangers of incest. They also passed down more simple rules to live by for children and adolescents. For example, by telling that it could be dangerous to move too far away from the settlement when a so-called mountain wanderer, a 'qivittoq', could capture you, or by allowing them to understand which plants you should not eat.

Cultural encounters

From the arrival of the Inuit in Northwest Greenland, they met with both the Tuniit (see p. 94) and the Norse. Meteoric iron was found in the area, which was an important trading commodity. Through cold forging, the Inuit processed the iron by hammering it into the desired shape. The Norse had several iron objects which they could exchange for other items, e.g. a ring from chainmail has been excavated from Nuulliit (see p. 103) and several artefacts have been excavated at Sermermiut.

These early cultural encounters are supported by the oral tradition, which reports of friendly encounters between the Inuit and the Norse, but also of disputes and even a tale of the last of the Norse of Nuuk Fjord who saw his kinsmen depart from Greenland.

From the mid-16th century to the 18th century, European expedition ships and whale hunting ships travelled along the west coast of Green-

Whale hunting was at stake, but since blubber was used to light up the living rooms of most European homes, demand was high. The baleen were used for the fine ladies' corsets that were the height of fashion at the time.

MANSELL COLLECTION/SHUTTERSTOCK/RITZAU SCANPIX

THE MAPPING OF GREENLAND

Different peoples have come to Greenland on various occasions. The first came from North America and spread over large parts of Greenland. Little is known of their mapping and exploration of the country but it must have been extensive. The first possible documentation comes with the geographer Pytheas (around the year 300 BC). He reached so far north that he was able to confirm his hypothesis that if you travel far enough north, you will eventually reach a place where the sun never sets in summer. Since then, there has been much discussion about how far north Pytheas actually came; was it Iceland or Norway, or did he get Greenland in sight? He called the country Thule. 'Pytheas' Thule' appeared on the world map around the year 140, when the Greek geographer and mathematician Ptolemy drew a world map based on meridians and latitudes, and Thule was placed around 63° north latitude.

THE MAPPING BY CLAUDIUS CLAVUS

An important milestone was when the Dane Claudius Clavus in 1427 drew the first map featuring both Iceland and Greenland. Clavus had studied Ptolemy's works, and his map was found together with Ptolemy's manuscripts. Clavus came up with a surprisingly good determination of Greenland's southern tip, and for the first time, he put the country on the world map under the Latin designation Gronlandia. However, he was wrong when he drew Greenland as being part of a continuous land mass with Russia.

MERCATOR'S WORLD MAP

After Clavus' map from 1427, the rendering of Greenland over the years remains a bit of a mystery. On his famous world map from 1569, Mercator placed Greenland surprisingly correctly and as an independent island. At that time, there was probably no one who had travelled in or collectively mapped the northern regions of the country. The knowledge base of the world map from 1569 thus remains a mystery.

MAPPING BETWEEN 1576 AND 1823

The rest of Europe grew increasingly interested in mapping the Arctic waters. Not least England was persistent in their search for a Northwest Passage – a sea route to the East round the north of America. In 1576, Martin Frobisher almost rediscovered Greenland, and in 1585-87, his fellow-countryman John Davis made three expeditions and made it all the way to Upernavik.

The exploration of the east coast was far more problematic due to the multiyear ice – or storis (great ice) – which blocked access to the unknown land. From the Danish colonies in the southwest, Peder Olsen Walløe made a journey by skin boat up the coast in 1751-53. In 1822, William Scoresby found the fjord that came to bear his name, and in 1823, Douglas Charles Clavering and Edward Sabine ventured further north and encountered a till then unknown Inuit tribe on Clavering Island.

MAPPING BETWEEN 1875 AND 1913

In 1875-76, a British expedition led by George Nares successfully managed to reach all the way to the Arctic Ocean and map the eastern part of the north coast. It was not until 1902-1904 that the 'Literary Expedition' led by Mylius-Erichsen was able to map the last part of the west coast. Meanwhile, a number of expeditions led by the American Robert Peary had increased the awareness of the northernmost part of the country, and in 1891, Peary was able to see for himself that Greenland is an island (see p. 20).

The inaccessible Blosseville Coast between Tasiilaq (formerly Ammassalik) and Ittoqqortoormiit (formerly Scoresby Sound) was later mapped by Georg Carl Amdrup and Ejnar Mikkelsen 1898-1900. The Swede Alfred Nathorst mapped the large King Oscar Fjord. The Denmark Expedition (1906-1908) put an end to the exploration of Greenland by exploring large parts of Peary Land. After Knud Rasmussen had established his station at the Uummannaq settlement (and later named it 'Cape York Station') in 1910, he embarked on a number of expeditions, the Thule expeditions 1912-1933, which, among other things, mapped the large fjords on the north coast and established the link to the east. With this, virtually the entire coast of the country was known.

In 1878, Arnold Diderich Jensen (Bildsøe) and in 1883, Alfred Nordenskiöld tried to access the ice sheet from the west. In 1888, Fridtjof Nansen chose to access the ice the opposite way at Tasiilaq and travel across the country to Nuuk Fjord. Johan Peter Koch followed the same path in 1912-1913 from Northeast Greenland to Upernavik, and it now became clear that the inner parts of the country were covered by ice.

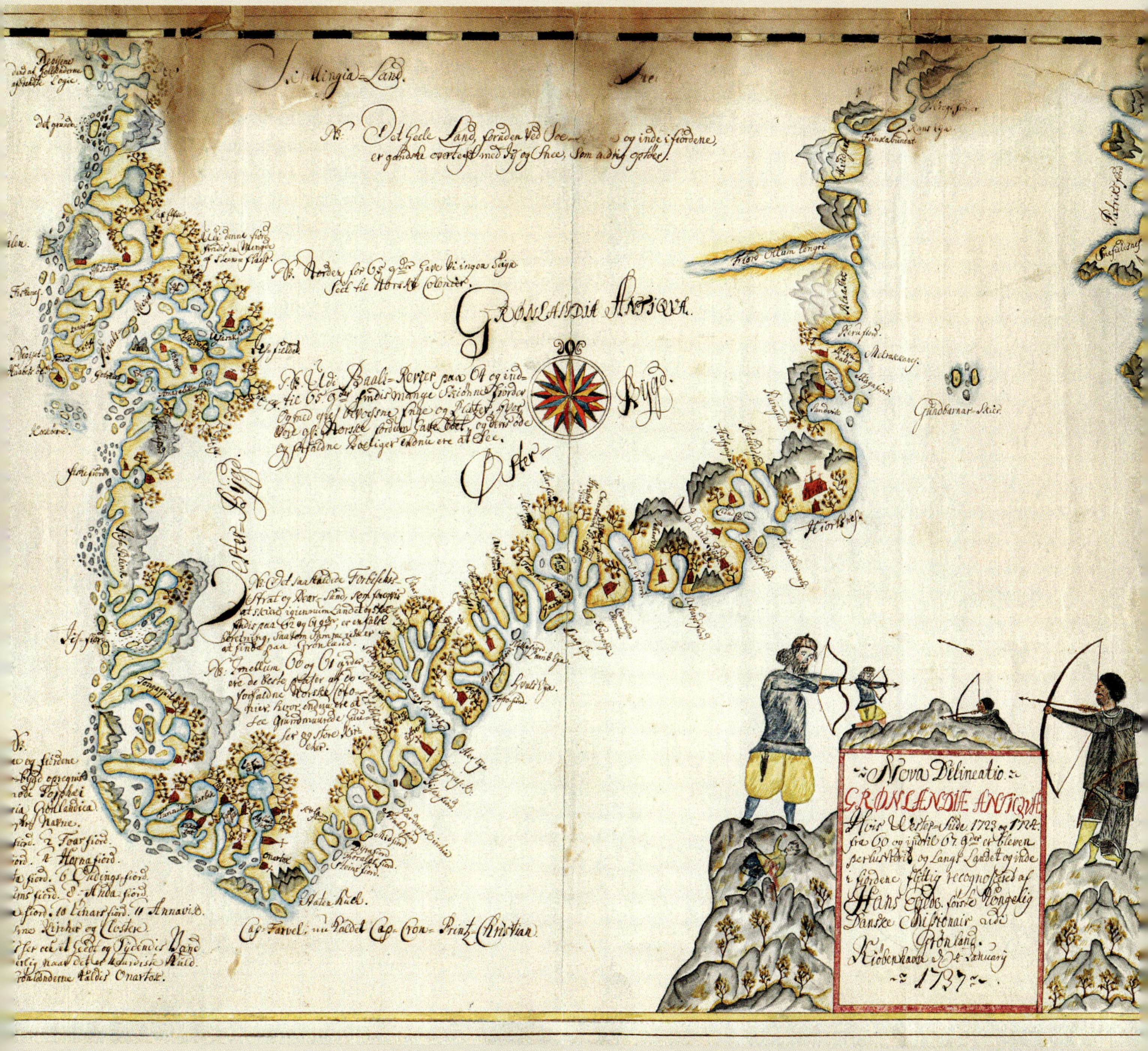

Hand-coloured map made by Hans Egede in 1737. The mapping of the coast and the deep fjords is based on Hans Egde's personal exploration, which was linked to his search for possible descendants of the Christian Norse. He did not discover any descendants, but the ruins of churches, etc. The map's indication of the ice sheet cannot be attributed to a proper survey. Scale approx. 1:2 000 000.

FOTOGRAFISK ATELIER, DET KGL. BIBLIOTEK

land. Trade opportunities quickly arose between the Europeans and the Inuit. Knives and needles of iron, but also glass beads, sugar and tobacco were traded in exchange for skins and walrus and narwhal tusks. The Dutch sailmaker knives were especially popular as they were made of metal and had the same shape as the Inuit women's main tool, the *ulu*, a knife used both to cut up and scrape skins.

Before long, the European presence became a driving force for long-haul travel, especially for the Inuit who travelled from the country's southern and eastern regions. According to the oral tales from East Greenland, the organised trading journeys could last for several years.

The skewed distribution of European commodities, as well as natural resources that characterise different regions in Greenland, was then traded internally among the Inuit. Qeqertarsuup Tunua (Disko Bay) had large supplies of baleen and whale oil, while there was a lack of wood, caribou skins and soapstone. Therefore, in order to obtain the best quality of soapstone for the production of lamps and pots, you had to trade with Inuit from the Nuuk area, while the area by Sisimiut was rich in walrus and caribou. In South Greenland, where the Inuit made more use of fishing, there was insufficient baleen to make fishing lines, while they themselves were rich in foxes and driftwood (see p. 105).

In 1636, Copenhagen shipowners were granted royal privilege to carry out whale hunting in Greenland, and they were officially encouraged to ab-

The lithograph from approx. 1860 illustrates the way of life at that time in Greenland. The top row of photos shows aspects of hunting life and the encounter with Europeans, whose presence is illustrated in particular by the whale hunting ships. The middle row photos show everyday life, for example, the summer skin tent and the winter peat hut, while the bottom row photos show refined hunting methods and hunting implements.

duct Inuits aged between 16 and 20, who were to be taught 'about the fear of God, the language and the art of academics'. The last abduction took place in Nuuk Fjord in 1654, presumably from the settlement of Ersaa at Sermitsiaq, a little north of Nuuk. David Danell abducted six Inuit, of which a boy jumped overboard, an elderly woman was put ashore, while Ihiob, Cabelou, Gunnelle and Sigjo were taken first to Bergen and then to Copenhagen.

An unknown number of Inuit were additionally abducted by British expedition ships, some of whom managed to escape after arriving in the British Isles. In 1682, a man in a kayak was seen rowing past Eday, and in 1684 another man in a kayak by Westray, both of these locations belonged to the Orkney Islands. In the early 1700s, an Inuit man drifted ashore at Belhelvie in Scotland in his kayak, but died shortly after arriving. These three episodes quickly turned into legends in Scotland, and the men were called *Finnmen*. Two of the kayaks are currently on display in Edinburgh and Aberdeen.

In the early 1700s, Inuit had considerable experience with the visiting whale hunting and expedition ships; from 1719 until 1721, between 50 and 100 whale hunting ships arrived in the country every year. In 1721 alone, 104 ships from the Netherlands and 14 ships from Hamburg were whale hunting by West Greenland.

THE COLONIAL PERIOD UNTIL THE WAR YEARS (1721-1940)

The Danish-Norwegian priest Hans Egede, together with family and crew, went ashore in Greenland on 3 July 1721. On an island in the mouth of Nuup Kangerlua (Godthåb Fjord) he established the Colony of Hope, and thus the Christian mission had come to the country.

LOUIS MOE/DANSK SKOLEMUSEUM/AU LIBRARY

Hans Egede sailed with the ship *Haabet* from Bergen in Norway on 2 May 1721 with his wife and four children to establish a colony in Greenland and run a mission from there. He tried to convert the Inuit i.a. with pictures, such as this one of the devil and hell. Colour lithograph on paper.

The location of the colony made it possible to meet many Inuit, as the great 'aasivik' area at Kangeq lay just south of this. The colony consisted of a relatively large farmhouse where all the colonists lived. Outside there were up to three cannons, a flag pole, a scale for blubber, some rowboats and workshops. The mission's view of the Greenlanders was not mild; they were considered to be pagan. Violence and threats were used in the work of converting them to the Christian faith.

Missionary work, trade and colonisation were closely linked. For example, Niels Egede, Hans Egede's son, served as both missionary, merchant and colony manager, and he had great success in incorporating new hunting methods, thus optimising both trade and earnings (see p. 117). The foreign merchant ships brought new goods, but they also brought epidemics, including smallpox, which took many lives. The colonisation meant that the Greenlanders effectively had no influence on development.

In 1905, Greenland went from being a mission field to becoming a deanery, and at the beginning of the World War II, the country was divided into 62 municipalities. The work to give Greenlanders greater co-determination over their country had started.

Hans Egede and the work for the mission service

Together with his wife, Gertrud Rask, their children Poul aged 12, Niels aged 11, Kirstine aged seven and Petronelle aged six, Hans Egede boarded the ship Håbet on 2 May 1721 and sailed out from Bergen to begin his journey to Greenland. Onboard were also the contractors and workmen who would remain in the

NUNATTA KATERSUGAASIVIA ALLAGAATEQARFIALU

Hans Egede's diary entry of 13 April 1722 states that 40 prominent women's boats with about 500 people sailed north past the colony at Kangeq. Each year about 100 families in umiaq went to the aasivik site Nipisat Sound near Kangeq to catch lumpfish. On 16 June 1723, Egede writes that the South Greenlanders and the local inhabitants »entertained each other and participated in games and play for several days and nights«. The watercolour from around 1860 was made by Aron from Kangeq and originally illustrated the tale of Kivioq.

country with them. The group counted a total of about 50 people.

Since 1711, Egede had proposed to King Frederik IV to establish a colony in Greenland and operate a mission from there, and in 1719 he received the backing of the King. The project received financial support from a group of merchants from Bergen, who founded Det Bergen Grønlandske Compagnie which would be responsible for voyages and trade. 300 rigsdaler had been allocated for Hans Egede's annual salary.

Hans Egede had prepared himself thoroughly for a mission among the unbaptised, but at his departure he still did not know whether the population of Greenland consisted of descendants of the Norse or 'real savages'. From 1261, the Norse had paid taxes to the Norwegian throne, and although no communication had come from the Norse since 1408, the Danish-Norwegian kingdom had maintained its power over the country. Moreover, Christianity had existed in Greenland in the past; Erik the Red's son, Leif, built the first church in the year 1000, and the country was a diocese in the years between 1124 and 1378 (see p. 100).

The construction of the Colony of Hope

The ship *Håbet* arrived at a small archipelago at Kangeq at the mouth of Nuup Kangerlua (Godthåb Fjord). A boatman from a Dutch whale hunting ship helped them into port, and on the last part of the journey they were joined by a lot of Inuit kayaks. Hans Egede set about the establishment of the provisional colony, the Colony of Hope, and at the beginning, the Inuit curiously followed the establishment and regularly lent a helping hand.

Hans Egede found no descendants of the Norse and devoted his mission to the Inuit. In Danish colonial administration sources, the Inuit are consistently referred to as 'Greenlanders'.

The mission's view of the pagan Greenlanders

Except during winter, when Greenlanders lived in their earth or peat huts at their winter settlement, Greenlanders lived a nomadic existence hunting for game. Up to 40 skin boats with a total of about 500 people took the long journey from the southernmost Greenland and southern East Greenland each summer to Qeqertarsuup Tunua (Disko Bay) to hunt whales. After wintering, the journey went south again. Along the way, they entered into marriages with inhabitants of other settlements and traded with other Greenlanders and European whalers.

A key element of the Greenlanders' belief was that not only humans, but everything in nature had a soul. It was believed that the actions of humans had an influence on the behaviour of the game and that a number of taboo rules had to be followed to ensure that the game would be caught. When the proceeds of the hunt became meagre or stayed away, an 'anqakkoq' – a shaman – was consulted who could help restore balance (see p. 157).

For Hans Egede, such a way of life was pagan, and he used threats and sometimes physical violence in the struggle to eradicate the shamans and drum song. Language problems made it even more difficult for Hans Egede to get the more than 200 Greenlanders at the small colony near Kangeq to accept the Christian message, and the first Greenlander was baptised on 1 January 1725.

Establishment of the colony in Godthåb

The project's economic foundation, the Bergenskompagniet, was not very successful either. The Greenlanders' brisk trade with the whalers made it so difficult for the trading enterprise to get the economy to balance that its activities ceased in 1726 and the king had to take over the trade.

The administration of cases regarding Greenland was moved from Bergen to Copenhagen in 1728. In the same year, Hans Egede left Håbets Ø (Island of Hope) and moved the colony to present-day Nuuk, which he gave the name Godthåb. King Frederik IV died in October 1730, and the heir, King Christian VI, commanded a halt to colonisation in 1731 and ordered all employees working with the mission and trade in Greenland to return home.

Hans Egede and his family, along with a few others, a total of 15 people, decided to stay in the country. The rest of the crew left the country, taking six Greenlanders with them. Shortly after their arrival in Copenhagen on 2 September 1731, the king changed his mind and instead urged the Danish-Norwegian citizens to form a new trading company in favour of colonisation.

In 1733, King Christian VI gave permission for the Moravian Brethren Mission to send three missionaries to Greenland. They were supposed to support Hans Egede in his missionary work, but were also allowed to work for their own cause. The Brethren Mission, which was based in Herrnhut, Germany, established the mission station Ny Herrnhut approximately half a mile from the colony of Godthåb.

Two of the six younger Greenlanders whom the colonists had taken to Denmark in 1731 were on the same ship as the missionaries. They had been allowed to travel home after the other four had died from smallpox. One died en route, while the other, Carl, only fell ill after arriving in Godthåb. By the time he died four months later, he had managed to pass on the infection. The smallpox epidemic ravaged at Godthåb in 1733-1734, and Hans Egede reported that only just over twenty baptised people had survived the epidemic.

Hans Egede saw a need for the establishment of more colonies in order to expand the mission and trade. On 25 July 1734, his son Poul Egede inaugurated a new colony, Christianshaab (today: Qasigiannguit), with his first sermon as a newly qualified priest. The colony was established by Jakob Severin, who since then also established the colonies of Jakobshavn (today: Ilulissat) in 1741 and Frederikshåb (today: Paamiut) in 1742.

Hans Egede and Seminarium Groenlandicum

Together with his son, Niels, and his two daughters, Hans Egede left Greenland in June 1736. He was burnt-out, depressed and disillusioned at not having managed to attract more Greenlanders to be baptised. At the same time, he was full of sadness and guilt for all those who had lost their lives during the smallpox epidemic, and he mourned the death of his wife who died on 21 December 1735.

In 1737 the Seminarium Groenlandicum was established in Copenhagen. The Mission College employed Hans Egede to lead the training of missionaries and advise the mission authorities on Greenlandic conditions. He wrote some textbooks to use for teaching in Greenland. When he resigned in 1747 he was replaced by his son Poul Egede, who since 1740 had served as his aide.

In 1744, Poul Egede published a translation into Greenlandic of the four Gospels and in 1750 a Greenlandic–Latin dictionary. In doing so, Greenlanders were given the opportunity to read the Christian message in their own language.

Relief portrait in plaster from the 18th century of Hans Egede and his eldest son, Poul Egede.

SVEND MALMBAK/ARKTISK INSTITUT, 1936

The distillery in what was then Godthåb, 1936. When the whale hunting ships brought in their catch, there was always work for the day labourers. In the distillery, oil was extracted from the blubber and poured into large barrels. Whale oil was an important export product and was a source of income for many families, and the processing plant was in operation until World War II.

Danish trade

In 1734, the king granted the Danish merchant and landowner, Jacob Severin, the exclusive rights to all trade and navigation in existing and future colonies in Greenland. This was to end the Greenlanders' trade with the Dutch whalers and ensure that the mission could continue. The continued trade between the Greenlanders and the Dutch resulted in a final showdown in 1739 in which, after being under fire for an hour from cannons on Severin's ships, the crew of four Dutch whale hunting ships surrendered and the ships were seized.

A shortage of capital forced Severin to give up his enterprise in 1749, and Det Almindelige Handelskompagni (the General Trade Company), whose owners consisted of merchants, landowners, and high-ranking officials, assumed the trade monopoly in 1750. By the time the Danish state took over the shares in 1774, Det Almindelige Handelskompagni (the General Trade Company) had set up another 12 trading posts and colonies.

On 1 January 1776, the Royal Greenland Trading Department (KGH) became an actual governmental management unit with its own executive board. In the same year, a Royal Decree came into force with a requirement for special permission to enter the country. This meant that since then, Greenland became more secluded from the outside world than most other European colonies.

From sea-based to shore-based whale hunting

The main game for the Greenlanders was the seal (see p. 160). Along with whale blubber, seal blubber was also the main export commodity for the changing trading enterprises. Due to the production of oil form seal and whale blubber, including for use in soap production and fuel for street lamps, until the early 1900s there was a high demand for blubber on the world market. The vast majority of the blubber came from seals that Greenlanders hunted from kayaks – and in the winter on solid ice with seal nets. Both Bergenskompagniet and Det Almindelige Handelskompagni had tried their hand at sea-going whale hunting, but competition, not least from the Dutch whalers, was too intense.

In 1759 when Niels Egede returned to Greenland to set up the colony of Egedesminde (today: Aasiaat), in collaboration with the Greenlanders, he realised his ideas on shore-based whale hunting. He also helped to move the colony of Holsteinsborg to the place where the town of Sisimiut is located today, and with the improved opportunities for whale hunting, this helped to create a good economy for the place. In cooperation with the Greenlanders, the sea-based whale hunting in 1781 was largely abandoned in favour of the coast-based. Besides his trading enterprise, Niels Egede did missionary work until he travelled to Copenhagen in 1782 and died shortly afterwards.

Other Greenlandic export products

Whale baleens, foxes, polar bears, down from the common eider, narwhal tusks and fishery products were also very popular, and trading took place by exchanging the Greenlanders' products for grocery goods.

In 1792, the so-called Greenlander book began to be used when The Royal Greenland Trading Department (KGH) kept accounts for each Greenlander by writing registering the revenue and subtracting the expenses. In 1805, Greenlanders were given a form of cash for the first time; credit notes which had values ranging from six shillings to five rigsdaler. From 1835, Greenlanders received cash for their goods.

There was coal in several places in Qeqertarsuaq (Disko) and by Uummannaq. Already from the 1780s, during the summer, the crew of the colonies mined so much coal that there was enough for a year's consumption in the colonies around Qeqertarsuup Tunua (Disko Bay). In 1924, the state established a coal mining town in Qullissat (see p. 268). The town had around 550 inhabitants in 1939, most of whom were Greenlanders, who all made a living as workers at the coal mine or by service occupations in the town. The coal was purely for local use.

From 1853, the Greenlanders traded graphite from Qaarsut and Niaqornat. In 1860, the supply to trading posts was almost 41.4 tons, giving Greenlanders a revenue of 7,300 rigsdaler. That was the equivalent of selling blubber and skins from more than 16,000 seals.

At Ivittuut, a total of 3.5 million tonnes of cryolite was mined from 1854 to 1987. Until 1894, the cryolite was mostly used to produce soda, and then increasingly to produce aluminium.

JETTE BANG PHOTO/ARKTISK INSTITUT, 1956

In 1968, the authorities decided to close the coal mining town of Qullissat. The approx. 1,400 inhabitants were scattered around Greenland. The last inhabitants left the town on Sunday 24 September 1972 on the coastal vessel Kunuunguaq. The abandoned town is visited every summer by former residents, and on 14 July 2015, they were joined by Queen Margrethe and Prince Henrik. A documentary book about the decision-making process and the human consequences, *Qullissat – byen, der ikke ville dø (Qullissat – the town that would not die)*, was published in 2013.

Trading stations and remote trading posts

While a unified settlement provided the best conditions for the work of the mission, a scattered settlement and thus larger hunting areas were in the interest of commerce. Consequently, small trading posts were established where hunters from the nearby settlements could deliver their catch. The first remote trading post, Eqaluk in the Uummannaq district, was established in 1787. Around the end of the 19th century, there were 23 remote trading posts in South Greenland and 271 in North Greenland. Communication between the districts was carried out by Greenlanders in the form of kayak and sled mail until the 1900s.

At the beginning, the managers of the remote trading posts were Danish contractors or unskilled Danes, who were generally married to Greenlanders. However, from the second half of the 19th century it was typically the so-called 'mixed' children of a Danish man and a Greenlandic woman who were employed to manage the remote trading posts. It was also mainly children of mixed ancestry who were educated or became employed in the trading posts and mission. In order to support the young people who came to be educated in Denmark, the authorities established a boarding house for Greenlanders in Copenhagen in 1880.

KUJATAA – FARMING ON THE BRINK OF THE ICE SHEET

In 1783, Norwegian Anders Olsen, his Greenlandic wife Tuperna and their family settled down as the first farmers after the Norse. They chose the Norse diocese of Garðar (Igaliku) as their settler place since it offered some of Greenland's richest grassland with good access to the surrounding resources. Although almost 150 years passed before Greenlandic sheep farming started in earnest, the distinctive farming tradition was founded by the descendants of Anders Olsen at Igaliku: grazing livestock – mostly sheep – which was near impossible and only sustainable through in-depth knowledge and intensive use of the resources of the sea and the land in the Arctic outfield.

The parallels to the farming and lifestyle of the Norse are obvious and natural. Therefore, the Norse and new Greenland farms always lie side by side on the 1,000-year-old green fields of the Arctic mountain landscapes. Due to the relatively late development of Inuit farming and early statutory protection of ancient monuments, the vast majority of Norse ruins are still preserved in the landscape. This unique circumstance, in which a near-intact fossil cultural landscape from the Middle Ages is superposed and maintained by a continuing Greenlandic farming tradition, was recognised in 2017 when UNESCO included the most central parts of Østerbygden (Eastern Settlement) on the World Heritage List under the title: Kujataa, Greenland: Norse and Inuit farming on the brink of the ice sheet.

ANINGAAQ R. CARLSEN/VISIT GREENLAND, 2017

Replicas of the longhouse and church at Brattahlíð – which today is the settlement of Qassiarsuk – as they might have looked at the time in the late Viking Age when, according to the sagas, Erik the Red settled as the first Norse in Greenland.

The first census took place in 1799 and showed a number of 5,122 Greenlanders from Uummannarsuaq (Cape Farewell) to Qimusseriarsuaq (Melville Bay). By 1901, the number of Greenland's total indigenous population had grown to 11,621. Of this, the population of mixed ancestry accounted for 65 % in North Greenland and 29 % in South Greenland.

The dependence on Greenlanders supplying goods for export and considerations for the spreading of the mission led to the continued work of constructing trading posts. In 1781, there were trading and mission posts along the entire west coast, from Upernavik in the north to Julianehåb (today: Qaqortoq) in the south. This development resulted in the West Greenland population becoming increasingly settled, and the purpose of the Greenlanders' traditional annual trade journeys between South and North Greenland disappeared.

Inspectorates and the Instruction

In 1782, West Greenland was divided into two inspectorates: a Southern Inspectorate from Sisimiut and southwards and the Northern Inspectorate with the colonies north of Sisimiut up to and including Upernavik. On 19 April 1782, the Royal Greenland Trading Department (KGH) issued a 15-chapter book of rules, *the Instruction*, which required the employees of the trading posts to always carry out their work with the aim of maximum earnings for the Royal Greenland Trading Department (KGH). They had to ensure that the Greenlanders caught the most possible game and that the products came to Denmark as soon as possible.

Although the Instruction was meant for the Royal Greenland Trading Department's (KGH) staff in Greenland, in practice it came to apply as a guide for the relationship between Greenlanders and Danes in general. The provisions were that the trading post staff had to discourage Greenlanders' trade with other Europeans and to limit the Greenlandic women's contact with wintering seafarers, colonists and other Europeans. Unlike in the past, the colonists were now only allowed to marry Greenlandic women of mixed ancestry following application. If a colonist was allowed to marry a Greenlandic woman, he had to commit to remaining in Greenland until his wife had died and the children had become adults.

Merchants had to make sure that Greenlanders did not get drunk from beer and stronger alcohol, and that they were not spoiled by coffee, tea, tobacco and other European goods. Having the same prices throughout the country when trading Greenlandic products should discourage competition among the merchants, and they were also the ones who had to enforce the very detailed regulations and penalties if Royal Greenland Trading Department (KGH) employees broke the rules. Many of the provisions of the Instruction applied right up to the beginning of the new arrangement in 1950.

The English Wars and the mission

By 1807, France, with Napoleon at the head, had conquered most of Europe. To ensure that Napoleon would not conquer and use the large Danish-Norwegian fleet to attack England, on 1 September 1807 the English launched an attack on Denmark and followed up with a bombardment of Copenhagen for the next two days.

The war, which lasted for seven years, also had noticeable repercussions for Greenland. The instability of the ship connections resulted in a shortage of goods. The Greenlanders lacked gunpowder and bullets for their hunting, and the trading basically came to a standstill because there were no goods to trade with. Many of the European employees of the Royal Greenland Trading Department (KGH) and some missionaries travelled back to Denmark or Norway. The epidemics, against which the Greenlanders had so little resilience, also caused problems. For example, a small community with a population of 429 at Uummannaq experienced a loss of 108 adults and 50 children in 1813.

France, an ally of Denmark, lost the war and in 1814 Denmark had to cede Norway to Sweden, but was allowed to retain its supremacy over the Faroe Islands, Iceland and Greenland. At the end of the war, of the Danish-Norwegian missionaries, only Bernhard Hartz at Ilulissat remained. The Greenlandic Catechists carried out the work of the mission throughout the war.

In 1815, the mission college chose to ordain the catechist Frederik Berthelsen of Maniitsoq as priest for all of South Greenland. He became the first priest of mixed descent. However, the 1835 Commission rejected a proposal that teachers' training colleges should also train Greenlandic priests. Members did not think Greenlanders were mature enough for a priesthood education. However, the state established two catechist seminaries in Nuuk and in Ilulissat in 1847 and 1848, respectively. Although the terms of the Danish mission had been fluctuating, by 1850 only a few unbaptised individuals remained on the west coast.

In 1894 Denmark set up a trade and mission station at Tasiilaq. The first baptism of East Greenlanders happened in April 1899, and in 1921 the West Greenlandic priest and missionary, Christian Rosing, baptised the last unbaptised East Greenlander. In the same year, the first royal visit to Greenland took place when King Christian X, together with Queen Alexandrine, Crown Prince Frederik and Prince Knud used the celebration of the 200th anniversary of Hans Egede's arrival to Håbets Ø (Island of Hope) as the opportunity to visit the country.

In 1934, the last adult was baptised in Thule. In 1910, Knud Rasmussen established a private trading station at Thule. Despite tremendous work and pressure from the committee behind the operation of the Thule Trading Station, the state did not take over until 1938, after which Thule continued as a colony.

For the Brethren Congregation, the English Wars did not have any noticeable repercussions. None of the missionaries had left the country, and supplies had come from Europe throughout the war. The Brethren Mission left Greenland in 1900, partly because of disagreements with the Danish mission and partly because their mission was complete. Before that, the Brethren Mission had gained great support in South Greenland. Among those baptised were more than 700 East Greenlanders.

Many of the East Greenlanders who rounded the Cape Farewell each summer to go on hunting and trading journeys took up permanent residence at the mission station Friedrichsthal, which today is Greenland's southernmost settlement, Narsamijit.

In 1925, 70 adults and children from the East Greenland colony of Ammassalik (today: Tasiilaq) moved north to Ittoqqortoormiit, where they were to help establish a new colony. Those who were due to move were told that hunting opportunities at Tasiilaq were no longer sufficient for the growing population. The main reason, however, was rather the dispute between Denmark and Norway over the right to East Greenland.

In 1931, Norwegian hunters occupied two areas in East Greenland. Denmark took the case to the Court in Hague, which ruled in 1933 that all of Greenland had come under Danish suzerainty upon its separation from Norway in 1814.

A group of people in umiaq docked with a ship, most likely at the East Greenland colony of Ammassalik (now Tasiilaq) around 1924. The following year, 70 people from hunting families, adults and children, were moved to what was then Scorebysund (now Ittoqqortoormiit).

ARKTISK INSTITUT, 1924

Reforms and Greenlandic co-determination

Europeans in Greenland were subject to Danish law, but there were no laws for Greenlanders. In 1828, the government decided that Danish law should also apply to the Greenlanders employed by the Royal Greenland Trading Department (KGH) or the mission, unless these were matters already described in the Instruction.

Colonisation had significantly altered the structure of society. New systems based on colonial trade and Christianity meant that Greenlanders effectively had no influence on the development of society. One of the actions that later gained great importance for the formation of public opinion in Greenlandic society was that the inspector of the Southern Inspectorate, H. Rink, procured a printing house for Nuuk in 1857 (see p. 122). Important Greenlandic personalities were managers at the printing company. The first one was teachers' training college teacher, Rasmus Berthelsen, then Lars Møller, who became qualified as a book printer in Denmark and managed the printing company right up to 1922, when journalist and author Kristoffer Lynge took over.

THE POLAR EXPLORER KNUD RASMUSSEN

Knud Rasmussen (1879-1933) was the greatest Danish polar explorer of the 20th century, matching the greats Roald Amundsen and Fridtjof Nansen. But unlike the his two Norwegian counterparts and other former polar explorers, his goal was not to reach the last white spots on the globe, but rather the people who lived furthest to the north. His expeditions came on the cusp of a new era, the age of decolonisation and democracy, and Rasmussen had a special background for making the meeting of cultures his mission in life. He was born and raised in Ilulissat where his father was a priest and his mother had Greenlandic ancestry. He therefore spoke Greenlandic just as good as he did Danish. Moreover, crucial to his development was the fact that he increasingly identified with his Greenlandic heritage. After his first expedition to Greenland in 1902-04, the so-called Literary Expedition, led by Ludvig Mylius-Erichsen (see p. 110), he stated: »I am, after all, a Greenlander and as such best suited for life in Greenland.«

The expedition had a long stay with an Inuit population north of the colonised Greenland. Here, he and Peter Freuchen established the Thule trading station in 1910 (see p. 169), which was to finance new expeditions, among other things. He went on a total of seven expeditions with the 5th Thule Expedition (1921—24) to Canada, Alaska and Siberia being the greatest and most important. The aim was in part to find the origin of the Inuit and their migration routes to Greenland, and, to the extent possible, to locate all remaining Inuit groups in order to acquire knowledge of their cultural traditions and religious conceptions. Before the expeditions, Rasmussen had collected legends and myths throughout the country. Across the areas visited, the expedition created an awareness of the Inuit as a cohesive nation with its own history, storytelling tradition and religion.

Rasmussen was not only a bridge builder between the Arctic peoples, but also between Denmark and Greenland. He wished to tie Denmark and Greenland as close together as possible. Through his books and lectures, he therefore sought not only to expand the Danes' knowledge of Greenland and the Greenlanders, but also to convey his great love of the country and its people. Because of his great charisma and great popularity in both Denmark and Greenland, he played an instrumental role in fortifying relations between the two countries both at the time and long after. Shortly before his death he managed to make the movie *The Wedding of Palo* in Greenlandic and with Greenlandic actors (see p. 220).

SVEND LAUGE KOCH/ARKTISK INSTITUT

Knud Rasmussen and Inûkitsoq study a map during the second Thule Expedition. The first two Thule expeditions used Inuit travel techniques, and both revolved around mapping. The first, in 1912, proved that Peary Land was not an island, while the second, in 1916-1918 mapped the coastline and fjords of northernmost Greenland.

JOHN MØLLER/ARKTISK INSTITUT, 1911

Greenlanders felt their views were not sufficiently listened to. The photo shows the first South Greenlandic provincial council assembled in Nuuk in 1911. Standing from left: Johannes Josefsen, Qaqortoq, Josva Kleist, Narsaq Kujalleq, Jens Hansen, Nanortalik, Jakob Hegelund, Paamiut, Otto Egede, Narsaq, Peter Rosing, Kangaamiut, Hans Motzfeldt, Qeqertarsuatsiaat, Gerhard Hansen, Alluitsup Paa. Seated from left: Nathan Lyberth, Maniitsoq, naalagaq Ole Bendixen, Carl Sivertsen, Sisimiut and John Møller, Nuuk.

Printing company, *Atuagagdliutite* and *Avannaamioq*

In conjunction with the printing company, in 1861 H. Rink founded the first Greenlandic language newspaper *Atuagagdliutit* (see p. 224). The newspaper included mentions of events around the world, serials with translations of foreign novels, translations of accounts of expeditions, and gradually also contributions from Greenlanders. The newspaper contributed to broader views out towards the world and a stronger sense of community across the otherwise relatively isolated and segregated colonial districts.

Another newspaper saw the light of day when the first issue of the North Greenlandic newspaper, *Avannaamioq*, was published in 1913.

The new governorships

In 1856, missionary and teachers' training college superintendent, Carl Emil E. Janssen, sent the missionary at the Moravian Brethren Congregation, Samuel Kleinschmidt, physician, Jacob Frederik Theodor Lindorff, and Acting Inspector of the Southern Inspectorate, H. Rink, a proposal for the Ministry of the Interior to establish governorships in all colonial district. After a trial period in South Greenland from 1857, the governorships began to formally operate in South Greenland in 1862 and in North Greenland in 1863.

The governorships were intended as a cooperative body between Greenlanders and Danes. The purpose was for Greenlanders to regain influence over and co-responsibility for their lives. However, Greenlandic points of view were often disregarded and cooperation in the governorships between Danes and Greenlanders failed. The main reason was that the economic interests of trade were often at odds with Greenlandic interests.

From mission site to deanery

The Law on the Church and School Service in Greenland of 1 April 1905 became the first law for Greenland.

The country went from being a mission site to becoming a deanery with its own rural dean under the Diocese of Zealand. The teachers' training college in Ilulissat was decommissioned and the teaching at the children's school was put into a system, with a lecturer in Greenlandic language employed as an advisor to the Minister for Ecclesiastical Affairs. Illiteracy was already largely non-existent in Greenland, since the founder of the mission, Hans Egede, had placed importance on people being able to read to be baptised.

In 1908, a new law was adopted for the governance of Greenland, which came into force in 1911. Trade and administration were now separated and had two juxtaposed leaders, a commercial manager for the Royal Greenland Trading Department (KGH) and a director of the new administrative body, the Administration of the Colonies of Greenland, respectively. Problems in making the scheme of juxtaposed leaders work in practice led in 1912 to the overall management passing to the director of the Administration of the Colonies in Greenland. Political responsibility for matters relating to Greenland came to the attention of the Minister of the Interior.

Municipal council, provincial council and districts

The governorships were abolished and instead the country was divided into a total of 62 municipalities, 26 in South Greenland and 36 in North Greenland, with each their municipal council. Two provincial councils were also established, one for South Greenland with 11 members and one for North Greenland with 12 members. All the members of the municipal councils were Greenlanders, who were elected by direct election among the local Greenlandic population. The two provincial councils were made up of people elected by municipal council members to represent their own region.

The inspectors for South and North Greenland were the chief executives of the provincial councils, however, without voting rights. With the introduction of municipal councils and provincial councils, Greenlanders were formally given the opportunity for the first time to discuss matters and formulate their positions without the risk of being trumped by the Danish colonial administration.

However, with the later criticism of municipal councils and provincial councils, it was said that they did not have a real influence on the final decisions on society development.

A revision of the Local Government Act of 1908 resulted in the Local Government Act of 1925 that brought about the introduction of 13 districts with each their district council. All the municipal council chief executives and all employees of the state in the districts, as well as the provincial council members of the regions, were members of the district councils. However, more Danes than Greenlanders was not allowed. In this way Greenlanders and Danes once again came to work together in the political bodies as they did with the governorships.

Headed for a new era

In the first half of the 1900s, the Administration of the Colonies in Greenland became aware of the potential for fishing in Greenland waters, and in the period from 1919 to 1939, 125 fishery stations were set up. The shift from hunting to fishing brought about major changes in the everyday life of Greenlanders. Many hunters moved with their family to the larger colonies, the women gained work in the production of fish, and the hunters acquired motor boats and rowboats. The number of rowing boats in South Greenland increased from 50 in 1880 to 743 in 1939 and in North Greenland from 33 to 673. Also further south in South Greenland, a new profession was becoming popular. The Administration established a sheep farming station in Qaqortoq in 1915, and the stock of sheep grew significantly – from a few hundred animals in 1915 to about 10,000 in 1936.

The population of West Greenland had grown from 11,190 in 1901 to 16,970 in 1938. Poor housing and hygiene conditions were seen as the main cause of the rapid spread of diseases, not least tuberculosis, which was the cause of 36 % of all deaths in the period between 1924 and 1933.

In the political world, the desire for greater influence on decisions grew for the Greenlandic population. Mathias Storch's futuristic novel *Sinnattugaq* from 1914 describes equality with Danes in a modern and well-educated Greenland in 2021. He also characterises the Greenlandic membership of the governorships as 'pseudo democracy'.

In 1939, the director of the Greenland Provincial Council, Knud Oldendow, invited four Greenlanders to Denmark to discuss Greenlandic conditions with the Danish Rigsdag's Greenland Committee between 1 August and 7 September 1939. However, the outcome of the negotiations did not have great effect as World War II broke out in early September 1939.

Greenland had been largely a country closed off from the outside world since 1776. This meant that Greenlanders had been cut off from the development that happens when a society has contact and trade with other countries and peoples. This situation changed drastically with the onset of World War II.

JETTE BANG PHOTO/ARKTISK INSTITUT, 1961

KGH's first self-service shop in Nuuk, 1961. During the development of a more modern Greenland, a modernisation of the supply service followed. In the self-service shops, brightly coloured shelves were filled with imported goods. During the 1960s, many of the old shops were converted into self-service shops.

THE WAR YEARS AND SUBSEQUENT DECOLONISATION

With the onset of World War II, Greenland became part of the United States' sphere of interest and several military facilities were established in the country. When Denmark was liberated in 1945, the connection could be resumed and in 1948, the Greenland Commission was established. The aim was for the Danish-Greenlandic cooperation to come up with proposals for reforms that would help to develop Greenland.

In 1953, Greenland was formally given equal status with the rest of the Danish kingdom, and the development that took place in the following years was governed from the Ministry for Greenland in Copenhagen. The development took place according to the Danish role model and often described subsequently as 'danification'. The focus was especially on business development, urban development, as well as education policy.

Greenland during World War II

On 9 April 1940, Denmark was occupied by Germany, and Greenland suddenly stood on its own two feet. Although the Danish government continued after 9 April, in the circumstances, Greenland was unable to receive directives, goods or new personnel from Denmark.

The President of the United States, Franklin D. Roosevelt, proclaimed that Greenland was part of the United States' sphere of interest and that the Danish government was thus, technically, hostile. The two governors in Greenland, South Greenland's Aksel Svane and North Greenland's Eske Brun, were the chief representatives in Greenland and could, by way of the applicable local government act for Greenland from 1925 and in exceptional cases, 'take such measures as the people's best interests may require.'

The primary objective of the governors during the war was to retain Greenland as part of Denmark and, moreover, continue the determined political line. In addition, the governors had to retain the support of the two Greenlandic Provincial Councils in order to continue the governance of Greenland. The two Provincial Councils had an advisory function on policy in Greenland, but Danish entitlement to Greenland was based on precisely trust and willingness to cooperate between Danish colonial rule and the Greenlandic population.

US military engagement

The Greenland policy aimed at developing society towards the opening of the country. One element of this was that over time Greenlanders had to have more and more influence over their own affairs, so that in the long term, society would be able to cope on its own. On this basis, a joint provincial council meeting was held at the beginning of May in 1940 at which the Greenlandic positions were presented. The meeting ended with two statements; one for King Christian X and one for President Roosevelt.

The message to the King expressed regret at the situation, while a gesture of friendship was given to the United States in the message to Roosevelt. Subsequently, diplomatic and trade relations were established with the United States and Canada to replace the lost connection to Denmark. Initially, it happened by both countries sending consuls to Greenland, and a US Coast Guard ship was deployed to patrol the waters.

The strategic importance of Greenland was highlighted during the course of the war. The 1941 Defence Agreement between the United States government and Denmark's envoy in the US, Henrik Kauffmann, ushered in the United States military involvement in Greenland, which is still in force. The agreement was reached by bypassing the Danish government, but Denmark's liberation government adopted the agreement as one of the first decisions shortly after Liberation in May 1945.

Cultural encounter and fraternisation ban

During the war, the United States set up a number of military facilities in various locations in Greenland, of which the South Greenland base in Narsarsuaq was the largest with up to 6,000 American soldiers. The population of Greenland at the time was roughly 20,000, so the relatively large American presence invariably had an impact on societal development. On the part of the colonial administration, the focus was on continuing the protection of Greenlandic society.

Gradually, a fraternisation ban was instituted and controllers were placed at the military installations. They were to ensure that there was no contact between the Greenlandic population and the American troops. On the part of the Greenlanders, however, there was a great desire for contact with the American troops. In many ways, the American presence was an eye-opener, partly through the Greenlandic population gaining insight into a different way of doing things, partly by seeing the USA's ability to implement initiatives, such as the establishment of the military bases, which led the thoughts of many towards the possibilities of a industrialisation of fisheries in Greenland.

Cryolite exports and increased passenger transport

During the war, the Greenlandic population maintained trust in Denmark, despite the broken connection and the American presence. Greenland's role in the Allies' struggle against the Axis powers was quite significant in several respects: Greenland provided the land for US military installations, including the runways that were the prerequisite so that the United States would send bombers to Europe. and through the Northeast Greenland Sled Patrol, later known as Sirius (see

The American Narsarsuaq Air Base in Greenland was established after the Defence Treaty of 9 April 1941. The base, built in the summer of 1941 and closed in 1958, had the military name Bluie West One. During World War II, it was the largest military base in Greenland, and was home to 6,000 people at its peak.

ERIK PETERSEN/RITZAU SCANPIX, 1952

KAJ SKALL SØRENSEN/ARKTISK INSTITUT, 1953

Ivittuut cryolite quarry. The quarry was established in 1856 and played an significant role for Greenland during World War II. It was the only open cryolite mine in the world, and as cryolite was used to make aluminium, it was essential in wartime manufacturing. The Americans therefore established the naval station Grønnedal to protect the quarry.

p. 311), German attempts to establish themselves in Northeast Greenland were thwarted. Meteorological observations in the waters off Northeast Greenland were also important as the weather in northwestern Europe is affected by the weather systems there.

The cryolite mine by Ivittuut in Southern Greenland was of special significance; it was the only one of its kind in the world, and the cryolite was used for aluminium production and was thus essential to the war industry. The cryolite from the mine was sold to the United States and Canada, and in this way the supply of goods for Greenland was financed.

The supply of goods had so far been shipped directly from Copenhagen to the colonial towns, after which the goods were sailed on to the smaller locations in the district. This meant that there was no significant contact between the colonial districts as there was no organised passenger traffic. During the war, due to the reduced freight capacity, transit ports were established from which smaller ships distributed the goods. It created increased contact between the different parts of the country, including through passenger transport.

The possibility of replacing personnel was poor and, in consideration of the situation, a centralisation of the administration was implemented, in which Nuuk became a kind of capital. The two provincial councils held joint meetings and decisions could be implemented faster than before.

The media during World War II

During the course of the war, several media were created in order for the population to be kept up to date on developments in the world: The Danish-language newspaper *Greenlandsposten*, which was later merged with the Greenlandic-language *Atuagdliutit*, today called *A/G*, opened up the dialogue between Danes in Greenland and those Greenlanders who could speak and understand Danish.

The radio service expanded and sales of radios increased quite significantly. So many people in Greenland got a sense that the world was getting closer and that Greenland was part of the world. Developments during the war gave Greenlandic society a new direction and a greater awareness in the population of what development was going to happen. Upon the liberation of Denmark in May 1945, the Greenlandic society was therefore quite different from the one that felt as if it was thrown into the deep-end in April 1940, when contact with Denmark was broken.

The initial post-war period

In May 1945, the announcement that Denmark was liberated was received with great joy in Greenland. The relations with Denmark could be resumed, and the Greenland National Council now saw the development that had occurred in Greenland during the war. Therefore, the idea was initially that conditions should return to pre-war status before any sweeping reforms were launched.

Around the turn of 1945/1946, the Greenland Provincial Council and representatives from the two Greenland provincial councils entered into the first negotiations on Greenland's future, but these did not result in major changes. Based on developments during the war, the Greenlandic representatives had a number of requests, but initially had to arm themselves with patience.

The Greenland Commission is established

In the summer of 1946, a Danish press delegation travelled to Greenland and reported on awful conditions in the country, including poor health conditions. Denmark was criticised for defaulting on commitments in Greenland, and the Greenland Provincial Council and the Danish government now came under pressure over the Greenland policy. The United Nations (UN) was created in 1945, and one of the organisation's objectives was for non-autonomous areas to change status from colonies to countries with autonomy or self-government.

From within, from the Greenlandic politicians, there was also pressure for a change in Greenland's colonial status. The retention of the country as part of the Danish kingdom was thus challenged from several sides. A new Danish government was formed in late 1947 led by Prime Minister Hans Hedtoft. Eske Brun, who had served as a bailiff in Greenland during the war, was employed in a newly created position in the Greenland Provincial Council.

Hedtoft and Brun jointly formulated a master plan for the future of Greenland, which was presented at the joint Provincial Council meeting in the summer of 1948. The plan was intended to strengthen ties and cooperation between Greenland and Denmark and contribute to the modernisation of society. The provincial councils gave their approval of the plan, and in the autumn of 1948 the Greenland Commission was established, broadly composed of Greenlandic and Danish members.

The task of the Commission was to propose reforms for the development of Greenland. As early as 1950, the Commission delivered its report, known as the G-50.

Greenland is formally given equal status with the rest of the Danish kingdom

A number of acts of the Danish Folketing followed on the basis of the Commission's report, including the centralisation of the provincial councils and of the administration. In terms of the constitution, Greenland was included in the revision of the Danish Constitution, and when the law came into force on 5 June 1953, Greenland was formally seen as an equal part of the Danish kingdom.

The process of the inclusion of Greenland in Denmark showed clear interests from both sides in maintaining the link between the countries. Denmark's interests had roots in the security-political trends of the time, including the relationship with the United States, while Greenland saw a clear advantage in continuing the long-standing relationship with a valued partner.

In 1951, a new defence agreement between Denmark and the United States regarding Greenland was entered into. Against that background, the United States continued its involvement in Greenland, this time with the emphasis on the Thule Air Base at Pituffik, which was established in 1951. The local Inughuit population of Thule was, for the sake of a planned expansion of the Thule base (see p. 408), forcibly relocated shortly before the Danish Constitutional Act came into force.

In the field of security and foreign policy, there was no involvement of Greenlandic politicians when decisions were to be taken in the early 1950s. At the same time, the pace of

MB/UN PHOTO, 1954

From left to right: member of the provincial council Augo Lynge, permanent secretary Eske Brun, member of the provincial council Frederik Lynge, and (behind) head of division at the Greenland Provincial Council, P.P. Sveistrup. Denmark informs the UN Committee for Non-Self-Governing Territories that Greenland has been incorporated into and given equal status with the rest of the Danish kingdom by the Constitutional Amendment of 5 June 1953. Denmark, consequently, ceases to report to the Committee.

development had increased significantly and, as a result of these actions, it soon became apparent that the population was finding it difficult to come to terms with the major changes that were happening.

Modernisation of society

The Greenland Commission's report in February 1950 provided the framework for the modernisation of Greenlandic society. The following years of development largely occurred with Danish conditions as a direct role model, and the policy used is subsequently often described as 'danification'. The governance took place from Copenhagen, from 1955 in the Ministry for Greenland.

One of the major tasks was the expansion of the infrastructure which was necessary for the establishment of an efficient business community. The earlier requirement that the administration of Greenland had to balance economically was waived, and in the following years the Danish state built ports, power stations, hospitals, schools, roads, waterworks, etc. From the outset, efforts were made to improve and renew the housing stock. Houses of stone and peat had to be replaced with larger and more hygienic dwellings (see p. 194), often constructed as self-construction projects and with advantageous construction loans.

Private initiatives and business support provide impetus

With the new arrangement, the monopoly disappeared, and thus the way was paved for private initiatives. The first years saw some new kiosks, but gradually shops with a stock appeared that could compete with the Royal Greenland Trading Department. In 1963, the first consumer cooperative was established and today Kalaallit Nunaanni Brugseni has outlets in several towns. Contractors, often stationed Danes, set up private enterprises and increasingly participated in the construction of homes and other buildings and facilities.

The socio-economy was to build on the exploitation of the country's resources, primarily fishing. Business support made it possible to invest in larger fishing boats and the number of fishermen increased, while the hunting profession gained less importance, except in the outermost regions. The first years focused on cod fishing, but as the cod population decreased, shrimp fishing gained increasing importance, and since then shrimp has been Greenland's main export product.

While private fishermen expanded the fishing fleet, there was virtually no private investment in factories and other facilities for the processing of the catch. Therefore, this task had to be undertaken by the state, and under the auspices of the Royal Greenland Trading Department (KGH), factories were built to process cod and shrimp.

Investments in open-water towns intensifies

In 1960, the state established the Greenland Committee of 1960, called G-60, which in a report published in 1964 proposed an intensification in investments in both infrastructure and production facilities. Investments were to be made primarily in the so-called open-water towns of Paamiut, Nuuk, Maniitsoq and Sisimiut, where it was possible to fish throughout the year and thus increase production.

Even before World War II, parts of the population had moved from smaller settlements and remote trading posts to primarily the towns. As part of the modernisation, a policy was implemented that would increase the moving from the small settlements. The Provincial Council supported this policy and the municipal councils were required to assess every year which of the inhabited settlements would be developed. These assessments became critical to whether investment should be made in infrastructure at the settlement and whether the population could receive funding for the construction of a new home. In this way, it pushed segments of the population to move to the settlements where the authorities wanted development.

The concentration policy was reinforced after 1960, in continuation of the plans for expansion in the open-water towns. In order to be able to house the many newcomers, concrete apartment buildings began to be constructed (see p. 197), but nevertheless during this period there was a constant shortage of housing in the towns.

Education policy after 1950

The education policy was also in focus. It had long been felt that good Danish skills and knowledge of Danish culture were a prerequisite for improving the level of education and thus to take over the positions that the Danes had held so far.

The new school law of 1950 opened the possibility that after the 2nd grade – initially in the bigger towns – pupils could follow a programme

At the shrimp factory in Aasiaat (Egedesminde), large quantities of shrimp are cooked, peeled and packed. Shrimp fishing was of major importance to the Greenlandic economy and provided new employment opportunities for the population, and from the 1950s, women increasingly entered the labour market.

VAGN HANSEN/RITZAU SCANPIX, 1965

REGITZE MARGRETHE SØBY/ARKTISK INSTITUT, 1965

School class in Kuvdlorssuak, 1965, about 300 km north of Upernavik. This year the school had 40 children in three classes, taught by a trained teacher and a reader. In Greenland, the number of pupils in primary school increased from about 4,000 in 1950 to about 10,000 in 1970. The answer to the teacher shortage was to employ Danish teachers on secondment, and for several years Danish teachers made up two-thirds of the trained teachers in Greenlandic schools.

where Danish was the language of instruction, although except for the subjects Greenlandic and religion. It would, among other things, qualify for admission to secondary school at the same level as a Danish lower secondary school leaving examination.

Between 1961 and 1976, more than 1,500 pupils were sent to Denmark after 6th grade, where they lived with Danish foster families and went to classes in Danish schools. During the same period, four boarding schools were established in Denmark, exclusively for Greenlandic pupils who would continue their schooling after the 7th grade.

Apart from the teachers' training college in Nuuk, no institutions of higher education were established until 1977, when the Iron and Metal School in Nuuk and the Building and Construction School in Sisimiut were established. Higher education had to be taken in Denmark, and in some places purely Greenlandic classes were set up, e.g. at the business school in Ikast and the pedagogue training college in Aarhus.

In Denmark, several associations for Greenlanders were formed in Denmark and, in several towns, institutions were established to pursue, among other things, the interests of people seeking an education. These still exist in Copenhagen, Aarhus, Odense and Aalborg as the Greenlandic Houses.

Population trends from 1950 to 1975

The many reforms meant major changes in demographic conditions. Along with improved housing standards, targeted action in the health care area meant tuberculosis gradually disappeared in the country. This caused a sharp decline in mortality and thus a sharp increase in the population, almost doubling from 1950 to 1970 from around 23,000 to just under 44,000 inhabitants.

Infant mortality declined dramatically, and child and adolescent groups made up an increasingly large proportion of the population. The average life expectancy increased from about 40 years in 1950 to more than 60 years in 1975. Previously, stationed Danes accounted for a few hundred, but with the new activities, thousands of Danes now came to the country. The so-called summer men were chiefly employed in construction and stayed in the country only during the busy summer period. Some had employment for several years before leaving the country, while others settled down because they had established a family or because they had decided to stay for other reasons.

In 1975, more than 9,000 Danes were living in Greenland, and they made up about 20 % of the population. At the same time, the number of Greenlanders living in Denmark also increased. Most were undergoing education and, for most part, returned to Greenland. Others got married to Danes and chose to stay in Denmark.

Negative consequences of development

The many reforms and investments in the 1950s and 1960s produced distinct improvements in material conditions. At the same time, a number of problems arose that can be interpreted as a result of segments of the population not becoming part of societal changes and ending up as passive bystanders to the developments from which they did not benefit.

A relatively large proportion of the population had no vocational training and thus ended up in a poor employment situation with low pay and risk of unemployment. A large part of this group were people with a traditional Greenlandic upbringing, especially people from the settlements or people who had moved from a settlement to a town. The symptoms were increased consumption of alcohol that could now be freely bought, increased crime and a rising number of suicides.

In political circles and organisations, there was an increasing dissatisfaction that the Greenlandic population did not adequately influence development and ended up as second-class citizens of society. The situation generated a renewed debate on the framework for development and demands for genuine equality.

The modernisation of society also meant the drawing up of legislation to ensure the legal rights of citizens, but in some cases the opposite occurred: The attempt to send 22 Greenlandic children on a year's stay in Denmark and then to orphanages in Nuuk in the early 1950s to make them 'exemplary citizens', the case of the legally fatherless children and more or less formal adoptions of Greenlandic children happened on the basis of a desire to assimilate Greenlanders, Greenlandic society and Greenlandic culture, but this had major negative consequences for those who were affected.

The impossible equality

A clear objective of the modernisation of Greenland was to improve the material conditions in the country, but at the same time on the part of the Greenlanders, there was an expressed desire to create equality between Danes and Greenlanders and that the Greenlandic population should have a real influence on their own circumstances. That, however, did not happen. The development was guided by the Danish authorities and overall decisions were taken in the Folketing and the Ministry for Greenland in ongoing negotiations with the Provincial Council.

The unsatisfactory situation was discussed in the Provincial Council in 1959 which, in a statement to the Folketing, called for real equality, and in the following years the problem surrounding equality was in focus. From the establishment of the first provincial councils, the state's chief representative in the country had been the chief executive of the provincial councils. The chief executive had approved the members' proposal for the agenda and been the leader of the meetings. This was now changed to allow the 1967 Provincial Council to elect the chair-

AAGE SØRENSEN/RITZAU SCANPIX, 1953

The Greenland ship Disko arrived in Copenhagen, carrying about 60 polio and tuberculosis patients to be dispersed to several treatment sites in the country. From 1951 to 1955, the number of tuberculosis patients who died was 886, representing 3.6 % of the population. In 1954, Queen Ingrid's Hospital (commonly referred to as Sana) was inaugurated in Nuuk, providing 211 beds exclusively for the treatment of tuberculosis. From 1959, all Greenlandic tuberculosis patients were treated in Greenland.

ANDREAS LUND DROSVAD/ARKTISK INSTITUT, 1948

The Johansen family, 1948. Kristian Johansen was the manager of Illorsuit from 1946, and in 1945-1950 he was a member of North Greenland's provincial council. Elisabeth Johansen was a member of the provincial council 1959-1975 and the only woman to have been a member of the provincial council. She was the first politician in Greenland to travel around giving election speeches, which was done by dog sled and boat. The couple's four children are: Astrid, Severin, Henrik Kristian and Lars Emil. The youngest, Ole, was not yet born.

man from among the members of the council.

The educated Greenlanders did not receive the same wages as their Danish colleagues, and the growing group of Greenlandic civil servants, e.g. teachers, telegraphists and priests, worked side by side with Danish colleagues under the same working conditions, but for smaller wages. They were now demanding the same wages as their Danish colleagues. Instead, the state introduced the so-called birthplace criterion which meant that everyone born in Greenland was remunerated on one scale, while workers who had been born in Denmark and were employed from Denmark received higher wages. The rationale was that wage conditions should be aligned according to productivity and in solidarity with other occupational groups. When Danish employees received more in wages, the argument was that it was necessary in order to attract labour to the country.

The birthplace criterion spawned fierce criticism among educated Greenlanders especially and, for several years, the Provincial Council demanded to have the scheme removed. However, the criticism was not supported by, for example, the Greenland Workers' Confederation (GAS), as equal pay between Danish and Greenlandic civil servants would create a large wage gap between the different occupational groups in Greenland.

In 1970, the Provincial Council had to recognise that the birthplace criterion affected only a smaller proportion of the Greenlandic population and that it was not possible to find another form that would be fairer. The acceptance of the birthplace criterion was at the same time an admission that equality was not possible in this area.

Women's equality

The introduction of women's suffrage in Greenland in 1948 meant that Greenlandic women had a more equal status than in the past, and more were given important societal functions and entered the labour market. Following this, more women expressed their desire for greater influence in society.

In line with the development process, discussions in the women's associations changed from topics such as family and children to also being about housing policy, education and health matters. Kalaallit Nunaanni Arnat Peqatigiit Kattuffiat (see p. 175) aimed to bring together the country's women's associations and contributed to a significant expansion of the importance of associations in the field of societal policy. Despite this, there were still comparatively few women in politics and business, but more women now took an education.

With the establishment of political parties in the 1970s, the function of women's associations as a platform for female candidates in politics disappeared more and more.

The desire for a Greenland according to Greenlandic conditions

Gradually, resistance also spread to other parts of the modernisation. In the mid-1960s, proposals for a new Greenlandic school law were discussed, one of the most important elements being the strengthening of teaching the Danish language (see p. 150). The proposal included a provision that parents could postpone teaching in Greenlandic until the 3rd grade. This met sharp criticism and a fear that such an arrangement would weaken the Greenlandic language.

In 1970, a conference was held in Sisimiut with the participation of Greenlandic politicians and representatives of a wide range of interest organisations with the aim of discussing the future of Greenland. Key topics were discussed, such as occupations, population concentration, school and education, language, social problems, etc., and set out guidelines which, when seen under one umbrella, required a more independent Greenland and policies that targeted Greenlandic conditions to a greater degree. In a special statement, representatives of Greenlandic youth stressed that Greenlanders should be considered a national minority and that they wanted a more politically independent Greenland.

Political trends

With the amendment of the Danish Constitutional Act in 1953, Greenland became, in principle, constitutionally equal to Denmark. In 1959, the Provincial Council invited the Danish authorities to set up a committee to adjust the development of society. The Provincial Council's invitation expressed a desire to establish a long-term objective. It was granted by the Danish government and led to the establishment of the Greenland Committee of 1960, with Greenlandic and Danish politicians and business representatives as members.

The Greenland Committee and the resulting debate

The Greenland Committee and its report became known by the designation G-60. The task of the G-60 Commission was to review the political, economic and administrative conditions. The G-60's plan was: '... to raise the political, social and cultural status of the Greenlandic population and boost their standard of living.'

In Greenland, a resistance to the consequences of the G-60 gradually emerged. The resistance was inspired by thoughts of national identity and international trends about 'free peo-

After the amendment of the Danish Constitutional Act in June 1953, Greenland had two representatives in the Danish Parliament. In September 1953, parliamentary elections were held for the first time in Greenland. The teacher and participant in public debate Augo Lynge received the South Greenland mandate, while the colonial manager Frederik Lynge received the North Greenland mandate.

POLFOTO/RITZAU SCANPIX, 1953

ples'. It would promote a more positive view of being Greenlanders, highlighting that Greenlanders were a characteristic people with their own culture, traditions and languages, and that this recognition was important in order for Greenlanders to meet Danes as equals.

There was now a pronounced desire to draw up their own political and administrative systems in accordance with the needs of the country and with greater participation of the Greenlanders themselves. The increased political awareness, including under the influence of simultaneous political developments in Denmark, the Greenlandic students' association and the Peqatigît Kalâtdlit association, influenced the societal debate. The principles behind the policy in Greenland were now turned towards, and thus also against 'danification'.

The concept of equality was redefined: In the same way that the Danes had to be understood based on their own prerequisites, the Greenlanders now had to be understood based on their prerequisites, and the natural consequence of this was demands for a development based on 'a more Greenlandic Greenland'.

Implementing this would require not only fundamental changes in the policy in Greenland so far, but also in the relationship between Greenland and Denmark. The 1972 decommissioning of the coal mining town of Qullissat (see p. 117) became the symbol of a failed modernisation of Greenland, where economic considerations were above the considerations of the population and where individuals felt that they had effectively no influence on their own affairs.

In January 1979, an indicative referendum on Greenlandic home rule was held. 70 % of votes were in favour, 25 % against. On 1 May 1979, Greenland's Home Rule became a reality. Within the framework of the Home Rule Act, Greenland assumed competence over its own affairs.

EBBE ANDERSEN/RITZAU SCANPIX, 1979

Enrolment in the European Community and the Home Rule Bill

Since Greenland was formally an integral part of Denmark (see p. 231), Denmark's desire to join the European Community also meant that Greenland was included in the application. However, while the negotiations with the EC were taking place, there was a change in Greenlandic views on relations with Denmark. A Greenlandic desire for a separate vote for Greenland was rejected by the government. The result in Greenland was a clear no vote to the EC, but when the Danish voters voted yes, Greenland had to follow Denmark into the European Community.

The oil crisis of the 1970s also spawned discussions about the right to the substratum in Greenland. These were also instrumental in political mobilisation in Greenland – a number of parties and associations saw the light of day. Greenlandic youth was a major factor in political mobilisation. For example, the rock band Sume put into words the critical attitude of many Greenlanders towards Danish politics and the desire for self-determination (see p. 217).

The new societal movements had a major impact on developments that occurred in the following decades. A new generation of Greenlandic politicians were elected into the political bodies in the early 1970s and, through their elected positions, proposals for a home rule in Greenland were formulated. The Greenlandic requirements were increased autonomy and participation in their own circumstances, control over the substratum, and that future development would take place, to a higher degree, on Greenlandic terms.

There were substantially different attitudes towards the ties to Denmark. From a desire for rapid independence, to the home rule arrangement being a necessary evil on the road to independence, and finally to Greenland remaining part of Denmark.

CLAUS BRUUN/RITZAU SCANPIX, 1979

The introduction of Greenland's Home Rule on 1 May 1979 was a big day of celebration. The Regent of the Kingdom, Queen Margrethe, arrives with Prince Henrik by helicopter at Godthåb, where they are welcomed by former chairman of the provincial council Lars Chemnitz and governor Hans J. Lassen.

HOME RULE (1979-2008)

The introduction of Greenland's Home Rule on 1 May 1979 meant a transfer of competences and responsibilities from Denmark to Greenland. This was in line with the terms for the country's development and occurred in continuation of ongoing negotiations between Greenland and Denmark. The desire behind the home rule arrangement was that Greenland should remain associated with Denmark, but that the Greenlandic community and the Greenlandic population should have a greater influence on their own affairs.

The structure of Home Rule

The law on Greenland's Home Rule stated in Section 1: 'The Greenlandic Home Rule attends to Greenlandic affairs within the framework of the kingdom.' Thus, under the law, Greenland's Home Rule had the authority and legislative powers for policy areas they had taken over.

The new status of Greenlandic society had great psychological significance. Greenland could set out its own course, thus defining its own society, but still in close association with Denmark. The adoption of their own national flag and the establishment of the National Day on 21 June, both introduced in 1985, have helped create a national identity.

The Supreme Legislative Assembly was now the Inatsisartut (Parliament), and the members of the Inatsisartut

FINN FRANDSEN/POLITIKEN/RITZAU SCANPIX, 2007

The Greenlandic flag Erfalasorput was raised for the first time on Greenland's National Day, six years after the introduction of Home Rule, on 21 June 1985. The flag was designed by the Greenlandic painter, graphic artist and former member of the Landsstyre, Thue Christiansen. In Inatsisartut 's (Parliament's) final decision, 14 members of Inatsisartut voted for the red and white flag with the circle, while 11 preferred a Scandinavian cross flag with the colours green and white.

had to appoint a Landsstyre (Naalakkersuisut), the government of the country. The Joint Greenland-Danish Home Rule Commission (1975-1979) had presented a plan for policy areas that could either remain a Danish responsibility, be divided between Greenland and Denmark or be taken over by Greenland.

Home Rule areas of responsibility

As early as January 1980, Greenland's Home Rule assumed responsibility for eight policy areas, including the governance arrangement, the church area, the teaching and cultural area and industrial relations. The judiciary/Greenland courts, police and prison service remained under Danish responsibility. However, locally trained police officers, magistrates and institution officers had to be used. The training of the groups took place in Greenland.

With each policy area that was taken over, a grant came from the state, scaled according to the expenditure on the area prior to the taking over. The Greenland treasury now obtained its income from the block grant and the taxation scheme, which was introduced before the introduction of home rule. In addition, the Danish State paid to have tasks performed in the Danish areas of responsibility in Greenland, just as foreign fishing licenses and taxes on certain goods provided revenue to Greenland's treasury.

The economic and political development

Prior to the introduction of home rule, on 4 April 1979, the first elections for the Greenland Inatsisartut (Parliament) were held (see p. 251). Four parties and a total of 95 candidates ran for office. Siumut received 46 % of the vote and the constituency division meant that the party gained 13 out of the 21 seats in the Inatsisartut (Parliament). Atassut gained the remaining eight seats.

Siumut's election issue was that Greenland should take over policy areas as soon as possible, as the party was sceptical about state funding of the respective policy areas prior to the handover. In addition, the withdrawal from the EC was a key issue (see p. 134) and, finally, much emphasis was placed on a more even distribution of investments throughout the country.

As the election's winner, Siumut formed the first Landsstyre, and the party's chairman, Jonathan Motzfeldt (see p. 256), became the first Landsstyre premier. In addition, the Landsstyre came to contain the following areas of government: one for settlements and outlying districts, one for business, one for schools and culture and finally one for the social area. Under the individual Landsstyre areas, directorates were established for individual fields.

A number of the civil servants who had worked in the administrative apparatus already built up under the Governor moved to the directorates. The creation of a Landsstyre area for settlements and outlying districts was, in its way, a break with the prior concentration policy, the 'danification' and thinking behind the modernisation after World War II.

The culture and development potential of the outlying districts were considered essential for society development and therefore needed to be supported.

First priorities for Home Rule

In order to achieve the overall and long-term policy objective of the Home Rule arrangement, it was necessary for Greenland to gradually become economically independent from Denmark. Home Rule's first priority was therefore to develop business so that exports of goods could offset the imports enabling the economy to move towards the ultimate goal of becoming self-sustaining.

As a starting point, the economic foundation was industrial fisheries, but relatively quickly the domestic fishing of Greenland halibut also gained significant importance for export income and hence the overall economy. However, the business policy was expanded into a multi-tiered economy with tourism, mineral resources and raw materials industries, as well as other land-based enterprises.

Strengthening the business sector

The staple industry was still fisheries and its derivative industries such as production and transport. Investment was made in both the private and publicly owned fishing fleets, as well as in production facilities with new factories and plants for the processing of catches. However, it was difficult to secure a stable income, partly because cod fisheries were failing and partly because of volatile prices on the international market. The production of other species such as Greenland halibut and crab was initiated, but with fluctuating success, and revenues from exports remained dependent on the shrimp fishery.

Sleeping tents belonging to the ruby mine at Qeqertarsuatsiaat/Fiskenæsset south of Nuuk. The Aappaluttoq mine (meaning red) is today operated by the Norwegian company LNS. For example, the rubies from here are used in the jewellery that the Government of Greenland's presented to Queen Margrethe on her 80th birthday.

PETER KLINT/POLITIKEN/RITZAU SCANPIX, 2014

It was necessary to develop other sectors that would be able to balance exports and imports. For example, this was to be done through the exploitation of the country's non-living resources, the mineral resources. Since then, a number of licences have been issued, partly for prospecting and partly for small-scale activities, such as the Nalunaq gold mine north of Nanortalik and a ruby mine at Qeqertarsuatsiaat.

The tourism sector required the development of the infrastructure of the country, for example, hotels, transport options, training of guides, etc. Greenland Tourism was established to target a new tourism strategy focused on cruise tourism as well as marketing both natural and cultural experiences.

Finally, the economy had to be strengthened by increasing local activities in other business sectors, such as construction, trade, manufacturing, service and consulting. In several instances, this was done with the introduction of cooperative ventures and other forms of co-ownership, not least in connection with the start-up of small production enterprises in the settlements.

In the latter part of the 1980s, Greenland ran into a severe economic crisis that came close to emptying its coffers and made it difficult to borrow funds for the projects that would develop the country. With the advice of the OECD, among others, a number of reforms were introduced. For example, a large part of the Home Rule's companies were identified as limited liability companies, although still with the Home Rule as the main shareholder. The crisis meant that Greenland and the Greenland business sector had to submit to market economy forces and that economic policy had to adapt to these.

Strengthening school and education policy

Up to the introduction of Home Rule, the school and education policy had a central role in the debate, and towards the end of the 1970s it had basically stopped sending pupils from primary and lower secondary school for stays in Denmark in order to strengthen Danish language skills and thereby the pupil's academic level. In 1977, a Higher Preparatory Examination course was established at Ilinniarfissuaq (teachers' training college) in Nuuk, and in 1983 the Higher Preparatory Examination course was established in Aasiaat. They were replaced in 1986 by the upper secondary education programme (GU), and at the same time an upper secondary education programme was established in Qaqortoq. In 2000, a technical upper secondary school was established in Sisimiut and the foundation for taking an academic youth education had been laid. In several towns, institutions were set up for vocational education programmes, such as the business schools in Nuuk and Qaqortoq, the food school INUILI in Narsaq and Imarsiornermik Ilinniarfik (Greenland's Maritime School) first in Nuuk, and later moved to Paamiut.

In 1984, the Ilisimatusarfik/Inuit Institute was created which changed its name in 1987 to Ilisimatusarfik/ Greenland University and was granted university status. In 2008, Ilisimatusarfik moved into the newly constructed campus, Ilimmarfik, on the outskirts of Nuuk and has since been given a number of new study programmes. Many higher education programmes in particular must still be taken abroad, often in Denmark, and others must be completed with periodic stays in Denmark.

The majority of the education programmes are in Greenland and the number of young people seeking an education is steadily increasing. The level of education has increased over time and the area is a high priority as a highly educated population is seen as a prerequisite for being able to cope on its own.

Strengthening culture and language

Cultural life also had a high priority during Home Rule, and in the early years many activities were initiated or supported. The language and Greenlandic culture became an important element in strengthening a common identity and a necessary prerequisite in the effort to build a national identity. The modernisation period had the 'danification' of Greenlandic society as one of its basic principles. The introduction of Greenland's Home Rule was based on a desire for a Greenland governed on Greenlandic terms.

Thus, 'Greenlandisation' was based on strengthening the Greenlandic language and Greenlandic culture. In relation to the language policy, there have been different views; Greenlandic children and young people have grown up since the 1950s under different language policies derived from different political priorities by changing authorities (see p. 149). From the political aspect in particular, it has been argued that being fluent in Danish allows better access to education programmes. The counter-argument has been that a fluency in the Danish language often undermines the child's learning of the Greenlandic language, and that this loss of language leads to a loss of identity.

Part of the work to ensure access to institutions that impart history, culture and art is precisely about helping to restore a Greenlandic identity. In 1981, a new legislation changed

TERKEL BROE CHRISTENSEN/RITZAU SCANPIX, 2011

Ilisimatusarfik, the University of Greenland, and Pinngortitaleriffik, the Greenland Institute of Natural Resources, are located on the outskirts of Nuuk between the suburb of Nuussuaq and the airport. At Ilisimatusarfik in 2020 there were four institutes with a total of around 750 students and 100 employees. In the background, Nuuk's landmark Sermitsiaq is shrouded in clouds.

the status of the existing museum to a national museum. The Act simultaneously provided the basis for the establishment of local museums in the towns (see p. 166). It became part of the foundation for increased knowledge of Greenland history and cultural history, which has since appeared in art, literature and design.

In 1984, actors established the project theatre Silamiut, which in 2011, by way of new legislation gained the status of The National Theatre/Nunatta Isiginnaartitsisarfia (see p. 210). The theatre is based in Nuuk, but since its early years, it has also given performances in towns and settlements throughout the country.

In 1997, in the heart of Nuuk, the Katuaq cultural centre (see p. 174) opened. Its cultural activities include theatre, concerts, films, exhibitions, talks, etc.

In addition, the Home Rule has actively supported literature, art and film.

The splitting up of the Danish government institutions

The growing political desire of the Greenlandic population for greater independence through the 1970s included a desire to do away with the two Danish government institutions, the Greenland Technical Organisation (GTO) and the Royal Greenland Trading Department (KGH). This desire was a result of the dominant influence of these companies on both the everyday life of individuals and overall societal

MORTEN RASMUSSEN/BIOFOTO/RITZAU SCANPIX, 2015

The coastal vessel *Sarfaq Ittuk* from Arctic Umiaq Line on its way through the Ilulissat Icefjord. The coastal vessel calls at the West Greenland coast from Qaqortoq in the south to Ilulissat in the north during summer. It serves a significant amount of the passenger traffic between the Greenlandic towns and settlements for tourists and the local population.

development. As that period also saw a political desire to create less powerful and more transparent organisations, the KGH and GTO split up into a number of independent organisations through the 1980s and 1990s.

The Royal Greenland Trading Department

Having managed the monopoly trade in Greenland since 1776, the Royal Greenland Trading Department (KGH) was in charge of the supply of goods, fish trading, fish processing and exports, transport infrastructure and postal service when Greenland's Home Rule was established in 1979.

In 1985, the Home Rule took over the part of KGH that oversaw the trade, production and export of seafood, and KGH's trading stations, production facilities and fishing fleet were taken over by Royal Greenland which was owned by the Home Rule.

From 1986, the Home Rule gained control of KGH's trading and transport activities and the name was changed Kalaallit Niuerfiat (KNI), and in 1993 KNI was split into several independent publicly owned limited liability companies, including Royal Arctic Line, the Pisiffik and Pilersuisoq retail chains and Polar Oil. In 2001, Pisiffik was acquired by private investors.

Parts of passenger service were privatised in 2006 as the Arctic Umiaq Line (AUL), but for economic and service reasons, privatisation had to be reversed as early as in 2007. Air Greenland A/S and Royal Arctic Line A/S now own AUL.

Greenland Technical Organisation

The former Greenland Technical Organisation (GTO) was established in 1950 as part of the Greenland Department under the Danish Prime Minister's Office, and from the beginning of the modernisation of Greenland, GTO was responsible for the technical planning and much of the societal operation. Thus, GTO was the owner of and operating organisation for electricity and water supply, roads, ports, telecommunications and feasibility studies, while construction tasks were contracted out to a number of mainly Danish contractors.

In 1987, the Home Rule took over GTO, which changed its name to Nuna-Tek, and in 1990 the national electricity, water and collective heat supply was spun off as an independent company under the Home Rule and later changed its name to Nukissiorfiit. Some of the other tasks, such as ports, major construction activities and administration were transferred to the departments of the Home Rule, while municipalities took over sewers, roads and municipal and local plans. Furthermore, Amutsiviit (Grønlands Værfter) was privatised.

In 1994, the Housing Administration was transferred from the Home Rule to the Home Rule-owned housing company INI A/S (see p. 283). In the same year, Tele Greenland A/S was spun off as an independent organisation and merged with Kalaallit Allakkeriviat (Greenland Postal Service), originally part of KGH, in 1997. Responsibility for electricity and water supply in the settlements was transferred in 1998 from the municipalities to Nukissiorfiit.

Grønlandsfly A/S was founded in 1960 by SAS and Kryolitselskabet Øresund with the aim of serving the US bases in Greenland. In 1962, ownership was extended to include the Greenland Provincial Council and KGH, and activities expanded with domestic passenger services. From 1998, it also offered transatlantic travel between Greenland and Denmark. Grønlandsfly A/S changed its name to Air Greenland A/S in 2002. In 2019, the Government of Greenland bought out the ownership interest of the Danish State and SAS, and Air Greenland A/S is now fully controlled by the Government of Greenland.

The struggle for indigenous peoples' rights

With the aim of highlighting the conditions of indigenous peoples – peoples who, as a consequence of colonisation, among other things, live as minorities in their country, often across national borders – the United Nations declared 1993 to be the International Year of Indigenous Peoples. Subsequently, the 1995-2004 decade was named the International Decade of the World's Indigenous Peoples. The aim is for the indigenous peoples to reclaim and secure the right to their own territories as well as language and culture.

In order to strengthen activities and promote dialogue between the indigenous peoples and the governments of their countries, a coordinating position was established under the UN, which was filled first by Henriette Rasmussen and later by Marianne Jensen, both former members of the Landsstyre. Since then, Greenland has played an active role in the United Nations Permanent Forum on Indigenous Issues, established in 2000, and in the adoption of the Declaration on the Rights of Indigenous Peoples in 2007.

The year 1993 was declared by the UN as the International Year of the World's Indigenous Peoples. At the official opening of this year's General Assembly in New York, Denmark was represented by members of the Greenland Home Rule.

MICHOS TZOVARAS/UN PHOTO

Inuit Circumpolar Council

Greenland's work in the ICC, the Inuit Circumpolar Conference, from 2002 Inuit Circumpolar Council, and in the United Nations has helped give the country a prominent position among the world's indigenous peoples and is seen by many as an example to be followed.

In the field of foreign policy, Greenland raised its profile early in the struggle for the rights of indigenous peoples. In 1977, at the Inuit Circumpolar Conference, Inuit from Greenland, Canada, the United States and Russia began a collaboration, the purpose of which was to strengthen the culture and society of the Inuit, and the ICC was founded in Greenland in 1980 as an NGO.

Through cultural events, political activities and sporting events, such as the Arctic Winter Games (see p. 176), the ICC has greatly contributed to strengthening awareness of a shared identity among Inuit. The ICC has also announced that the indigenous people want to be called 'Inuit' and not 'Eskimo', which is a term that has been forced upon them from the outside and is perceived as patronising and discriminatory. In addition, the organisation has worked to improve social, economic and political conditions in the Arctic region.

UN Secretary-General Ban Ki-moon presents the 2007 Mahbub ul Haq award to Canadian Inuit activist and former ICC President Sheila Watt-Cloutier in recognition of her work raising awareness of the impacts of climate change. The award is given to people who have made a special commitment to improving people's development opportunities.

MARK GARTEN/UN PHOTO, 2007

ICC Greenland has positioned itself as a prominent Inuit NGO that has attracted international attention, not least in the field of environmental policy and the development of a common Arctic environmental strategy. The ICC received awards for its environmental policy work from the United Nations in 1988 and the Nordic Council in 1996.

Foreign policy initiatives

The Home Rule Act ruled that foreign policy matters, including defence and security policy, were governed by the Folketing and the Danish government. On the other hand, the Danish Government had to consult the Home Rule before entering into agreements with other countries that came into contact with Greenlandic interests, and the Home Rule was given the opportunity to be represented in negotiations of particular interest to Greenland. Since then, Greenland has pursued an active foreign policy in several fields and created the Foreign Office in 1994, which subsequently and under different designations has been responsible for the country's foreign policy contacts and initiatives.

Greenland's withdrawal from the EC

The desire for home rule was strongly linked to Greenland's involuntary membership of the EC, and after the introduction of home rule, negotiations began with the EC on withdrawal. The result was an association agreement whereby Greenland was associated with the EC as one of the 'Overseas Countries and Territories', ensuring free movement of people to the EC countries, duty-free access for Greenland products to the European market and a number of aid schemes. As a quid pro quo, the EC demanded a fisheries agreement negotiated by Greenland's first diplomat representing the Home Rule in Brussels. In an indicative vote in 1982, 53 % – headed by the anti-EC movement ANISA – voted in favour of the Community, and in 1985, Greenland withdrew from the European Community. The agreements with the EC, today the EU, have been renewed several times.

Determination of catch quotas

In the following years, Greenlandic representatives, on behalf of the Danish Government and with the special authority of the Minister of Foreign Affairs, concluded an agreement with Iceland and Norway, establishing common rules for capelin fishing, and agreements with the Faroe Islands on the hunting and conservation of seals and small whales. Representatives from Greenland have since participated in Denmark's delegations at the negotiations of the International Whaling Commission, IWC, which sets, among other things, annual catch quotas. For a period of time, the International Whaling Commission was unable to make decisions on quotas because of disagreements over the size of quotas and widespread international requests for total conservation of whales.

In 2007, the EU imposed a common position on its members in the IWC, creating conflicts between Denmark and the rest of the EU, and between Greenland and Denmark. In 2013, since the IWC could not reach an agreement, Greenland introduced quotas for whale hunting in its own waters, contrary to Denmark's wish not to violate existing conventions.

PEER PEDERSEN/POLFOTO/RITZAU SCANPIX, 1982

Demonstration in the run-up to the referendum on EC membership in January 1982. The opposition to the EC was instrumental in mobilising Greenland politically, and after home rule was introduced in 1979, the move to leave the EC began.

The Igaliku Agreement

The US presence in Greenland developed in parallel with the changes in the international political situation. In 1991, as part of an agreement with Denmark, the United States withdrew from the base at Kangerlussuaq and ceased to use the airport at Kulusuk. The agreement also provided for the establishment of a committee with representatives from both Greenland, Denmark and the United States.

When the US government submitted plans in the late 1990s for a missile defence that also included Thule Air Base, the Inatsisartut (Parliament), through the Foreign and Security Policy Committee of the Folketing, demanded that any agreement to replace the 1951 Defence Agreement should be made with Greenland as an equal partner. If this did not happen, Greenland would not be able to accept a new agreement.

In 2004, Greenland, Denmark and the United States concluded the Igaliku Agreement, named after the settlement of Igaliku, where the final agreement was signed by the foreign ministers of Denmark and the United States and the foreign affairs member of the Landsstyre. In addition to the defence agreement itself, the agreement also provides for expanded cooperation between the United States and Greenland in areas such as research, education, energy, health, the environment and tourism.

The Igaliku Agreement was an important step in Greenland's efforts to fill and expand the framework for Home Rule, and the agreement helped to initiate work on formulating and realising the ideas of self-government.

Self-Government Commission (2004 to 2008)

At the beginning of the new millennium, the first stages of the next step for Greenland began. After 20 years

IVARS SILIS/AFP/RITZAU SCANPIX, 2004

In 2004, Josef »Tuusi« Motzfeldt, member of the Landsstyre, reached an agreement with Per Stig Møller, Danish Minister of Foreign Affairs (left), and Colin Powell, US Secretary of State (right). Among other things, the agreement gave the US the green light to upgrade the radar in Thule for missile defence. The signing of the agreement took place in Motzfeldt's home settlement Igaliku, hence the name Igaliku agreement.

of Home Rule, the planned policy areas had been taken over. Moreover, much had happened since 1979: an increasing number of Greenlanders took up education and Greenland had taken over a wide range of societal functions in the public and private sectors, which meant, among other things, that the number of Danes summoned gradually decreased. There is no denying that society was ready for a new step with less dependence on Denmark and thus could look forward to a greater degree of independence. Consequently, at the turn of 1999-2000, the Greenland Landsstyre established the Greenland Self-Government Commission, considering that there was an increased need to review Greenland's position within the Unity of the Realm and the Danish Constitutional Act. Furthermore, the terms of reference provided for an economic analysis of the relationship between Denmark and Greenland. The Commission's work resulted in a report which was submitted to the Landsstyre in 2003. Both the Landsstyre and the Inatsisartut (Parliament) accepted the Self-Government Commission's report.

Self-Government Commission process

The next step was a joint Greenland-Danish Self-Government Commission to consider the proposal for the introduction of self-government in Greenland. The terms of reference were signed by the Greenland Premier and the Danish Prime Minister on 21 June 2004 in the 25th anniversary year of the introduction of the Home Rule.

The Commission, which was composed of members of the Greenland Inatsisartut (Parliament) and members of the Danish Folketing, was to work according to the terms of reference given by the government and the Landsstyre. The Commission's job was to produce a report to consider and propose how the Greenland authorities could assume further powers where constitutionally possible.

In addition, the Commission was to draw up proposals for a new arrangement concerning the economic situation between Greenland and Denmark.

Greenland's constitutional position and the self-determination right of the Greenland people under international law were also to be examined.

Finally, the terms of reference established that it is for the people of Greenland to decide whether Greenland wants independence. Any implementation of independence must be made on the basis of an agreement made within the framework of Section 19 of the Danish Constitutional Act.

Three working groups were established to address specific topics on subsurface resources, economy and business development, and matters of state and international law. Elections to the Folketing in February 2005 and to the Inatsisartut (Parliament) in November of that year, as well as the formation of the government following the November 2007 Folketing election, resulted the replacement of some members of the Commission, including the Chairman of the Commission.

The Commission drew up draft legislation on a new arrangement which also took into account policy areas already taken over by the Greenland authorities under the Home Rule Act for Greenland. In 2008, the joint Greenland-Danish Self-Government Commission completed its work and delivered its report and draft legislation for a self-government arrangement in Greenland.

Self-Government is introduced

The referendum was preceded by an extensive awareness campaign based on the Self-Government Commission's report. Printed matter was distributed door-to-door, the newspapers ran information campaigns, and public meetings were held in many places in the country.

At public meetings, the content of self-government was presented and citizens could ask questions of politicians and public officials. Some of the meetings were televised via nationwide media, and radio and television broadcasts with subsequent debate programmes were produced. Information and debate meetings were organised at educational institutions and widely communicated through the Commission's website.

The indicative referendum was held on 25 November 2008. Voters were asked to answer 'Yes' or 'No' to the question: 'Do you want self-government to be introduced in Greenland with the content and conditions outlined in the Greenland-Danish Self-Government Commission's draft self-government act?'

With a high turnout of 71.96 %, the result of the referendum showed that a clear majority was interested in Greenlandic self-government: 75.54 % of all the votes were in favour the introduction of self-government, 23.5 % against.

The result of the referendum was submitted to the Inatsisartut (Parliament) and on that basis, a request was forwarded to the national authorities. Then the Folketing confirmed the self-government of Greenland by the Act on Greenland Self-Government.

POPULATION TRENDS FROM 1805 TO 2021

Since 1805, occasional censuses have been carried out in West Greenland, although the first years should probably only be used tentatively. West Greenland was divided into South and North Greenland and included the area from Cape Farewell to Kraulshavn (Nuussuaq) in the Upernavik district.

In the almost 100 years from 1840 to 1938, the Greenlandic-born population more than doubled from 7,877 to 16,969. During the same period, the European part of the population increased from 251 to 383 persons, a growth which was mainly due to the increasing number of workers at the cryolite quarry in Ivittuut (see p. 39), while the number of Europeans in the colonies was modest.

The population at Ammassalik first came into contact with Europeans in 1884, and in 1901 there were 431 Greenlanders and ten Europeans registered in the district. The settlement at Scoresby Sound was established in 1921, mainly by families from Ammassalik. In 1938, a total of 1,071 Greenlanders and eight Europeans were registered in the two districts, mainly due to population growth in the Ammassalik district.

North-western Greenland north of Qimusseriarsuaq (Melville Bay) only formally became part of Denmark when the state took over the Thule trading station in 1938. In 1938, 271 Greenlanders and six Europeans were registered in the district.

Post-war modernisation (see p. 128) was accompanied by an almost explosive population growth. In one decade, the population grew by 35 %, with migration from Denmark being greater than the increase in the Greenlandic population, and from 1947 to 1967 the total population grew from just over 20,000 to almost 44,000. A population growth that, due to the Danish centralisation policy, took place almost exclusively in the towns.

The figure shows the demographic trends from 1805 to 2021, the period up to 1901 being for West Greenland only.

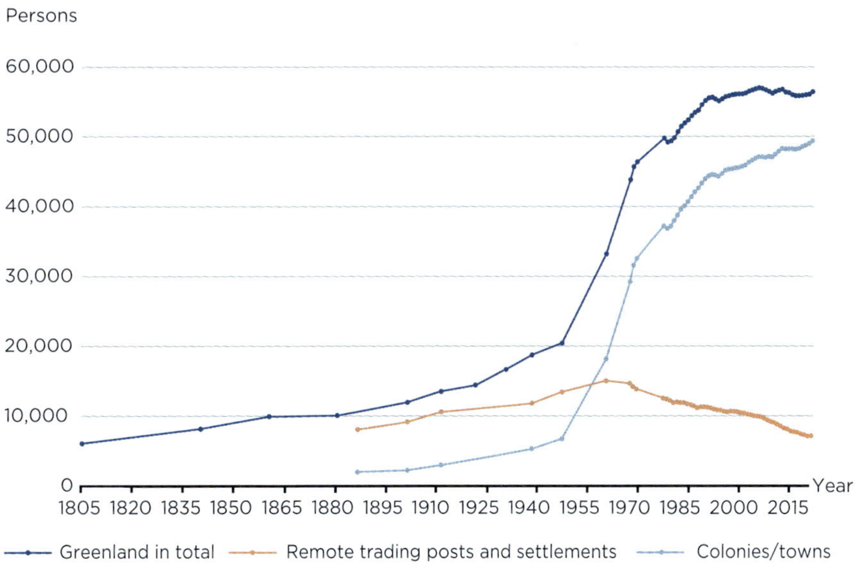

Figure 3. Population development 1805-2021. The figure shows All of Greenland, Remote trading posts and settlements and Colonies/towns.
The period 1805 to 1947 includes only the population born in Greenland. The Ammassalik district is included in »All of Greenland« only from 1901, Scoresbysund (Ittoqqortoormiit) from 1925 and Thule (Qaanaaq) from 1938.

FOR 1805-1938: BERETNING VEDRØRENDE GRØNLANDS STYRELSE 1942; 1947: GRØNLANDSKOMMISSIONEN BETÆNKNING 1950; 1960: STATISTISK ÅRBOG 1962; 1968-69: STATISTISK ÅRBOG 1971; 1977-2021: GRØNLANDS STATISTIK STATISTIKBANK.

CULTURE

JULIE EDEL HARDENBERG, 2017

Julie Edel Hardenberg, *Ineriartorneq/ Progress*, 2017. The work is one of a series of seven debate-provoking works created by Hardenberg in response to a commission from the Greenland Reconciliation Commission: To show reflections of Greenland and promote dialogue across language and ethnicity.

The ancient myths, nature, history and traditions play a key role for Greenlandic culture. Since the Greenlandisation of the 1970s, there has been a growing interest in reviving traditional cultural expressions, which were disappearing, and at the same time there has been a keen interest in strengthening the existing culture. However, some artists are turning their eyes more to global challenges and international currents.

Photo pages 146-1457. In Tasiilaq, a woman hosts a kaffemik in her home, where she tells about the East Greenlandic women's costume she is wearing. Making the costumes is a painstaking craft, and you often get help from family and friends to make the different pieces.

MADS PIHL/VISIT GREENLAND, 2015

The traditional drum dance and song are now seeing renewed interest. In 2021, the tradition was included in the UNESCO World Heritage List of Intangible Cultural Heritage.

Music has generally been of great importance for Greenlandisation, including the rock band Sume, who sang in Greenlandic in the 1970s on topics dealing with Greenlandic themes and at the same time had a critical perspective on the overly dominant Western influence on development (see p. 217).

Another example of retention of traditions is the hunting culture. Life as hunters is closely related to identity, community and cultural rules that help maintain the balance between man and nature. Although occupational hunting today has limited importance to the overall economy, it is still considered an integral part of the culture.

While new materials and architectural trends have been adopted buildings, they are still being adapted to culture, nature and the climate challenges.

Among the earliest visual artists is Aron from Kangeq, who in the mid-

19th century depicted Greenlandic legends and myths in watercolours. In 1972, the first art education, Grafisk Værksted (Graphic Workshop), opened. In newer visual arts, a movement is seen away from the national narrative, and the artist now tend to address colonial and post-colonial issues.

The same trend is seen in fashion and design: Where previously, mainly products that referenced Greenlandic culture were created, the gaze is now turned to more global trends – while using the symbols of the Inuit.

The oral storytelling has been an important element of social life, and from the 19th century many of the tales were written down. Where literature in the past most often contributed to the love of Greenland and national consciousness, a new development emerged in the early 2000s, addressing different themes and breaking down taboos.

Cultural dissemination has played an increasingly larger role, and by the end of the 1900s there were cultural history museums in most Greenlandic main towns. Nunatta Katersugaasivia Allagaateqarfialu, the Greenland National Museum & Archives, established in 1966, is the largest in the country. In 1995, Greenland's first art museum opened in Ilulissat, and in Nuuk, it has been possible to visit the Nuuk Art Museum since 2005. In addition, most main towns have settlement halls or cultural centres; in 1997, the Katuaq cultural centre opened in Nuuk.

LANGUAGE

The Greenlandic language is a branch of the Inuit languages spoken in a wide area of the Arctic: from west Inuit in Alaska and Inuit at the eastern tip of Siberia, as well as Inuit in Canada and Greenland. The area spans some 4,000 km and yet the entire area is surprisingly linguistically homogeneous, with many phonological and structural similarities. The number of Inuit speaking Inuit languages is less than 100,000, about half of whom speak Greenlandic.

Greenlandic is the official language of the country, and Danish can be used in public contexts. Public authorities and private companies with at least ten employees must develop a language policy.

The latest language policy act in Greenland was adopted on 19 May 2010. The aim is to provide a clear framework for the country's language policy, including for linguistic integration. This must be done by ensuring that the Greenlandic language is complete and capable of supporting society as a whole, as well as by strengthening the language and its development as a native and second language. At the same time, the language policy must address and remove existing language barriers and promote language acquisition.

The Greenland Language Council is tasked with monitoring the development of the Greenlandic language, finding authoritative Greenlandic designations for imported words (replacements) and approving terminology.

The three main dialects

The three main dialects, West Greenlandic, North Greenlandic and East Greenlandic, are quite distinct. Towns and settlements each have their local dialect, and if you hear a Greenlander speak, you can often tell which town or settlement they come from.

There is no legislation governing the oral language, everyone can use their own dialect, but in writing people are required to use standard writing and must follow the orthographic rules. Yet, more Greenlanders often write in their own dialect; for example, East Greenlanders and North Greenlanders often write dialect on social media.

The IT-80 football team, founded in Isortoq in 1980, competes here in the annual coastal championships held in Tasiilaq. Regardless of the language, the volume is turned up as fans cheer and roar for goals.

MADS PIHL/VISIT GREENLAND, 2013

Map of the areas pertaining to the three significantly different Greenlandic dialects.

Figure 4. Language development 1984-2003 based on data from the 2007 Survey of Living Conditions in the Arctic.

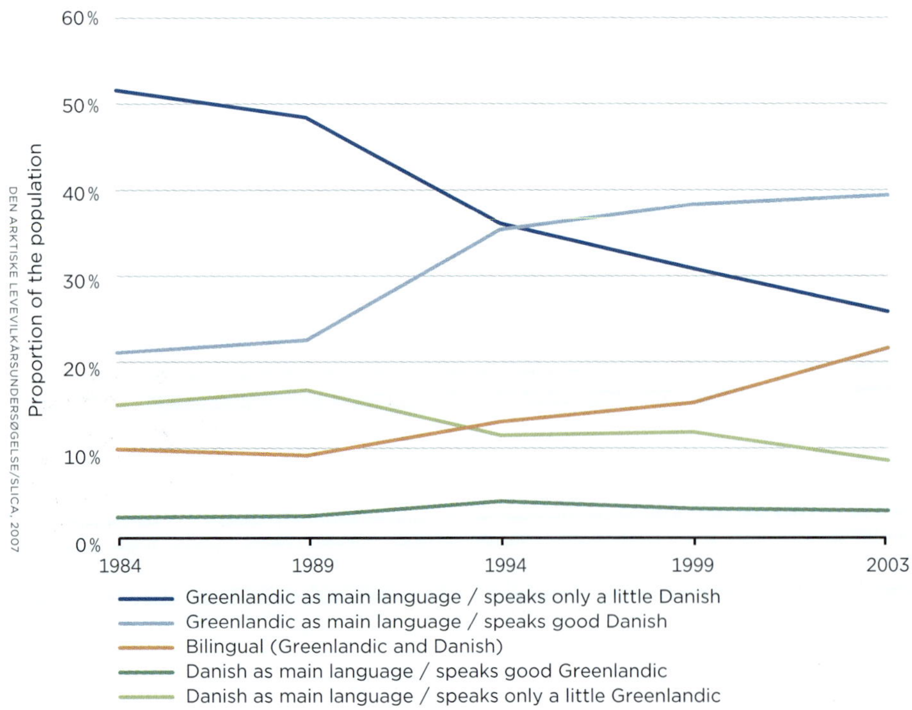

Language use until the mid-1900s

From Hans Egede's arrival in 1721 until World War II almost all inhabitants spoke Greenlandic, including the missionaries. The missionaries managed to translate the religious language into Greenlandic, and the first dictionary, published by Hans Egede's son Poul Egede, came out already in 1750.

In the catechist seminaries in Godthåb (Nuuk) and Jakobshavn (Ilulissat), established in 1847 and 1848, respectively, teaching was conducted in Greenlandic. Danish was one among several subjects, and in the 1860s, Danish was taught five hours a week.

In 1851, seminary teacher Samuel Kleinschmidt published a Greenlandic grammar which described the language on the basis of morphology. At the same time, he developed the first actual orthography, which until 1973 was used as standard orthography. The transition from the old to the new orthography took place gradually, and there are still reminiscences of the old orthography, which can be used, for example, in the Greenlandic personal names. The new orthography is phonemically oriented, meaning that the spoken and written language have come closer each other.

From the early 1900s up to the 1950s, many authors wrote in Greenlandic, and the first novel in Greenlandic, *Sinnattugaq*, written by rural dean Mathias Storch, came out in 1914 (see p. 225).

The use of language from 1950 to the 1970s

From around 1950, a vigorous Danisation of Greenlandic society occurred (see p. 128). Since there was a lack of trained Greenlandic teachers and Greenlandic teaching books, it was necessary to understand Danish in order to pursue further education. A new school act in 1967 introduced an option to postpone Greenlandic teaching until the third year of school, so as to give children better opportunities to learn Danish.

During the 1960s, the school structure approached that of Denmark. After the 6th grade, Greenlandic children were commonly sent to Denmark for one year to improve their language skills. Many who grew up at that time speak Danish without any problems, but for a large part, this resulted in inferior skills in written Greenlandic.

The use of language from the 1970s up to the present day

The introduction of Home Rule in Greenland (see p. 135) brought significant changes to the language policy. The Greenland Home Rule Act states that the Greenlandic language is the main language and that at the same time the Danish language must be thoroughly taught. It must be possible to use Greenlandic and Danish in all public contexts. The strengthening of the Greenlandic language was ensured in the school law of 1979, in which Greenlandic became the language of instruction from the 1st grade and Danish became the first foreign language. From that day, it was compulsory to teach Danish as a foreign language from the 4th grade and English from the 7th grade.

The priority of Greenland teaching has improved the mother tongue skills of those speaking the Greenlandic language. A language study from 1999 found that 69.1 % of the population spoke Greenlandic, 15 % were bilingual (Greenlandic and Danish), while about 15 % only spoke Danish. A study from 2003 of the language of 14 to 17-year-olds found that 94.6 % of young people could speak Greenlandic, while 73.3 % could also speak Danish.

In the two language studies, the respondents had stated their language skills themselves. If you look at them objectively, the figures may turn out to be different, but nevertheless, it is evident that more and more Greenlanders speak both Greenlandic and Danish.

The only both quantitative and qualitative language study was conducted in Qaqortoq in 2001. The study participants included people born in 1959, who had grown up in a time of great focus on the Danish language, and young people born in 1982 and 1983 who had grown up under the Home Rule during the Greenlandisation. The results showed that those who grew up during the Danisation were bilingual, but that many of them are not very good at writing Greenlandic. The young people who grew up during the Greenlandisation speak mainly Greenlandic, and many of them do not speak Danish very well.

The language and the education problem

For some, the increasing number of monolingual young Greenlanders has meant inferior educational opportunities, because although teaching in primary schools is primarily in Greenlandic, higher education is primarily conducted in Danish.

The language of instruction in upper secondary schools is Danish, except in the subject Greenlandic, and almost all higher education requires Danish skills. There are several reasons for this: There is a lack of Greenlandic teachers, the vast majority of teaching books are in Danish, and in many disciplines no specific Greenlandic terminology has been developed.

Only two higher educational programmes do not require Danish skills: the carpentry training programme and the decentralised social worker training programme. Both training programmes are aimed at adults with practical experience.

The degree programmes at the Ilisimatusarfik University are primarily conducted in Danish. However, if the teacher speaks Greenlandic, teaching can be conducted in Greenlandic.

FINN FRANDSEN/POLITIKEN/RITZAU SCANPIX, 2007

First day of school in Nuuk. The Greenlandic school system is based on the Danish system, and the first school year, preschool, can be compared to the Danish preschool class.

The language of administration

Although the language policy act from 2010 state that the Greenlandic language is used in public contexts, much of the language of administration is still Danish.

The Inatsisartut (Parliament) rules of procedure stipulate that negotiations should be conducted in Greenlandic, but that members of Inatsisartut and Naalakkersuisut (government), who do not speak Greenlandic, may speak Danish with interpretation into Greenlandic, and that it must be possible to follow negotiations in Danish. All documents relating to Inatsisartut and Naalakkersuisut are translated to be available in both Greenlandic and Danish. Therefore, there are many translators and interpreters in the Government of Greenland, and well-educated Greenlanders are assigned interpreting tasks along with their normal job.

With this amount of translations, translators and interpreters are significant language developers. Unfortunately, they are not always familiar with the specific terminology in individual fields, and this may lead to imprecise translations that cause confusion. The Language Council and the Language Secretariat investigate and develop the Greenlandic language, including in relation to terminology.

FOUR TALES ILLUSTRATED BY ARON OF KANGEQ

NUNATTA KATERSUGAASIVIA ALLAGAATEQARFIALU

For this drawing, Aron writes: »Qivaaqiarsuk disposes of some of the many women by letting them be taken by the tide on a large skerry off the houses at Kangeq.«

THE TALE OF QIVAAQIARSUK

Fertility is central to Inuit culture and is a recurring theme in the ancient storytelling tradition, not least in the tale of Qivaaqiarsuk (Qivaaqi means sculpin) told by Aron of Kangeq, among others.

One of the focal points is fertility, both of the landscape and of women. Qivaaqiarsuk is always lucky in his hunting endeavours, but he has to provide for many women, and he is unscrupulous if they become more of a nuisance than a help.

The drawing of Qivaaqiarsuk shows how he kills many women in one go; he lures them out to gather seaweed at low tide, and when high tide comes, he lets them drown.

THE TALE OF ULAAJUK

Innersuit (plural; singular: innersuaq) are fire beings often called beach spirits because they live under the beach. Those who live highest up are friendly. They are often the kayakers' helping spirits and protect them when they go hunting. Those who live deeper down have no noses. These innersuit should be avoided. They are jealous of humans and may imprison them, cut off their noses and turn them into innersuaq.

In this story, the young man Ulaajuk has been out hunting and has met an innersuaq from deep below. The innersuaq forces Ulaajuk to follow him, cuts off his nose and holds him captive. Ulaajuk invokes his helping spirit, who comes to free him. As they row away together, the helping spirit says »Hurry up and turn around, your nose is coming flying after you!« and Ulaajuk's nose comes flying and returns to its place.

NUNATTA KATERSUGAASIVIA ALLAGAATEQARFIALU

Here Aron notes: »Ulaajuk gets his nose cut off and is made into innersuaq.«

NUNATTA KATERSUGAASIVIA ALLAGAATEQARFIALU

Aron's caption is as follows: »While hunting caribou in Qooqqut, Mattiina dreamed that they descended from the sky to give him a message: Christian Renatus and Poorkaaputsi.«

THE TALE OF QOOQQUT

A tale from the 1850s featuring a powerful man who refuses to get baptised. He does not understand all the talk about salvation, but he understands what the angakkuit are saying. After a long time, however, he changes his mind, and in baptism he is given the name Mattiina. On a caribou hunt in Qooqqut in Nuuk Fjord, Mattiina dreams that two angels descend from the sky. They sing beautifully and then tell him that one angel, previously called Poorkaaputsi by its loved ones, has no body but has become pure soul, while the other, Christian Renatus, still has a body.

Christian Renatus was one of the early Christian catechists and also one of the good storytellers. This tale shows, among other things, how features of European culture are incorporated into the Inuit universe.

THE TALE OF KAASSASSUK

Kaassassuk is the name of the main character in a widely told tale among all Inuit. In East Greenlandic, the name means water beetle. It was believed that when the beetle lives at the bottom of dark lakes, it cannot be seen by death. When parents gave their boys this name, they did it to protect their lives in a society where it would be critical for the family if the provider died. The oral tradition of storytelling often portrays love for the childless and fatherless who lived without protection.

Kaassassuk was an orphan and bullied by both children and adults, but after enduring many hardships, he ventures out into the wild where he draws strength. Having gained incredible strength, he takes terrible revenge on all who have been mean to him.

NUNATTA KATERSUGAASIVIA ALLAGAATEQARFIALU

The caption to this picture is as follows: »The other children make fun of Kaassaasuk and wash him in the snow, and some of them pee on him.«

TRADITIONS AND TALES

The oral storytelling has been a key element of Greenlandic social life. The listeners were entertained with strange, eerie and supernatural events that were always told as true stories. The tales are closely related to the original religion, way of life and morality, and deeper understanding of the tale thus presupposes knowledge of the culture.

The tales challenge culture, and at the same time they challenge the listener's understanding of it. Even when the tale does not moralise, it puts existential matters into words, and this way, the language and the listener's perception of key cultural concepts are developed.

The missionaries and early collection of tales

The tales first noticed by the missionaries were typically those which explained where different animals come from and why they are as they are. Often the missionaries saw these tales as signs of backwardness.

In the late 18th century, Europe began to take interest in both their own and others' adventures, myths and legends. In the 1820s, the Greenlandic storytelling tradition met positive interest from the missionary Peder Kragh and his North Greenland catechist Wittus Steenholdt. They took the initiative to write down and transcribe various texts, mostly the traditional oral tales, which were popular and circulated around for reading aloud. The crucial role of Bible reading in the Lutheran mission and the efforts of the catechists meant that almost anyone could read.

H. Rink and the first printed work

In 1852, civil servant and geologist Hinrich Johannes Rink came to South Greenland, where he was first colony manager and in 1855 was appointed inspector of South Greenland. H. Rink reported that Danish colonisation was heading towards a fatal end. The population was unhappy, became increasingly passive and was declining as a result of hunger and epidemic (see p. 119).

In Copenhagen, H. Rink received a positive response to his ideas of writing down, collecting, printing and distributing the oral tales. With a wish to strengthen and preserve indigenous culture, this work began, and it became a key element in the building of a Greenlandic nation. Respecting the peculiarity of the language, he drew attention to the importance of ensuring that the dialect and spelling of storytellers remained their own and were not corrected.

In the introduction to the book *Eskimo Tales and Legends* (1866), H. Rink writes how he perceives the decay of Greenland culture and what, in his opinion, has gone wrong: The strangers have not understood that the small Greenlandic communities had a civilisation; without an actual social life, the Greenlanders could never have survived. The staple functions of community life were undertaken by an angakkoq (shaman) who, with his special abilities, helped the people of the small communities. During the drum song, he was able to leave his body and soar elsewhere (see p. 210). The angakkoq had a connection to the beyond and knew how things worked, what you could or could not do, he could advise and heal.

To undermine the authority of the angakkuit, the mission started to work hard against them. In the opinion of H. Rink, this was a fatal mistake as it removed the core element in society and thus broke down the culture: '*One could also dare say that if they are abolished, and no corresponding heads of the nation are put in their place, it will eventually perish.*'

Much of H. Rink's *Eskimo Tales and Legends* was written down by Aron of Kangeq and Jens Kreutzmann of Kangaamiut, who had also both made illustrations. Among Aron's tales (see p. 152), there are many dealing with the meeting with the Norse. Aron of Kangeq writes flawlessly and elegantly, and one senses a self-censorship trying to make the tales more decent, while Kreutzmann is more outspoken.

H. Rink had also got hold of some tales from Labrador, and Peder Kragh sent his old collection, which was also included. Later, H. Rink published the collection in English.

Greenland expeditions and documentation

The collection of the tales mainly started for the Greenlanders' sake, but soon it became commonly recognised that Inuit culture was disappearing, or at least changing radically, and now the driving interest became to document the culture before it was lost. When in 1883-1985, Gustav Holm led the Umiak Expedition to East Greenland (see p. 342), where the expedition wintered and established contact, he collected as the first tales.

From the early 1900s, Knud Rasmussen travelled on expeditions with dog sleds, reaching almost all areas inhabited by the Inuit (see p. 121). Along with his own personal records, he industriously collected tales and documented the Inuit way of life, including in areas that had only very rarely contact with people from outside. However, it is not clear how accurate his renderings are.

At the same time, a number of West Greenland missionaries started to make great efforts to document life in East Greenland. First, the mis-

Linguist and anthropologist William Thalbitzer spent a year in the newly established trading post of Ammassalik in 1905-1906. He collected several sound recordings of the East Greenlandic language using a phonograph. He took many photographs, especially portraits, including this of the Angakkoq Ajukutooq Attakkaat, later christened David.

WILLIAM THALBITZER/ARKTISK INSTITUT, 1906

sionary Christian Rosing (see p. 182) arrived in 1900, in 1906, he published the book *Tunuamiut,* describing East Greenland culture. In 1921, Danish eskimologist Thalbitzer published a number of songs, poems, incantations and stories from East Greenland.

Audio and visual recordings of tales

The first audio recordings from Greenland came in 1906 with the Austrian Rudolf Trebitsch. Released in 2003, they include hymns, poems, legends, sagas and myths in short lyrical forms, as well as rhythmic play with words and nonsense.

A few years after, in 1909, Norwegian music ethnologist Christian Leden came to Uummannaq and Qaanaaq, and in 1912, he travelled to the Uummannaq-Upernavik areas. On both voyages, he recorded cuddle songs, satirical song and drum songs (see p. 210).

Since French ethnologist and polar explorer Paul-Émile Victor collected 500 songs in East Greenland in 1935, France has had a particular interest in the area. Together with anthropologist Joëlle Robert-Lamblin, he has also published some tales.

In 1934, for the first time, a cinema audience could experience a film, *Palos Wedding* (see p. 220), with sound and in the East Greenlandic dialect. It was written by Knud Rasmussen, who himself played a key role during filming.

A third pioneer was eskimologist Erik Holtved, who in Thule 1935-37 and 1946 – 47 recorded almost 200 tales and songs, and transcriptions of those were published in English. Holtved's categorisation of the tales provides insight into the diversity: Myths and legends, fabulous beings, epic tales, fairy tales, children's songs, fables and animal tales. The remaining, almost half, which he calls historical legends, are subdivided into several types of legends about resurrection, killing, revenge and taboos.

Throughout her life, the writer Mâliâraq Vebæk (see p. 226) collected folklore from South Greenland, which she wrote down and published, and she also recorded tales and songs.

The East Greenlander Isaiah Kuitse probably became the last traditional storyteller. Teacher Niels Grann and

INUIT WAY OF LIFE

KONRAD NUKA GODTFREDSEN, 2012

Reincarnation was part of the Inuit way of life, and this included humans and animals. When a name was given to an infant, the characteristics of the name and any helping spirits would follow.

IInuit tales played a crucial role as a cohesive force that told the Inuit how the world was created and how everything was interlinked. People were connected to each other by family ties, but also through a name soul. Each human being consisted of a body (timeq), a soul (tarneq) and a name (ateq). If a person died, it was therefore important that the next newborn child was named after the recently deceased, so that the name – and its special abilities and amulets – could live on. This tradition still thrives to some extent in present-day Greenland.

However, humans were not the centre of Inuit life, for humans were merely one of many relatively equal beings, and therefore everything had an inua, which is a life essence or inherent being. The word itself translates as 'its essence', and inua was found in all living creatures (humans and animals), natural elements such as rocks and the wind, in a stone and even in laughter, hunger and sleep.

Just as religious writings prescribe good and bad morals, oral tradition served to pass on vital lessons and admonitions for socially acceptable behaviour. In addition, the tales gave rise to social gatherings and togetherness both during the long, dark winter nights and during the annual aasiviit (summer camps, see p. 213) where people from near and far gathered.

MALIINA, ANINGAAQ AND SASSUMA ARNAA

One of the recurring tales among Inuit in Greenland, northern Canada, Alaska and as far east as Siberia is the tale of Maliina and Aningaaq. They were siblings but the boy Aningaaq fell deeply in love with his sister, Maliina. When his sister to her horror realises that she has been sleeping with her brother in the dark, she flees the settlement in shame. She manages to grab a piece of glowing moss from the blubber lamp, her brother does the same, but in his moss there are only embers. In their flight, they both rise to the heaven, and so we see in the sky the Moon (Aningaaq) chasing the Sun (Maliina), though he will never catch up with her.

Another common story is about Sassuma Arnaa (in early literature called Mother of the Sea or the Sea Woman). In this tale, a (according to the father) marriageable daughter is presented with a number of suitors, whom she rejects as she is able to support herself, i.e. go hunting. However, her father will not tolerate this, so one day he pushes his daughter over the side of his boat. As she clings to the boat, he cuts off her fingers, so that she sinks to the bottom of the sea, while her severed fingers transform into the myriad animals of the sea and she becomes 'Mother of the Sea' (Sassuma Arnaa). Since then, when hunting went wrong or settlements experienced famine, it was attributed to man's rule-breaking, and it was now the task of an angakkoq (spiritual adviser, see p. 158) to make a spirit journey to Sassuma Arnaa in an attempt to appease her and negotiate with her to restore the balance of the world.

This anorak belonged to the four-year-old boy Tertaq, and it features no less than 80 amulets sewn on the outside and inside. The amulets, which include the skull of a wolverine, a whole raven, polar bear claws and more, were meant to protect the boy from spiritual attacks, but some of the animals' unique properties could also infect the boy. It was unusual, however, for a costume to have so many amulets, so Tertaq's father understandably demanded a good price from polar explorer Knud Rasmussen: the polar explorer's pencil, which could »capture stories in lines«, and a lock of his hair. In 1924, when the costume was collected, not all Inuit could read and write.

ROBERTO FORTUNA/NATIONALMUSEET, KØBENHAVN

CHRISTIAN SØLBECK, 2020

The Inuit tattoos are known from historical sketches and appear on the well-preserved Qilakitsoq women (see p. 106). Most of the tattoos are found on the female body as amulets, as the woman has to go through menstruation and potentially dangerous births. They could also be mythical references; for example, lines across the second finger joint symbolise Sassuma Arnaa, who, it is said, had these finger joints cut by her father. The tattoos were created by poking soot from the blubber lard lamp into the skin with a needle or by pulling a soot-soaked thread under the skin. In recent years, the tradition has been revived, predominantly women getting tattoos on their hands, forearms, neck, thighs, lower legs and, more rarely, on the face. The woman to the left is explaining that the four dots on the back of her neck symbolise her four siblings, while the lines across the second joint of the fingers of her right hand and the dots on the joints closest to her knuckles symbolise her children. The woman to the right explains that she has on her hands the symbol of the Mother of the Sea and some amulets for her husband and son. The dots keep the evil spirits out.

ANGAKKOQ, AMULETS AND TUPILAK

On a spirit journey to the unseen world, the angakkoq had to draw on the spiritual strength of its helping spirits (toornat). These spirits had been overcome by angakkoq during its apprenticeship, when spirits could be negotiated with. However, if an angakkoq apprentice had to give up its training, it could become an ilisiitsoq: one who used its understanding of the spiritual world for misdeeds. People without spiritual training, on the other hand, had to protect themselves with amulets and tattoos.

The amulets had a dual function. They were meant to protect the wearer from harm, but at the same time the wearer could acquire a desired ability from the animal which the amulet consisted of. The snowy owl gave you better eyesight, while the polar bear gave you strength and the common eider made you a faster kayaker. If you were born with ailments, your survival depended mainly on whether the settlement had plenty of food resources and whether you were spiritually protected and empowered.

One feared creature was the tupilak, created to harm another human. A tupilak could consist of many different parts from animals and possibly hair from the person you wanted to hurt. When the creature was awakened, its creator would command its mission.

Whereas the original tupilak was a mischievous creature, today the beautifully carved tupilak is sold in souvenir shops as a lucky charm, and thus the object now has a new meaning (see p. 204).

TABOOS AND RITUALS

The Inuit rituals offer deep insight into the Inuit way of life. In addition to a host of taboos (see p. 104) surrounding life transitions such as menstruation, birth and death, the Inuit performed a number of rituals when interacting or even negotiating with the game or animals. The hunt was not attributed to the hunter's years of training and skills, but rather to the persuading of the animal to 'give itself' to the hunter, his family and the settlement. The hunter achieved this by being moral and hard-working, by making the hunting tool so that the animal could recognise itself in it (see p. 198) and by the hunter showing respect after the death of a seal, for example, by offering the seal fresh water to drink.

A lifetime of abiding by these rules would see a good hunter in the land of the dead with an abundance of game. The disrespectful and greedy hunter, on the other hand, went to the land of the dead where he had to catch butterflies with his mouth because his hands were tied behind his back – a cruel fate from an Inuit perspective.

French linguist Nicole Tersis filmed him. The footage appeared in 2016, and in parallel, a transcript and an analysis of the gesturing were included in the accompanying book *Paroles et gestuelle. Un conteur inuit du Groenland oriental/Words and gesture. An inuit storyteller in East Greenland*. Kuitse's episodes and scenes alternate between suspense, revenge, hopelessness and comedy, which he plays with great intensity.

Ulrikke Oodaaq of Qaanaaq is in the movie *I remember …* from 2002. Karen Littauer is behind the filming of storytellers from the districts of Ammassalik, Thule and Upernavik. Among other things, Oodaaq tells to the camera a story about a mysterious childhood experience that happened one day she was home alone. The oral tale is supported by gesturing, for example, the hand creates an illusion of a wall in the house, both before and after it is mentioned verbally. It shows her ability to recall the moment, and the audience has to wait for the few words that tell what the gesturing shows.

INUIT HUNTING CULTURE

For many people in Greenland, hunting, whale and seal hunting is associated with 'the good life'. Hunting means meat supply, food, hunting traditions, skin processing and sailing trips in the fjords. Although hunting may have different purposes, it supports good relationships between family and friends, animals and nature. The money earned from the catch is often part of the sharing economy, where meat is distributed in social networks or sold in local markets. For many families, this local meat supply is an important source of fresh nutritious produce, and it is a focal point in Greenlandic cuisine.

In addition to meat, game brings skins and bones that are used in home pursuits, national costumes and arts and craft. Hunting is also the purpose of many sailing trips, hikes and summer camps – and it sets the stage for interesting and insightful tales in good company. In addition, hunting is deeply connected to traditional Inuit hunting culture and knowledge, which has been passed on from generation to generation. Thus, hunting is interrelated with Greenlandic culture, identity, and community.

Hunting as an occupation today

All over Greenland, people go hunting and fishing in their spare time, making sure that the family's chest freezers are well stocked. Approximately 5,000 citizens hold a license for spare time hunting. For the approximately 2,000 occupational hunters in Greenland, hunting takes up most of working life, and it forms a significant part of the income of hunting families. Some hunted animals can be traded for their meat or skin, which is sold locally, nationally or internationally. The hunting profession is costly in terms of spending on work equipment such as boat and motor, gasoline, rifles, cartridges, nets, hooks, harpoons and work clothes. Overall, the income from

THE TALES OF ANAQORSUAQ AND QIVAAQIARSUK

Hans Hansen and Hans Holm from the Greenlandic radio station KNR broadcast tales by contemporary storytellers in the 1950s and 1960s. They are still broadcasted on KNR, and there are still many listeners to the two stories about Anaqorsuaq and Qivaaqiarsuk as well.

The life of Anaqorsuaq, the great hunter, is recounted by Leopoldus Ringsted, called Puustusi, from Kapisillit. It takes the form of a national epic poem, a classic heroic tale, consisting of two parts: The first part is about Anaqorsuaq's upbringing in East Greenland, while the second part recounts his later life in West Greenland.

Growing up in East Greenland, Anaqorsuaq always hears about 'Pisissarfik', a large mountain at the head of Nuuk Fjord in West Greenland. It is a place with an abundance of game: seals, caribou, fish. His dream is to make the long journey there and settle.

Anaqorsuaq goes there, and in this harmonious world he has a son with whom he is happy and goes hunting, but occasionally he makes high demands on his son's skills. One day while hunting, his son is pulled down by a seal, but Anaqorsuaq does not intervene to save him from drowning. Puustusi comments that the father's extreme sternness shows that this tale is true. An untrue tale would only show Anaqorsuaq's good side.

Atypical of traditional tales, however, this one is highly idealised, largely conflict-free, the gallery of characters are helpful, and the protagonist's skills almost make Anaqorsuaq superhuman. As an utopian idea and pilgrim tale, this tale is unique. At the same time, it is clear that Puustusi is excited about talking about the greatness of the past. Puustusi considered himself a Christian, but his tale emphasises the humanity of the pagans.

The story of the childless catcher Qivaaqiarsuk is told by Simon Nielsen from Qoornoq. Qivaaqiarsuk's wife is barren, and during a period when the seals do not come to shore, he gives her some mattak to eat. When this kills his wife, Qivaaqiarsuk takes out her stomach, inflates it and dries it. He then rows out far from the shore, and three times he finds an ice floe further and further from the shore against which he strikes the bloated stomach. Only the third time does it burst, and the seals begin to rush in droves towards the shore, while Qivaaqiarsuk himself risks his life in the storm at sea.

Qivaaqiarsuk puts his life on the line; he manages to defy and control the supernatural forces and thus gets the seals back. His worship of fertility, that of women and of the sea, is his ultimate purpose.

JONATHAN NACKSTRAND/AFP/RITZAU SCANPIX, 2019

Cleaning fish in Kulusuk. There is a lot of fish in the waters around Kulusuk, especially cod, but the settlement does not have a factory, and the only possibility to buy fish in the district is in Kuummiut, some 40 km to the north.

the catch is modest, and a hunter's domestic economy often relies on a partner's fixed monetary income from employment. Occupational hunting is therefore often combined with fishing and tourism, depending on season, supply and demand.

Although occupational hunting today is of less importance for the national economy, the profession still plays an important role for the labour market in North and East Greenland, as well as in settlements in South Greenland, where there are fewer job alternatives. Analyses of the informal economy of hunting have shown that hunting makes a significant socio-economic contribution to Greenlandic society, primarily in the form of self-sufficiency and food security.

The earliest Inuit hunting culture

Greenland has evolved with the hunting culture. Since the first immigrations, Inuit have made a living from hunting on land, on the sea, on the ice and in the fjords. Through time, marine mammals, terrestrial mammals, fish and birds have been the basis of food and clothing, light and heat, tools and means of transportation, such as dog sleds, kayaks and large skin boats.

The sustenance of families and settlements has relied heavily on favourable hunting conditions and successful catches. Hunting has therefore traditionally been a key focal point in the organisation of Arctic communities, where roles and responsibilities have been defined by hunting activities. Hunting itself has most often been carried out by men, and the preparation of meat and skins has been the responsibility of women.

Meat, blubber and skins were – and are – distributed according to special rules. A *piniartorsuaq* (great hunter), who has extensive hunting expertise and skills, has had a special responsibility for sharing his catch and the surplus from the hunt with those in need. These roles and relationships served to provide food security and strengthen overall welfare. In addition, cultural rules, such as the special ways in which the hunter and the hunter's wife should show respect for the hunted animals, ensure favourable hunting conditions and maintained the balance between humans, animals and nature.

Hunting culture during and after the colonial era

The role of hunting in Greenlandic society and culture has obviously changed through historical processes, including the trade with European whalers in the 17th and 18th century and in particular with Denmark's colonisation. Until the early 1900s, the Danish colonial strategy was based on the outcome of hunting.

The colonisation was financed by trade with skins and blubber. In this process, the catch from hunting was partially commercialised, and in the colonial discourse, 'hunting' became 'the national profession.' As revenue from the blubber trade went down in the late 1900s, the first steps were taken towards changing to fisheries (see p. 123). Subsequently, the colonial power considered the hunting profession worthy of preservation for the sake of in-kind economy and culture.

Broadly speaking, these political developments meant that fishing became the trademark of modernisation, while the Inuit hunting profession was increasingly categorised as cultural heritage. These historical processes have undoubtedly influenced the development of hunting culture in Greenland and are also reflected in Greenlandic politics today, where hunting is often articulated as cultural heritage and takes up an increasingly smaller role in development plans.

In spite of these developments, Greenlandic hunting practices escape any rigid categorisation of hunting as a relic of the past, which is opposed to fisheries. Hunting, in contrast, relates to the lived life, which is defined by dynamic and diverse hunting practices and purposes. Hunting is not just about cultural preservation and need-based food supply. It is associated with countless aspects of Greenlandic living, and it encompasses culture, vocation, originality, traditions, economy, way of life, identity, community, and local sovereignty. Hunting also relates specifically to the animals of the land and the sea, their availability, the cycle of seasons and *sila* (weather and climate) – and it is made possible by passed down and acquired hunting skills and an in-depth knowledge and conditions of nature, the seasons and weather. There is no one way of being a hunter but many, and these vary from place to place.

Hunting and the hunting profession is managed by the Department of Fisheries, Hunting and Agriculture under the Government of Greenland (see p. 266). According to legislation, the Department is required to manage the utilisation of living resources based on scientific advice from Pinngortitaleriffik (Greenland Institute of Natural Resources) as well as commercial and socio-economic needs. Emphasis is also placed on including the local knowledge of hunters. This principle sometimes proves difficult to follow, because biological research and hunters' observations of animal populations may be contradictive.

Hunting on land

Caribou, arctic fox, mountain hare and muskox are the main *nersutit* (terrestrial mammals) hunted on land. Birds such as grouse and little auk are also hunted on land, whereas thick-billed murre, black-legged kittiwake and eider ducks are hunted at sea or on the islands where they breed. The populations of terrestrial mammals have historically fluctuated more

EMIL HELMS/RITZAU SCANPIX, 2021

Harbour porpoise tails for sale in Nuuk. Whale meat is a popular food, and the skin of the whale – mattaq – is a highly prized delicacy, often served raw in small cubes or prepared in a variety of ways for festive occasions such as kaffemik.

DOUG ALLAN/NATURE PICTURE LIBRARY LIMITED/RITZAU SCANPIX, 2012

Near Siorapaluk a man catches little auk (alle alle) with an ipoq, a handmade net on a long stick. In general, there are strict rules on how birds can be caught and strict quotas.

than those of marine mammals. Caribou populations, in particular, have seen dramatic fluctuations. In East Greenland, caribou have gone extinct due to climatic changes.

Hunting of caribou and muskoxen

Among the animals hunted on land, the caribou holds a special status. In many places in West Greenland, summer hunting is associated with holidays, spending time together with the family and connection to ancient traditions and places. The vast inland area near Kangerlussuaq in West Greenland, Angujaartorfiup Nunaa, is a caribou hunting area where families from coastal towns and settlements have hunted for generations and where tent sites are passed down from generation to generation. These days, the caribou hunt begins on 1 August, and many families camp out even before the hunt sets in. Most summer hunting grounds are close to rivers or estuaries, or other places where arctic char, polar cod and ammassat (capelin) are also fished. Summer hunting is important to many – both vocational as well as spare time hunters – because it fills the freezers, and because the community in nature builds up strength for a long winter.

Although animals in Greenland are primarily thought of as 'wild', caribou and muskoxen carry a history of human interference. In the 1950s, domestic reindeer were introduced from Norway to South Greenland and to a farm in Nuuk Fjord (see p. 324), and in the 1960s, 27 East Greenland muskoxen were moved to Angujaartorfiup Nunaa (see p. 45).

The muskoxen bred so much that animals have since been translocated from here to other areas in West Greenland. With the muskoxen, new forms of hunting followed. In winter, commercial hunters from Sisimiut and Maniitsoq come to Kangerlussuaq to hunt muskoxen for their meat and hides, and new economic opportunities have emerged: A small-scale wool production industry supplies national and international markets with muskox yarn. In addition, trophy hunting is a new form of hunting that takes place in the Kangerlussuaq area and elsewhere in Greenland. The muskoxen have become coveted game for foreign hunters travelling to Greenland for an Arctic hunting experience. Trophy hunting is different from traditional Greenlandic hunting practices, and it has provided a new economic opportunity for hunters and occupational hunting, which generates cash earnings while regulating the population of old bulls.

Hunting at sea

Many different *miluumasut* (marine mammals) are hunted in Greenland. These include various species of whales and seals, as well as polar bears (see p. 68). Since the animals of the sea are the most stable source of proceeds, they have always been of great importance for Greenlandic living, and ensuring good relations between humans and the animals of the sea has been particularly important. This is illustrated in the Central Greenlandic tale of *Sassuma Arnaa* (Mother of the Sea, see p. 156), who holds back marine animals when humans lose respect for animals and nature and forget to follow the prescribed rules of conduct. She only releases marine mammals again when humans have managed to appease her by combing her hair. In a way, *Sassuma Arnaa* was respected as an authority of marine resources.

Today, hunting at sea, like all other hunting, is managed by the Government of Greenland in consultation with biological advice from Pinngortitaleriffik (the Greenland Institute of Natural Resources). As marine mammals migrate in international waters and cross national borders, the advisory services are provided in collaboration with international organisations, such as the International Whaling Commission and the North Atlantic Marine Mammal Commission. In particular, the hunting of walruses, polar bears, narwhals and beluga whales is under the watch-

Walrus hunting near Qaanaaq. The walrus is the largest seal species in the Arctic, and a full-grown male weighs around 1,200-1,500 kg, so it takes a lot of effort to drag home a caught animal. Quotas for walrus were introduced in 2006 and the walrus hunting season in Northwest Greenland runs from 1 October to 30 June.

LOUISE MURRAY/SCIENCE PHOTO LIBRARY/RITZAU SCANPIX, 2015

ASHLEY COOPER/NATURE PICTURE LIBRARY LIMITED/RITZAU SCANPIX, 2008

Seal skins hung to dry in Ilulissat. For a lot of hunters, sealing, selling seal meat and trading in seal skins is a major source of income. Great Greenland purchases seal skins in a number of settlements and towns, where the purchased seal skins are frozen and sent to the tannery.

ful attention of biologists, authorities and international environmental organisations. Since the mid-1900s, the hunting of certain marine animals has been regulated via quotas. This regulation represents a new seasonal rhythm that hunters must adapt to.

Seal hunting

Seal hunting, in particular, has played a significant and distinctive role throughout history, both in pre-colonial Greenland and in the colonial period. The seal has been the primary game in the Arctic because it has covered the basic daily necessities with its meat, blubber, skins and bones. Therefore, the seal has a special place in the Inuit culture, identity and world view.

With Denmark's colonisation, seal hunting also became key to the colonial trade that commercialised it and capitalised on skins and blubber, which were attractive in the European market. Seal blubber was sold as oil for lamps and soap production in Europe until the blubber was replaced by mineral oils in the late 1900s. The sale of sealskins on the international market continued through the colonial period and is an important contribution to hunters' income today.

Whereas, sealskin trading was formerly undertaken by the Royal Greenlandic Trading Company, it is now undertaken by the government-owned Great Greenland A/S, which trades, tans, and markets sealskins, nationally and globally. However, since the 1960s and 1970s, the market has been threatened by the Western World's anti-seal hunting campaigns, which brought about a market collapse in the 1980s with devastating consequences for Arctic hunting economies. The market has never recovered, and sealskin has since been subsidised by the Home Rule and Self-Government. The anti-seal hunting movement flares up again at times, and in the 2000s a campaign ended with an import ban on sealskin products in the EU.

The history and evolution of seal hunting reflect that hunting culture arises and changes through various processes in which indigenous, colonial, local and global realities come together, clash and intertwine.

Future perspectives for the hunting culture

While it has become harder to subsist on hunting alone, the number of spare time hunters has increased in

LOUISE MURRAY/SCIENCE PHOTO LIBRARY/RITZAU SCANPIX, 2015

Breathing hole hunting near Qaanaaq. Hiding behind a camouflage sail, a hunter slowly sneaks up on a seal sunbathing next to its breathing hole.

recent years. It points to the fact that a life of hunting, despite hardships of the work, is highly valued. Climate change is increasingly affecting hunting conditions. However, some hunters highlight that the hunting culture has always adapted to the changing forces of nature, but that quotas and political regulation can create uncertainty about the future. Other hunters say: As long as there is life, animals and ice in Greenland, there will be hunting.

FOOD CULTURE

Greenlandic food is characterised by coming directly from nature. *Neri* make up the root of the Greenlandic word for *eating*. The word is close to the Greenlandic word *neqi* – meat. That is no coincidence. In pre-colonial times, food in the Greenlandic sense was synonymous with meat. While it is easy to think that it has been a monotonous diet, wildlife provides the basis for a very varied diet, both in terms of nutrition and taste. Variation is especially formed by the course of the year, as each season brings new hunting opportunities and limitations, and at the same time the country's vast geographical area provides access to different foods in the different areas. Therefore, a person from Nanortalik will describe Greenlandic food quite differently from a person from Uummannaq or Ittoqqortoormiit.

Diverse use of individual animals

Each season and each place have great variety if you know about nature and animals. A caribou, for example, consists of more than roast and pieces for soup. A caribou can be the source of very diverse dishes. In addition to the meat itself, a caribou can offer delicacies such as refreshing fly maggots from the inside of the animal's skins in winter, organs of high nutritional value and, not least, the most important Greenland vegetable – the caribou's half-digested stomach contents. The same applies to the other important quarries. For example, a walrus also offers mussels if you look inside its belly. And all animals, from fish and seals to seabirds and eggs, can be sources of diverse taste experiences when thoroughly fermented.

Throughout time, the geographical and seasonal food has been influenced by the possibility of going shopping at the supermarket. Today, rice, onions, potatoes and apples are thus important elements of traditional Greenlandic food, even though all of them but the potatoes need to be imported.

Cooking

In the past, as well as today, cooking of the food most often consists either of boiling or no cooking at all. Eating the food raw, dried, frozen or fermented has been important for the nutritional value of food, and it remains the most common way to eat much of the Greenland produce.

Today, many people consider a cold buffet consisting of mattak (whale skin), large and small dried fish and raw or fermented seal blubber, a fermented boiled cod with boiled potatoes, a bit of dried caribou and perhaps some apple boats the epitome of Greenlandic food. These foods will also often be the ones you see at the festive meals – the kaffemik.

Kaffemik

The kaffemik is a central part of Greenlandic culture, although it is a phenomenon that was introduced with the colonisation. Kaffemik is held on special occasions such as birthdays, weddings, christenings, confirmations, first days of school and funerals. At kaffemiks, people open their home

Cafe A21 in Narsaq serves food almost straight from nature: dried caribou meat, whale meat, blubber, cod liver, dried cod and turnips. A few traditional foods are available in supermarkets, but most require knowledge of the natural environment. Many families spend the summer hunting and collecting food for the year's festivities, so the best Greenlandic dishes can be offered to kaffemik guests.

METTE BENDIXEN/RITZAU SCANPIX, 2009

to everyone, even those they do not know. In return, guests are expected not to take up the chair and coffee cup for too long so that space can be made for as many people as possible. Guests are invited to a strong sensory experience, where especially the senses of smell, taste and sight are stimulated by fine cuisine and an always impressive number of cakes. The literal meaning of kaffemik is 'give me coffee'. Each family has their own way of hosting a kaffemik. No one does it the same way, and often the host gets new inspiration for the food from other kaffemiks. This is precisely what makes kaffemiks special. They are renewed and show the trend in society.

The elderly recall that in their childhood kaffemiks used to start quite early in the morning. The day in advance, it was the task of the children to grind the coffee, which was often made in a large saucepan. If anything was served beyond the coffee, it was Kalaallit kaagiat, the traditional Greenlandic cake, which was made in a coal oven that reportedly gave the cake an amazing flavour.

As with other parts of the food culture, kaffemik is closely related to hunting and nature. Nowadays a lot of time and energy is spent going caribou or muskox hunting with the aim of being able to serve a good roast at the kaffemik. Many families spend their vacations inside the fjords smoking the best looking trout, to be able to serve it at the family kaffemik. For many people today, the kaffemik is the main link to Greenlandic food.

Diet and sustainability

These years, diet and, in particular, a sustainable diet are high on the agenda. For the vast majority, that means less meat and more plants. Of course, such a diet is not easy to introduce in an area of the world where many of the vegetables have to be transported from afar, and where the culture is based on traditions surrounding the hunting and eating of animals. A sustainable diet in Greenland therefore looks different than it does in many other places in the world. For Greenlanders and other Inuit in the Arctic, the sustainable diet is illustrated in one of the most famous myths, the myth of Sassuma Arnaa – Mother of the Sea (see p. 156). The myth teaches us that if we do not treat our fellow creatures and our shared habitat with care and consideration, then the riches of nature disappear.

MUSEUMS OF CULTURAL HISTORY AND HERITAGE

Most main towns and some settlements have museums of cultural history. In addition to the Greenland National Museum, there are 23 museums along the coast, including two art museums. Most of the local museums are staffed by a museum manager and assisted by a few other employees.

Nunatta Katersugaasivia Allagaateqarfialu (The Greenland National Museum & Archives)

Founded as a national museum in the mid-1960s, the museum moved to its current location in the Colonial Port of Nuuk in 1978 after the old warehouses in the area had been renovated. The following year, with the introduction of Home Rule, it was included under the Greenland Home Rule.

The National Museum merged with the national archives in 1989, and the institution was given the official name of Nunatta Katersugaasivia Allagaateqarfialu. In addition to respon-

WERNER FORMAN/RITZAU SCANPIX, 2011

This distinctive storage bag from Ammassalik probably dates from around 1900. The sewn-on skin ornament inspired the logo used by Nunatta Katersugaasivia Allagaateqarfialu/Greenland National Museum & Archives since 1978.

ANINGAAQ R. CARLSEN/VISIT GREENLAND, 2020

ANINGAAQ R. CARLSEN/VISIT GREENLAND, 2020

The entrance to the museum area in Sisimiut is adorned with a whale's jaw as a testimony to the importance of whale hunting in the town's history. The jaws are from a baleen whale caught in the early 1900s. The museum area is characterised by a number of historic buildings from the colonial period. The museum has also built a reconstructed peat house, which gives an insight into colonial housing conditions for the Greenlandic population around 1900.

sibility for most of the Museum and Archives Act, the National Museum also has responsibility for the day-to-day administration of the Preservation Act. The museum is the legal authority for the management, monitoring and protection of all cultural heritage interests in two of Greenland's UNESCO World Heritage sites: the farming and cultural landscape of Kujataa (see p. 118) as well as the Inuit hunting ground of Aasivissuit – Nipisat (see p. 371).

With more than 70,000 registered items of cultural history, it is the largest museum in the country. A large number of items and collections of cultural and natural history are also kept in Denmark, including the collection of human material, collection of archaeologically excavated animal remains and bones and collections of fossil finds.

Local museums

A predominant part of the local museums have been established on the initiative of the townspeople – out of a desire to preserve, disseminate and collect artefacts. The museums are often located in the old part of the town and are housed in former churches, colony manager residences and other historic buildings.

Upernavik Museum

The museum in Upernavik was established and furnished in the early 1950s in the old church, which the district council had used as a chamber. In 1979, buildings in the old town were handed over to the municipality and the area was zoned for cultural purposes. After a new restoration, in 1997 the buildings were put into use as part of the museum. The buildings included the old church, the old shop, built in 1864, the manager's residence from 1832 (originally built at Imerissut around 1777), the cooper's workshop from 1848, the old school from 1911 and the vicarage from 1866.

The cooper's workshop was converted into an artist's residence in 2006, and since then, the residence has been visited every year by artists from all over the world. The municipality is responsible for the operation and selection of artists, and every visitor has handed over a work to the museum. Some artists have also held exhibitions or workshops for school children or the town residents.

Qaqortoq Museum

On 5 March 1972, the then Julianehåb Museumsforening opened its first exhibition in the old blacksmith's shop. In the mid-1970s, Anda Høegh built and furnished a traditional peat house in the backyard of the museum. In the autumn of 1997, the municipality took over the old colonial manager's residence from 1804 that was subsequently renovated, and in the spring of 1998 the museum moved out of the blacksmith's shop. The official opening of the museum in the colonial manager's residence took place on 21 June of that year.

Ilulissat Museum

Locals were behind the establishment of the first collections in Ilulissat in the late 1960s. The museum had its first board on 5 October 1972, and the association applied to the municipality to use the old vicarage where Knud Rasmussen was born in 1879. This was

FILIP GIELDA/VISIT GREENLAND, 2019

Maniitsoq Museum is housed in four listed colonial buildings from the second half of the 19th century. This beautiful two-storey house is the main building in which part of the collection is displayed. The timber houses originated in Norway, called »laftehus« and were used as living quarters for tradesmen. A great advantage of this type of house was that it could easily be dismantled and sailed to Greenland, where it could be rebuilt quickly.

where the premiere exhibition opened on 22 April 1973. A peat shed was built and the old whale oil refinery was moved from the port area and placed here. A hunting and fisheries department was inaugurated on 23 July 1994. Today, the museum today has several buildings with public access.

Narsaq Museum

The wish to have a museum in Narsaq initially came from local politicians. A museum committee was appointed in 1973 to work for the establishment of a museum.

The Narsaq Museum consists of several buildings located in three different locations in the town. Until the 1950s, the main building from about 1850 served as the town's first and only grocery shop. Furthermore, the museum has an extensive stone collection in a building constructed in 1848 as a storeroom and cold store. In addition, there is a residential house from around 1900 as well as a 1920 house that was originally used as a remote trading post manager's residence and later became Narsaq's first post office. Associated with the museum is also the town's first power plant from around 1950, a reconstruction of the town's last peat house as well as a memorial room for the priest Henrik Lund.

Sisimiut and Kangerlussuaq Museum

The association Sisimiuni Katersugaasivik was founded on 14 March 1974. The items collected were placed in Gammelhuset, the former assistant's residence from 1756. In November 1984, Gammelhuset opened after restoration. Today, the museum is housed in a number of historic buildings from the colonial period: the 18th-century whale hunting lodge, the missionary residence from 1759, the blacksmith's workshop from about 1900, the Bethel Church from 1775, the old shop from 1825, the Halfway House from 1844 and the colonial manager's residence from 1846. Furthermore, a peat house rebuilt in 1992 is attached to the museum.

In 2015, the local council decided to let the small privately started Kangerlussuaq Museum become part of the Sisimiut Museum, which therefore now has the name Sisimiut and Kangerlussuaq Museum.

Maniitsoq Museum

The Maniitsoq Museum opened on 25 October 1974 in the former town treasurer's office on Imeqvej. Prior to 1974, the museum committee in Maniitsoq had used venues like the settlement hall for exhibitions. The four old colonial buildings from the second half of the 19th century, which today house the museum, were slated to be dismantled in 1970 at their original loca-

ANINGAAQ R. CARLSEN/VISIT GREENLAND, 2020

Qaanaaq Museum, hunter's equipment. From the beginning, the local museums have shown a tremendous local commitment, reflected in the collection of local artefacts. This helps to define the shared history of a local area.

In addition to collecting knowledge and objects related to whale hunting in days gone by, the museum in Paamiut intentionally presents the recent fishing industry, which has had and still has great significance for the town. The photo shows two of the museum buildings.

ANINGAAQ ROSING CARLSEN/VISIT GREENLAND, 2019

tion on the waterfront to make way for an expansion of the town's fish processing plant. At the request of the local council, a project was prepared to rebuild the four buildings.

The buildings were an assistant's residence made of timber, an old shop, a half-timbered warehouse and a cooper's workshop and a workshop building made of stone. They were re-erected at Illunnguit in Kirkegårdsdalen in the late 1970s, and in early 1977, the first buildings were ready for use.

Aasiaat Museum

The Aasiaat local council established a museum committee in 1977, which would carry out the work of establishing a local museum in the town. Today it is based in the house of late member of the Danish Folketing, Frederik Lynge. Employees were hired on 1 May 1981, who were to be in charge of the registration of the collection and general operation of the museum. In 1989, the museum was given two more buildings, which are used as exhibition room and storage facility.

Qaanaaq Museum

The establishment of a museum in Qaanaaq started when a museum association was founded in 1979. The purpose was to collect items from the ancient culture and to try to get a building for the collections. The association's wish was that Knud Rasmussen's house could be moved to Qaanaaq and furnished as a museum. The house was originally erected in 1910, when Knud Rasmussen and Peter Freuchen founded the trading station in Cape York. The building was dismantled and re-erected in Qaanaaq during 1986 and inaugurated in March 1987. Since 1987, Qaanaaq has had its own museum of local history, and especially during the 1990s exhibitions were built with a special focus on the local area.

Qasigiannguit Museum

The museum is housed in Qasigiannguit's oldest buildings, dating from the town's founding in the first half of the 18th century. In 1981, the Royal Greenland Trading Department handed over the old telegraph manager's residence, the town's old salute cannons and flag bastion to the municipality. The telegraph manager's residence was taken into use as the first museum building. From 1988 also the manager's residence was handed over and is now the museum's main building. The so-called Poul Egede's House (the oldest European wooden building in North Greenland) underwent listing and renovation in the 1990s, and it was then used as an exhibition room.

Over the years, several deeds on buildings have been handed over to the museum, and the building stock has doubled with the inclusion of three more buildings – a peat house erected at the citizens' initiative, a self-built house and a storehouse of a former paint business. In line with the museum's development focusing on citizen participation and interactive dissemination, all three buildings have been included in the museum's activities.

The storehouse of a former paint business is used by the museum as a workshop for the municipally initiated historical interactive reenacting program, Nunaqarfik Uumassusilik/ Living Settlement, working in the workshop and in the open air with the dissemination of the everyday life of summer in the pre-colonial Inuit culture of Qeqertarsuup Tunua (Disko Bay) (see p. 90).

Paamiut Museum

Around 1980, the museum began its operation, and today there are exhibitions in two colonial period buildings as well as storage facility and workshop in several other houses. In the former colony manager's residence,

MADS PIHL/VISIT GREENLAND, 2016

Evening light over Nanortalik Museum and a section of the old town. Most of the local museums are centrally located down by the water. Like other local museums, the museum makes a point of presenting the ancient Inuit hunting culture. The hunting culture was in sharp decline from the 1950s with the increasingly dominant role of fishing. This decline acts as a catalyst for the creation of more local museums.

various local works are displayed, including carvings, photos (including slides) and music (sheet music). Moreover, the building has an interior consisting of locally made furniture. Dinghies, kayaks, tools and models are exhibited in the former shop building. One of the very few whale hunting costumes from traditional whale hunting in Greenland is one of the museum's largest treasures.

On the first floor of the same building there is a geological exhibition with rock samples from all over the country as well as cooper's tools, tubs and barrels.

Nanortalik Museum

A museum committee was formed in Nanortalik on 31 October 1977, and the municipality made the old hospital available. From 1981 to 1982, the old remote trading post manager's residence was restored and handed over, and it became the first home of the Nanortalik Museum, which inaugurated the building on 1 August 1982.

The Nanortalik Museum is set up as an open-air museum in the old colonial port. The museum includes almost all of the buildings, most of them being used as exhibition rooms. The open-air museum includes replicas of peat sheds, summer camp sites, skin boats with covered stands and kayak stands which can be used by the kayak association, as well as a kayaking roll practice ground. In addition, the museum has skin clothing, which is used at cultural events as well as at events for tourists.

Uummannaq Museum

The municipally owned Uummannaq Museum of local history opened on 21 July 1988. The opening featured Edvard Kruse's collection of old hunting implements, an exhibition on the German polar explorer and geologist Alfred Wegener, an exhibition on the now abandoned settlement of Maarmorilik, various tools from European whale hunting and a kayak exhibition.

Over several years, writings with varying contents were printed from the museum, such as historical surveys of buildings from all over the municipality, interviews with elderly residents, reports of archaeological surveys and special writings on the history of the settlements. All of these were produced and copied in the museum office and subsequently distributed to the population.

Ammassalik Museum

The Tasiilaq Museum Association held its first general meeting on 18 November 1981. In August 1990, the museum had its opening exhibition in the old church from 1908 which houses the Ammassalik Museum. In addition to the church, the museum consists of a chapel, a photo museum, a potato house and a peat house. The museum's main building is listed. The museum has a significant collection of items that shows the cultural heritage of East Greenland.

Qeqertarsuaq Museum

The original museum building is the old inspector's residence, later the bailiff's and doctor's residence, erected in 1840. Established in 1992, the museum was given another building in 2005, in which a so-called cold museum is designed for kayaks, boats and sledding equipment, as well as other major exhibits.

The collections are quite comprehensive and include originals and copies of Jakob Danielsen's watercolours, a fine collection of carving works, finds from several archaeological sites in the area, costumes and clothing, hunting implements, a cooper's and joiner's workshop, the bailiff's full-dress uniform, measuring instruments from and a picture exhi-

ANINGAAQ R. CARLSEN/VISIT GREENLAND, 2020

Harpoon gun at Qeqertarsuaq Museum. Around 1860, the harpoon gun was invented and, with the steamship, heralded a new era in whale hunting. It was now possible to hunt blue and fin whales, both of which are relatively fast species that had not previously been conspicuous prey.

REBECCA GUSTAFSSON/VISIT GREENLAND, 2016

Visitors to the Nuuk Art Museum explore an exhibition of works by members of KIMIK, the Association of Artists in Greenland. Located in a former church, the museum has permanent and temporary exhibitions.

bition on the Arctic Station (see p. 80) as well as an exhibition on whales.

A particular focus of the museum is the whale hunting period, the time of the town as the 'capital' of North Greenland and the life and work of Jakob Danielsen. In addition, there is a small room where the museum has a memorial room for writer and priest Otto Sandgreen.

Nalunaarasuartaateqarnermut Katersugaasivik

The Greenland Telemuseum opened on 20 May 2004 in the old radiotelegraph station in Qaqortoq. It is a historical museum, which aims to present the main thrust of telecommunications development in the country. The station is the only one left today of the first four radiotelegraph stations, inaugurated in 1925. From the beginning, it served as the main station, and all the telegraph traffic out of and into the country (except for Tasiilaq) passed through this station in the early years.

Kangaamiut Museum

The Kangaamiut Museum opened on 27 May 2011 with the exhibition *Images from Kangaamiut* by photographer Preben Harkamp. The museum focuses on arts, crafts and tales from and about Kangaamiut, and the current exhibition includes paintings, birdskin wall hangings and carved figurines made of ivory and soapstone. The artists behind the works exhibited all have ties to the Kangaamiut: Esra Berthelsen, Aron Berthelsen, Mads Kreutzmann, Peter Rosing and Lauritz Larsen.

The museum, which has been a department of the Maniitsoq Museum from the beginning, is housed in the old merchant manager's residence as well as a small stone house.

Nuutoqaq — Nuuk Local Museum

Located in Nuuk's beautiful colonial port, the museum is housed in and old boatyard. The museum opened in 2019, and the permanent exhibitions take the visitors back to the 1950s onwards where Nuuk was on its peak on becoming the capital. In addition, the museum has changing exhibitions.

Other local museums

Narsarsuaq Museum was founded in 1991 by local forces and an airport service that provided a building. The original collection consisted of a series of documents and photos from the Americans' time at the Bluie West One airbase (see p. 125) as well as items from the same period. Since its opening, material has been continuously collected which illustrates the history of the airport, and the emphasis has also been the Norse settlement in South Greenland.

The Ittoqqortoormiit Museum opened as a museum on 4 September 1997 but has never had an actual museum manager. In 1995, the Home Rule transferred the former office of Greenland Technical Organisation (GTO) to Ittoqqortoormiit Municipality. Changing employees from the municipality administration are in charge of the duties at the museum. From the beginning, the exhibitions have included furniture from a home in the settlement of Ittaajimmiit, photos from the first year of the town, as well as paintings made by the first doctor.

In Kulusuk, a private initiative has turned into a good attraction for both the place's own residents and guests. The museum is founded by and is based on a private collection. In 2012, a small building was restored and used as a museum, and it is still in this building you can see the collections of a family.

Art museums

In addition to Greenland's largest art museum, the Nuuk Art Museum, there are smaller local art museums in Nuuk and Ilulissat, and in Niels Lynge's House, many of his paintings are on display.

Nuuk Art Museum

Since Svend Junge's arrival in Greenland in 1959, Svend and Helene Junge purchased art, literature, stamps and more. It was the family's wish that this collection, which around 2003 was one of the largest private collections of Greenlandic art, should benefit the public. In 2005, they opened a private museum, and in 2006, they donated the collection and exhibition building to the then Nuuk Municipality. Today, Nuuk Art Museum is Greenland's largest art museum.

The collection at the Nuuk Art Museum includes sculptures, video art, watercolours, drawings, graphics and approximately 500 paintings. Most of the collection is made up of the gift from Svend and Helene Junge, including works by Emanuel A. Petersen as well as figurines made of soapstone, bone, wood and tusk.

New purchases of art were also made, and the museum has works by Pia Arke (see p. 202), Anne-Birthe Hove, Arnannguaq Høegh, Inku Silis-Høgh and many more.

Ilulissat Art Museum

Greenland's first art museum opened on 15 August 1995 in a colony manager's residence built in 1923. During the first years, the art museum was part of

To the left: Ilulissat Art Museum, interior. The art museum has an extensive collection focused on the painter Emanuel A. Petersen, who fell in love with the fascinating light variations of Qeqertarsuup Tunua (Disko Bay). Today, the museum has a varied art collection, and it has become firmly rooted in the local environment. To the right: Ilulissat Art Museum was opened as a museum in 1995 in the old colony manager's residence from 1923. The building, which is listed, was designed by architect Helge Bojsen Møller.

FILIP GIELDA/VISIT GREENLAND, 2019

FILIP GIELDA/VISIT GREENLAND, 2019

the Ilulissat Museum, from which the art museum took over an art collection with emphasis on the painter Emanuel A. Petersen. In 2004, a museum manager was hired for the art museum and the two museums were separated. Since then, the art museum has gained a strong entrenchment in the local environment; the fascination with early foreign artists has dimmed, and instead more emphasis is placed on contemporary art.

Niels Lynge's House

In 1990, the then Nuuk Municipality bought the estate of Niels and Karoline Lynge. The priest, poet and artist Niels Lynge died back in 1965, and since then, the widow Karoline lived in the house until her death in 1989. When the municipality took over the house, it was clear that the entire property and its contents were a piece of cultural history in their own right.

In cooperation with Nunatta Katersugaasivia, the contents were registered and the house furnished to be able to receive visiting guests. The house opened in 1995 as an exhibition venue, and it was the first museum owned by Nuuk Municipality.

Cultural centres and settlement halls

Most main towns have local settlement halls which are also used for other cultural purposes such as music venues, concert events and theatre. In February 1997, an independent cultural centre opened in Nuuk, Katuaq, which organises art exhibitions, conferences and meetings and concerts with artists from all over Greenland and the Arctic as well as the rest of the Nordic countries.

Taseralik, the cultural centre in Sisimiut, offers many kinds of experiences for everyone in the town, including films and talks, music, theatre and art exhibitions with both professional and popular actors.

The Sermermiut cultural centre in Ilulissat also has varied cultural events.

ASSOCIATION ACTIVITIES AND VOLUNTEERING

Greenland has a wealth of association activities borne by volunteers. In particular, there are many sports associations and activities for the socially marginalised groups, including especially children.

Social media make it possible to create and maintain contact across the vast geographic distances. Associations make extensive use of these platforms, also when it comes to fund-raising activities for victims of natural disasters or calls for demonstration.

The football pools and Lotto funds support non-elite sports as well as talent and elite sports. In addition, associations may apply for project funds for non-profit purposes.

Costs for the construction of sports facilities and cultural centres are covered jointly by the Government and municipalities.

The earliest associations

The country's first association, the Christian association Peqatigiinniat, was founded in Maniitsoq in 1907 and spread to the whole country. The purpose was to debate, on a Christian basis, everyday topics such as hygiene

ADAM MØRK/SCHMIDT HAMMER LASSEN ARCHITECTS, 2017

Katuaq, Greenland's Cultural Centre, is located in the centre of Nuuk. With its undulating façade and prominent skylight, the building symbolises the northern lights. Katuaq is an oasis in the middle of the town where people gather around art and culture. The house helps to display Greenlandic culture, as well as mediating cultural cooperation between Greenland and the rest of the world.

ANINGAAQ ROSING CARLSEN/VISIT GREENLAND, 2020

Football field in Qeqertarsuaq. In a number of towns, gravel fields have now been replaced by artificial grass fields. Football is played in indoor halls during the winter, but during the summer the outdoor fields are used around the clock.

and the resources of the country. The associations held two kinds of meetings, the spiritual ones that pertained to Christian life, and the worldly ones that dealt with the material circumstances of people.

The Aqissiaq sports association was founded in 1931 by Augo Lynge. During World War II, Lynge took the initiative to form the youth organisation Nunatta Qitornai. Nunatta Qitornai published the magazine *Inuusuttoq*, the goal of which was to revive the young people who were encouraged to take initiative and not hold back on new thoughts.

Everyone lived in small spaces, and the need for gathering places for enlightenment work grew. During the 1920s and 1930s, settlement halls were built in most towns. The settlement halls were all run on a voluntary basis and based on settlement hall associations. The settlement halls were frequently used for business, political and enlightenment purposes. This was the place of new ideas, of theatre and films, talks, dancing and partying.

Kalaallit Nunaanni Arnat Peqatigiit Kattuffiat (Union of Women's Associations in Greenland) was founded in 1948. The purpose was to provide mutual assistance, work for proficiency in domestic affairs, arouse interest in public work and to work for the maintenance of Greenlandic skins.

Sports associations

The first four sports clubs were formed in the 1930s in Qaqortoq, Nuuk and Maniitsoq. Ten years later, 11 sports clubs had been established, and by 1956, there were a total of 36. The goal in all sports was the Greenland Championships, held at fluctuating intervals of up to a few years.

The Sports Confederation of Greenland (GIF) was founded in 1953. There are about 130 associations with more than 9,000 members under the auspices of GIF. In the 1970s, sports halls were erected in the large towns, and the many clubs joined together in specialist federations. The year of its establishment is commonly indicated by the name, e.g. UB-68 (football), KS-69 (skiing), B-67 (badminton). Today, GIF has nine special federations.

Aalasa (get moving) is the motto of the confederation that has set itself the goal that by 2030, Greenland should be the world's most physically active country. This must be done, among other things, by strengthening the volunteer culture. The Make May Healthy initiative contributes to *this. Make May Healthy* is a movement facilitated by local volunteers, and the goal is to get as many people as possible to be physically active. Sports clubs, gyms, swimming baths, businesses, volunteers and others can set up events where attendance is free of charge. Social media and local newspapers are used for communication.

The various sports clubs organise club championships, regional championships and Greenland championships. Behind each event, volunteers work hard organising the event, finding accommodation, catering, etc.

The improved travel options have made it possible to participate in Danish championships, world championships and the Winter Olympics. However, the Unity of the Realm means that Greenlandic participants in most sports compete under the Danish flag. Since 1989, Greenlandic athletes have participated in the Island Games, the Olympic games for island communities, where competitors compete in 12-14 sports.

An athlete performs a 'one foot high kick' at the 2016 Arctic Winter Games in Nuuk. Arctic Sports has been done by Inuit for centuries. It is very audience-friendly and the competitions are followed with great interest.

MADS PIHL/VISIT GREENLAND, 2016

Arctic competition disciplines

Physical and mental strength was needed to hunt and survive in the Arctic. The training, which was mostly in the form of games, is today included in competitive sports. The Arctic Sport Greenland association holds national championships and participates in circumpolar competitions. Today, dog sledding and kayaking are competition disciplines organised into associations.

Football

Kalaallit Arsaattartut Kattuffiat (the Greenland Football Association) was founded in 1971, and the association counts 76 local clubs. This makes football the primary organised sport. Many towns have outdoor gravel fields, and several artificial turf fields have been established in recent years.

In the years after 1979, many mini halls were built, making it possible to play indoor football, among other things.

STEEN OLSEN/ARCTIC CIRCLE RACE, 2018

During the three-day Arctic Circle Race at Sisimiut, people camp overnight in tents. Temperatures can drop to -30ºC, but if tired skiers stick their heads out of the tent, they're often rewarded with the sight of the stunning northern lights.

Arctic Circle Race

The Arctic Circle Race, ACR, is known as the world's toughest ski race. 160 km must be completed over three days whatever the weather. The race, which first took place in 1998 and has since been held annually, attracts many people, both nationally and internationally. All the logistics around the race are handled by around 250 volunteers.

Marathons

Marathons and extreme races are held in several towns throughout the summer and autumn. Events are organised by local sports clubs or associations.

Associations within culture, nature and society

An old tradition of holding summer camps, aasiviit, where there was room for political debate, music, art, talks and practical activities, was resumed in the mid-1970s. The Aasivik movement (see p. 212) was based on the efforts of volunteers, and only in 1980 did the Home Rule begin to provide funding. An annual Aasivik was held until 1996, and the camps resumed in 2012.

Cultural festivals and theatres have been given institutional settings with cultural centres in Nuuk, Ilulissat and Aasiaat (see p. 174) and a national theatre in Nuuk (see p. 210).

Nature and environment

The introduction of self-government, the transfer of responsibility for nature and the environment, as well as subsoil and mineral resources, combined with the freedom to work towards independence, have been the backdrop for new voluntary initiatives. Citizens demand to be involved in decision-making processes, and several new non-governmental organisations, NGOs, have emerged. The World Wildlife Fund, WWF, has a branch in Nuuk, which is also home to the nature conservation associations Ocean North, Friends of the Nuuk Fjord and Avataq as well as an active group, Plastic not so fantastic. The Urani Naamik/No to Uranium society has several local departments. Knowledge of radioactive

EMIL HELMS/RITZAU SCANPIX, 2021

The environmental organisation Urani Naamik handing out flyers in Nuuk, April 2021. Uranium is found in several places, especially in South Greenland. The debate for or against extraction has been going on for decades. In November 2021, Inatsisartut (Parliament) passed a law banning uranium mining projects.

substances is collected and disseminated, and efforts are made to promote citizen participation. The society is behind a series of demonstrations against mining projects aiming to extract uranium.

Transparency International

In 2011, Transparency International opened a branch in Greenland. Transparency International Greenland works to actively prevent corruption and promote transparency, accountability and integrity. The association wants to ensure that economic growth benefits the whole society.

Greenland Guide and Scout Association

The Greenland Guide and Scout Association was founded in 1973, but the guide and scout movement has been active since the 1940s. In 2007, the Greenland Guide and Scout Association had 500 members, after which a decline in activity levels was seen. Today, the guide and scout movement, helped along by increasing access to social media, is restoring its active groups.

Voluntary social work

Over the years, the disadvantaged people of society have called for volunteering. A number of associations founded abroad today have independent local branches in Greenland. Similarly, associations have emerged in Greenland.

In 2012, the Self-Government established MIO, the National Advocacy Center working for Children's Rights, and in 2011, UNICEF opened a branch in Greenland. Both collaborate with the voluntary organisations.

Arnat Suleqatigiit

Arnat Suleqatigiit (women working together) emerged with inspiration from the Danner movement in Denmark in 1981 as a cross-party group set up with the aim of investigating the problem of violence between cohabitants.

JENS ASTRUP/RITZAU SCANPIX, 2016

The biggest Greenlandic cultural event outside of Greenland is 'Greenland in Tivoli', held every year on 31 July and 1 August. Here at the 40th anniversary in 2016, children from Uummannaq Music are getting ready to go on stage.

The need for a shelter for battered and endangered women soon became clear, and in 1983, the first women's shelter opened in Nuuk. Soon thereafter, women in Ilulissat, Maniitsoq, Sisimiut and Paamiut joined forces and formed local branches of Arnat Suleqatigiit as well as shelters. The shelters were predominantly based on volunteer labour. In Nuuk, the association dissolved in 2001 due to a lack of volunteers.

The Association of Greenlandic Children

The Association Greenlandic Children originated in Denmark in 1924. The background was poverty and tuberculosis, which made a big impression on a visiting delegation from the Danish Rigsdag. The association established and ran homes for orphans and children affected by tuberculosis. Later, nurseries came about.

Following the introduction of the Home Rule, the buildings were handed over and the association was changed to a project and development organisation. Today, the association is in charge of mentoring projects, the Najorti children's expert scheme and summer camps for children. The association has offices in Nuuk and Copenhagen. All projects are run by volunteers.

Meeqqat Inuunerissut /Better Lives for Children

The association Meeqqat Inuunerissut/Better Lives for Children is actively working to improve conditions for children. The association focuses particularly on the right to one's own body and the prevention of child sexual assault. The association was established around 2010. It is run by volunteers and conducts nationwide activities.

Nanu Meeqqat/Nanu Children

Through nationwide concrete projects focusing on vulnerable and socially marginalised families, Nanu Children works for increased social responsibility in Greenland. Moreover, the association distributes Christmas aid and organises Christmas parties for families with children.

Natteravne/Unnuarissat

Natteravne/Unduarissat aims to provide a greater sense of security. In the evening and night, adults walk around the towns with the purpose of being visible to children and young people who can contact them. Unnuarissat has existed in Nuuk since 2000 and in several large towns for shorter periods of time.

Save the Children

Meeqqat Ikiortigit – Save the Children Kalaallit Nunaat was founded in 2013. Save the Children is an international organisation and the starting point for the work is the United Nations Convention on the Rights of the Child. The goal of Meeqqat Ikiortigit is to empower Greenland's children, with a particular focus on promoting their well-being and development. The association speaks out in the public debate and runs a number of projects.

Red Cross

The Danish Red Cross has worked in Greenland since 1949, when they set up three orphanages. In the 1970s, the first first aid instructors were trained and local departments were established in several towns.

Established as an autonomous department in 1992, Kalaallit Røde Korsiat provides relief aid in both international and national disasters, trains first aid instructors and runs recycling shops in a number of towns.

NIELS NIELSEN/NATTERAVNENE I NUUK, 2019

The Nightwatch/Unnuarissat in Nuuk
Nightwatch groups do not intervene, but by their presence they contribute to the safety of children and young people at night. The Nightwatch groups are popular and are welcomed by the children, with offers of sweets and a chat.

ARKTISK INSTITUT, 1984

The Ikiu youth section aims to help socially marginalised young people carve out a meaningful life. One of the ways to achieve this is to build a culture of volunteering among young people.

Neriuffik

Disease sufferers are increasingly being treated in Greenland, however, some patients are referred for treatment in Denmark. In connection with illness, the language barrier can be challenge in that information is needed in one's own language. Some associations are actively working on this. The biggest one is Neriuffik, a nationwide organisation dedicated to the fight against cancer. Neriuffik is an autonomous division of the Danish Cancer Society and has departments in most towns. Grants are awarded to cancer sufferers and their relatives, informational materials are prepared and supportive talks are offered to cancer sufferers.

Nunatsini Inuit Innarluutillit Kattuffiat

People with physical and mental disabilities are covered by the UN Convention on the Rights of Persons with Disabilities and have its own spokesperson, Tilioq. Associations for the blind, deaf and people with autism have joined together in the Disability Organisations of Greenland, Nunatsinni Inuit Innarluutillit Kattuffiat (NIIK).

Qaamaneq

In 2019, the association LGBT (lesbian, gay, bisexual, transgender) Greenland was formed. The association presents itself as a network for young people who would like to promote and work for diversity. The association uses the media and social media to raise awareness, and it carries on the tradition of the annual Gay Pride held for the first time in 2010 in Nuuk and is now conducted in several towns.

Zion Church in Sisimiut was built in 1926 to replace the so-called Blue Church, Bethel Church from 1775, which is today a museum. The Bethel Church was the first church in Greenland that was built by the Greenlandic congregation. It was shipped from Copenhagen as a kit and paid for with 60 barrels of whale blubber. Zion Church also ended up being too small; it was rebuilt and expanded in 1984.

RELIGION AND RELIGIOUS COMMUNITIES

Christianity came to West Greenland on 3 July 1721, when Hans Egede, together with his family and a large entourage, settled on an island at the mouth of Nuup Kangerlua (Godthåb Fjord) (see p. 113). He came as a royally appointed missionary to reconnect Norway and Greenland and re-incorporate the country into the Danish-Norwegian Empire by imparting the Evangelical-Christian faith to the inhabitants. From the mid-13th century, Greenland, along with Iceland and the Faroe Islands, had been part of Norway, but by the early 15th century, contact between Norway and Greenland had ceased (see p. 102).

The West Greenland missions

During the 18th century, the Danish-Norwegian king's envoys laid out a number of towns evenly distributed along the west coast of Greenland from Upernavik in the north to Nanortalik in the south. From these new settlements, Christian teachings were widely spread. A Greenlandic written language was established and locals learned to read and at the same time were taught the Christian faith. Around 1810, almost all West Greenlanders were baptised and had learned to live a Christian life, attending church and reading *the New Testament* and Luther's Small Catechist in their native tongue.

The German-speaking Moravians (the Moravian Brethren) worked alongside the Danish-Norwegian state mission and, with the king's permission, they started their missionary activities in Nuuk in 1733 (see p. 115).

The Herrnhut established their first and largest mission station in Greenland, Neuherrnhut, 2 km south of the state mission colony of Godthåb (Nuuk). Here, traditional Greenlandic peat houses with imported European windows and chimneys are seen near the mission station on the far left of the photo. Lithograph by Lars Møller after a watercolour by Hinrich Johannes Rink, 1863.

METROPOLITAN MUSEUM OF ART, NEW YORK. LSA/ALBUM/RITZAU SCANPIX

ANINGAAQ R. CARLSEN/VISIT GREENLAND, 2021

The church in Qaanaaq is one of the northernmost church buildings in the world. The inauguration was in 1954, a year and a half after the population of the settlement of Uummannaq – close to Mount Dundas in the valley of Pituffik – was moved 130 km north to Qaanaaq in connection with the establishment of Thule Air Base. The church was extended in 1988 and has the world's northernmost pipe organ, a gift from Thule Air Base given in 1987. The altarpiece is a painting made for the church and shows Jesus in the company of Greenlandic children.

Over the next 130 years, they established a total of six mission stations in the Nuuk area and southwards all the way down to Cape Farewell (Uummannarsuaq). After 1864, relations between Denmark and Germany became tense, and against this backdrop, the Moravians abandoned their presence in Greenland. To the sorrow of the communities, they left the country precipitately in 1900. The Danish state mission took over the mission stations of the Moravians, and the communities were incorporated into what was soon to become the Church of Greenland.

Missions in East Greenland and North Greenland

In East Greenland, a mission station was established in 1894 on the island of Ammassalik. This was done on Danish initiative with the priest Frederik Carl Peter Rüttel as the first missionary in the place. His successor became the West Greenlandic catechist Christian Rosing, who had a degree from Ilinniarfissuaq (Godthåb College of Education). Christian Rosing was ordained as a priest in 1904 and then served as a missionary in Ammassalik until 1922. He baptised the last East Greenlanders in 1921.

A newly founded Danish ecclesiastical association, the Committee on Ecclesiastical Affairs in Greenland, established a mission station in Thule district (north of Cape York) in 1909 under the direction of the West Greenlandic catechist Gustav Olsen, who had been ordained as a priest that same year. He baptised the first Inughuit in 1912 and did not leave Thule until 1921. The last adult Greenlander in the district was baptised in 1934. Fearing the great power of the United States, it was not until 1937 that the Danish state dared to assert its sovereignty in the territory by taking over the church, school and medical services.

The mission in East Greenland and Thule made a great impression on the already Christian West Greenland community, and news from these areas were eagerly read, not least Christian Rosing's book *Tunuamiut* on the pagan customs of the East Greenlanders, which was published by Ilinniarfissuaq in 1906.

The Church of Greenland

The Church of Greenland came into being in 1905 when the Danish Parliament passed the Greenland Church and School Bill. During the negotiations in the Danish Parliament, it was perhaps for the first time officially stated that Greenland was no longer a 'pagan' country, but that the country's church was part of the Evangelical Lutheran Church in Denmark. From then on, Greenland was therefore no longer a mission field, and the priests were no longer to be called missionaries.

The former mission districts became parishes, a concept taken over from the Faroese Church. Today, Greenland is divided into seventeen large parishes.

At the head of the Church of Greenland was from 1905 a rural dean who was under the supervision of the bishop of the Diocese of Zealand, later the bishop of Copenhagen. From 1969, the Greenland dean was given the title of national dean.

The Danish Constitutional Act of 1953 emphasises that the Constitution of Denmark applies to all parts of the Kingdom of Denmark. It has the obvious consequence that the Evangelical Lutheran Church in Denmark is also the Church of Greenland.

When Greenland was given home rule in 1979 (see p. 135), the church was the first thing to be transferred, and the national dean became a vice-bishop. Not until 1993 did Greenland become an independent diocese with its own bishop. When Greenland gained self-government in 2009, the Greenland Church Act of 2010 established that the Church of Greenland, Kalaallit Nunaanni Ilagiit, was assigned to the Government of Greenland and at the same time would be part of the Evangelical Lutheran Church of the Unity of the Realm. About 95 % of the Greenlandic population are members of Kalaallit Nunaanni Ilagiit, which is funded by the Greenland Government. No church tax is paid in Greenland.

The country's seventeen parishes are grouped into three deaneries: Provsteqarfik Avannaa (North), Provsteqarfik Qeqqa Tunulu (Middle and East) and Provsteqarfik Kujataa (South). Each parish has a main town and one or more settlements. In total, the Church of Greenland has 67 church buildings with the Annaassisitta Oqaluffia (Church of the Savior) in Nuuk as the cathedral (see p. 188).

There are 25 priestly offices in Greenland, including the bishop and three deans. The Church of Greenland also makes one to two priests available to the Greenlandic community in Denmark. Long-term vacancies are common, especially in the outermost regions. In addition, there are approximately 40 part-time catechists with a short education and about 15 so-called 'readers' who are catechists with no education. There are organists at roughly half the churches.

Queen Margrethe's arrival for the ordination of Greenland's new bishop, Paneeraq Siegstad Munk, at Hans Egedes Church in Nuuk on Sunday, 10 October 2021. To the right of the Queen is former Dean, Isak Davidsen, who was knighted with the Order of Dannebrog in 2018. Hans Egede's Church was consecrated in 1971, the 250th anniversary of Hans Egede's arrival in Greenland. In the background, on the wall, are portraits of Hans Egede and his wife Gertrud Rasch. It has the only carillon in Greenland.

CHRISTIAN KLINDT SØLBECK/RITZAU SCANPIX, 2021

Most priests in the Church of Greenland have a bachelor's degree in theology from Ilisimatusarfik supplemented by a pastoral theological course organised by the church. In addition, the church also conducts organist and catechist education.

The catechists

For centuries, the catechists have played a big and important role both in the church and in the community. They originally acted as the mission priests' helpers, but have also often had ample opportunity to act more or less independently. There is no doubt that a significant part of the overwhelming success of the Greenlandic mission is due to the fact that the mission very early on allied with Greenland converts, leaving much of the mission work to them.

The catechists perform a number of pastoral functions in combination with teaching obligations and work assignments, which in Denmark are left to the parish clerks. Each parish spans large geographical distances that make it impossible for the priest to carry out all duties. The local catechists can therefore independently, under the supervision of the priest, hold church services without communion, conduct baptism ceremonies and funerals and teach candidates for confirmation. They do not undertake weddings and confirmations, nor are they allowed to hold communions without special permission from the bishop.

Before 1950, the church and school services in Greenland belonged together, and Greenland's dean was the school's chief executive. The catechists received thorough education at Ilinniarfissuaq and, alongside their ecclesiastical functions, often also acted as school teachers. When the church and school were separated,

Confirmation day in Nanortalik, South Greenland. The special Greenlandic costume is worn on this day of celebration, just like on other religious holidays: The men in white anoraks and black trousers, the women in intricately designed dresses, kamiks and beaded collars in complicated patterns.

MADS PIHL/VISIT GREENLAND, 2016

the responensibility for the catechist was transferred to the church and is today just a short course education.

Congregational representation

The Greenlandic counterpart of the Danish parochial church council are so-called congregational representations, which consist of elected representatives of the congregations in all towns and settlements, in addition to the local catechists and priests. According to Greenland Church Act, the congregational representations must 'provide for good conditions for the service of the Gospel' and must be consulted before a decision is made on filling a vacancy with a priest in the parishes concerned. Along with the priests of the diocese, the congregational representatives also have the right to vote in electing a bishop.

JØRGEN MALLING/ARKTISK INSTITUT, 2008

Interior of the church in the East Greenlandic settlement of Kulusuk (population 215), consecrated as a school chapel in 1924. The church was partly built from ship timbers from a shipwreck with the help of surviving crew members. The votive ship, seen to the left of the photo, was also made by one of the shipwrecked sailors.

Congregational life

The Greenlandic ecclesiastical tradition has two origins: the Danish state mission, and the German-speaking Moravian mission (see p. 115). The period of state pietism in Denmark formed the basis of the Greenland mission, and the later more rationalist religious trends have affected Greenland less than the rest of the Unity of the Realm. The country is therefore characterised by a special spirituality and liturgical tradition. At the ecclesiastical feast and in connection with the major life events – baptisms, confirmations and weddings – there is a tradition of wearing the national costume to church.

From the beginning, both the state mission and the Moravian mission placed great emphasis on speaking the Greenlandic language. At an early stage, Greenlandic hymns were composed, first by the foreign-language missionaries, later also by Greenlandic poets. Today, the church therefore has a rich repertoire of well-known Greenlandic hymns, translations of hymns from especially Denmark and Germany and hymns originally composed in Greenlandic.

Peqatigiinniat

The Christian association, Peqatigiinniat (those who want fellowship), is the country's oldest association. It emerged as a revivalist movement in 1907 under the inspiration of the Home Mission and Grundtvigian movements in Denmark and quickly spread throughout the west coast.

Members meet for Bible reading, joint prayer and hymn singing in settlement halls and private homes and also do voluntary social work. This purely Greenlandic language movement has had great importance in building national consciousness and church life, but has less impact in the present than it has had in the past.

The Missions to Seafarers

Indenlandsk Sømandsmission – the Danish Missions to Seafarers – is a Danish organisation that has worked in West Greenland since 1969, and currently operates three seamen's homes and one hotel. The seamen's homes serve as both hotels and Christian drop-in centres from which the words of God are preached. Based on the Missions to Seafarers, the Danish Christian movement, Kristeligt Forbund for Studerende (KFS), has started a student mission in Nuuk in recent years.

EMIL HELMS/RITZAU SCANPIX, 2021

The Catholic community in Greenland belongs from an organisational perspective to the Catholic diocese in Copenhagen. The only Catholic church building in Greenland is Christ the King Church in Nuuk, an important ecclesiastical and social meeting place for the growing foreign population, particularly of Filipino origin. Catholic masses are also occasionally held in other towns.

Religious communities outside the Evangelical Lutheran Church

With the amendment of the Danish Constitutional Act in 1953, Greenland had freedom of religion, and a number of Christian and non-Christian religious communities emerged outside the Evangelical Lutheran Church. Some of them disappeared again.

The Catholic Church

Catholic Christianity came to Greenland around 980 with the Icelandic speaking Norse who disappeared during the 1400s. In more recent times, the Catholics had missionary activities from the 1930s, and in 1972 they were able to consecrate their own church building in Nuuk, Christ the King Church. There are approximately 300 Catholics in the country, more than 80 % of whom are of Filipino descent.

The Pentecostal Church

Since 1953, evangelical free churches have worked in Greenland, primarily the Evangelical Free Church of Denmark, the Apostolic Church and Pentecostal communities in Denmark, Norway and Sweden. These three denominations joined forces in 2000 under the name Inuunerup Nutaap Oqaluffia (The New Life Church), which today is a Pentecostal church with some 13 communities around Greenland with approximately 500 adherents.

The Evangelical Christian Community
This religious community was established in Nuuk in 1970 by missionaries from the main free church in the Faroe Islands, Brøðrasamkoman, which is rooted in the British revivalist movement The Plymouth Brethren. Known for their mission house Ebenezer, the community purchased a larger building in Nuuk in 2021, which was named Naalakkap Jiisusip Oqaluffia (Church of the Lord Jesus). In addition to the ecclesiastical activities, since 1975, the Evangelical Christian Community has run the Naalersitaq summer camp at the head of the Ameralik Fjord to the delight of many families with children. The community is financially supported by fellow believers in the Faroe Islands and also conducts missionary activities elsewhere in Greenland. The community has approximately 100 adherents.

The Baptist Church
A Baptist priest from Independent Baptists in the United States has been conducting missionary activities in Ilulissat since 2007. Today there is a church building and a small community, Ilulissani Baptistit Oqaluffiat (the Ilulissat Baptist Church), supported financially by a group of American Baptist churches.

The Salvation Army
The Salvation Army commenced their activities in Nuuk in 2013. In addition to the ecclesiastical activities, the movement has extensive social activities.

The Seventh-day Adventist Church
The Seventh-day Adventist Church erected a church building in Nuuk in 1958 and was active in Greenland until 1992. Today, the former Adventist Church is an art museum.

The New Apostolic Church
The New Apostolic Church was established from 1989 in three towns in Greenland. Today, the religious community has only very few followers; occasional church services are held by an itinerant priest from Germany.

The Church of Jesus Christ of Latter-day Saints (Mormon Church)
The Church of Jesus Christ of Latter-day Saints had missionary activities in Greenland from 1998, but stopped their activities in 2014.

Jehovah's Witnesses
Jehovah's Witnesses have been active in Greenland since the 1950s. Today, the religious community has about 150 members and is represented with kingdom halls in five towns.

Bahá'í
The Bahá'í religious community came to the country in the 1950s and today has about 150 followers spread across eight towns and settlements, as well as a meeting house in Nuuk established in 1974.

BUILDING CUSTOMS AND ARCHITECTURE

Ever since the first Inuit immigrated, building customs and architecture have been characterised by life in a harsh nature. From the dwellings of early nomadic people to modern standard houses, weather conditions, rocky ground and permafrost have been challenges to overcome, and past times are reflected in today's modern architecture.

Paleo-Inuit, Independence and Saqqaq

In the Arctic cultural landscape, 4,500-year-old traces of early Inuit dwellings are still seen. Both building custom and architecture were adapted to the nomadic way of life and the Arctic climate. For thousands

The abandoned mining town of Qullissat was once one of Greenland's largest towns. When the mine closed in 1972, the residents were forced to leave, and the town is now used only as a holiday resort by the families who used to live there.

LINDA KASTRUP/RITZAU SCANPIX, 2015

CHURCH BUILDING

REBECCA GUSTAFSSON/VISIT GREENLAND, 2016

In the middle of snow-covered Nuuk rises the cathedral's tower-adorned facade. In the multitude of buildings, the Ombudsman's service residence can be seen, the large grey building that hides the long side of the church. The red two-storey building to the right of the church was the home of Greenland teachers' training college from 1847 to the early 1900s. Since then, the house has served as the residence of the national dean, vice bishop and bishop.

All towns and settlements in Greenland, even the smallest ones, have a church. The churches are very different; in the towns you finds large, often architecturally valuable churches, exquisitely decorated with altarpieces and other adornments. In the settlements, with a few exceptions, small towerless, red-painted wooden school chapels from the 1900s dominate, often decorated with only one or two edifying images. In many places, the church room still serves as a schoolroom during the week, and the altar is hidden behind a folding door or rolled aside. On Sundays, the school desks are pushed away and the room is made ready for the churchgoers.

CHURCHES IN NUUK

Nuuk has two large churches, the cathedral consecrated in 1849 and Hans Egede's Church consecrated in 1971. In the suburb of Nuussuaq, space has been reserved for a third church, but it is not known when it will be built.

Annaassisitta Oqaluffia, Greenland's cathedral

Annaassisitta Oqaluffia, Greenland's Cathedral, is the Church of Our Saviour in Nuuk. The elevated north-south-facing church (23 × 12 m), located north of the old town, is clad in red-painted, lap-boarded planks, and the pitched roof is covered with shingles. A white-painted portico with a triangular gable, columns and stately staircase forms the entrance. Above this is a small tower with slender, copper-clad pyramidal roof and a clock from 1928, which was the year of the town's 200th anniversary. Four dials are inlaid in the tower, so that you can see the time from all corners of the world.

The white-painted church room is panelled, and to both sides of the aisle are white-painted wooden benches. Instead of an altarpiece, there is a large crucifix flanked by two kneeling angels made of plaster of Paris. The altar silver, dated 1722, was donated by King Christian VI. On the south wall hangs a frequently reproduced painting by J.E.C. Rasmussen, which depicts the kneeling Hans Egede in a snow-covered landscape. Behind the church is a listed cemetery, where, among others, hymn writer Rasmus Berthelsen is buried.

Hans Egede's Church

A detached light-filled church at the southern end of the town towards Skibshavnen. The church, which has white walls and no tower, was designed by Ole Nielsen, who had previously designed the church in Aasiaat (1965). On the back wall hang pendant portraits of Hans Egede and his wife Gertrud Rask painted in Bergen in 1718 by an unknown artist.

THE EARLIEST CHURCH BUILDINGS

The first real church buildings outside Nuuk were erected in 1758 (Paamiut), Bethel Church in Sisimiut (1775) and Zion Church in Ilulissat (1779-83). Of these three 18th-century churches, the first one has been demolished, and the church in Sisimiut has become part of Sisimiut Museum. Only the church in Ilulissat is still in use, although it has been altered several times, most extensively in 1904 and 1929-31, respectively.

The eastern section of the building was initially used as a hospital, but in 1904 this part was merged with the rest of the church. In 1929-31, the large wooden building, which was located directly on the ground close to the water's edge, was dismantled, partially renewed and moved 50 m up in the hinterland. The Zion Church (22 × 10 m) is a wooden church built of heavy logs with exposed corner posts. The round-arched windows are painted white, and the chancel section now faces east. The church's high pitched roof is clad with shingles, and in the middle of the nave rises a large fleche with a pulled back upper floor and a pyramidal roof, which was originally covered in lead. However, during the war in 1807-14, the lead roof was replaced and the lead was distributed among the town's hunters who desperately needed it for bullets.

Inside, the church is clad in boards with a flat ceiling carried by two rows of white-painted wooden columns. The altarpiece with baptismal font and kneeler is adorned by a copy of Bertel Thorvaldsen's *Christ*. On the south choir wall hangs the painting *The Sacrament* (150 × 115 cm), donated by the Mission College in 1785 and originally used as an altarpiece. One of the distinctive features of this painting is that it renders the words of the sacrament in Greenlandic in an early version.

CHURCHES BUILT IN THE 19TH CENTURY

Time has been hard on the churches that were erected in the main towns during the 19th century. This is largely due to demographic change; the rapidly growing population meant that the churches became too cramped and therefore were demolished and replaced by larger churches one by one. However, the church in Upernavik from 1882 has survived as

KATHRINE KJÆRGAARD, 2003

Peter Gemzøe's picture from 1836 shows a nobly dressed Jesus reciting before his disciples the words, »Fear not, little flock! for it is your Father's good pleasure to give you the kingdom« (Luke 12:32 according to the contemporary translation). The picture by Gemzøe was bought by polar explorer Knud Rasmussen's great-grandfather, merchant Hans Mossin Fleischer, and donated to Uummannaq old church, where it served as an altarpiece. In 1935, it was transferred to the new church, where it is placed on a side wall.

a museum, while the church in Qasigiannguit from 1889 has been used by the scout movement since 1969, and for some time also as a place where you could 'park' your children.

Of all the 19th-century town churches, only the beautiful and minimalist Church of Our Saviour from 1832 in Qaqortoq is now used for its original purpose. The red-painted wooden church has a green, half-hipped roof with an octagonal fleche that has been added later. Inside, the ceiling is supported by two parallel rows of columns, which are painted light grey with red fields at the top. The altarpiece *Jesus Calms the Storm* was made in the late 1800s by the highly popular Anker Lund. Not far from the altarpiece hangs a lifebuoy from *Hans Hedtoft*, the only trace of the large ship, which shortly after having left the port of Qaqortoq wrecked and disappeared in the Denmark Strait off Cape Farewell (Uummannarsuaq) on 30 January 1959, there is also plaque with the names 95 people who died. The church has two large paintings, *the Baptism of Jesus* and *the Supper at Emmaus*, the latter possibly by Adam Müller, as well as three portraits of Bone Falch Rønne, Hans Egede and Poul Egede. Rønne founded the Danish Missionary Society (1821), which initially worked mainly for Greenland, and he donated the church to the then colonial town.

CHURCHES BUILT IN THE 20TH CENTURY

Of the large number of churches that have been built in the 20th century – in many cases to replace demolished churches from the previous century – it may be mentioned that from the first half of the century new churches were built in the settlement of Alluitsup Paa, and in Paamiut, where a red wooden church in Norwegian stave church style was built in 1909, as well as Sisimiut, which in 1926 got a red-painted church with room for no less than 400 church goers. The church in Sisimiut is built on a cliff overhang near the old church from 1775, with which it engages in an impressive aesthetic-architectural interaction.

A new church was erected in Qeqertarsuaq in 1914, and in 1935, a quite unusual church made of hewn boulders with a massive 15 m high square tower with pyramid tip was erected in Uummannaq. In the church hangs an outstanding painting in a gilded frame with Greenlandic inscription, *Jesus With His Disciples At The Sea Of Galilee*. The painting has recently been identified as the work of Eckersberg-apprentice Peter Gemzøe (1836).

The Danish(-Norwegian) dominance in church building and decoration remained virtually total until the mid-1900's when new ideas emerged also here.

Egede's Church, Aasiaat

The first significant example of this change is the church in Aasiaat. In the early 1960s, it was decided to build a new church to replace the old one from 1901, which had become too small. Danish architect Ole Nielsen was chosen, and the Greenlandic artist Jens Rosing was chosen to be in charge of the decoration.

With its low outer walls of almost all black-painted wood and its high, shingled, hipped roof, which protrudes above the flimsy wooden bell frame to the west of the church, Egede's Church is a significant landmark in the townscape, especially from the seaside. However, the real sensation at the consecration on 29 August 1965 was the interior of the church. A large, high-ceilinged church room with white brick walls and natural-coloured woodwork, where your gaze is naturally drawn towards the altarpiece – two white brick plinths with a wooden board in natural colour – no matter where you stand. Above you can see Jens Rosing's *The Creator and the Creature*, a grand relief made of glazed ceramic inserted into a wrought-iron frame and divided into four equal fields (231 × 118 × 5 cm).

Elijah Church, Maniitsoq

Jens Rosing's altarpiece heralded a style that has since been replicated, first in Maniitsoq (1981), then in Tasiilaq (1989) and in Qaqortoq (1996). In Elijah Church in Maniitsoq, which was built in 1980-81 based on the design of Bo Jørgensen in collaboration with Holger Mørch-Sørensen and Lise Sonne Nielsen, the inside of the church was decorated by the young artist Aka Høegh (see p. 202). Above the communion table made of round natural stone gathered from all over the parish, cast in cement and covered with a thick plank of driftwood, Høegh has placed a cross, also of driftwood, decorated with Lapland rosebay. The artist has stated that by using driftwood and Lapland rosebay, she wants to signal that Christianity arrived in Greenland by boat just like driftwood, and that it – like Lapland rosebay – has taken root throughout the large Arctic island. The baptismal font is a roughly hewn granite stone from Atammik on which Høegh has carved five fish painted in blue. The silver baptismal basin was donated in 1862 by Dowager Queen Caroline Amalie, who had great veneration for Greenland.

Tasiilaq Church

When a new church was to be built in the East Greenland capital of Tasiilaq in 1989, it was designed by the Danish architect Holger Jensen, and Aka Høegh was commissioned to decorate the church. The altar decoration is a large crucifix (132 × 106 × 6.5 cm) formed by sawn-out driftwood and painted with five red and black circles. Above this is placed a crucifix outline formed by narwhal tusk. The baptismal font is a large polished tree root of Siberian larch (height 79 cm, largest diameter 100 cm), washed ashore off Aka Høegh's hometown of Qaqortoq.

Gertrud Rask Church in Qaqortoq

Gertrud Rask Church in Qaqortoq was built in 1973 based on Ole Nielsen's design as an auxiliary church to the Church of Our Saviour from 1832. A 4 m high and 6 m wide painting made in 1996 by Kîstat Lund (see p. 201) dominates the north wall of the large white-painted brick church room. *Let the little children come to me* is painted with acrylic and airbrush directly on the wall. It shows a barefoot Jesus Christ wearing a blue robe placing his blessing hands on two Greenlandic little girls. The girls are in national costume and stand surrounded by other children and with their watchful mothers in the background. Above the magnificent scenery shines a golden arc of light.

The altarpiece, which almost disappears under the weight of Kîstat Lund's enormous wall painting, was made by Marie Haagen-Müller (1973) and shows 47 Greenlandic plants painted with great accuracy.

KATHRINE KJÆRGAARD, 2003

Jens Rosing's relief *Skaberen og skabningen* (*The Creator and the Creation*) shows the universe as a testimony to God. In the upper left field, the Savior is seen with a golden halo around his head and hands in a blessing gesture. In the next field you can see the sky with the Charles Warbler and the North Star on a cobalt blue background. The bottom left shows the sea and its bounty, symbolised by four codfish, and next to that earth, land with angelica, a well-known Greenlandic plant. In the same field, an almost invisible spider is also seen – a allusion to the town's name (Aasiaat: 'the place with spiders').

Paleo-Inuit lived in skin tents with a living area divided in two by a central mid-passage and a hearth. Later, a peat wall was added, where the floor area and entrance were below ground level.

INGE BISGAARD, 2022

of years, animal skins were used as a cladding material for walls, ceilings and plank beds; fur side against fur side, the skin also served as insulation material. Driftwood was used to support the roof structure, while the walls were made of stone and peat. The snow was a seasonal building material, and for heating, blubber and driftwood were used.

Early Inuit houses were mostly overlooking the sea. The floor plan of the dwelling was circular, the structures, lashed tent poles covered with animal skins, were light.

The dwellings had a stone-paved mid-passage and central cooking facilities with two side platforms. The hearth was bordered by thick flagstones placed upright around a horizontal slab that served as the base, and the heating took place with heated stones. Structurally, the circular dome shape is the most stable structure in all sorts of weather, and at the same time it is the fastest shape to heat.

Paleo-Inuit, Greenlandic Dorset

About 1,500 years later came the next big wave of immigration, bringing with it a building culture which was developed as the climate grew colder and snow volumes increased. With the colder weather, three different dwelling types were developed, depending on season.

The winter dwelling had peat wall covered with skins on the inside and with posts of driftwood/whale bone. A paved mid-passage structure stretched from entrance to rear wall. The lowered living area was square to oval and divided into three units; the mid-passage area with a central hearth, work table and a storage room for fuel, etc., and a sleeping area and working area on each side of the mid-passage area. These dwellings thus had a three-level structure.

In the spring, when daylight returned, and before the thaw set in, time had come to take off to the hunting grounds. The roof of the winter dwelling was removed to give the dwelling a much-needed vent, and when the people returned for the winter, a new roof made of driftwood, animal skins, and then peat and stone was installed.

The early spring voyages to the hunting grounds required the construction of temporary snow houses along the way. The suitability of snow as a building material varied greatly from place to place.

By summer, resources were more dispersed, and it was time to split up into family groups and head out to the hunting grounds near the archipelago. The settlements were carefully selected near a protected harbour and favourable hunting conditions. In all areas of the country, the summer dwelling consisted of a mobile skin tent bordered by a low rock mound that offered protection against the wind.

The layout of dwellings with a central mid-passage became the common architecture for almost 4,000 years and disappeared around the year 1350.

The Inuit culture

The last major immigration from the north happened around 1150 with the Inuit culture that had developed a zoned dwelling adapted to a colder climate. Gradually, vessels and implements had been developed which enabled whale hunting, and the large animals were a source of new building materials as well as oil for heating.

Building customs were closely aligned with the season, hunting opportunities, and generations' accumulated knowledge of what worked in different weather conditions.

The winter dwellings of the precolonial Inuit culture varied in their architecture according to the regional conditions, but the architecture was much the same as in the past, though now with a level division of living area, floor area and entrance area.

Sometimes a small extension used as a kitchen annexe was found just inside the entrance. Plank beds were covered with skins laid on mats made of whale baleen strips, preventing the skins from getting wet.

Lighting and heating of the dwelling consisted of blubber lamps with a wick of moss, over which the food was also prepared. In the facade, the dwellings had a window that was covered with a stretched expanse of stitched gut skins.

Traditionally, each summer people gathered at selected summer gathering places, aasiviit (see p. 105), or festival houses. The traces from the gathering places show rectangular building facilities, which vary from 8 m to 43 m in length and with a width of about 5 m. The building orientation is north-south with the longitudinal axis parallel to the coastline and thus views towards the sea. The stone blocks walls are up to 1 metre high, and the entrances are located at the gables.

The buildings had a 1 metre wide mid-passage marked by flagstones placed edgewise – in some places mid-passages were filled with pebbles and then covered with flagstones. The buildings could house up to 30 families arranged in pairs opposite each other throughout the length of the building. Nothing indicates that the buildings were roofed, and presumably, they were used for ceremonial events when weather allowed.

The building structures of the Inuit culture evolved from a cloverleaf shaped floor plan, landscaped alone or in clusters, to types with a rectangular, square or even trapezoidal outline.

The later house types, such as the longhouse and the square house, were given a saddle roof used especially on the east coast, where large amounts of snow traditionally fall.

The longhouse, which can be more than 20 metres long, shows a substantial break with both architecture and social patterns. The floor plan of the longhouse resembles the layout of the summer gathering facilities with a long row of family apartments.

The architecture of the longhouse is believed to be rooted in the 18th-century whale hunting industry. First,

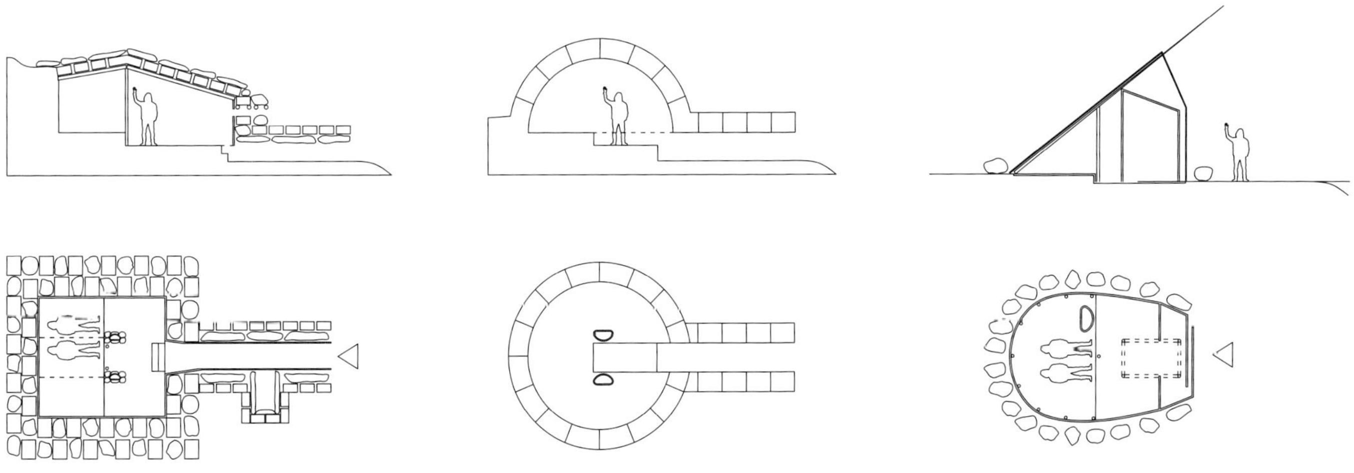

Winter dwelling — Winter/spring dwelling — Summer dwelling

Paleo-Inuit developed three different types of housing adapted to winter, winter/spring and summer, respectively.

INGE BISGAARD, 2022

Dutch ships arrived each year in February or March to conduct whale hunting on the West Coast and exchange European consumer goods for whale and seal blubber, furs and walrus and narwhal tusks. They started to gather in large settlements at the colonies and created housing communities, which had advantages, both in terms of housekeeping and whale hunting. It was no longer necessary to go on long hunting trips, and as a result, the nomadic way of life gradually disappeared and urban communities started to form.

The Norse

With the Norse immigration in the Viking Age from the late 900s to about the 1400s (see p. 94), settlements were established in Greenland by a northern European people. The Norse mainly settled as farmers, and the settlements were usually located at the head of the deep fjords and valleys. The sizes of the farms were very diverse, and ranged from large building facilities to single farms.

Building customs and architecture after colonisation

The pattern of settlement after colonisation gradually changed from a dynamic to a more stationary life form. As the country developed, the settlements grew from being small colonial sites to form actual urban communities. With the missionaries' settlements from 1721 (see p. 113) and the subsequent arrival of traders, the first colony on an island in the mouth of Nuup Kangerlua (Godthåb Fjord) became a place where people came from afar to trade.

The trade included all public construction, such as churches, schools, shops, official residences, hospitals, fish and salt warehouses. The house types included stone houses, timber houses and board-covered half-timbered houses, all inspired by Norwegian-Danish building customs.

The early colonial buildings were formed and carved so that they were ready to be assembled on the building site without the use of nails and fittings. The building foundation consisted of natural stone without binder, which prevented the timber from rotting. Therefore, in many public buildings erected before 1900, only wooden dowels and nails were used to assemble the main structures, while hand-forged nails were used for fixing cladding, brackets, etc. When a building no longer served a function in the country, it was carefully disassembled, moved and reassembled to serve a new function.

From peat house to wooden house

At the onset of colonisation, the peat house and longhouse were still in use, but with the advent of the mission and trade, the traditional dwelling gradually changed.

The most common family home of the early colonial era was a clone of the Norwegian wooden house and the Greenland peat house. The house, a one-room dwelling with a living area, kitchen and plank bed, evolved over the years. First, glass windows were installed in the façade, then the long house mid-passage was replaced by a shorter and taller storm porch, and then one or more of the outer walls were made of boards. Eventually the house had a high pitched roof, a saddle roof just like the houses of Europeans. Greenlanders' peat huts, skin tents and longhouses were slowly replaced with these wooden houses.

Many of the new self-built houses were of very poor quality compared to the traditional building customs. As a result of inadequate guidance in building wooden houses, the self-

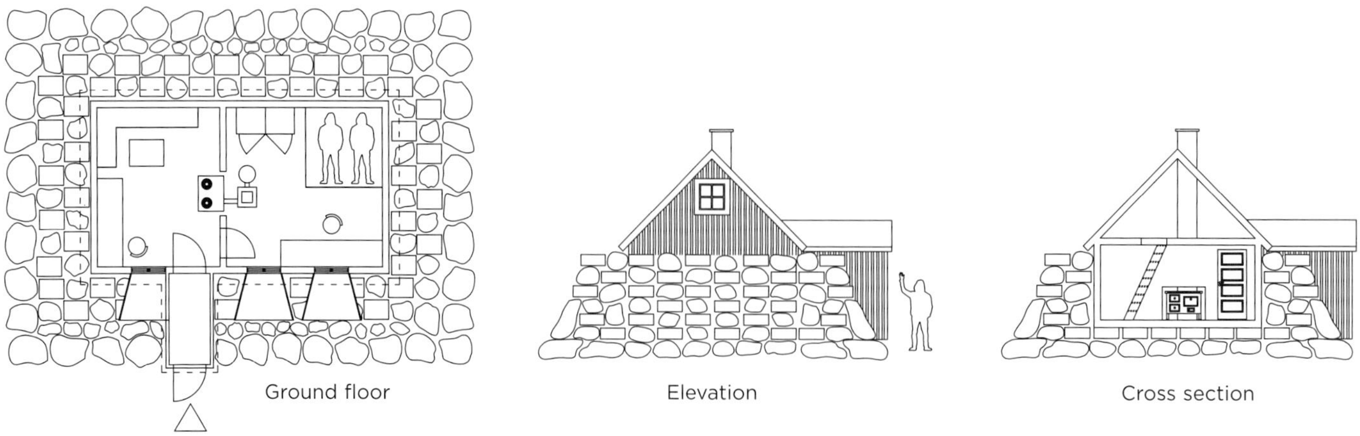

The most common family home of the early colonial era was a copy of the Norwegian and Greenland peat house.

INGE BISGAARD, 2022

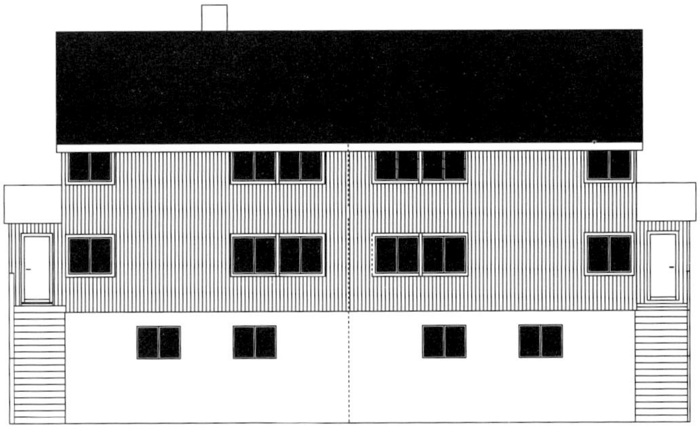

Elevation – facade

Elevation – gable

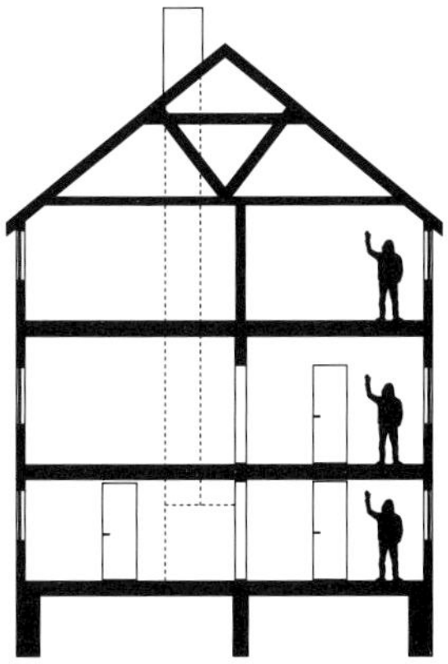

Cross section

built houses were cold and draughty and generally in poor condition.

The most widely used house paint in early times was the cod liver oil paint, which dates back to the Norwegian cod liver oil recipe. In the Greenlandic variant, seal blubber was used as a binder and colour pigments were added along with crushed resin. The house colours of the colonial period are mainly red ochre, ochre-yellow, white, blue, green, black and brown. Based on the colour of the houses, the various public buildings can be roughly identified.

Stone wall houses

At the beginning of the 19th century, a distinctive building custom emerged in South Greenland, where the Danish-Greenlandic house took on a new look. The peat wall was replaced by a sheer stone wall. The chairman of the trade, the Norwegian Anders Olsen, moved to Igaliku with his Greenlandic wife and their children to run a sheep farm, and here they built their house from the stones from the ruins of the Norse. With small distinctive stone walled houses made of the red Igalikusand stone, Anders Olsen created a distinctive built heritage which has been partially preserved in the settlement.

Modern materials from the United States

During World War II, Greenland was cut off from Denmark (see p. 124) and became an important staging point for Allied air transport from the United States to England. The missing supplies from Denmark were replaced with new ones from the United States.

During World War II, US airbases were built with the best developed technical aids, and in the aftermath, self-building gained new breeding ground. New materials such as concrete, plywood, corrugated sheets, insulation, paint, etc. became available, and at the same time, the number of peat houses was further reduced.

The years after 1950

Better housing was on top of the agenda of the Greenland Commission of 1950 (G-50). The new houses had to be of good quality, both in terms of construction and insulation. The living space was to be larger, and the kitchen, living room and toilet room had to be separated. Houses had to be heated by a coal stove or tiled stove, but there were no water installations.

Denmark devoted more money to an improved infrastructure, and the population became more concentrated in the towns. With the abolition of the monopoly trade, several shops gradually appeared in the townscape. Under the Ministry of Greenland, the Greenland Technical Organisation (GTO) was established, bringing together the country's construction and operation activities.

During the war, Greenland had a reasonably good economy, based mainly on the extraction of cryolite and cod fishing. American air bases and buildings were constructed in several places in the country. This sketch shows the former trade inspectorate in Nuuk.
Inge Bisgaard, 2022

This resulted in frenetic building activities in the years after 1950. Quayside facilities, water supply plants, warehouses, power plants, shops, hospitals, schools, sanatoriums, fisheries facilities and official residences were constructed. Fishing and sheep farming intensified. Scattered settlements were the main obstacle to the financial gain of fishing, and by tempting the population with lenient loans for building houses, the population was pooled into larger urban areas.

The GTO carried on the house colour traditions of the 19th and 20th, but now the different colours were used for the residential houses, giving the settlements life and variety and helping to identify who lived where. The picturesque urban areas with the house colours red, ochre yellow, green and blue can be seen in many levels in the hilly landscape.

The years after 1960

A new committee, the G-60, was established, and its overall goal was to depopulate the settlements and concentrate the population in four open-water towns, Paamiut, Nuuk, Maniitsoq and Sisimiut.

Standard house construction developed rapidly, and by the mid-1960s, houses were erected in many places with one and a half floors. They could be provided with central heating systems, hot and cold water installations, bathrooms and utility rooms/boiler rooms. Urban settlement plans were now intended to accommodate as many homes as possible, though for fire reasons with due distances.

Building in the Greenland landscape presented great challenges, and for reasons of establishing access roads, drainage conditions and suitable land plots, many single-family homes had disproportionately high concrete foundations.

The early house types had a small entrance hall, but later the classic storm porch was added as a wind lock or cold trap. A dream house in the 1960s consisted of separated kitchen, living room and toilet, a habitable attic, storm porch and entrance hall, and water and a gas stove installed. Due to prevailing wind and snow conditions, the inward-opening exterior door was located in the south or north facade. A good view of the sea was – and still is – weighted almost as highly as the living room and kitchen.

Due to the hilly rock landscape and subsurface permafrost layers, it is a major economic challenge to install water supply and sewer lines in scattered single homes. If you wanted these facilities, it had to be done through rationalised construction.

The following years, many multi-storey concrete element buildings were constructed. In 1965, the longest block of flats of the Unity of the Realm, colloquially called Block P, was erected in Nuuk. The 200 m long five-storey balcony access building was Greenland's first prefabricated building. By the standards of that time, the building had the comforts of modern times: light, air, rubbish chute, running water and heating, modern toilets and kitchens. The following years, many prefabricated buildings were constructed along the coast, though none as voluminous as Block P.

In the late sixties came a new series of standard houses in four different sizes, based on different principles than before. The standard house went from half-timbered structures to frame structures. The drawback of the houses was that they were too tight, often resulting in rot and mould in wall and roof structures.

The years after 1970

Around 1970, the high-density low-rise buildings started to leave their mark. The up to three storey high buildings were adapted to terrain conditions and emphasis was placed on variety and good common areas.

While urban house construction was in full swing, the opposite was the case for improving the standard of housing in the settlements. In the early 1970s, a self-build house was introduced, which was supplied as an assembly kit with drawing illustrations in Greenlandic, allowing anyone to build their own house. The self-build house became very popular, especially in the settlements, and over the years, it was revised and modernised.

The building regulations took into account local conditions, for example that houses on the east coast are exposed to severe Piteraq storms (see p. 283) and must therefore be designed to withstand these.

In 1977, the country's first act on land use, urban development and housing development was adopted. The act was based on the prior legal

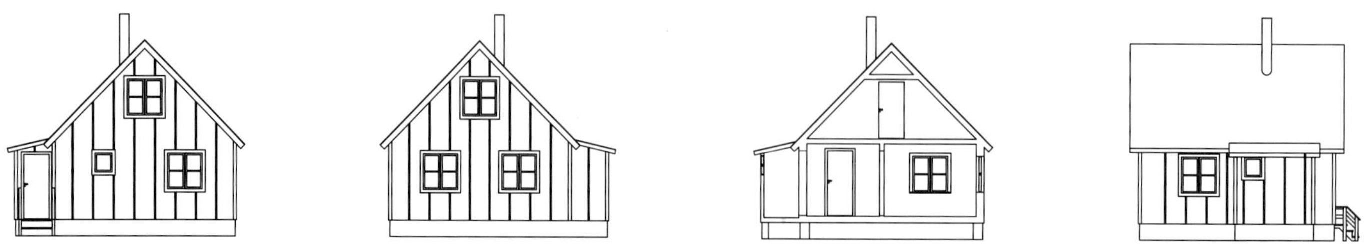

In the 1970s, the Greenland Technical Organisation introduced enhanced self-built houses, which hunters in the settlements could buy and build with financial aid.

INGE BISGAARD, 2022

MARINE GASTINEAU/RITZAU SCANPIX, 2018

Australian graffiti artist Guido Van Helten created this work at Block 5 in Nuuk in 2014. The subject is a portrait of the hunter Pooŋojorteq taken by W. Thalbitzer in 1906.

condition that land cannot be purchased and that permission must be obtained to take an area of land into use. A land planning committee was established and it was agreed, among other things, that Greenlanders, through the land planning committee, should have a greater influence on the land management of the country. In the ensuing decade, municipally owned service houses were erected in most settlements.

The purpose of the service houses was to provide people of the settlement with access to any facilities they might lack in their homes, such as laundry facilities, shower facilities and facilities for large events. At the same time, the service houses had housekeeping workshops and more.

The years after 1980

The cooperative housing movement saw the light of day in the mid-1980s. The interest was based on a desire to ease the pressure on the public construction funds for housing. In addition, the idea as to increase private savings, create more resident influence and responsibilities and to reduce the cost of operation and maintenance. With thoughts of social community and common facilities, the cooperative housing idea fits nicely with the Greenlandic building customs.

Greenland's first housing company was created in 1994, and it today manages much of the country's rental housing. The homes are either multi-storey buildings or high-density low-rise buildings. Each home is provided with kitchen and cooking recess as well as shower, toilet and storage room.

The years after 2000

The country's first capital strategy was presented in 2016. It forms the basis for the development of the capital. Nuuk is experiencing increasing relocation from the coast and the infrastructure is undergoing a massive expansion. The capital strategy sets out for Nuuk to be a sustainable pioneering town in the Arctic. Renewable energy, which has so far been hydropower plants, needs to be developed, and other options, such as solar cells, hydrogen and geoenergy, waste heat from industrial and waste incineration, must be exploited and expanded in the future.

In the urban landscape, rows and rows of high-rise buildings sprout up. The large panoramic windows often face each other, and only a few overlook the town and sea. The building culture is the most visible part of cultural heritage. Despite the acquisition of other external materials and architectural trends, the building custom has been continuously adapted to Greenlandic culture. It forms the framework of a life in a harsh nature and has deep roots. Today's modern cooking islands and sociable kitchens are not new in traditional Greenlandic kitchens.

VISUAL ARTS AND CRAFTS

The word for art, *eqqumiitsuliorneq,* is first seen in 1871 in a dictionary by Samuel Kleinschmidt, where it translates to 'one who often does uncommon things: an artist'. The word can be translated directly into *making something remarkable* or simply *art*.

The Greenlandic concept of art is first seen in the middle of the 19th century. Until then, traditionally, beautifully ornate objects were made, and the decoration itself has been entrusted with a function; it was believed that the decoration made the thing better and that a seal would rather be hunted by a beautiful harpoon than by an ugly harpoon.

From the middle of the 19th century, the functionless art appeared on the scene, and as the art community has

The work *Bag maskerne* (*Behind the masks*) by Aka Høegh, 2008. Her works include paintings, sculptures and installations. Since the 1980s she has been one of the driving forces behind Greenlandic visual art. Høegh was involved in establishing the Grafisk Værksted (Graphic Workshop), later the Greenland School of Art, and sat on the board of the Foundation for the Establishment of Greenland's National Gallery of Art.

IVARS SILIS, 2021

Photo from the exhibition Lights ON – Lights OFF by Bolatta Silis-Høegh (2014). Through the works, Silis-Høegh expressed the pain she felt when Inatsisartut lifted the zero tolerance on uranium in 2013. The majority of the works in the exhibition were large self-portraits. A selection from the exhibition is included in the Greenland National Museum's art collection.

RUBEN BLÆDEL, 2014

grown stronger, the desire and need for art institutions has also grown. From this time, there has been a tendency for Greenlandic artists to have had someone who has taken on some kind of mentoring role, including by sourcing outside materials and buying up the artist's works. Before 1970, the artists were predominantly male, but then a shift occurred and today, the majority are female.

Visual arts

Today, there are small municipal art museums in Nuuk and Ilulissat. Both are based on former private collections, are created on private initiative and are today run by the municipality. Greenlandic artists also frequently exhibit their works in Greenland's cultural centre Katuaq in Nuuk, in the cultural centre Taseralik in Sisimiut, in local museums and in some companies offering the opportunity to make exhibit in their premises.

Exhibitions outside Greenland are usually in Denmark at institutions that primarily work with Greenland and not with art, for example in De Grønlandske Huse (the Greenlandic Houses) in Denmark. It is a challenge that art is viewed from a Greenlandic perspective and efforts are being made to create a national art museum.

Art in Greenland has always been and still is such a small field that it is borne by quite a few people both at the institutional level and at the artist level. The art very much relies on people and is therefore difficult to put into periods and trends. Today, the Greenlandic art world is made up of just over 25 professional artists, and this is probably the largest number ever.

Early visual arts – anthropological collection

In the transition from functional decoration to functionless art, the concept of *eqqumiitsuliorneq* arose, and in this early phase, art consisted primarily of drawings, watercolours and prints.

One of the first visual artists was Aalut Kangermiu, Aron from Kangeq (see p. 152). His works include watercolour, wood carving and drawing, but also written texts. Aalut Kangermiu was a catechist of the Moravian community in Kangeq and could both read and write Greenlandic, and then he was a hunter. He fell ill with tuberculosis, and during one of the periods when he was bed-bound, the Greenland explorer and civil servant H. Rink requested Aalut to write down Greenland legends, myths and ways of life and *'Especially, moreover, anything that could serve as amusement or entertainment for Greenlanders'* as stated in the announcement that H. Rink sent out to all of West Greenland in 1858. The request included both writings and pictures.

The invitation testifies that the purpose was to gather knowledge for the Greenlanders themselves. Aalut's works were therefore a kind of commissioned works with a defined purpose; the collection of local history was to help build a national identity (see p. 154).

Aalut Kangermiu shares history with, among others, Jakob Danielsen in North Greenland, who also created art from his bed, and in East Greenland, it was Kârale Andreassen who depicted everyday life, hunting, shamans' journeys and myths. They all relied on good connections and materials from abroad, and none of them

GREENLAND PAINTERS

Greenlandic art is influenced by artists who have travelled and explored the country for extended or short periods of time. They are often referred to as expedition artists or Greenland painters, and a lot of the art still being made in Greenland is rooted in this tradition.

The first Greenland painters were sent by the Danish State in the early 19th century. They travelled to make a visual documentation of the expeditions, and they often served as teachers for Greenlandic artists for short periods of time. These mainly Danish painters often created romantic depictions of nature, and to draw attention to the Danish colony in the north, their works were exhibited in several places in Europe.

Artists in the period before Hans Lynge and Jens Rosing were particularly inspired by these painters, and they picked up on their style.

Greenlanders on their summer voyage from 1878 by J.E.C. Rasmussen is owned by the Nuuk Art Museum and was purchased by the couple Svend and Helene Junge Pedersen shortly before the museum opened in 2005. In the background of the work is Nuuk's landmark, the Sermitsiaq mountain.

NUUK ART MUSEUM, 2017

were trained in the art profession. The works were not necessarily collected because of their artistic qualities, for example very naive drawings by Andreassen were collected from when he was only 13 years old. The works thus lie in a field of tension between anthropological collections and art. No women were registered in visual arts until around the mid-1900s.

From landscape painting to abstraction

Aalut Kangermiu was followed by a growing number of artists, and in terms of materials, the works extended to include oil painting on canvas. It was still primarily catechists, priests, teachers and other relatively well-off Greenlanders who created the pictures. Nature was the predominant motif, and this genre includes Steffen Møller, Peter Rosing and Niels Lynge. It continued to be mainly men trained and employed in professions other than art.

After this time, when nature painting took centre stage, a transitional period followed characterised by larger abstractions and more impressionistic art. There was more experimentation with the motifs, and the first female artists began to establish their presence. In Kîstat Lund's landscapes, the goal is no longer to reproduce the seen or to create a tale; the works stand as an interpretation of moods and play with colours. Another significant profile from this period is Thue Christiansen, who is known as the artist behind Erfalasorput, the Greenlandic flag (see p. 14).

Only in the latter half of the 1900s did artists begin to educate themselves in the field. There was no such thing as actual academic education, but more and more were taking courses, including at the Royal Danish Academy of Fine Arts, including sculptor Simon Kristoffersen, who in the years around 1967 had a short stay at the Academy of Fine Arts in Copenhagen and was awarded the Academy's silver medal for outstanding artistic work.

Danish artist Bodil Kaalund was in Greenland for the first time in 1968, and she came to serve as a lecturer and mentor to a number of Greenlandic artists, among others, she helped artists such as Hans Lynge, Aka Høegh and Anne-Birthe Hove make contact with art academies.

Kaalund has been behind a number of international exhibitions by Greenlandic artists and published the largest work to date on Greenlandic art in 1979. In addition, she co-initiated the establishment of Grafisk Værksted (Graphic Workshop) in Nuuk, and she was active in the work surrounding the founding of a national art museum.

The Art School and institutionalisation of art

The first Greenlandic art education, Graphic Værksted (Graphic Workshop), opened in 1972. The school, today called Eqqumiitsuliornermik Ilinniarfik, Greenland Art School, was initiated by Hans Lynge and Bodil

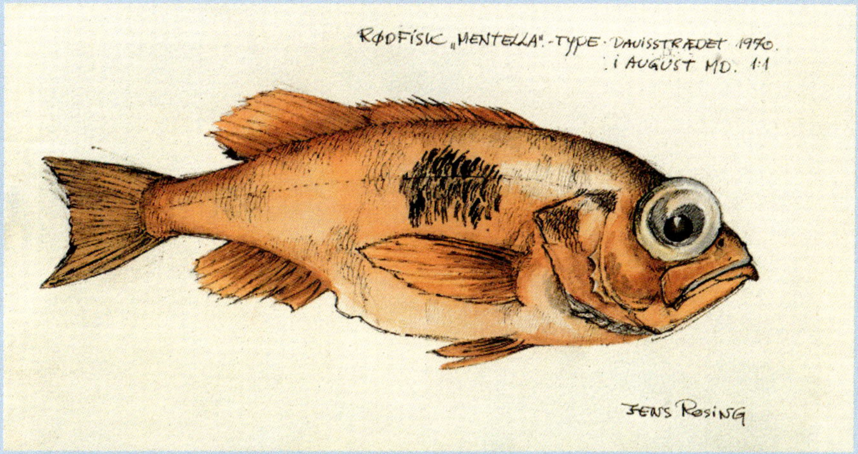

JENS ROSING, 1970

Redfish, »Mentella« Davis Strait 1970, Jens Rosing.

HANS LYNGE AND JENS ROSING

Hans Lynge (1906-88) and Jens Rosing (1925-2008) stood out from previous Greenlandic artists in that they did not depend on having good connections, they were able to obtain materials themselves and were able to create the art they wanted. They both graduated from the academy, and art was their only profession.

Inspired by his father, Niels Lynge, Hans Lynge too was interested in nature romanticism. His works are characterised by distinct brush strokes, large colour surfaces and abstraction. The motif is often nature, people, legends and myths, but unlike previous works, these works are more reflections and memories than actual tales. Lynge liked finding inspiration outside Greenland, and Greenlandic artists still find themselves in this tension field between the local and the international spheres.

Lynge and Rosing's professionalisation allowed them to do decorating assignments (see p. 191), book illustrations and exhibitions. It marked the beginning of a national awareness of Greenlandic art, which, among other things, fostered the idea of professionalising it further. Focus was turned to the value of having an own environment versus being trained in the European way of thinking and expression taught at international academies.

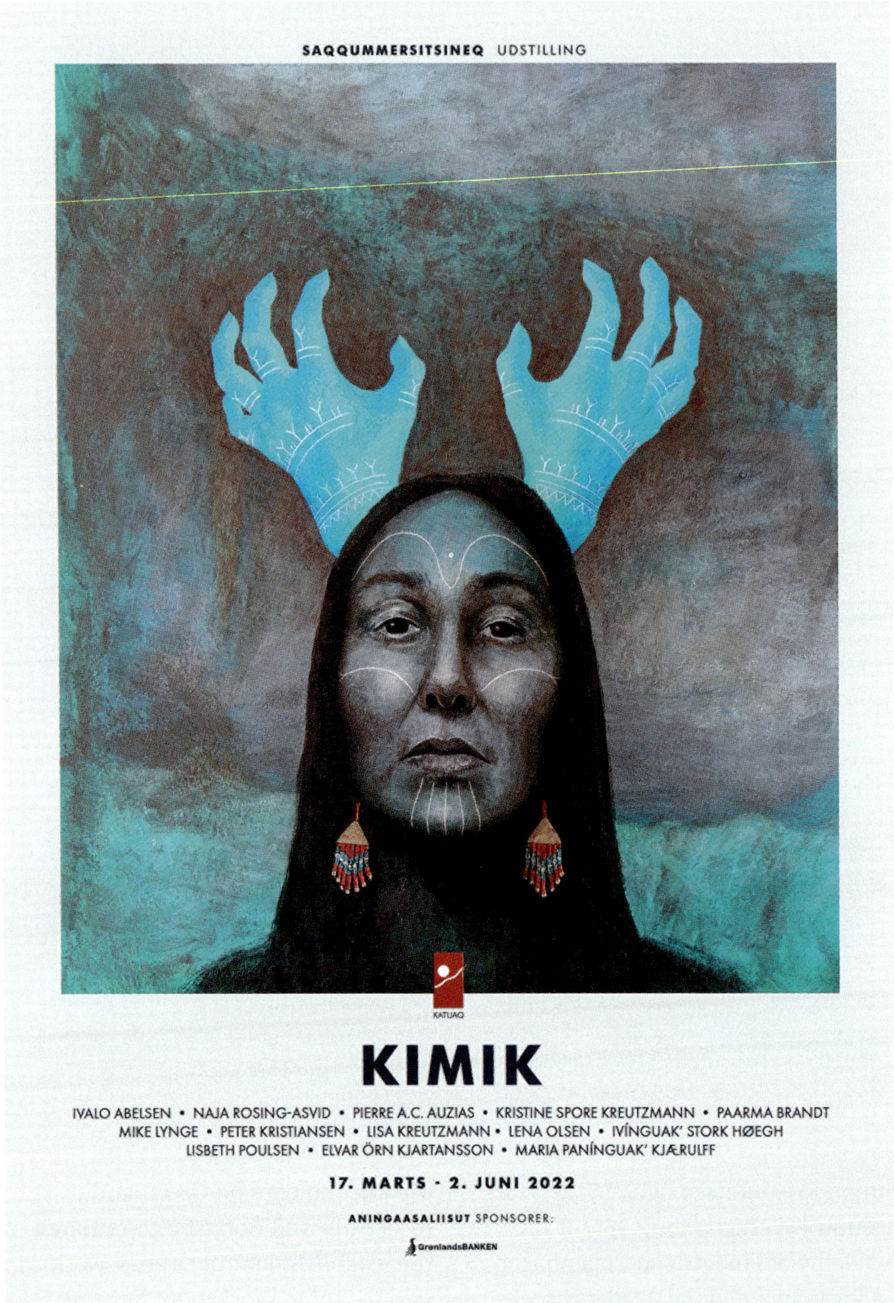

KIMIK. ASSOCIATION OF ARTISTS IN GREENLAND. 2022

Exhibition poster for the 2022 spring exhibition by the Association of Artists in Greenland, KIMIK. The artist behind the design is Lisa Kreutzmann with the work *Tuttu Inoorroortoq*. KIMIK has existed since 1995 In 2022, it was comprised of 22 professional Greenlandic artists.

Kaalund. They argued that a graphic workshop was a natural further development of the tradition of carving and incising scenes. The goal was the same as in the time of Aron from Kangeq (see p. 152), namely to create works that came out to the people and were of use in society. There was a clear desire to be inspired by traditional Inuit culture and thereby create a visible Greenlandic identity in art.

Since then, almost all Greenlandic artists have been students and often also teachers at the school. One of the first and most well-known people is the artist Aka Høegh. From her childhood, she dreamed of becoming an artist, but the Home Rule refused to let her study at the Academy of Fine Arts in Copenhagen, but without being enrolled, Høegh managed to follow the courses for a number of years. In the following years, it became significantly easier for young artists to graduate abroad, and both Arnannguaq Høegh and Anne-Birthe Hove graduated from art academies. Both came to play a major role up through the 1980s and 1990s, becoming the start of a sustained wave of academic-trained artists, the majority of whom remain women.

The field grew with artists such as Buuti Pedersen and Miki Jacobsen, who in 1995 were co-founders of KIMIK, the Association of Artists in Greenland. They helped to ensure that the Greenlandic artists had a mouthpiece. They were negotiated for annual spring exhibitions, and together could apply for funds for major joint projects. It also resulted in a number of releases, including *Grønlandske nutidskunstnere* (Greenlandic contemporary artists) by Camilla Augustinus from 2004 and KiMIK's own 2015 anniversary book, as well as a series of exhibition catalogues for exhibitions outside Greenland, e.g. *The Red Snowmobile*, a joint exhibition held at The North Atlantic House in Copenhagen in 2005.

Debate-provoking art and art after the introduction of home rule

Danish-Greenlandic artist Pia Arke graduated from the Academy of Fine Arts in Copenhagen in 1992. Her graduation project, the essay *Ethno-Aesthetics*, outlines the expectations for art when it is perceived as ethnic and Greenlandic. Arke here questions what makes art Greenlandic. Should you be able to tell from the motif that it is Greenlandic, and if so, what would that say?

An artist like Frederik 'Kunngi' Kristensen, who created non-figurative works, often saw his works being rejected because they were not Greenlandic enough. Should art in Greenland have a local connectedness to have a local relevance, or can it be global and yet native to Greenland?

A lot of art continues to be created with Greenlandic motifs, but it seems more like an inspiration drawn from the everyday life of the artist rather than a desire to create something that underpins a national narrative.

The motifs may be, for example, block of flats, myths or nature, but the works may as well deal with current plans for mines at Kuannersuit – Kvanefjeld in Narsaq, as seen for example in the works of Bolatta Silis-Høegh (see p. 199) and Ivínguak' Stork Høegh. The colonial past also plays a large role in, for example, the works of Julie Edel Hardenberg (see p. 148). This is seen in her work *Rigsfælles-skabspause* (Unity of the Realm Pause), a straitjacket created from the Danish and Greenlandic flags.

Since 1990, a number of artists have worked with conceptual art, such as installations, interventions, performance and video art. Specifically, Jessie Kleemann has created some attention-grabbing performative works. Together with Danish artist Asmund Havsteen-Mikkelsen, Inuk Silis Høegh created the *Melting Barricades* project, a major performance project that toyed with the idea of creating of a Greenlandic military. The project involved personnel carriers, uniforms and recruitment in the streets of Nuuk and Copenhagen.

Crafts

Two main trends are seen in Greenlandic crafts: one that is closely related to the visual arts, and one that leans towards applied art.

In the first and more recent trend, you will often find artisans who create both visual arts and crafts. Many are members of the Association of Art-

A propaganda poster from Inuk Silis Høegh and Asmund Havsteen-Mikkelsen's *Melting Barricades*, a combined performance and exhibition project in which the artists created a fictional Greenlandic military.

MADS PIHL/VISIT GREENLAND, 2013

TUPILAK

The tupilak, a small figure carved in bone or tusk, is one of best known Greenlandic artistic expressions, but it is quite a new thing.

The word 'tupilak' originally denotes the evil spirit you could cast on a person who you would like to harm (see p. 158). The spirit has been illustrated by Kârale Andreassen. It was not until around the year 1900 that figures depicting the spirit began to be created, and these are the figures we know today as tupilak: grotesque figures carved in the bone and tusk that often also incorporate a number of animal elements. Very few of the artists are named, and the figures are often stated as works by an unknown artist.

There are a number of references to the figure among Greenlandic visual artists, for example Anne-Birthe Hove, who in her work *Portrait* creates a figure-like creature in the same way as a tupilak.

A number of attempts have been made to provide supplementary training in and raise the quality of this type of handicraft, but none of the initiatives have been successful.

Tupilak in walrus tusk carved by Gedion Qeqe from Tasiilaq.

ists in Greenland, KIMIK, e.g. Buuti Pedersen, graduated from the School of Decorative Art, today the School of Design at the Royal Danish Academy of Fine Arts. In addition to the visual arts, she has created a number of glass sculptures, decorations with baptismal fonts, jewellery series, etc. Another example is Kristine Spore Kreutzmann, graduated from the Royal Danish Academy of Fine Arts' Glass and Ceramics School at the island of Bornholm. Kreutzmann makes both experimental ceramic works and cups and dishes for the home. Maria Panínguak' Kjærulff works in the field of tension between design and visual art. Her series *Inuit Lady*, a series of watercolours and prints, depicts small ulus (knives) that have been turned into women with kamiks, mittens with orchids and other decoration. The series stands out from her otherwise large and expressive works, and it has become a standalone design line with a very different expression from her other works.

The part of the handicraft that lean more towards applied art includes figurines, jewellery and the like, often made of bone, stone and tusk. The only place in Greenland to train in this field is Kalaallisuuliornermik Ilinniarfik, the School of National Clothing Education in Sisimiut, which works to create the Greenlandic regional costumes.

Simon Kristoffersen is one of the people who have historically had a great importance, especially for the soapstone figurines. Father and daughter Aron Kleist and Cecilie Kleist have established a presence in the field of bone carving work, and artists still refer to their works.

Today, especially Kristian Fly, Kim Eriksen and Oline Binzer are worth noting. All of them create polar bear heads for quick sale, but at the same time immerse themselves in large-scale experimental unique works.

FASHION AND DESIGN

For many years, the fashion and design industry has been characterised by clothing and jewellery that refer to Greenlandic culture. However, young designers are increasingly turning away from the use of cultural symbols and instead referring to more global trends.

In the years leading up to 1979, when home rule was introduced, young Greenlanders wore anoraks and Greenlandic jewellery to show their national pride. Today, Greenlandic cultural symbols still characterises much of the fashion clothing that is now being mass-produced in foreign factories.

Fashion and design are being used to market the country. The designers use their Greenlandic identity and culture to express themselves, and their products do not only target the Greenlandic market.

From fashion to tradition

Greenlandic skin clothes have changed considerably over the past 500 years, partly due to contact with Europeans and partly because of Greenlanders' own approach to fashion. Ancient tales, myths and diaries of Europeans tell of how the women sewed clothes for their impending meeting with other compatriots at summer camps (see p. 212), and how they also swapped creations.

The whalers and the Norwegian-Danish mission brought European fashion accessories such as beads, lace, embroidery and fabric. The women used it for their costumes – and no two were the same. The more resources one had, the more decorated the costume.

There are three regional costumes: arnatuut in Thule, tunumiutuut in East Greenland and kalaallisuut in West Greenland. The latter, due to contact with the Norwegian-Danish mission since 1721, has changed the most, and the presence of European whalers from the Netherlands and Great Britain in particular since the 15th century characterised the West Greenland costumes.

Several designers has drawn inspiration from the West Greenlandic women's costume. Since the 1930s, it was possible to buy fabric for dresses or buy ready-made dresses, but it was not until during World War II that a majority of women began to wear American and European clothing. The women's costume became a national costume worn at holidays and special events, and its shape has not been changed since the 1940s.

Clothing design

Some 50 years after Greenlanders began to wear western clothes, the first major collections by Greenlandic designers were put into production. Many of their designs rely on Greenlandic natural and cultural history, but there are also lines that completely or partially distance themselves from that style.

Isaksen Design

The first Greenlandic fashion designers to launch collections were Rita and Nickie Isaksen, who founded Copenhagen-based Isaksen Design in 2002. Mother and daughter, both residing in Denmark, often draw inspiration from Greenlandic nature, culture and history, including cultural

For the Nordic Fashion Biennale 2014, Austrian-American artist duo Cooper & Gorfer collaborated with fashion designers from Greenland, the Faroe Islands and Iceland to create theThe Weather Diaries collection. Designer Bibi Chemnitz co-created *The Fifth Daughters* (left) and *Ena with Eyes Shut* (right), which also feature beaded collars sewn by Nikolaj Kristensen. The series was published as a book with the title *The Weather Diaries*.

THE FIFTH DAUGHTERS, 2014 / COOPER & GORFER

ENA WITH EYES SHUT, 2014 / COOPER & GORFER

MICHAEL TONSBERG/ISAKSEN BY NICKIE ISAKSEN, 2020

As in this 2020 collection, Isaksen Design draws inspiration from traditional Inuit tattoos. Isaksen Design was the first Greenlandic designer to attend the Copenhagen Fashion Fair in 2006, where American, Nordic and Canadian buyers discovered them. Isaksen Design was sold for several years on New York's Fifth Avenue.

INUIT QUALITY CLOTHES OF GREENLAND, 2020

In 2019, Inuit Quality Clothes launched their first T-shirt 'Save Our Arctic'. By creating clothes that focus on climate change in Greenland and the Arctic, the collection emphasises the importance of preserving Greenland's nature.

symbols such as the women's knife ulu, tupilaqs (see p. 204) and Inuit tattoos. The collections have been well received in both Greenland and Denmark. Isaksen Design makes fashion clothes and jewellery both for men and women and children's clothing.

Bibi Chemnitz

Designer Bibi Chemnitz founded her brand in 2005 in Denmark, and today she is based in both Nuuk and Copenhagen. Her design draws on the two widely different universes: Greenland's harsh nature and the cobbled alleys of Copenhagen. The collections, which are aimed at both men and women, are called *streetstyle* and are sold to both young Greenlanders and internationally. One of Bibi Chemnitz's hallmarks is the Greenland flag, Erfalasorput, in black-and-white and the silhouette of Greenland. It is also used as a logo in her outdoor collection.

Anita Høegh

In 2008, Anita Høegh set up a yarn production and opened a shop in Sisimiut, Qiviut, selling yarn from muskox and sheep wool. In 2013, she opened another shop in Nuuk, and the two shops sell only what the employees design and produce. With Anita Høegh in charge, the shops have their own wool workshop and purchase muskox skins directly from the hunters. The production company focuses on nature's own raw materials and sells shawls, beanies, sweaters, wristlets and other garments made of muskox and sheep wool. In addition, Qiviut also sells quality outerwear sewn from sealskin, which is intended for people who spend a lot of time on outdoor activities in winter, such as dog sled drivers and hunters.

Louise Lynge Berthelsen

In 2010, Greenlandic seamstress and designer Louise Lynge Berthelsen opened Nuuk Couture. Initially, she sewed to orders, but after a few years, she began designing and mass-producing fashion clothes for women, and later, she started to focus on men and children. She is the first Greenlandic designer to have a collaboration with Kalaallit Nunaanni Brugseni (Greenlandic supermarket chain) and sold children's clothes with inspirations from the West Greenland costume.

Louise Lynge Berthelsen, who comes from a large family in Nuuk, has a grandmother who is a national costume seamstress. Her designs are based on the narrative of the West Greenland women's costume and pay respects to both the seamstresses and the costume. With the slogan 'Sapinngilagut' (we can do it), she has also created a street wear collection aimed at young people. She is also behind a jewellery collection that refers to the West Greenland beaded collar.

Inuit Quality Clothes of Greenland

Inuit Quality Clothes of Greenland was founded by Dorthea Rødgaard from Qaqortoq and Hanne Kristiansen from Aasiaat. Both women own clothes shops, and in 2012, they found that there was a lack of clothing designed by Greenlanders, and they therefore decided to make a brand with which the Greenlandic fashion wearers could identify with.

With a silhouette of a Greenlandic woman with a hairtop and a Greenlandic man included in the word *Inuit*, graphic design fashion clothes and outerwear from Inuit Quality Clothes of Greenland have now been produced for the last nearly ten years. The clothes are promoted by singers Julie Berthelsen and Simon Lynge on social media.

Bolt Lamar

Bolt Lamar is a graphic design brand created by two anonymous young men. The name of the brand is a merger between two international stars, Olympic gold sprint winner Usain Bolt of Jamaica and American rapper Kendrick Lamar. The brand started as leisure project for two friends, and because it became so popular, they chose to mass-produce their designs on T-shirts, hoodies, beanies and caps.

Bolt Lamar can be seen as a backlash to the use of the Greenlandic cultural symbols in clothing collections. When the surnames of the two world stars are merged, the brand symbolises globalism and that as a Greenlandic designer you can pick from all over the world.

Jewellery design

Jewellery designers and entrepreneurs who mass-produce jewellery also use traditional cultural symbols. Isaksen Design, Nuuk Couture and Inuit Quality Clothes of Greenland also produce jewellery as part of their designs. As one of the first Greenlandic entrepreneurs, Duuden Design from Qaqortoq started to pro-

Jewellery artist Kim Kleist Eriksen's company is called Eriagisaq. It means »something you don't want to lose«. He sells jewellery made from caribou antlers and walrus tusk.

KIM KLEIST ERIKSEN, 2018

duce acrylic glass jewellery. In addition to the mass-produced jewellery, there is also jewellery made from the bones, roof, claws and tusks of animals. The jewellery points back to the Greenland aarnuaq, an amulet whose power was supposed to protect the wearer.

Many jewellery designers are inspired by the Greenlandic culture and nature. The autodidact Kim Kleist Eriksen of Sisimiut cuts various motifs out of narwhal and walrus tusk. Also Gedion Qeqe from Tasiilaq, who in 2018 was appointed as cultural ambassador for Kommuneqarfik Sermersooq, makes popular jewellery from caribou antlers, tusk and stone. Both artists interpret different symbols from Greenlandic tales and myths in their jewellery.

Aviaaja Kleist Burkal also draws on Greenlandic symbolisms. She graduated in 2015 as a jewellery designer from the Institute of Precious Metals in Copenhagen and also has a degree in art and design from Italy and Austria. Her jewellery has references to the South Greenland storis (great ice) and the raven.

THEATRE AND DANCE

The performing arts are based on a strong storytelling tradition, drum dance and mask dance. These traditions are rooted in social, cultural and spiritual rituals. The drum and mask dance are two different old traditions.

The dances give a glimpse into a Greenland that was both then and is now rooted in presence with family and friends, thrills and laughs and myths and mystique. Both mask dance and drum dance (see p. 210) were disapproved of by the Christian missionaries who considered the dance a pagan ritual. This meant both were under threat and close to disappearing altogether in West Greenland. The prevalence of Christianity was strongest in Middle and West Greenland, which is the background to the fact that many existing songs and drum dances originated in the North and East. The mask dance, on the other hand, originates predominantly from East Greenland, where it was secretly kept up.

Mask dancing is an approximately 4,000-year-old tradition that has been used especially during the winter time as entertainment. The dance is impromptu and occurs in the meeting with the audience. It varies regionally and spans three overarching themes: to poke fun, to scare and to pay court. Thus, the mask dance reflects both the sexual drive, the humour and the frightening. Children and adults alike are struck by the delight of horror, laughing at the absurd movements and grotesque appearance.

In the period 1819 to 2011, the theatrical scene as we know it from the West evolved from early amateur theatre to a genre-borne and more professional scene. With the establishment of the Silamiut Theatre in Nuuk in 1984, the country got its first professional theatre, and in 2011, this stage transitioned to being the national theatre.

Theatre

Theatre, as it is known from Western cultures, came to Greenland in 1819-1820 when Peder Kragh, a missionary in Aasiaat, translated B.S. Ingemann's Bible drama *The voice in the wilderness*, which is about John the Baptist. The main content was known by the Greenlanders from *The New Testament*, but it was not until 1874 that the script was printed. Birgit Kleist Pedersen points out that Greenlandic theatre history can be divided into four phases: In the first phase, the period 1820 to 1920, amateur theatre was primarily performed, and the plays were mainly translated.

MARTIN LEHMANN/RITZAU SCANPIX, 2021

Mask dance at the National Theatre of Greenland in Nuuk. The dancers are Connie Kristoffersen and Hans-Henrik Suersaq Poulsen.

GERTH LYBERTH/NUNATTA ISIGINNAARTITSISARFIA, 2017

In 1977, the Tuukkaq Theatre created the show INUIT, which has been revived several times and is considered a Greenlandic theatre classic. This photo is from 2017, when the School of Acting at Greenland's National Theatre revived INUIT.

During the period 1820 to 1920, mainly amateur theatre was performed in Greenland, and the plays were mainly translated. Then, around 1930, different genres started to be seen within Greenlandic acting, including plays by Pavia Petersen and Hans Lynge. The third phase was initiated in 1975 with the establishment of the Tuukkaq Theatre in Fjaltring near Lemvig, and it ran until 1984. It lasted until 1984, when the Silamiut Theatre was established.

In 2011, the Silamiut Theatre became the national theatre – the National Theatre of Greenland – which is also home to the National Acting School.

Tuukkaq Theatre

The Tuukkaq Theatre has played a special role in reviving, developing and preserving the mask dance. The theatre in Fjaltring was founded by Reidar Nilsson, and its purpose was to train Greenlandic actors as well as to conduct performances on subjects drawn from Greenlandic culture. This started a new phase of the performing arts.

Tuukkaq's focus was not the traditional word theatre, but rather physical theatre, singing and music. The focus was on old cultural practices and its renewal, both in form and content. The actors worked with body, movement, singing and masks as the essential expressions, and there was very little focus on the spoken word.

The theatre functioned both as a theatre and educational institution; both were experimental and characterised by the continuation of the mask dance. The Tuukkaq Theatre existed from 1975 to 1994, and today it serves as a living museum and visitors' centre in Fjaltring.

Silamiut Theatre

Performing artists who had graduated from the Tuukkaq Theatre travelled back to Greenland and estab-

GERTH LYBERTH, NUNATTA ISIGINNAARTITSISARFIA, 2017

The performance Angutivik//The Greenlandic Man was a co-production between Nunatta Isiginnaartitsisarfia and Teater freeze Productions. It played in Greenland and Denmark and won Klaus Geisler a Reumert Award for Male Ensemble Role of the Year. Premiere in 2017.

In 2017, Greenlandic dancers came together to create the performance Arctic Dreams. Antero Hein was videographer, Maliina Jensen producer, Alexander Montgomery-Andersen artistic director. Dancers and choreographers were Madelaine Gordon Graadahl, Ruth Montgomery-Andersen, Maliina Jensen, Antero Hein, Amanda Darfelt and Alexander Aarø. Also four dance aspirants from Nuuk: Ivi Luna Olsen, Maria Lyberth, Miyuki Daorana and Ululinannguaq Olsen.

ANTERO HEIN, 2017

lished the Silamiut Theatre in Nuuk in 1984. Alongside physical theatre, Silamiut produced classical theatre based on both original Greenlandic material and on works from the outside. The theatre also set up musical productions.

In 1984, Simon Løvstrøm (Mooqqu) became the manager of the theatre, taking over Tuukkaq's role as theatre school and running it until 1997, when Silamiut, together with Norwegian director Indra Lorentzen, helped to open the Katuaq cultural centre with the performance *Qasapi's last day*.

Nunatta Isiginnaartitsisarfia – The National Theatre of Greenland

The idea of the National Theatre and the work on a theatre act was set in motion after a theatre seminar held in 1992, and the Theatre Act was passed in 2010. On 1 January 2011, the act was introduced, and this made the Silamiut Theatre the national theatre, Nunatta Isiginnaartitsisarfia. In 2013, the theatre moved to Industrivej in Nuuk, where it is still housed.

The National Acting School opened in August 2012 with Makka Kleist as the director of Nunatta Isiginnaartitsinermik Ilinniarfia. The National Theatre and the School of Acting form a unified whole which has three main tasks: to produce, develop and make accessible Greenlandic performing arts to a wide audience, to train actors and to support and develop amateur theatre in Greenland. In 2018, Varste Mathæussen became the director of the school, followed by Ruth Montgomery-Andersen in 2019 and Vivi Sørensen in 2021.

The School of Acting is a two-year education that accepts five students every two years. The school opened in 2012, and since then a total of 18 actors (2021) have graduated.

Amateur theatre

N.A.I.P (Nuummi Aliikkusersuisartut Isiginnaartitsisartut Peqatigiiffiat) is Nuuk's amateur theatre association, established by Kristian Egede in 1959. He was a noted entertainer, who toured all over Greenland in the 1960s and 1970s. In addition to theatre, N.A.I.P. is a cultural institution which preserves and develops the vernacular use of folk dance, kalattuut (see p. 215), mask dance and drum dance, and is also known for radio theatre. The association is open to everyone and helps to ensure that these cultural elements are handed down from generation to generation and that they are accessible to the local community.

Throughout time, amateur theatres have existed in several locations in West and East Greenland. Before the sports hall came into being, these theatres were often housed in the local community centres.

In recent times, the best known and established amateur theatre is Pakkutat in Sisimiut. In 1992, under the direction of Valerius Nielsen, the theatre represented Greenland at the Arctic Winter Games and later at the Nordic Amateur Theatre Festival, where in the mid-1990s they received the award for best performance. East Greenland's Ortu Ignatiussen also helped establish and maintain interest in theatre in Tasiilaq.

Nunatta Isiginnaartitsisarfia has had an important task in supporting and developing amateur theatre. In 2015, 2016 and 2017, the National Theatre held amateur theatre seminars, and in 2018 an amateur theatre festival called Kiinnat Festival.

Dance

The original dance traditions of Greenland are mask dance and drum dance, and from the late 16th century, when European seamen brought dance and fiddle music, the foundation of the somewhat younger tradition kalattuut was created.

Within European genres, there have been several dancers who have participated in performances and held workshops since the 1960s; among them, Ann Crosset has toured both with Nyt Dansk Danseteater (New Danish Dance Theatre) and subsequently, in 2015 and 2018, with Company B of The Royal Danish Ballet.

Establishing a dance community with trained dancers in ballet, modern, standard dance, hip hop and other dance genres began with Greenland's hosting of the 2002 Arctic Winter Games (see p. 176). During the period 2003 to 2011, dance lessons were offered by Dida Jensen and Niels Berglund for children and young people in Nuuk.

Qiajuk Studios opened a dance studio in 2011 with lessons in ballet, vertical dance, modern, urban styles and jazz. This was now possible because young dancers had returned to Greenland after graduating abroad.

MAKKA KLEIST

Actress and storyteller Makka Kleist (b. 1951) has been a driving force in the work to develop the Greenlandic theatre. Kleist grew up in Qullissat but was sent to Denmark to attend the 6th grade and finish school. As a newly qualified primary school teacher, Makka Kleist joined Tuukkaq Theatre in Denmark in 1978 (see p. 209) and was one of the driving forces and worked as an actress, director and playwright until 1983. It became the beginning of a long career, and she has subsequently lived in e.g. Canada and Norway, worked as a stage actor, had film roles and also engaged in children's theatre and has written and translated plays.

In 2003, Makka Kleist returned to Greenland, where she, among other things, has worked at the Silamiut Theatre and been active in the work to establish a national stage and a drama school in Greenland. When The National Theatre/Nunatta Isiginnaartitsisarfia opened in 2012, Makka Kleist became head of The National Acting School until 2018 when she returned to Denmark.

GERTH LYBERTH/NUNATTA ISIGINNAARTITSISARFIA, 2016

Makka Kleist in the performance *Arnatsialaat* (*The Rez Sisters*). Tomson Highway's play was first performed in Greenlandic in 2016. It was a graduation performance for four student actors who performed alongside professionals including Kleist.

DRUM DANCE AND DRUM SONG

Drum dance and drum song have traditionally had social, religious and cultural significance; they have been used as entertainment, storytelling, flirting, as part of spiritual ceremonies and to settle disputes. The songs have been passed down from generation to generation, and the performance of drum song and dance can be a mix of music, dance and acting. They can be performed by both men and women.

ANNE BANG/ARKTISK INSTITUT, 1980

Aasivik in Niaqornap Kuuani, 1980 (near Qoornoq). The photo shows Tobias Tukula from Kulusuk in East Greenland, who also was at the Aasivik in Qullissat in 1977. Note the relatively large East Greenlandic drum.

THE HISTORICAL BACKGROUND

The first Greenlandic frame drums (qilaat) and fragments of such drums date back more than 4,000 years and have been found in men's houses in Northwest Greenland. It is therefore assumed that the tradition came to Greenland with the Inuit.

The Inuit frame drum comes in many varieties, from the relatively large East Greenland drum to the small and slender North Greenland drum. The construction also varies from region to region, but basically the drum is made of wood or bone, over which is stretched a membrane of animal skin or hide, which resonates when the underside of the frame is struck with a wooden or bone drumstick. During the performance, the person will traditionally stand with his or her legs slightly apart and make a series of body movements while singing and drumming. The drum was an important part of the Angakkoq's spirit ceremony or spirit journeys.

When the Danish-Norwegian mission arrived in West Greenland in 1721, drum song and dance were associated with pagan culture and was more or less outright prohibited. Therefore, the main knowledge of the tradition comes from East and North Greenland.

DRUM DUEL

Drum dance and drum song have been used in the East Greenlandic drum duel. Here, disputing parties could sing accusations and perform light violence against each other in front of a choir made up of the disputing parties' families. In this way, the parties were able to resolve conflicts as an alternative to violence and vendetta.

The drum duel was the Inuit court system where disputes were settled by the verdict of the audience after the disputing parties had sung and danced their songs against each other. If a person became so angry at the audience's laughter that he was unable to complete the duel, he would automatically lose.

The drum duels were usually performed at the large aasiviit, the large summer camps at Sermermiut, Taseralik, Kangeq, Uunartoq and Aluk, to which the Inuit travelled long distances to meet, trade and find potential new partners.

ANNE BANG/ARKTISK INSTITUT, 1980

Aasivik in Niaqornap Kuuani, 1980. The summer camps were political and cultural rallies. The photo shows two women with frame drums. In the original drumming contest, opponents had to be of the same sex; in other situations, the frame drum could be used to spot someone of the opposite sex.

RENEWED INTEREST IN THE TRADITION

In the wake of the Greenlandisation from the 1970s and onwards, there has been a growing interest in drum song and dance as an indigenous Greenlandic musical culture. The tradition has undergone a revival, consisting partly of traditional performance of collected drum songs, partly of new compositions and integration of drum song and dance into globalised music genres.

Danish ethnomusicologist Michael Hauser recorded drum songs in the Thule region in the 1980s, where the tradition was disappearing. He wrote down and analysed the songs and was able to demonstrate that many of the songs have a complex structure, while others are simpler. The latter originates from later immigration from the West, the last in the 1860s.

Robert Hivshu Peary, who grew up in Thule and is trained in traditional drum dance, has been working to revive the tradition. In 2010, Michael Hauser published *Traditional Inuit Songs from Thule Area*, a work of over 1,500 pages, made in collaboration with Robert Hivshu Peary, among others. In 1975, ethnomusicologist and composer Poul Rovsing Olsen released songs by Egon Sikivat, a well-known drum singer from Kuummiut in East Greenland.

In 2021, drum dance and song (qilaatersorneq) was entered on UNESCO's list of Intangible Cultural Heritage. One of the aims is to ensure a living cultural heritage. Greenland is the first country in the Unity of the Realm to have a cultural expression accepted on this list. The inclusion on the list is not only about Greenlandic drum dance and song but also about the use of the frame drum by other indigenous peoples in the Arctic.

Alexander Montgomery-Andersen and Maliina Jensen have secured development and Greenlandic entrenchment of these dance genres, and they have taught 200-300 children a year, both in the capital and in other towns. Together with the individual municipalities and cultural centres, Katuaq, Bårdar Dance Academy in Oslo, Norway, and NAPA (Nordic Institute in Greenland), Greenland's dance community has hosted the national dance festival OuterVision in the years 2010-2017.

National and international cooperation

The performing arts have always been based on cooperation between different institutions and bodies. The Katuaq cultural centre has supported artists and has served as producers for performances by various institutions. The Nordic Institute in Greenland, NAPA, has supported the genesis of the Katuaq cultural centre as well as numerous performances. By serving as host and producer of several productions, NAPA has helped to ensure a Nordic presence in the Greenlandic theatre and dance community, including through cooperation with The Royal Danish Theatre, The Royal Danish Ballet, Jo Strømgren (the performance Polaroid) and other artists.

The dance community has benefited from the stages of Katuaq and the National Theatre. Young artists have been given the opportunity to create and show their own productions, which has included collaborations with local actors, sound and lighting designers, as well as collaborations with Nordic and international theatres established and directed by Sami and Indigenous peoples in Australia, Canada and Alaska.

As the dance community has grown, dance performances have been created both featuring NAPA, The National Theatre, foreign artists and local youth talent. The driving force in these productions has been the Greenlandic dancers Maliina Jensen and Alexander Montgomery-Andersen as well as Ruth Montgomery-Andersen.

MUSIC

The earliest Greenlandic music, drum song (see p. 210), is still alive – both by virtue of the performance of collected songs and through new compositions where the drum song is integrated into global music genres. Since European whalers brought instruments and tunes with them, there has been a tradition of integrating the new and putting it in a Greenlandic context.

In recent music history, this is seen, among other things, both in the 1970s rock of Sumes and in the rap music of Tarrak.

Choral and hymn singing

With Hans Egede and later the Moravian mission (see p. 210), choral singing became part of the Greenlandic music culture. Both missions were Protestant, and the early hymn repertoire has therefore been Greenlandic versions of the European repertoire.

Starting in the mid-19th century, Greenlandic composers, such as Rasmus Berthelsen, began to build a repertoire of original hymns and national songs that today play an important role in the Greenlandic hymn and choral tradition. For example, Berthelsen's Christmas hymn *Guuterput Qutsinnermiu* is an important part of Christmas traditions, and it is an example of the special slowly undulating singing style inuttooq that is distinctive of much Greenlandic choral singing.

There has been a tradition of spontaneous polyphony at church services, where the congregation could sing four-part chorale harmonisation to a number of hymns. Although this tradi-

Rapper Josef Tarrak-Petrussen gained a lot of attention in Greenland and Denmark with the song *Tupilak* in 2016. Using samples from a well-known song of the same name by the group Mariina (1992), he again broached the Danish-Greenlandic theme in popular music.

MARTIN LEHMANN/RITZAU SCANPIX, 2014

ARKTISK INSTITUT, 1880

The accordion is ready, and a group of festively dressed young people wait in the late 1800s for the first kaavikulooq, or polka, to begin. At the settlements along the coast, a flat rock was used as a dance venue for centuries. Gradually, trade and crafts led to the construction of buildings that provided opportunities for indoor dancing. At some settlements, you can still find flat rocky areas, worn down by kamik soles.

KALATTUUT – GREENLANDIC STYLE

Foreign whalers were the first to bring European instruments, music and dance to Greenland. The Inuit traded with the whalers, and the cultural meeting laid the foundation for the dance tradition kalattuut, 'Greenlandic style'. It was probably mainly the fast-paced two-part *reel* that was played and danced in 1585, which is when it is first heard of on Greenlandic soil.

From 1721, more and more Danish and Norwegian officials were sent out, and some of them brought melodies and dances from their homelands. The local traditions therefore differ quite a lot, but also have clear common features.

Kalattuut can be danced as quadrille, circle and line dances. There used to be considerably more dances than there are today, but since there has been no tradition of writing down notes and steps, many have been lost.

The two most popular kalattuut, polka and two-step, came to Greenland during the 1850s. They are played and danced enthusiastically in all parts of the country, both for national costume performances and often as a festive feature during gatherings and celebrations.

Particularly the violin and the two-row accordion have been the instruments of choice; today the music is often recorded.

Two distinctive features of kalattuut are the incredibly fast pace and some gender-specific steps. The men dance very fast stomping steps, while the women have gliding steps.

tion is not upheld everywhere today, choral singing is an important part of national and local culture, and there are a number of choirs that regularly record and tour in Greenland and abroad.

Vaigat music

Vaigat music was inspired by American country and western music that became accessible in Greenland during World War II in conjunction with US base radio and ordering catalogues. Some of the songs are Greenlandic versions of this repertoire. The music is named after the Vaigat Strait between Nuussuaq Peninsula and Qeqertarsuaq (Disko), and the genre is closely tied to the abandoned mining town of Qullissat (see p. 117), where it originated and abounded.

The genre emerged in the 1950s when the local orchestra in Qullissat, under the direction of Jens Hendriksen, was given the opportunity to purchase a variety of instruments, including a *pedal steel* guitar, which came to play a significant role in the genre's sound.

After the authorities closed Qullissat in 1972, the music has gained political significance in subsequent criticism of the shutdown, but originally it was entertainment and dance music.

Rock'n'roll

In the 1960s, rock'n'roll music came to Greenland. There were a number of bands within the genre which, among other things, competed in music competitions in Nuuk. The best known band from that time is The Eskimos, which played at a regular basis at Aleqas Kaffebar (Aleqa's Coffee Shop) and in the settlement hall in Nuuk in the mid-1960s. Through the record label Kajak owned by Grønlands Indkøbsring, the band released two EPs in 1966. The Eskimos had Barselaj Danielsen (guitar and singing) and Peter O. Petersen (drums) as recurrent figures. The band disintegrated in 1966 when Barselaj Danielsen went to Denmark to study.

The Eskimos sang primarily in English or Danish, and their repertoire consisted partly of songs from internationally known rock bands, partly of music from the kalattuut repertoire exposed to the rock'n'roll instruments, and in addition they had compositions of their own.

The other bands of the 1960s also stuck primarily to the rock'n'roll repertoire.

In 1966, The Eskimos released their two EPs, recorded in Nuuk as a two-track and released by Grønlands Indkøbsring through the Kajak record label. The repertoire consisted of a mix of world-famous rock'n'roll hits and kalattuut interspersed with electric guitar, bass and drums.

ANDREAS OTTE, 2022

Sume and the Greenlandisation of popular music

As the home rule movement of the 1970s gained ground, the band Sume was the musical spearhead of the movement's politics. With Malik Høegh and Per Berthelsen in the front, the band released their first album *Sumut* in 1973. Their rock music received considerable attention, with the lyrics being in Greenlandic and dealing with Greenlandic cultural themes, and because the band made use of visual and auditory symbols of Inuit identity. Furthermore, the band's lyrics often contained explicit criticism of Danish influence in Greenland. Sumes' musical approach, appropriating a global music genre and putting it in a Greenlandic context, remains prevalent today.

Sumes' first two albums were created in collaboration with the socialist and anti-imperialist publishing house DEMOS in Copenhagen, and in the 1970s, other Greenland-language releases also came out at the publishing house, such as Rasmus Lyberth's

1975 debut album *Erningâ (To My Son)* and Juaaka's album *Kalaaleqatikka*, both of which genre-wise are more folk music than rock.

The socialist and anti-imperialist line from Sume was upheld by rock bands, such as Inneruulat and Piitsukkut, but from the mid-1980s, the band Zikaza as well as artist Ole Kristiansen became some of the most trend-setting names on the Greenlandic music scene with top-selling albums such as *Miki Goes to Nuussuaq* (1986) and *Isimiit Iikkamut* (1989), respectively. These two names had a less confrontational approach to Danish influence in Greenland and focused more on general human themes, such as love and grief. However, the Greenlandic context still had a strong presence through language, symbols, the use of legends and nature as metaphor in the texts.

SUSANNE MERTZ/BAM/RITZAU SCANPIX, 1974

Sume is the biggest rock band in Greenland's music history. But the band was primarily active in Denmark, where the band members were studying. Here from left: Malik Høegh, Emil Larsen, Per Berthelsen and Hjalmar Dahl.

Pop music and women at the front

Greenlandic rock music did not feature female performers in the 1970s and 1980s, but in the Vaigat music and the softer end of the pop-rock spectrum, there were female singers, among others Birthe Olsen, who sang alongside her brother Per Berthelsen on some popular releases in the 1980s.

Female lead singers took on an increased importance in the 1990s, initially with the band Mariina having Marina Schmidt at the front. Together with Hans Lange, she achieved great success with the 1992 album *Utaqqivunga*. Mariina was a role model for Nina Kreutzmann Jørgensen, who initially came forward with the children's release *Arnajaraq* (1997) together with Julie Berthelsen, but became more influential in the group Qulleq in the late 1990s.

Along with names like Kimmernaq Kjeldsen and Tupaarnaq Mathiassen, Nina Kreutzmann Jørgensen and Julie Berthelsen remain some of the best known names in Greenlandic pop.

Rap

Rap came to light with Nuuk Posse in the early 1990s. The group rapped with humour in both Greenlandic, Danish and English. In 2003, the band Prussic politicised the genre by using rap as a mouthpiece for members' own experiences of neglect. Since then, large quantities of underground rap have been produced in Greenland.

In 2016, the genre again saw wide popularity with rapper Josef Tarrak-Petrussen. With the song *Tupilak*, based on a sample of Mariina's number by the same name, he resumed Sumes' criticism of Danish influence in Greenland and Danes' racist prejudices towards Greenlanders.

Recent music and international aspirations

Since Sume, the Greenlandic music scene has been mainly nationally oriented, with the songs largely being in Greenlandic and dealing with local themes. However, certain bands and artists have managed to gain attention outside Greenland, including especially Sume and Rasmus Lyberth. In addition, Angu Motzfeldt, Simon Lynge, Nive Nielsen, Nanook and Small Time Giants, among others, have gained attention in Denmark and internationally.

From the 2010s, electronic music has played an increasing role on the Greenlandic music scene through artist Uyarakq, who also produces rap music, and with Da Bartali Crew, which has achieved great popularity within the genre. In addition, the band Inuk has been at the top of the sales charts with Greenland-language pop-rock.

The music industry

In the 1970s, desire arose for record companies and recording studios dedicated to a wave of new music in Greenland. From its creation in 1976, it was the record company ULO that reigned in Greenlandic music. Throughout time, more and more record companies appeared. The most popular releases during this period sold over 10,000 copies.

In the 1990s, development peaked with five active record labels. Eventually, stores had large stocks of unsold CDs, as fewer and fewer purchased CDs due to illegal copying and internet distribution. Of these companies, Atlantic Music, created in 1990 by Ejvind Elsner, is the only one active in 2021. The company has survived by publishing music along with the sale of musical instruments, rental of music systems and holding of festivals.

In the first part of the history of the record industry in Greenland, the record company oversaw the whole process of recording and publishing music. Musicians were transported and accommodated with per diem during

At the Danish Melodi Grand Prix in Jyske Bank Boxen in Herning on 23 February 2019, Julie Berthelsen and Nina Kreutzmann Jørgensen came in second place with song no. 7: *League of Light*. The number was mainly in English, but partly in Greenlandic. Subsequently, DR had to publish their rating system due to complaints about errors. It turned out that Julie and Nina had received the most audience votes, but were rated lowest among the finalists by the jury.

HENNING BAGGER/RITZAU SCANPIX, 2019

the recording period, and the record company had responsibility for marketing, sales, distribution, print and cover. In 2020, when the most popular releases sold about 3,000 copies, Atlantic Music either acted only as a distributor of artists' own recordings, or the company kept spending down. Recording time has become shorter, and the musicians typically arrange for part-financing of the release themselves, e.g. by obtaining funds. Along with the decline of the established music industry, during the 2010s, artists started to release their music themselves on the digital media. By 2020, 16 out of 24 releases were recorded and released by the artists themselves.

The slow transition to a more digital music industry is linked to population figures. Releases must sell about 50,000 downloaded songs or be streamed 2 million times to make the same profit as 1,000 CDs sold. Therefore, there is only limited access to Greenlandic music online. The most popular bands or artists can still maintain part-time or, in a very few cases, full-time earnings through concert activities.

+ MICHELLE FONTAINE/NIVE AND THE DEER CHILDREN, 2016

Nive Nielsen debuted in 2009 with the album *Nive Sings!* The core of the group was Nive Nielsen and her then partner Jan de Vroede, who had also played with Chilly Friday. An international collective of musicians, The Deer Children, provided musical backup. Her second album *Feet First* was released in 2016. Nive Nielsen has gained a lot of attention internationally, and she is also active today as an actor in international productions such as the American drama series *The Terror* (2018).

FILM

The first several years of producing films with Greenland as the theme, the productions were foreign, and the perspective was usually stereotyped and exotising. Both in feature films and documentaries, however, this focus has changed; where previously you saw Greenland from the outside, you now increasingly see the country from the inside. The focal point is no longer to the same extent colonial history; it has changed to general human experiences. After the turn of the millennium, a young Greenlandic film industry has flourished, and one of many ambitions is to create purely Greenlandic films.

The early years – foreign representation

The first films shot in Greenland were foreign/Danish productions. Often they drew on exotic stereotypes of indigenous peoples or showed the country as a productive colony under Denmark. These documentaries aimed to present Danish sovereignty to audiences at home and abroad at a time when tensions with Norway were growing over territorial claims on the east coast of the island (see p. 120).

Two films produced from the early 1920s, *Kongeparrets Grønlandsrejse (King Christian X and Queen Alexandrine visiting the Faroe Islands, Iceland and Greenland)* (1921) and *Den store Grønlandsfilm (the Great Film of Greenland)* (1922), a silent film about the Fifth Thule expedition 1921-1924, led by Knud Rasmussen and Peter Freuchen, showed Greenland as alien, exotic and as a successful colonial endeavour.

Film history's first feature film with Greenland as its theme was Wal-

SKANDINAVISK TALEFILM, NORDISK TONEFILM, NORRØNA FILM/DET DANSKE FILMINSTITUT, 1930

In the 1930 film *Eskimo*, Austrian Paul Richter plays the carefree brother-in-law who inadvertently ends up in Greenland, where he falls in love with an Eskimo girl Ekaluk, played by Norwegian Ada Egede-Nissen. *Eskimo* was the first sound film produced in Denmark.

ter Schmidthässler's silent film *Das Eskimobaby* from 1918, starring Asta Nielsen as Ivigtut, who meets a young polar scientist, Knud, whom she follows to Berlin. The film thematises exotic Greenland versus Western culture and depicts Nielsen as a nature child who causes trouble in a philistine family. The film shows only sporadic familiarity with Greenland and is full of misguided stereotypes.

In 1930, Georg Schnéevoigt's Danish-produced film *Eskimo* came with Norwegian/Greenlandic speech and with Greenlandic actors as extras, created on the basis of polar explorer Ejnar Mikkelsen's novel *John Dale* (1921). The film stages Greenland as the country where you experience life as significant, purposeful and healing.

Documentaries

From the first short film, *Driving with Greenland Dogs,* by director Peter Elfelt, recorded in Fælledparken in

PALLADIUM/DET DANSKE FILMINSTITUT, 1934

THE WEDDING OF PALO

Paalup nuliarsarnera (*The Wedding of Palo*), which premiered in 1934, was the first East Greenlandic-language film. Knud Rasmussen (see p. 121) wrote the script, the German documentarist Dr. Dalsheim was behind the camera.

The film is about falling in love and the road to marriage. The process is presented as more than a ritual, and it shows how those involved handle some of the conflicts that a wedding can bring about. At the same time, the film is an example of how Knud Rasmussen, through his commitment, found many ways, including the medium of film, to promote Greenlandic culture.

In *The Wedding of Palo*, Knud Rasmussen is assisted by the purely East Greenlandic cast, which, together with the stunning scenery, is what makes the film succeed.

The film premiered three months after Knud Rasmussen's death and thus became a memorial to him.

1897, many documentaries have been made by Danes about Greenland. In particular Jørgen Roos and his fascination of Knud Rasmussen and Greenland resulted in the Danes getting good knowledge of the country, not least with Per Kirkeby and Aqqaluk Lynge's: *And the Authorities Said Stop* (1972) about the decommissioning of the coal mine and thus the town, Qullissat (see p. 117). Another documentary that describes Greenland as a developing country is Lene Aidt and Merete Borker's *Women workers in Greenland* (1975).

From the beginning of the new millennium, a series of documentaries are seen that are increasingly critical of former taboos in Greenland's history. As Greenlanders play an increasing role in the productions, the focus shifts from seeing Greenland from outside to experiencing it from within.

One of the Danes who has a critical eye on colonial power is Ulrik Holmstrup with *Hingitaq – the Outcasts* (2004), who, like Roos and Lynge's films, depicts one of the black chapters of colonial history, the forced relocation of an entire population in 1953 after the establishment of the Thule Air Base (see p. 127). Karen Littauer made the commemorative documentary *I remember … Tales from Greenland* (2002), in which 14 Greenlanders tell of their childhood and youth and give a portrait of a country and its people.

Sasha Snow's *Arctic Crime & Punishment* (2002) depicts a North Greenlandic community; violent crime is balanced between traditional lifestyles and a contemporary conception of crime. Laila Hansen's *Inuk Woman City Blues* (2002) approaches traditional themes about the legacy of colonialism, alcoholism and rehabilitation from the perspective of women in Denmark. With *The Land of Human Beings – My Film about Greenland* (2006), the director Anne Regitze Wivel shows a multifaceted portrait of a country where the debate on political independence and identity is discussed. Young Greenlandic artists are in the cast, including Angu.

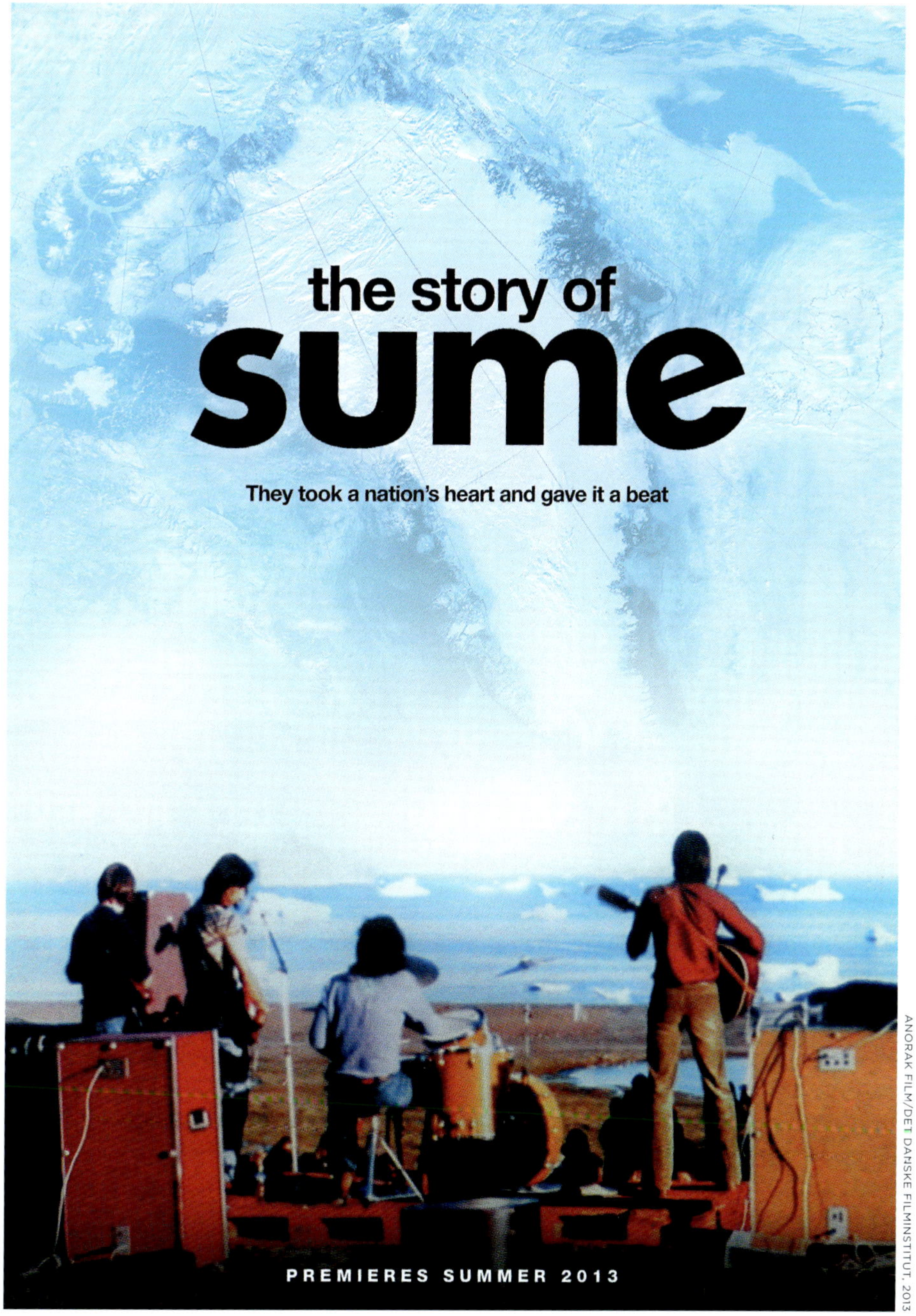

ANORAK FILM/DET DANSKE FILMINSTITUT, 2013

A reviewer of the film *Sume – the sound of a revolution* wrote 'The guitar is often mightier than the sword'. Sume's powerful melodies and songs became the soundtrack to a national campaign for freedom. The film takes viewers through the past and present with music performances, political speeches and interviews with well-known Greenlanders who witnessed this song-political revival. The red loudspeaker, as on the poster, is used as a constant prop and symbol.

NORDISK FILMS KOMPAGNI/DET DANSKE FILMINSTITUT, 1956

Qivitoq (*Mountain Wanderer*) is a tale of hardship, of scorned – and newfound – love and of a society forced to change. The mountain wanderer is portrayed as someone to be feared and as someone who sets out to die in solitude. The film was shot in Greenland in the summer of 1956 and clearly shows how vulnerable humans are in Greenland's magnificent nature. In 1957, *Qivitoq* was nominated for an Oscar in the category 'Foreign language film'. On the poster, the Danish actors are mentioned by their real names, while the Greenlanders are mentioned by the name of the character they played.

The television documentary *The Escape from Greenland* by Poul-Erik Heilbuth (2007) stirred fierce debate in Greenland over unilaterally focusing on the dark sides of society, with many calling it unethical and wrong. Another 2011 documentary, Sarah Gavron's *On the edge* (2011), gave a different positive portrait of an isolated settlement which survives against all odds.

The documentary film *Sume – the Sound of a Revolution* (2014) with screenplay by director Inuk Silis Høegh, Emile Hertling Péronard and Ine Urheim breaks with the past. The title alludes to the band Sume (see p. 217) that was the musical soundtrack to the struggle for home rule 1979. In an alternation of historical clips, music and interviews, the film depicts Greenlanders' longing for freedom, co-determination and break with colonial power.

Danish-Greenlandic collaboration is seen in *The Raven and the Seagull* by Lasse Lau (2018), which shows Greenlandic identity's break with the prejudices and clichés about themselves. *The Fight for Greenland* from 2020 by Kenneth Sorento and co-producer Jørgen Chemnitz depicts the attempts by young Greenlanders to create a new and better Greenland. Among other things, attention is focused on the independence debate and the relationship with Denmark.

The documentary *Pilluarneq Ersigiunnaarpara* (*Happiness Scares Me No More*) from 2020 by Nivi Pedersen focuses on life after sexual assault as part of Killiliisa, Naalakkersuisut's strategy against sexual assault.

The years of collaboration

In the years from the mid-1950s and to the late 1990s, it was especially the collaboration between Denmark and Greenland that characterised Greenlandic films. *Qivitoq* (The *Mountain Wanderer*) from 1956, directed by Erik Balling, focuses on a familiar theme about Greenland as the country where the traveller can find himself again after defeat in Denmark – and at the same time the film highlights the difficult local transition from the hunting to the fishing profession. *Qivitoq* has a Danish and Greenlandic cast, including Greenland's first film star, Niels Platou.

The collaboration continued with Palle Kjærulff-Schmidt's *Tukuma* with screenplay by Palle Kjærulff-Schmidt, Klaus Rifbjerg, Josef Tuusi Motzfeldt and Greenlandic actors from the Tuukkaq Theatre. The theme is familiar; personal crisis, suicide attempts and the journey to Greenland in a search for identity.

Self-representation and collaborative production

The 1980s and 1990s mark the beginning of Danish-Greenlandic collaborative productions that deal with both Greenlandic and more universal themes. *Heart of Light* from 1998 by Jacob Grønlykke with screenplay by Grønlykke and Hans Anthon Lynge marked a milestone: It was filmed in Greenland, the language was Greenlandic and a number of well-known Greenlandic actors were in the cast. The drama deals with a family tragedy against the backdrop of colonialism, alcoholism and the search for identity in the wild nature.

The collaborative films paved the way for the first generation of Greenlandic filmmakers, and their works began to appear at the turn of the millennium. The films had more and more Greenlandic cast members: writers, directors, producers, actors. The topics reflected that the young generation wanted more than colonial history.

Films such as Inuk Silis Høegh's *Sinilluarit* (1999) and *Eskimo Weekend* (2002) broke with depictions of Greenland as a troubled or exotic country. Instead, these short films show an ordinary youth weekend filled with jealousy, romance and happy end. A particularly Greenlandic dispute is felt when two young rivals 'outsing' each other as in the traditional singing duels (see p. 210).

The feature film *Hinnarik Sinnattunilu* (2009), directed by Angajo Lennert-Sandgreen who also plays the lead role, has become cult among Greenlandic children. The film is produced by Aka Hansen and Malik Kleist in their own production company Tumit Production. *Nuummioq* by Torben Bech and Otto Rosing also came in 2009, and like *Hinnarik Sinnattunilu*, it marks a break with the focus on the past and past inherited problems seen in *Heart of Light* and in the French-Greenlandic collaboration *Inuk* (*Le Voyage d'Inuk*) by Mike Magidson (2010). On the visual and thematic level, *Nuummioq* breaks with the stereotypical notions of Greenland and individually describes general human events such as love and sickness.

Between the series of Greenlandic films, another Danish film is seen with a critical view of Greenland's colonial history, *The Experiment* from 2010, directed by Louise Friedberg from a screenplay by Friedberg and Rikke de Fine Licht. The film tells the story of 22 Greenlandic children who, as part of a Danish-Greenlandic experiment in 1951, were sent to Denmark to be raised to be elite Greenlanders (see p. 130). The film, inspired by Tine Bryld and Per Folkver's interview book *I den bedste mening (With the Best of Intentions)* (1998), led to demands for Denmark to apologise.

Young film industry

The first Greenlandic youth horror movies, the hit *Qaqqat alanngui* (2011), is directed by Malik Kleist. The film easily measures up to American horror movies, and arguably there are many international references, but the horror also draws on Greenlandic aesthetics and cultural history: The murderer, who is a qivittoq (mountain wanderer), one by one murders the young people out in the Greenlandic nature.

Scene from *Heart of Light* starring Rasmus Lyberth and Vivi Nielsen. Rasmus is devastated and has already lost his self-respect when tragedy strikes: His eldest son kills several young people and commits suicide. To regain the respect of his family, the settlement and himself, Rasmus sets out on a hunting expedition, and he returns home a hunter and a human being.

ASA FILM PRODUCTION/DET DANSKE FILMINSTITUT, 1998

ADAM LYBERTH/POLARAMA GREENLAND, 2018

ULANNAQ INGEMANN/POLARAMA GREENLAND, 2018

The singer Anori is played by singer and actor Nukâka Coster-Waldau. She grew up in Uummannaq, trained at the National Theatre School in Copenhagen, and appeared in many Danish and Greenlandic films. Anori, which translates as wind, is a dramatic love story; the death of Anori's lover takes her on a mythical and sensual journey, with Greenland's magnificent scenery and the New York skyline acting as a visual backdrop.

By the same director, Malik Kleist, came *Unnuap Taarnerpaaffiani* (*When The Darkness Comes*) in 2014. Also this film is completely Greenlandic in terms of director, production, cast and language. Although the horror elements are worthy of an international horror film, however, there is a special Greenlandic vibe with monsters from Greenlandic mythology. One of the newest feature films is Pipaluk Kreutzmann Jørgensen's film *Anori* (2018) which she has written, produced and directed. The film blends a tragic love story with old Greenlandic myths, and featuring Nukâka Coster-Waldau as Anori, the film has a series of dreamlike sequences set in Greenland and views to New York.

With these films, a young film industry has also blossomed up with the creation of Tumit Production, film.gl and Polarama Greenland, which aim to professionalise the local film industry and create entirely Greenlandic produced films. The film language is now more varied with looks at both past, present and future.

LITERATURE

Literature has evolved from the anthropological narratives collected by the colonial power over a modernism that discussed, among other things, Greenlandic identity, nationality and traditions in the face of new currents. Increasingly, thematics have become more individualistic, just as Greenlandic writers are increasingly breaking taboos and discussing their own identity.

Time up to 1930

The pre-colonial era was characterised by orally handed down myths and legends intended to enlighten and to preserve tradition and language. With colonisation came the inspiration for collecting these myths and legends, and from 1857, South Greenland's inspector, H. Rink, published the *Kaladlit Oqalluktualliait/Grønlandske Folkesagn (Greenlandic legends), which consisted, among other things,* of manuscripts by Aron of Kangeq and Jens Kreutzmann of Kangaamiut (see p. 154).

In 1921-1925, Knud Rasmussen published *Myter og Sagn fra Grønland (Myths and Legends from Greenland),* collected during The Literary Expedition in 1902-04 and the Fourth Thule expedition in 1919 (see p. 212).

Wanting to help build the country, H. Rink set up the first book printing house in Godthåb in 1857, publishing the first book, *Pok, in 1857,* and in 1861, he founded the first Greenlandic newspaper *Atuagagdliutit* (see p. 122). From 1861, the chief catechist Rasmus Berthelsen was from 1861 editor of *Atuagagdliutit.*

The literature that could now be published by the printing house was characterised by religiosity and patriotism.

Modernism

The first published authors

Among the pioneers were Josva Kleist and the poet priest Henrik Lund, who in 1912 wrote the lyrics for *Nunarput utoqqarsuanngoravit*, adopted as the national anthem of Greenland in 1916, with music by Jonathan Petersen. In 1921, Lund published *Erinarssûtit,* a folk song book that has contributed to national consciousness.

The Act on the Greenland Provincial Council of 1911 gave the population new requirements for the country's development. The population and writers participated in the debate on Greenlandic identity. Representatives of the progress were seen in the first two Greenlandic novels: Mathias Storch's *Sinnattugaq* in 1914 (*En Grønlænders Drøm (the Dream)*, 1915) and Augo Lynge's *Ukiut 300-nngornerat* in 1931 (*300 år efter (300 years later)*, 1989). They criticise their own time, demand change and produce a vision of the future in which Greenland has developed into a society full of wealth and equality.

Romantic nationalism

From 1930-1950 onwards, the well-educated elite, with outside inspiration, wrote poems, songs, plays and novels. The great changes that occurred made it natural to strengthen national consciousness and seek a cultural and identity standpoint in romanticised depictions of bygone days, the dream of change and confidence in one own's strengths. Frederik Nielsen wrote Greenland's first collection of poems *Qilak – nuna – imaq* (1943) and several novels, including *Tuumarsi* (1934) and *Ilissi tassa nunassarsi*. Pavia Petersen wrote the first contemporary comedy, *Ikinngutigiit* (1934) and posthumously *Niuertorutsip pania* (1944) about the life of the daughter of a mixed marriage. With *Ersinngitsup piumasaa* (1938), the all-round artist Hans Lynge depicted the pre-colonial era when codes of honour force two brothers into an unhappy killing.

Mental decolonisation and breakup

After the amendment of the Danish Constitutional Act in 1953, Greenland was incorporated into the Danish kingdom (see p. 127). The Greenlanders adopted a wait-and-see and not always enthusiastic attitude, and the following years of literature was marked by protest against the upheaval of society.

JOHN MØLLER/ARKTISK INSTITUT

Teachers' training college students and their teachers: Terkel Møller, Hans Motzfeldt, Jonathan Petersen, Gustav Olsen, Christian Rosing, Schultz-Lorentzen, pastor Nicolai Edinger Balle, Rasmus Berthelsen, Niels Lynge, Sechmann Rosbach, Jørgen Brønlund and Josva Kleist. Nuuk. April, 1899.

Ole Brandt is an icon of Greenlandic literature, not least because of his novel *Qooqa* (1971), which helped to give Greenlanders cultural dignity and identity by showing the skill of their ancestors. Brandt was a teachers' training college teacher and founded the Greenland Writers' Association, of which he was also chairman. Here he is attending a Christmas party in Aasiaat, ca 1953-1955.

NIELS ERIK THYGESEN/ARKTISK INSTITUT

Many kept focusing on the past, as seen, for example, in the novel *Nunagssarsiak* (1954) by Ullorianguaq Kristiansen, which tells the myth of the struggle of the Greenlanders and Norse. Otto Rosing's novel *Taseralik* (1955) pays tribute to the ancestors and describes colonial Greenland and the annual summer camps of the mid-19th century.

The debate over centralisation and depopulation of the settlements is seen in Ole Brandt's 1971 cult novel *Qooqa*, which urges compatriots to fight against development and to believe in their own experience. The authorship of the priest Otto Sandgreen was based on source material collected from elderly people in the outermost regions. The main work, the novel *Isi isimik kigullu kigummik* (1967) (*An eye for an eye and a tooth for a tooth*, 1987), describes soul loss and blood feud in an East Greenlandic family of shamans. Sandgreen continues this line with his biography *Georg Quppersimaan: Taamani Guutimik nalusuugama* (1972) (*Min eskimoiske fortid (My Eskimo past)*, 1982), in which he tells, via the memoirs of Georg Quppersimaan, of the life of an East Greenlander before his conversion to Christianity. With the collection of poems *Nalusuunerup taarnerani* (1965), Villads Villadsen described in lyrical form both East Greenland and the clashes of the past between the Norse and Inuit.

From the late 1960s to the home rule in 1979 – and the time thereafter – a number of angry young men put the resistance into words. The protest poetry that followed in the wake of the student revolt in Europe was led by Greenlandic students in Denmark, such as Aqqaluk Lynge, Moses Olsen and Aqigssiaq Møller. They embarked on a new and central theme, homesickness, and took a critical perspective on Denmark's presence in Greenland. The style was confrontational with a keynote of anti-colonialism and anti-imperialism. The themes revolved around the unfortunate repercussions of the new arrangement, and concepts such as identity crisis and alienation emerged.

Poetry as a weapon in the political struggle is seen in Moses Olsen's short story *Aamma uagut taamaappugut?* from the collection of poets and short stories *Allagarsiat* (1970), which portrays the consequences of modernisation and a subdued people. Aqqaluk Lynge satirises what the Danes have accomplished in the poem *Ode til Danaiderne (Ode to the Danaids)* from the collection of poems *Tupigusullutik angalapput (The Veins of the Heart to the Pinnacle of the Mind*, 1982). He sees the Danes as power seekers who discard the Greenlandic people and practice inhuman humanist imperialism. Lynge clearly addresses the Danes, just like Jens Geisler and Aqigssiaq Møller.

Individualist writers and global influence

In 1979, Greenland gained home rule, and the authors gloomily described the change of times, but now also focused on individual issues. In 2010, Kristian Olsen Aaju wrote Greenland's first crime novel, *Kakiorneqaqatigiit*. Hans Anthon Lynge published *Umiarsuup tikinngilaattaani* (1982), *Allaqqitat* (1997) (*Bekendelser (Confessions)*, 1998) and co-authored the screenplay *Heart of Light* (*Qaamarngup uummataa*) (1997) (see p. 223).

One of the first female writers, Mâliâraq Vebæk, focuses on the oppression of women and the problems that can arise from the Greenland–Danish culture clash in his *Bussimi naapinneq* of 1981 (*Historien om Katrine (The story of Katrine)*, 1982) and *Ukiut trettenit qaangiummata* from 1992 (*Tretten år efter (Thirteen years later)*, 1997). With Grethe Guldager Thygesen's *Kalaaleq arnaq* (1984) and Dorthe Nathanielsen's novel *Aani* (1986), the depiction of female life takes a place in literature.

Although views and literary expression could still be trenchant, the period leading up to and following the introduction of self-government in 2009 became more diffuse and characterised by more voices. The diversity of literature now allowed for taboos to be broken, individualism and avant-garde art. Jessie Kleemann caused outrage with her avant-garde and provocative poem collection *Taallat* in 1997.

New currents with individual issues and tackling taboos in society are seen in Vivi Lynge Petrussen's *Naalliutsitaanerup pissaanerani* (2001) on domestic violence. Kelly Berthelsen describes sexual abuse, alcoholism and substance abuse with *Misigissutsit: Taallat* in 2000 and in *Tarningup ilua* in 2001 (*En sjæls inderste kammer (The innermost chamber of a soul)*, 2002).

Otto Steenholdt embarks on the horror genre of qivittut/mountain wanderers with *Inuillisimasup ikioqqunera (Et råb om hjælp (A cry for help)*, 2001/2014). With the novel *Tarrarsuummi Tarraq* in 1999 (*Saltstøtten (The pillar of salt)*, 2000) and the short story collection *Uumasoqat* in 1992 (*Det andet dyr (The other animal)*, 1998), the provocateur Ole Korneliussen brought phenomenology, existentialism and human being into literature. The lyricist Marianne Petersen, nominated for the Nordic Council Literature Prize in 1993, adds burlesque and grotesque humour to her poems.

Many writers paint, write and make music, e.g. the visual artist Frederik 'Kunngi' Kristensen (see p. 203) is also known for his poems.

Sørine Steenholdt's debut book *Zombieland* was nominated for the Nordic Council Literature Prize in 2016. In *Zombieland,* Steenholdt serves up a societal critique in powerful short stories that cannot be ignored. The gloomy universe is supported by expressionistic illustrations by the young artist Maja-Lisa Kehlet Hansen.

MAJA-LISA KEHLET (ILLU. COVER) OG ULANNAQ INGEMANN (PORTRÆT) /MILIK PUBLISHING, 2015

Individualistic issues and breaking taboos

In 2012 and 2014, NAPA (Nordic Institute in Greenland), Nunatta Atuagaateqarfia (Central Library of Greenland), NUIF (Nuuk Youth in Progress) and Milik Publishing organised the short story competition Allatta! for young people between the ages of 15 and 25. It brought a breakthrough in Greenlandic literature with two short story collections and several spin-offs, including a novel by Niviaq Korneliussen, *HOMO sapienne* (2014), about finding your gay identity, a short story collection by Sørine Steenholdt, *Zombiet nunaat* (2015) (*Zombieland*) describing vulnerable people and abuse, and a short story and poem collection by Pivinnguaq Mørch, *Arpaatit Qaqortut* (2015), depicting existential themes of life and everyday challenges. As the first author, Niviaq Korneliussen was successful in her complaint against the refusal of support from the Danish State Art Foundation and wrote her second novel, *Naasuliardarpi* (2020) (*Flower Valley*), following a grant. The novel, which in 2021 won the Nordic Council Literature Prize, describes grief at witnessing suicide and feeling different.

Many of the young writers are so secure in their identity that they can negotiate it or criticise parents, society and their compatriots. Taboos are broken and the view is individual with local issues, but also international and innovative both in style and content.

MEDIA

With a population size of only just over 56,000 people living widely scattered, the few nationwide media have much significance. However, the lack of a media policy means that news coverage is given low priority.

In Greenland, there are two nationwide media houses, the public service station KNR, which produces radio, television and web news, as well as the media house Sermitsiaq/AG, which publishes two weekly newspapers, web news and a few maga-

MARTIN SYLVEST/RITZAU SCANPIX, 2021

Niviaq Korneliussen will receive the Nordic Council Literature Prize in November 2021 for the novel *Naasuliardarpi Blomsterdalen* (*Flower Valley*) at The Royal Danish Playhouse, Copenhagen.

zines, as well as the household-distributed free newspaper *Nuuk Ugeavis*. In addition, there are a few local newspapers, local television and local radios stations, which operated on a voluntary basis and mainly in Nuuk, Sisimiut, Ilulissat, Maniitsoq and Qaqortoq.

KNR – public service

The first television programmes broadcast in Greenland were recorded from Danish television on VHS tape and were broadcast via the Godthåb closed TV networks from 1966. The association was paid by its members and based on volunteering. In 1981, there were 25 associations, but a wish existed for real television that did not depend on Danish television programmes being tapped, transported by ship, selected and broadcast by the local associations.

Following the introduction of home rule, the cultural area was transferred in 1981 (see p. 136), including also broadcasting. Home rule now legalised the illegal privately based tappings.

Simultaneous television broadcasting began on 1 November 1982. In the years before, there were endless discussions of economy, culture and language. Politicians and media professionals thought that once the technology and economy were in place, it would gradually become possible to produce Greenlandic broadcasts that could meet the public service obligations.

The prospect of national television was characterised by ambivalence: Idealists focused on the fact that entertainment would keep young people off the streets, that it would benefit language skills and strengthen education and cultural work. Critics feared that the foreign influence would undermine the Greenlandic language and national identity. A 1975 television report showed that a language distribution was envisioned with 85 % self-production/versioning and 15 % non-versioned. However, the figures turned out to be the other way around, often to the criticism of the population.

Over the years, airtime increased steadily: In 1983, KNR aired an average of four hours and 20 minutes per day, of which 17 minutes were self-production. By 2011, the airtime had expanded to 13 hours and 52 minutes, of which one hour and 27 minutes were self-production – with current news and current affairs programmes such as *Qanorooq* (news) and *Tamassa*, in-studio visits, entertainment and debate.

The fact that KNR-TV's annual broadcasts consisted of an average of about 15 % self-production was a matter of resources. A comparison with the Danish public service channel showed that it was impossible to expect more in a large and sparsely populated country. It was not until 2014 that there was 100 % TV coverage everywhere.

In 2013, KNR reorganised from flow channel to niche channel. The mix of Danish and Greenlandic programmes was superseded by self-produced programmes. Danish programmes were now broadcast live from DR and TV 2. Channel and programme selection could be decided by television consumers themselves, and additional TV packages could be purchased in more places.

Many new programmes have come into being, and Greenland TV has a modern multi-platform interface, but with fierce competition from payment channels and internet.

Media standpoint in a democratic society

In December 2010, a media working group established to shed light on the development perspective of Greenlandic media released a statement that said, among other things, that the media must tie the country together and reflect all of society to include all types of audiences. But the limited funds and resources available for distribution and travel prevent the media from fulfilling their role optimally.

The terms of reference of the working group intended to create a diversity of free media that could involve citizens in the free democratic process. The media should meet the need for versatile and engaging experiences, raise the level of information and, not least, reflect Greenlandic language and culture. But the expected political debate did not materialise.

After more than a decade, the report can still form the basis for a discussion that addresses the development perspective of the media.

Just as in the rest of the world, many Greenlandic young people pick up most of the news from social media, and not from the productions of the nationwide media. A glance at the content of the nationwide media shows that their primary target audience is not the young generation. The pattern uncovered by the media working group in 2010 is largely unchanged: Children and young people were and are the group that gets the least attention in the media picture.

Media landscape today

The Greenlandic media have few resources, the advertising market is limited and there is a shortage of professional staff and specialised knowledge. In 2017, there were only about 35-40 full-time journalists to produce all kinds of products – from news and cultural programs to investigative business journalism. That picture has not changed. Yet both Sermitsiaq/AG and KNR have online newsdesks, which bring daily news.

The *Sermitsiaq/AG* and *AG/Atuagagdliutit* newspapers, which are the only nationwide newspapers, each have their editorial staff of about five journalists, with about half being bilingual (Greenlandic) journalists and the other half being Danish journalists recruited from Denmark. The same applies to KNR's news department Radioavisen, whose content is broadcast in both Greenlandic and Danish. Furthermore, KNR has a television news department, Qanorooq. Five days a week, Qanorooq broadcasts news from the coast and short news from the outside world with an editorial staff of only five or six bilingual journalists. Television news are subtitled in Danish, allowing everyone regardless of language to keep up.

It is not only journalists who make up the professional workforce in the media. The nationwide newspapers have graphic designers and versionists, as well as a photographer. Likewise, KNR has AV assistants, technicians, super users and production assistants.

The media and the language issue

Versionists have an indispensable function in media production, as both Danish journalists and some Greenlandic journalists write articles in Danish and have them translated into Greenlandic.

Within virtually all public and private companies, educational establishments, etc., the working language is Danish, so although Greenlandic is the official main language, the everyday language is strongly influenced by the Danish language, which means that the journalist's sources are very often in Danish This requires journalists to master both languages fluently, and it can be a barrier for young people wanting to train to become a journalist.

On 24 April 2018, elections were held for Inatsisartut, the Parliament. On 18 April, KNR held a live TV debate about self-government.

JONAS OLUFSON/RITZAU SCANPIX, 2018

SOCIETY AND BUSINESS

IVARS SILIS/NUKISSIORFIIT, U.A.

What and who sets the framework for developing the individual, businesses and society in Greenland? The overall framework is set by the Danish Realm, the Government of Greenland and, not least, by the global trends in the political and economic markets.

Basically, the social structure and political system as well as administrative structures in Greenland are based on Scandinavian/Danish norms and ways of organising. The significant differences lie in the fact that it is a completely different country with a completely different people and a culture that was beyond comparison with the Scandinavian or Danish ones. In addition, the population is only about 56,000, which is of course a limitation in itself, not least in a very extensive country of about 2.2 million km^2 with a scattered settlement pattern which impedes all public efforts.

In essence, Greenlandic communities are operationally so-called insular communities where it is necessary that all societal functions, such as electricity supply, school, healthcare, shop, etc., be found in all communities, almost no matter how small they may be.

The Qorlortorsuaq hydropower plant, built in 2004-2007, supplies energy to the two towns of Narsaq and Qaqortoq. The plant is part of the Nukissiorfiit public utility, whose supply responsibility covers 17 towns and 53 settlements. Nukissiorfiit is a net managed company owned by the Government of Greenland. Previously, assignments in the field of energy were carried out by the Greenland Technical Organisation (GTO).

Photo pages 230-231. At the Government of Greenland's experimental farm, Upernaviarsuk Experimental Station, agricultural experiments are conducted, and students at Upernaviarsuk Agricultural School are taught here.

MADS PIHL/VISIT GREENLANDW, 2015

The question is often: What could the Greenland society be compared to? Most often, it is compared with the Danish society, regardless of the aforementioned fundamental differences. In short, the living conditions and many other societal conditions in Greenland are not comparable to those in Danish society.

Greenland is more comparable to all current and former colonies of Western countries – and, of course, not least comparable to other Arctic, isolated and small communities. In all colonised societies, from north to south and from west to east, people grapple with the same social, personal and economic problems. There are approximately 375 million people on the globe who identify themselves as indigenous peoples. These are most often people who have been split up into different states and island communities when borders were drawn between the states. What characterises these communities is that they exercise self-determination to varying degrees, but the colonising state still has jurisdiction over crucial matters.

A number of publicly owned infrastructure companies, such as Tusass (formerly Tele Post Greenland), Air Greenland, Arctic Umiaq Line, Royal Arctic Line, KNR (Greenland Radio), town buses and privately owned taxis and broadcasters provide a physical and digital infrastructure that is accessible to the vast majority of places.

Climate change is the cause of much of the renewed attention on Greenland and the rest of the Arctic. In the new world order Greenland has again been brought into play against old and new enemies in the world for the United States. During COP26 in Glasgow on 1 November 2021, Premier Múte B. Egede stated that Greenland will join the Paris Agreement, which was adopted at COP21 in 2015. Greenland is thus positioning itself as a climate-friendly country where oil and gas exploration has stopped. Critics believe it will cut off the country from foreign and much needed investment, while Naalakkersuisut is confident that a green profile for Greenland is more alluring to potential foreign investors. Renewable energy sources, such as wind and hydropower and the sun, are seen as great potentials for the country's own supply of energy, but also as a basis for stored energy and as a supply to energy-intensive industry.

The Danish State and its institutions in Greenland perform a number of civil and military functions. In addition to the legal system and the upper courts, this includes the enforcement of the borders of the kingdom and the fisheries inspection, as well as 31 other more or less essential responsibilities.

Lack of education remains a crucial Achilles heel for the development of society. Regardless of steady progress, there is widespread criticism of the results at virtually every level of the education system. In primary school, both the children and teachers grapple with problems that relate more to life outside school than actual schooling. In the institutions of higher education and in the trades, the dropout rate of students and pupils is high. There is also free admission to Danish educational institutions, and students can obtain Greenlandic educational aid if they take courses deemed eligible by the Self-Government, or they can choose to get SU (Danish state education grant).

The low degree of formal education, and perhaps, as a result, also the low level of attention to international relations, shape the public debate and thus also set part of the political agenda. At the same time, it makes it necessary to call in a relatively large number of formally trained workers from outside. Recent statistics show that in 2019, young people 'outside the system' in the 'youth target group' (16 to 25-year-olds) amounted to just over 3,000 persons. That is young people who are not in education or employment.

THE UNITY OF THE REALM AND THE DANISH STATE

The formal connection between Greenland and Denmark came about when in 1380 Denmark and Norway formed a union. That was the starting point when, from 1721, Danes and Norwegians resumed colonisation (see p. 113). When Denmark was forced to cede Norway to Sweden in 1814 after the Napoleonic Wars (see p. 119), trade and administration of Greenland and the Faroe Islands were transferred to Copenhagen.

After World War II, the Danish government asked one of the country's most prominent jurists, Professor Alf Ross, to assess the changes planned with regard to the position of Greenland and the Faroe Islands. He took the opportunity to describe the concept of the 'Community of the Realm' as legally meaningless, because a 'community' requires the parties to have equal status. Ross therefore believed that the term 'unity of the realm' better described the relationship. Nevertheless, in daily practice, the concept of the 'Community of the Realm' forms the framework for relations between Denmark, Greenland and the Faroe Islands – and for the discussions on what the relationship should be like.

The constitutional status of Greenland and the Faroe Islands has been contested in separate ways as a result of widely differing history in terms of belonging to Denmark. The Faroese have insisted on describing

THE GREENLANDIC INSULAR COMMUNITY

MADS PIHL/VISIT GREENLAND, 2016

The settlement of Aappilattoq in Kujalleq Kommunia has just under 97 inhabitants (2021). The isolated settlement is just over 50 km from Greenland's southernmost point, Uummannarsuaq (Cape Farewell).

Greenland is characterised by its geographical expanse, corresponding to the distance from the Shetland Islands to North Africa, combined with an Arctic climate and a population of just over 56,000. With just 0.03 inhabitants per km^2, Greenland is thus the most sparsely populated country in the world, while covering a geographical area larger than Central Europe where more than 300 million people live. Accordingly, Greenland society and its business structure differ in a number of areas from the countries with which Greenland is often compared, such as the Scandinavian countries and Iceland.

A CHALLENGING TRANSPORT INFRASTRUCTURE

Besides the ice sheet, which covers eight-tenths of the country, Greenland is also challenged by sea ice where the multiyear ice along the east coast prevents boats from reaching Tasiilaq and the district's towns and settlements from November to mid-May. Ittoqqortoormiit usually receives only two annual supply ships. Similarly, the multiyear ice can periodically complicate – or even make impossible – sailing in South Greenland. In North Greenland, the districts of Uummannaq and Upernavik can today only be reached by boat from May to December, Qaanaaq receives only two annual supply ships, and in winter sea ice causes delays or ship cancellations in Qeqertarsuup Tunua (Disko Bay). To this should be added the challenges related to cold, hurricanes, large temperature fluctuations, precipitation and drought.

Due to climate, permafrost as well as the wide and deep fjords, and because several towns and settlements are located on islands, establishing long road connections is complicated and extremely expensive. These are challenges, which also exist in other desolate areas such as Alaska, Arctic Canada, Iceland and Siberia, where a coherent road network has been established to some extent. However, for Greenland, the challenges, combined with the very small population, have meant that no inhabited places are connected by road to other places, apart from some sheep farms in South Greenland.

Therefore, transport between Greenland's 17 towns and 54 settlements takes place by boat or aircraft, and in winter combined with snowmobiling and in North and East Greenland also by dog sled. In some places the sea ice is used in winter to drive between towns and settlements by car and truck. The transport structure means that it might take one or more days to get from one town to another, and then you still have to factor in weather delays and cancellations. So travel time can be long, even between the towns and settlements that lie the closest to each other.

GREENLAND FUNCTIONS AS A COMMUNITY OF 71 ISLANDS

The transportation challenges mean that residents are dependent on the facilities and services available in the individual place and that all facilities and services must basically be established and operated as a so-called insular community that has no connections to other towns and settlements. For instance, all towns and settlements have their own electricity and water supply without any backup from other parts of the country, with the exception of Qaqortoq and Narsaq, which have a joint hydropower plant.

At the same time, the insular community and transport infrastructure mean that there is nowhere where it is possible to commute to work elsewhere on a daily basis. This limits the labour market for each citizen to the individual settlement – town or settlement – and, in terms of fishing, the profitable radius of action is the individual fisherman's vessel.

THE MARKET ECONOMY MECHANISMS ARE CHALLENGED

All towns and settlements are small. Even the capital of Nuuk, with more than 19,000 residents, is, in an international context, a very small town, and the country's smallest settlement has less than 20 residents. Accordingly, the average size of Greenland towns and settlements are more comparable with those in, for example, parts of Alaska, Arctic Canada and Sápmi (Sameland).

The size of the towns and settlements and the complex and costly transport infrastructure challenge the market mechanism in a number of vital areas because in many towns and settlements, no customer base exists for competing providers of a given product group or service. This creates natural monopolies that can challenge price formation unless regulated by the government.

Moreover, several towns and settlements are so small that there is no customer base for a single provider, even within such vital areas as retail. This issue can either lead to vi-

HENRIK FORSBERG

Greenland placed on top of Europe in the same scale to illustrate size and distances.

tal services not being offered or that the government will have to take over tasks that in a market economy would be provided by private operators. The alternative is for the government to pay a private company in whole or in part in subsidies to provide the service, as is the case for example in the Self-Government service contracts with private operators.

A LARGE PUBLIC SECTOR

Consequently, in trying to maintain Greenland as a welfare state, the government owns a number of companies that carry out societal tasks in those places and/or commercial areas where the customer base and geographical and climatic conditions do not allow the tasks to be solved on market terms. In the large towns, some of these tasks are provided on market-like terms. However, the wish for and political priority to maintain a dispersed settlement community with access to welfare services (albeit at different levels) would not be possible without the government-owned companies and, for example, the same prices on energy, water and telecommunications services.

In general, the insular community entails a large number of challenges, for instance that each town and settlement must have its own school, health centre, electricity and water supply, shop, etc. This means that several of the government-owned companies have many 'branches' for a small customer base to ensure that, for example, all settlements have someone who maintains the power plant and water works. It also means that the public administration will inevitably be much larger than in countries where, for example, children can ride by bus to school in the nearest major town.

SETTLEMENT PATTERN

The settlement pattern is rooted in history, but it is also prerequisite for exploiting local natural resources due to the size of the country and the large distances.

Fish and shellfish account for about 95 % of the value of total goods exports. Fisheries can be roughly divided into the trawler-based offshore fishery, where it is less important where the trawler has its home port, and inshore fishery which contributes with over 40 % of the total catch value (2018). For large parts of the inshore fisheries, the distance to the fishing grounds is crucial, for instance when it comes to longline fishing for Greenland halibut from the sea ice in North Greenland. It is therefore necessary to maintain a number of small trading places and thus towns and settlements.

The settlement pattern has been a source of debate since the Danish modernisation after World War II (see p. 128), and although the settlement pattern is dynamic and constantly evolving, it is a framework condition for Greenland where distances are substantial. For example, it is about 1,200 km from Qaanaaq to Ilulissat and about 1,500 km from Qaanaaq to Nuuk. Accordingly, the size of the population and the insular community set the framework for the structure of the society.

The coastal passenger ship *Sarfaq Ittuk* at Kangia (Ilulissat Icefjord). *Sarfaq Ittuk* sails between Narsaq and Ilulissat, calling at most towns and a few settlements along the way.

MADS PIHL/VISIT GREENLAND, 2014

themselves as a separate nation that has never accepted the transfer of sovereignty to Copenhagen. Greenlandic criticism has been focused on whether the formal integration of Greenland in 1953 was based on a free and informed consent that lived up to United Nations' standards for decolonisation. The objections have pointed in particular to the fact that Greenlanders were not presented with UN alternatives to integration: independence and free association.

Politically, the Unity of the Realm is often referred to as equal, but the official Danish interpretation remains that the home rule and self-government arrangements in the Faroe Islands and Greenland, respectively, act as regional authorities that have been handed over power from the state. It is generally agreed that it will be politically impossible for Denmark to unilaterally abolish the arrangements. Some independent jurists believe that the arrangements have been given constitutional status, even though they are not mentioned in the Danish Constitutional Act, and that they are protected by the fact that under international law the Greenlanders constitute a people with the right to self-determination. Equality therefore consists of a *right* to independence, although it is manifested differently. Since the Self-Government Act of 2009 describes a procedure for how Greenland declares its independence, discussions have since concentrated on foreign and security policy, where equality is interpreted differently in Greenland and Denmark.

MADS CLAUS RASMUSSEN/RITZAU SCANPIX, 2021

Premier of Naalakkersuisut, Múte B. Egede (IA), third from left, among colleagues at the 73rd Session of the Nordic Council at Christiansborg. The Faroe Islands, Greenland and Åland are working towards increasingly independent status in the Nordic Council and the Nordic Council of Ministers.

Greenland's self-government is far-reaching in comparison with the overseas territories of other states. Self-government covers the vast majority of internal affairs and can be extended to virtually all other areas, even those traditionally considered core tasks of a state, such as police, legal system, foreigners and border control. The only thing maintained by the Danish State is that the constitution, citizenship, Supreme Court, foreign policy, security policy, defence policy and monetary policy should remain under Copenhagen as a prerequisite for being able to speak of a unitary state. The official Danish response to demands for increased self-determination in these areas has usually been that it would be in breach of the Danish Constitutional Act. However, the interpretation of the Danish Constitutional Act has changed over time, most spectacularly when, after all, it proved to be possible to take over mineral resources and, in principle, the police and legal system. The same movement has unfolded in relation to the involvement of Greenland and the Faroe Islands in foreign affairs. For some observers, the development underlines that interpretation of the Danish Constitutional Act in relation to the Unity of the Realm is the result of a political rather than a purely legal process.

Greenland and the Danish Royal House

According to the Danish Constitutional Act, the Danish Kingdom is a constitutional monarchy. The monarch is the head of state for all parts of the kingdom, including for Greenland and the Faroe Islands. Where the relationship with Denmark and the Danish state is a controversial subject, then generally in Greenland there is great devotion and respect towards the Danish royal family.

From the beginning of colonisation, knowledge of the Crown became widespread in Greenland, both through the posted missionaries and merchants and through oral accounts from Greenlanders such as Pooq, who had visited the royal family in Denmark. The notions of the distant king were arguably obscure: Was he the powerful, wealthy and punishing master or the loving, protective family father? The king played an important role as head of the Lutheran Christian Church, and the Greenlandic congregations have since then prayed for the king and royal house at church services.

As the first regent, King Christian X visited Greenland with his family in 1921. For most Greenlanders, it was the first meeting with royalty of flesh and blood. Later royal visits with accompanying kaffemik and celebrations have helped build emotional bonds and more personal relationships between Greenlanders and the royal house.

The close ties are strengthened by intertwining the formal and the emotional, e.g. in the following ways: Queen Margrethe II herself presented the Greenland Home Rule Act in 1979 (see p. 135) and the Self-Government Act in 2009 at ceremonies in Greenland. Greenland shows its recognition of the monarch and royal house through official flag-flying days where royal birthdays are celebrated. The members of the royal house are depicted on Greenlandic stamps, often wearing Greenlandic national costumes. In 2000, the Home Rule decided to name a remote area in northernmost Greenland after the crown prince, Kronprins Frederiks Land. It can be seen as a gesture of appreciation for the crown prince's appropriation of Greenlandic culture when, in 2000, during his four-month dog sled journey, he travelled 2,800 km in northernmost Greenland. Accompanied by former members of the Sirius Dog Sled Patrol and with the assistance of hunters from Qaanaaq, he travelled from Qaanaaq to Daneborg. The crown prince's twins were given the Greenlandic baptismal names Ivalo and Minik in 2011. By many Greenlanders, it was considered conciliatory and significant when in 2015, the royal couple went to Qullissat with the Royal Yacht Dannebrog to see with their own eyes the town which was closed by the authorities in 1972 (see p. 117). Queen Margrethe II, in her televised New Year's speeches, has continued and developed her father's tradition of sending warm and acknowledging greetings to Greenland. From the first New Year's speech in 1972, the Queen's greetings were addressed to 'the people of Greenland' several decades before the Self-Government Act's official recognition of the Greenlanders as a people.

The Greenlandic members of the Danish Folketing

Since the amendment of the Danish Constitutional Act in 1953, Greenlandic voters, interests and views have been directly represented with two members in the Danish Folketing. Siumut has more or less held one of the Greenlandic mandates since the party was created in 1977. The second mandate was held by Atassut until 2001, when IA took over. As opposed to the Faroese members of the Danish Folketing, more than twenty years passed from when Greenland had its own representation until a Greenlandic member joined a Danish party (Lars Emil Johansen became a member of SF – the Socialist People's Party – in 1975). After this, Greenlandic members have most often belonged or cooperated more loosely with Danish parties, most continuously Siumut and the Danish Social Democrats. From 2001 to 2011, the members of the Danish Folketing from Siumut and IA, formed a North Atlantic party together with the Faroese Republican.

KELD NAVNTOFT/RITZAU SCANPIX, 2021

Queen Margrethe II giving the New Year's speech in 2021. The Queen often complements the speech's traditional greeting to Greenland with Greenlandic handicrafts on the desk.

MP Aaja Chemnitz Larsen, elected for IA in Greenland since 2015, during the Prime Minister's Question Time, 2021.

PHILIP DAVALI/RITZAU SCANPIX, 2021

The Greenlandic members of the Danish Parliament have mostly dealt with political issues which have been particularly relevant to Greenland. In particular, they have used the Folketing's question time and committee work to check the Danish government's administration of policy areas still under Danish jurisdiction. As members of the Danish Parliament, Greenlandic politicians have also had the opportunity to influence Danish politics, not least in those cases where they have been instrumental in building a parliamentary majority: After the 1971 elections, Moses Olsen and Knud Hertling were decisive for forming a Social Democratic government. In 2007, for some time, it looked as if the Greenlandic members could become crucial to the centre-right government for getting enough votes in a case involving the possibility of asylum families to live outside asylum centres. In such situations, the reasonableness of Greenland mandates being included on an equal footing has been contested. In recent years, Greenlandic politicians have been part of spending agreements on several occasions, although their mandates have not been necessary for a majority, for example in connection with the defence agreement and in connection with the establishment of a pool of funds for North Atlantic projects on the Budget.

Politics and administration of the Unity of the Realm

Politically and administratively, the Unity of the Realm is made up of a number of rules, institutions, practices and contacts between politicians and officials that tie Greenland and the Faroe Islands together with Denmark. At this more concrete level, the Unity of the Realm is not so much a triangle as two bilateral relations between Denmark and Greenland and Denmark and the Faroe Islands, respectively: The Danish Folketing has a Greenland Committee and a Faroe Island Committee. The Faroe Island home rule and the Greenland self-government might be similar, but they have been negotiated separately and at different times: Block grants are agreed and calculated in two separate processes, and different national legislation and international agreements may or may not be put into force with different adjustments for Greenland and the Faroe Islands, respectively. As far as Greenlanders and Faroese have coordinated, the contact has been mainly been undertaken by individual politicians. In addition to the Danish Parliament, the work of the Nordic Council and Council of Ministers, the Arctic Council, the Arctic Parliamentary Conference and especially the Western Nordic Council have helped to bring the parties together.

Since 1983, the so-called Summits of the Realm have been one of the few institutions to bring together the three political parties of the Unity of the Realm. Since 1995, the Prime Minister of, the Prime Minister of the Faroe Islands and the Greenland Premier of Naalakkersuisut have, by turns, hosted biannual or annual meetings where the three top politicians speak together behind closed doors without officials. Often the venue has had a symbolic value, either nationally or for the host, and the meeting is surrounded by a short or long programme with spouses and sightseeing giving the impression of an official state visit. The nature of the meetings has alternated between family gatherings and strained arm wrestling about current issues.

The Summits of the Realm are prepared by officials. For a number of years, a regular series of meetings was formalised between the manager of the Greenland Home Rule and the Permanent Secretary of the Prime Minister's Office, but due to the intensified

MADS CLAUS RASMUSSEN/RITZAU SCANPIX, 2021

After the Summit of the Realm with Danish Prime Minister Mette Frederiksen (S) and Faroese Prime Minister Bárdur à Steig Nielsen (Sambandspartiet) at Marienborg on 10 June 2021. Premier of Naalakkersuisut, Múte B. Egede (IA), to the right, discusses new roles for Greenland in the Arctic Council and the coordination of the foreign policy of the Unity of the Realm.

flow of foreign and security policy matters in recent years, the series of meetings has been supplemented by ongoing contacts. In June 2021, a committee of ministers from all three parties to the Realm was established and the Prime Minister at the head of the table was established in an attempt to bring the foreign policy of the Unity of the Realm into unison. Both with regard to matters of state and cooperation in issues under Greenlandic jurisdiction, contact between ministries is most often on an ad hoc basis. A secretariat, a steering group or the like with representatives of both the Self-Government and the Danish authority is often formalised to facilitate concrete Danish efforts in devolved areas (e.g. social affairs).

The High Commissioner

The High Commissioner in Greenland, as the supreme representative of the Kingdom, is the heir to an office which has existed since 1782 under various designations and with changing tasks portfolio. Initially, the *inspectors'* authority was primarily the administration and staff of the Royal Greenlandic Trading Company. From 1925, the two *governors* were chaired each their provincial councils in the North and South, supervisors of district councils and municipalities and at the same time heads of the administration of justice. After 1950, the *Governor* of Greenland was the bailiff for the technical area, housing, family law, education, health and, until 1965, the police. At the same time, the Governor chaired the Provincial Council. However, from 1967, the Council elected their own chairman. From 1979 to 1992, the newly established Home Rule Government gradually took over most of the administrative areas.

Consequently, changing High Commissioners have undertaken increasingly diminished administration, but at the same time developed a new role as a link between the Government of Greenland and the national authorities. Specifically, this is realised by the fact that the High Commissioner:

- ensures communication between Danish authorities and Naalakkersuisut
- keeps the Prime Minister's Office up to date on developments in Greenland through reporting
- is involved in the planning and carrying out of visits by members of the Royal House, the Folketing, the Government, etc.
- proposes candidates for royal distinctions
- ensures that parliamentary elections, as well as any referendums, determined by the Folketing, are held in Greenland and that it is possible to vote by post in Greenland for elections in Denmark, the Faroe Islands and to the European Parliament
- has national laws and other legal regulations valid for Greenland translated into Greenlandic and publishes them on its website.

In addition, the High Commissioner deals with a number of family law matters such as separation, divorce, adoption, determination of child and spousal support, name changes and custody agreements. Added to this are supervision of funds held in trust for legally incompetent persons, application for free legal representation, verification of public certificates for use in the legalisation process by the Ministry of Foreign Affairs of Denmark and citizenship proceedings. The High Commissioner also settles complaints about registration in the national register.

Justice area

The police, criminal services and courts, apart from the Supreme Court, are among the areas that Greenland can take over under the Self-Government Act. Despite independence being on the agenda of almost all Greenlandic parties, taking over the legal system has not received much attention. Although the Self-Government has appointed a minister as responsible for the justice area, the Danish Minister for Justice is the person in charge, and it is the Danish Parliament, which after consulting in Inatsisartut, adopts laws and amendments in this area. The Danish Minis-

OTHER GOVERNMENT INSTITUTIONS

A number of government authorities have staff in Greenland to manage areas still under Danish jurisdiction:

Naviair, which provides air traffic management services over Greenland.

The Danish Meteorological Institute's (DMI) weather service provides knowledge about weather, sea and ice conditions.

The Centre for Occupational Injuries, which administers the Workers' Compensation Act in Greenland, is a centre under the Danish Labour Market Supplementary Pension and the Danish Labour Market Insurance.

The Working Environment Authority in Greenland is, as part of the Danish Working Environment Authority, a board under the Ministry of Employment and oversees the Greenlandic occupational health and safety legislation.

The Maritime Authority in Greenland, which is part of the Danish Maritime Authority under the Danish Ministry of Economic and Business Affairs, undertakes tasks relating to safety on board, safety at sea, ship registration, etc.

GEUS, the Geological Survey of Denmark and Greenland, has had a small office at the Greenland Institute of Natural Resources since 2013 and assists authorities and companies with advice and specialist knowledge on issues relating to mineral resources and climate.

EMIL HELMS/RITZAU SCANPIX, 2021

The police in Nuuk is housed in the older parts of the administration complex, which was constructed in 1966 to house the administration of the Governor and the secretariat of the Provincial Council.

GRØNLANDS POLITI, 2021

The new 49-foot police vessel launching and naming ceremony is Aasiaat on 23 July 2021. *Kimik* is to support the police district's efforts at sea and in the surrounding towns and settlements in emergency situations and with police tasks.

ter for Justice is the person in charge of this area, and it is the Danish Folketing, which, after consulting in Inatsisartut (Parliament), adopts laws and amendments in this area.

Police

The task of the police is to maintain security, peace and order. Greenland forms one police district under the jurisdiction of the Ministry of Justice in Denmark. The police district is headed by the police commissioner, who must hold a Master's Degree in Law.

The office of police commissioner is located in Nuuk along with the central administration of police and is organised into three pillars:

- The prosecution service.
- The executive pillar, which, in addition to internal tasks, is also responsible for citizen contact regarding the issue of various permits. In addition, the Police College organisationally falls under the executive pillar.
- The police pillar, which is in charge of the national emergency operations centre as well as the 18 police stations in police regions North and South.

The largest staff group of the police are resident police staff trained at the Nuuk Police College. This group, which at the end of 2020 included 88 employees, will be supplemented by dispatched or assisting police staff from Denmark. In addition to policing duties, police officers also appear in court as prosecutors on behalf of the prosecution service, just as the police carry out a variety of civilian duties, including issuing passports, driver's licenses, motor registration and alienation and paternity cases. In the settlements, the police are represented by municipal police officers, who receive quite short training. By the end of 2020, the settlements had 59 municipal police officers.

Courts

The district court in most cases forms the first instance and consists of one district judge and two judges appointed by the municipal council. The district courts follow the municipal boundaries, except for the Qaasuitsoq District Court which cover the two northern municipalities: Avannaata Kommunia and Kommune Qeqertalik. The district judges are not attorneys at law, but they have completed special district judge training with a duration of two years and three months.

Prior to a major legal reform in 2008, there were no educational requirements for district judges, they were merely required to be of good character and meet the conditions of eligibility to the local council. After the reform, the district judges are employed in permanent positions. In order to strengthen the quality of the administration of justice, a legal court has been established, the Court of Greenland, that takes care of legally complicated cases as the first instance. It is also the Court of Greenland that is responsible for the guidance and training of the district judges. The High Court of Greenland is the court of second instance and exclusively an appeal court. The High Court consists of one judicial county judge and two lay judges appointed by Inatsisartut (Parliament). The defence in the district courts is in most cases conducted not by attorneys at law, but by authorised defence counsels who have undergone a defence counsel training. The office of defence counsel is a sideline occupation. In the legal courts, the Court of Greenland and the High Court of Greenland, attorneys at law appear as defence counsels.

Prison and probation service

The prison and probation service is responsible for the enforcement of the measures imposed by the courts: supervision, community service and imprisonment. The Prison and probation Service also administers people detained by the police and conducts pre-trial personal investigations. The Prison and probation service in Greenland has its own director in Nuuk, who reports to the Directorate of Prisons and Probations in Copenhagen.

Imprisonment takes place in so-called institutions of delinquency. The first penal institution was erected in Nuuk in 1967 and had room for 18 people. The main principle of the penal institution was that the convicts should maintain attachment to the surrounding community through employment in the town, they should be re-socialised and supported for a future existence free of crime.

Throughout the 1980s and 1990s, the number of inmates increased to a level well above the other Nordic countries. Today, there is room for 154 inmates in Ilulissat, Aasiaat, Sisimiut, Nuuk, Qaqortoq and Tasiilaq. Despite the intention of openness and contact with society, the institutions of delinquency have evolved over the years into closed institutions, with few inmates noticing the principles of the open institution. In the institutions of delinquency, there are few opportunities for activity, and re-socialising offers are limited.

With the legal reform, semi-closed wards were introduced in all institutions of delinquency, and a closed institution was built in Nuuk which widely has the characteristics of a prison. Inaugurated in 2019, the new closed institution, will, among other things, house persons given an indeterminate sentence who had so far been sent to serve in the Institution of Herstedvester.

The Defence

According to the Self-Government Act, defence policy belongs under Denmark and cannot be devolved to Greenland. Formally, the Danish Defence undertakes the military defence of Greenland. In effect, the task has since World War II been undertaken jointly by the United States and Denmark. The United States has significant military interests in Greenland, especially because the island lies on the shortest air route between North America and Russia and Asia.

JOINT ARCTIC COMMAND

Danish Defence's tasks in Greenland and the Faroe Islands are handled by the Joint Arctic Command (JAC) with units provided by the Navy, the Air Force and the Special Operations Command in particular. JAC was created in 2012 by a merger of the Greenland Command and the Faroe Islands Command. The capabilities and personnel of the Danish Defence vary depending on the season, but on average around 250-300 people are deployed in and around Greenland.

The Joint Arctic Command typically has the following capabilities:

- 6 dog sled teams under the Sirius Dog Sled Patrol
- 1 Challenger surveillance aircraft
- 2-3 Thetis inspection ships with helicopter
- 2 Knud Rasmussen inspection vessels
- 2 land-based Air Greenland helicopters on standby
- Periodically a frigate.

The Joint Arctic Command has the following stations:

- Liaison element Thule with a liaison officer at Thule Air Base
- Station Kangerlussuaq in the civilian Kangerlussuaq Airport
- Station Daneborg in Northeast Greenland, base for the Sirius Dog Sled Patrol
- Station North in Northeast Greenland
- Station Mestersvig in Northeast Greenland
- Liaison element Faroe Islands in Tórshavn
- Station and patrol service in Greenland at Aalborg Airport
- Station Grønnedal in South Greenland.

MADS PIHL/VISIT GREENLAND, 2016

Defence assistance in ice-breaking, rescue operations and the like is appreciated, especially in settlements and outlying districts, while some independence advocates in Nuuk complain about the visibility of the Defence. The inspection ship Thetis in Tasermiut Fjord with a Lynx helicopter in the air.

IDA GULDBÆK ARENTSEN/RITZAU SCANPIX, 2019

Today, the Thule Base is restricted to a few hundred American soldiers, most for six months at a time, and a few more civilian employees from Denmark, Greenland and the United States. The Americans have created a golf course on top of the Dundas mountain in the background.

The military presence of the United States has been linked to scandals several times, such as the forced relocation of the local inhabitants in connection with the expansion of the Thule Air Base in 1953 (see p. 408) as well as the secret stationing of American nuclear weapons in Greenland during the first half of the Cold War. Since the end of the Cold War, US military interest in Greenland has been more limited, but the Thule Air Base, now the United States' only military facility, forms part of the US missile defence and warning system and is thus an important part of the US nuclear deterrent.

Since the late 2010s, the United States has again shown a greater interest in Greenland. This is due to increasing American fears about the presence of China and Russia in the Arctic. The United States has therefore placed greater demands on Denmark's military presence in Greenland.

The Danish Defence asserts the sovereignty of Denmark, monitors the territory and helps protect Greenland militarily as a complement to the US defence of the North American continent. In addition, the Defence serves as a coast guard, carrying out tasks of great importance to Greenlandic society, especially search and rescue service, fishery inspection, maritime pollution prevention, hydrographical surveys and other civil society support tasks. A large part of these tasks are solved together with other authorities, for example, the search and rescue task is carried out in cooperation with the Greenland Police and Naviair, while environmental monitoring tasks are solved in cooperation with the Self-Government.

The Danish Joint Arctic Command carries out the tasks of the Defence in Greenland and the Faroe Islands (see p. 243).

The Defence has established the Arctic Emergency Force, an organisational structure that makes it possible to use the resources of the entire Defence, Home Guard and Emergency Management Agency in the Arctic in disaster situations or for sovereignty enforcement.

Defence expenditure in Greenland and the Faroe Islands amounts to approximately DKK 1 billion, but in response to the new security policy situation and in particular the interpretation of it by the United States, Denmark has announced new investments in the Arctic, initially to monitor airspace and ship traffic. In 2019, Denmark and Greenland reached an agreement allowing the Defence to continue using the airport at Kangerlussuaq, regardless of whether the Self-Government moves civil traffic to new airports at the main towns.

The headquarters of the Joint Arctic Command in Nuuk houses DMI and Naviair staff in Greenland in order to be able to coordinate and share knowledge, particularly in the context of crisis and rescue situations, environmental monitoring and sovereignty enforcement. In addition, the Americans established a consulate in Nuuk in 2020, reflecting the renewed interest of the United States in Greenland. So far, the consulate has an office in one of the buildings of the Joint Arctic Command.

Foreign and security policy

As regards foreign and security policy, the Unity of the Realm formalises a number of procedures for the involvement of Greenland and the Faroe Islands and how they can act independently. The framework is set in the Danish Constitutional Act and (by agreement between Greenland and Denmark) in the Self-Government Act as well as in various laws on interaction between government and parliament. Thus, according to the Self-Government Act, Greenland can 'negotiate and conclude international law agreements' with foreign states and international organisations 'on behalf of the kingdom' within cases involving 'devolved policy areas'. However, if Denmark is directly involved or is a member state of the international organisation in question, or if *both* Greenland and the Faroe Islands are involved, the state takes over. In such cases and in other foreign affairs conducted by the Danish Government, a number of principles for the discussion and involvement of Greenland (and/or the Faroe Islands) have been established. Greenland has won this influence particu-

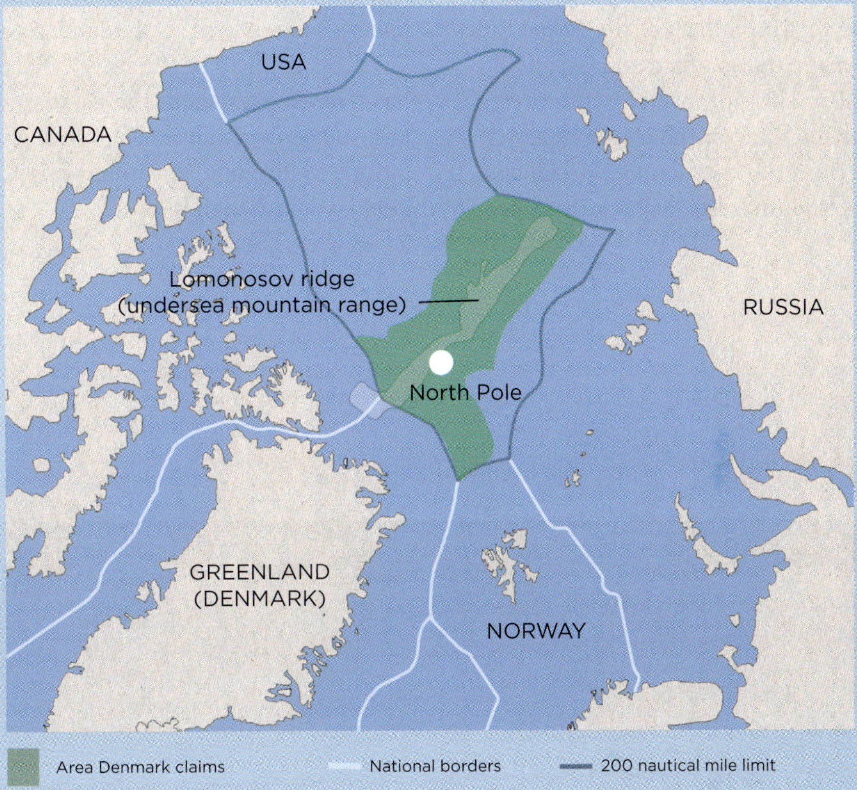

EFTER KORT AF PER JØRGENSEN GADS FORLAG, 2017, *DANMARK OG KOLONIERNE: GRØNLAND – DEN ARKTISKE KOLONI*

Map of the Arctic with the area claimed by the Kingdom of Denmark on behalf of the Unity of the Realm. The Geological Surveys of Denmark and Greenland (GEUS) has been the project manager of the continental shelf project for the purpose of collecting and interpreting data in order to submit claims to the Commission for the Limits of the Continental Shelf.

THE CONTINENTAL SHELF PROJECT

Launched in 2002, the continental shelf project is to document the Unity of the Realm's rights to the seabed and thus the right to exploit the resources that lie further out than the exclusive economic zone of 200 nautical miles.

Denmark ratified the UN Convention on the Law of the Sea on 16 November 2004 on behalf of the Unity of the Realm and then had ten years to make claims for extended rights to the continental shelf. This marked the beginning of a collaboration with the participation of Danish, Faroese and Greenlandic institutions.

Based on i.a. seismic measurements and seabed samples, claims have been made for rights to the subsoil in two areas near the Faroe Islands and three areas near Greenland. Between 2009 and 2014, a total of four reports on this were submitted to the Commission for the Limits of the Continental Shelf – including a claim for an area north of Greenland of approx. 895,000 km2, including the North Pole.

larly through three episodes involving the EU, the US, and the recent great power attention to the Arctic:

When Greenland left the EC (see p. 142) in 1985 and during the same period took over the fisheries policy, international negotiations on fishing quotas became a key task. However, the first diplomats of the Home Rule foreign affairs office and Greenland in Brussels purposefully interpreted the mandate of the Home Rule Act to cater to Greenland's 'special commercial interests' as broadly as possible. Today, the EU's relationship with Greenland is defined by three agreements, two of which are continuously renegotiated. These are the Greenland Treaty following Greenland's exit from the EC in 1985 and fisheries and partnership agreements. As long as the EU is satisfied with the fisheries agreement, Greenland, as an 'overseas country or territory' (OCT), has duty-free access to the EU market. Today, Greenland has representations in Brussels, Washington DC, Reykjavik and Beijing.

The Greenlandic diplomats are formally part of Danish diplomacy, but in practice report to Naalakkersuisut and the Department of Foreign Affairs in Nuuk. The representation in Copenhagen, which also houses professional staff from departments and government agencies, falls within the Premier's Office and the head of representation is part of the Diplomatic Corps in Copenhagen.

In parallel, the relationship with the United States has helped drive the expansion of the foreign policy competence of home rule and self-government authorities. In response to a number of cases related to the Thule Air Base, a Foreign and Security Policy Committee was established in 1985 where Inatsisartut (Parliament) members could be in-

In May 2021, US Secretary of State Anthony Blinken (D) visited Kangerlussuaq, where a press conference was held in Air Greenland's large hangar. Naalakkersuisut was represented by the chairman, Múte B. Egede (IA), and nalakkersuisoq for Foreign Affairs, Pele Broberg (N). Danish Foreign Minister Jeppe Kofod (A) also attended.

ÓLAFUR STEINAR RYE GESTSSON/RITZAU SCANPIX, 2021

Greenland is Denmark's ticket to Arctic co-operation. In 2017, The Crown Prince did the opening address at a conference on the UN Sustainable Development Goals in the Arctic. Three Greenlandic ministers, including Nalakkersuisoq for Foreign Affairs, Suka Frederiksen (S), spoke – while her Faroese colleague declined in protest at the restrictions which Denmark places on the foreign policy of the autonomous territories.

JENS DRESLING/POLITIKEN/RITZAU SCANPIX, 2017

formed in confidence, and in 2004 the gained insight and influence on foreign and security policy was formalised in the Itilleq agreement, the provisions of which are found in the Self-Government Act.

Most recently, changes in global climate and shifts in global economic and security balances have turned attention to the Arctic. This has given Greenland more opportunities to establish links outside Denmark. In 2008, Greenland and Denmark, along with the four other Arctic coastal states, signed the so-called Ilulissat Declaration. In the declaration, governments confirmed that they would resolve any disagreement over new borders in the melting Polar Sea amicably and according to UN procedures. At the same time, it reaffirmed the willingness separately and in established cooperation forums, primarily the Arctic Council, to take responsibility for the environment and development, so proposals by, among others, global environmental organisations for an international treaty to protect the momentum lost by the Arctic.

In plans for the financing of future independence, China, following the takeover of the mineral resources area by the Self-Government in 2009, has been high on the list of potential investors. In response to this, in 2020, the United States has opened a consulate in Nuuk that, with the help of American consultants and cooperation programmes, is tasked with increasing the integration of Greenlandic economy and society with North America. At the same time, the US sees the modernisation of bases in Russian Arctic as a threat to the Thule Air Base. It therefore wants increased investment in the defence of Greenland, including with obvious civil/military 'dual use'. In doing so, more and more of the 'internal affairs' of the Self-Government have taken on a security policy dimension and this puts the division of competences in the Unity of the Realm under pressure. If the state insists on deciding as soon as anything is *also* a security policy issue, it will roll back Self-Government. Conversely, it contradicts the interpretation of the Danish Constitutional Act requiring Danish monopoly to conduct security policy if the Self-Government insists that the state must not interfere.

SELF-GOVERNMENT

In a referendum held in Greenland on 25 November 2008, with a turnout of 71.96 %, a majority of 75.54 % voted in favour of the introduction of self-government (see p. 143). With this, the Folketing adopted the Act on Greenland Self-Government to take effect on the National Day on 21 June 2009. The term 'Self-Government' is colloquially used in several ways. This includes the Self-Government Act that established the transition from home rule to self-government. Self-Government can also refer to the central administration, Naalakkersuisut (government), the political system consisting of Inatsisartut (Parliament) and Naalakkersuisut and/or the entire Greenlandic public sector. Thus, reference can be made interchangeably to aspects of constitutional law, organisation of administration or governance – or to self-government in more abstract terms. In Greenlandic, *Namminersorlutik Oqartussat* is used officially, meaning literally 'those who govern themselves', but the word for authority (pisortat) is also used to refer to the public sector.

KELD NAVNTOFT/RITZAU SCANPIX, 2009

Chairman of Inatsisartut, Josef »Tuusi« Motzfeldt has just been presented with the Self-Government Act by the Queen. In the background, Danish Prime Minister Lars Løkke Rasmussen wearing a white anorak.

Self-Government is very much a continuation of home rule and the institutions created and developed since 1979 (see p. 135). But the Self-Government Act marks a development that makes Greenland one of the autonomous areas with the widest competences and self-determination in the world, without being formally sovereign. The Self-Government Act, uniquely for Danish legislation, as a preamble – a short, introductory text. This essentially acknowledges and redefines the population of Greenland as a people under international law with the right to self-determination. Since World War II, and the worldwide decolonisation, international law developed under the auspices of the United Nations but it was not until 2009 that Greenlanders became recognised by Danish law as a people in their own right.

Status as a people under international law is a condition for secession and state formation, and the Danish acknowledgement followed decades of work on the rights of indigenous peoples under the Inuit Circumpolar Council (ICC, see p. 142) and UN auspices. Under the law, a referendum in Greenland, where a majority votes for independence, can start the process towards an independent Greenland.

Furthermore, the Self-Government Act states that Naalakkersuisut and the Danish government are equal parties and that Greenlandic is the official language of Greenland (see p. 149). With the Self-Government Act, Greenland was allowed to take over a number of new policy areas which could not be taken over under the Home Rule Act (see the box on p. 249). However, unlike the takeover of policy area under the Home Rule arrangement, financial resources are not included in the form of increased block grants from Denmark. While the areas allowed to be taken over under the Home Rule Act were taken over sooner than expected, the process has been slower under the Self-

BRODDI SIGURDARSON/UNITED NATIONS, 2018

Greenlandic politicians, diplomats, and the Inuit Circumpolar Conference have played an essential role in the UN's work on indigenous peoples' rights.

POLICY AREAS ELIGIBLE FOR TRANSFER UNDER THE ACT ON GREENLAND SELF-GOVERNMENT

LIST I: THE SELF-GOVERNMENT MAY DECIDE TO TAKE OVER:

a) industrial injury compensation
b) the remaining areas under the health care area
c) the road traffic area
d) the law of property and obligations
e) the commercial diving area

LIST II: SUBJECT TO NEGOTIATION, THE SELF-GOVERNMENT MAY TAKE OVER:

1) prison and Probation Service
2) passports
3) police and prosecution service as well as related parts of the administration of criminal justice
4) administration of justice, including the establishment of courts of law
5) criminal law
6) The immigration area and border controls
7) law of capacity
8) family law
9) succession law
10) law practice
11) the weapons area
12) radio-based maritime emergency and security services
13) the radio communications area
14) the company, accounting and auditing area
15) the food and veterinary area
16) aviation
17) intellectual property law
18) copyright
19) shipwreck, wreckage and degradation of depth
20) safety at sea
21) ship registration and maritime law matters
22) charting
23) the buoyage, lighthouse and pilotage area
24) marine environment
25) financial regulation and supervision
26) the mineral resources area*
27) working environment(*)
28) meteorology.

* Transferred
(*) Transferred, however, only with regard to offshore work (oil drilling etc.).

Government. Shortly after the introduction of Self-Government, one and a half area was transferred, namely part of the occupational health and safety area (offshore work) and the management of the mineral resource area, as the ambition was for extraction of mineral resources to be a significant step towards a self-sustaining economy.

The first chapter of the Self-Government Act establishes the tripartition of power, placing legislative power with Inatsisartut (Parliament), the executive power with Naalakkersuisut and the judicial power with the courts (which are so far under the jurisdiction of the state). The vast majority of legislative work and governance concerning Greenlandic affairs takes place under the auspices of Inatsisartut and Naalakkersuisut. However, including policy areas not yet transferred, the separation of powers is more complicated, since Inatsisartut and Naalakkersuisut are systematically involved in special legislation adopted by the Folketing for Greenland when national legislation is put into force for Greenland, and by international agreements of special significance for Greenland. In relation to foreign policy and the relationship between Danish and Greenlandic law, a number of grey areas remain. Although the Danish Constitutional Act describes a unitary state, the Unity of the Realm has some features that are more reminiscent of a federation. The political system has developed since the 1970s, and although Greenland is a relatively young democracy, the institutions are stable.

The electoral system

The 31 members of Inatsisartut are elected for a maximum term of four years. The turnout at Inatsisartut elections has remained stable at 70-75 % in this century, but fell to around 66 % in the April 2021 elections. Interest in national politics is somewhat higher in terms of turnout than in municipal elections (approx. 60 %) and Folketing elections (approx. 50 %). All of Greenland forms one constituency in Inatsisartut elections, and the electoral system aims at proportionality, so that a party receives approximately the same proportion of mandates as the proportion of votes at the national level (specifically, the d'Hondt method of proportionate distribution of mandates is used). There is no formal minimum percentage of the votes necessary for a party to be represented in parliament, but a mandate costs about 1,000 votes or 3 % of the total votes. The system encourages voters to distribute mandates across quite a few parties, as known from multi-party systems in Scandinavia, among others. In the last few elections, the proportion of women who were elected directly has fallen to about 30 %, but the figure fluctuates quite a bit during each term, depending on which alternates are called upon.

On each party slate, candidates are nominated equally, i.e., the party mandates are awarded to the candidates who obtain the most personal votes, and not by priority party list. The votes for each party often gravitate towards a single vote catcher. He or she thus gets votes for not just his

CHRISTIAN KLINDT SØLBECK/RITZAU SCANPIX, 2021

Parliamentary sessions traditionally begin with a service at Nuuk's Cathedral, Our Saviour's. After the service, the speaker of Inatsisartut and the premier of Nalakkersuisut lead a procession to the Inatsisartut chamber in the administration building.

or her own seat, but carries one or more colleagues. The result is that a number of politicians are elected to Inatsisartut with quite a few personal votes (e.g., just over 50 votes were enough to be an alternate after the 2021 elections). In this way, it can appear somewhat random whether it is one or the other who is assigned a mandate – which you might then take with you if you change parties. Compared to other countries, far more people vote for a specific candidate rather than for a party at large.

Alternates are quite often called upon, for example, if a member asks for leave from Inatsisartut after being elected to Naalakkersuisoq (minister), the Folketing or municipal council, in connection with international delegation trips, sick leave or emigrating from Greenland. Dual mandates with a seat in both national and municipal politics or the Folketing are widespread, but often up for discussion, and it varies over time whether or not the respective parties allow dual mandates.

Originally, the Inatsisartut elections act ensured broad geographical representation in that eight districts were guaranteed one or more seats. One of the main arguments for amending the act with effect from 1999 was that less local politics was wanted in national politics. Instead, Inatsisartut was to take its cue from the overall interests of the country. That purpose has not been fully achieved. Looking at the candidates' personal votes, many have a clear geographic base. Debates, negotiations and decisions are still to a certain extent characterised by local issues and regional interests, not least in relation to infrastructure and the fishing industry. A change to the electoral system is regularly proposed by different parties, for example by reintroducing local constituencies.

CHRISTIAN KLINDT SØLBECK/RITZAU SCANPIX, 2018

Greenlandic election campaigns take place on many platforms: on posters, at election booths, on Facebook, in letters to the editors of newspapers *Sermitisaq* and *A/G* and with a caravan of candidates meeting voters at televised meetings in the larger cities.

The party system

The three old parties of the 1970s were traditionally relatively easy to pinpoint ideologically: Inuit Ataqatigiit (IA) far left, inspired by anti-imperialist movements with Marxist rhetoric. Siumut was the broad social democratic people's party. The conservative-liberal Atassut was the moderate party, both in terms of economy and in relation to Denmark.

Over time, three features have characterised the evolution of the party system. Firstly, the IA has gradually grown from a marginal position – both in terms of size and political content – to being equal to Siumut as a party fit to lead government. Secondly, the big right-of-centre party, Atassut, has lost some voter appeal. The ideological place and voter base are largely taken over by Democraatit (Democrats). Thirdly, over the years, a number of new parties have been launched, at times based on local interests or personal projects.

Most new parties have had limited lifespans, others have managed to take root. A number of breakaway members from existing parties have created new ones, several of which are based on fundamental features of the Siumut movement, such as Nunatta Qitornai (NQ) as the party of the big plans and a lot of show and Naleraq (N) as the party of the little man, fishermen and hunters. The social liberal Demokraatit was also originally founded by a breakaway member from Siumut, while Suleqatigiissitsisut (the co-operative party) was formed by former members of the Democrats about 15 years later.

No party has ever managed to become elected again if first voted out of Inatsisartut. In the 2018 elections, as many as seven parties were elected, but the two newest failed to become re-elected in 2021.

Figure 6 gives an overview of the Greenlandic party system, although

Figure 5. Distribution of seats in Inatsisartut elections since 1979. Since 1995, there have been 31 seats in the Inatsisartut. Siumut remains in the centre, IA has emerged from the left, while Atassut's place on the right side of the spectrum has been gradually and partially taken over by Demokraatit. They are all challenged by small new parties.

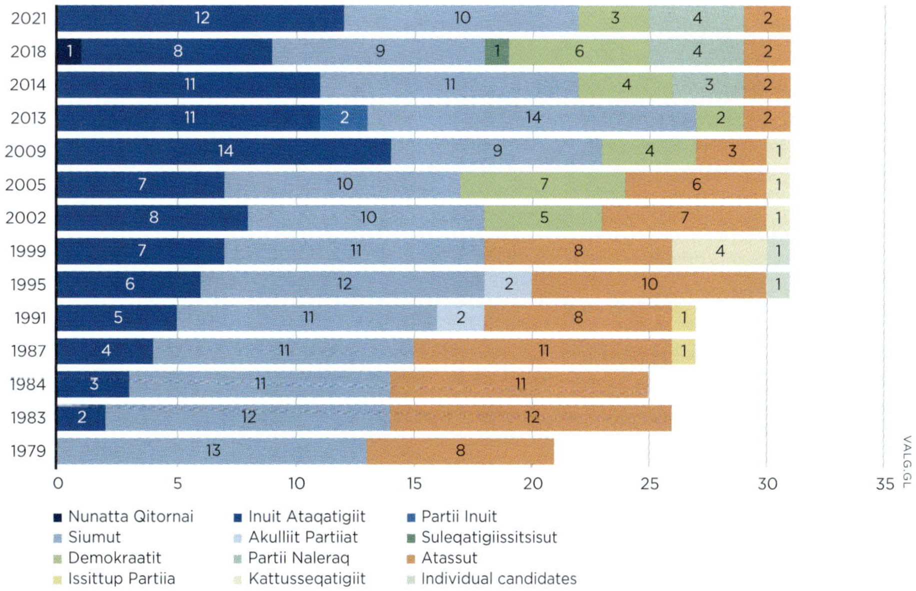

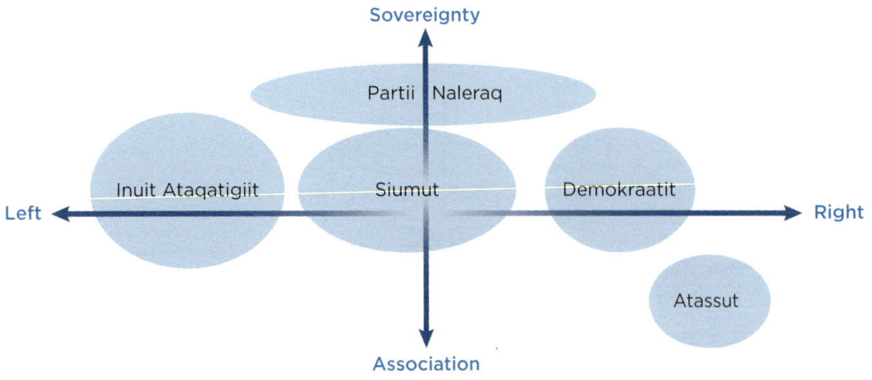

Figure 6. The political parties in Greenland roughly distribute themselves across a traditional right/left spectrum and a separate axis concerned with sovereignty from Denmark or not.

RASMUS LEANDER NIELSEN OG ULRIK PRAM GAD, 2022

complications in terms of both the axis of independence and the right-left axis make it difficult. In terms of the axis of independence, most of the parties have a spectrum of views, and most of the parties have moved over time. IA has gone from being *a* secessionist party to underline the value of a close pragmatic relationship with Denmark, like in other formal forms. For some periods, Siumut has been busy distancing itself from Denmark, but at all times, the party has also included politicians taking a pragmatic approach. Most recently, Atassut has returned to its original starting point as a proponent of the Unity of the Realm, after a number of years as a supporter of independence. Demokraatit has moved toward independence, only with considerably less haste than Naleraq.

In terms of the right-left axis, there are several newer parties that designate themselves as social liberal and are advocates of privatisation or a smaller public sector. Arguments for a smaller public sector, however, also exist on the left wing. For some parties, privatisation means selling large public monopoly companies to private investors (in practice mainly foreign companies due to lack of Greenlandic capital). For others, it means dissolving monopolies to make way for small local initiatives. Several parties talk about redistribution to address economic inequality, but some emphasise reducing regional inequality, others want to prioritise settlements and small towns, and again others have the unskilled and underemployed in the concrete blocks of towns in mind. Siumut has recently started to move back towards the left and Atassut simultaneously towards the right, while Naleraq's original starting point focusing on hunters and fishermen in recent years has been complemented by more liberal politicians and ideological thinking. The IA and Democrats have more urban voters, while Naleraq and Siumut voters are found in smaller settlements along the coast. Atassut was formerly a Nuuk party, but now primarily picks up votes in towns such as Ilulissat and Maniitsoq.

CHRISTIAN KLINDT SØLBECK/RITZAU SCANPIX, 2021

The atmosphere in the Inatsisartut Hall and on social media can be harsh, but voters are relatively forgiving, so it is possible to make a comeback. Greenlandic politicians also need to be able to deal with each other, even after a disagreement. The photo is taken during a break from KNR's party leaders' debate before the 2021 elections.

Coalition formation

After the April 2021 elections, for only the second time in Greenlandic political history, IA could take the post of premier of Naalakkersuisut. And in 2022, Siumut entered for the first time as a junior partner in a coalition. Siumut has otherwise sat almost continuously as the leading party since the birth of the party system in the 1970s (see p. 251). Siumut has formed coalitions with ideologically very different parties and has frequently switched coalition partners several times during a parliamentary term. Siumut, as a broad popular movement, has been home to many political currents, local interests and strong personalities. The premier of Naalakkersuisut and the leader of Siumut has most often been one and the same person, and often premiers have had almost the same amount of work keeping their own party together as they have had for their coalition partners. Coalition breakdowns happen relatively often as a result of political disagreements over concrete issues, matters of principle or shifting balances of power internally within Siumut – or as a result of one of the smaller parties having lost patience with Siumut's leadership, course or lack of the same. Most governments have been based on majority coalitions; however, there have been minority governments in the 1980s and for most of the 2018 – 2021 parliamentary term.

A settled tradition has developed in which post-election coalition formations or government transformations are accompanied by more or less verbose political coalition agreements. However, large parts of the documents consist of adherence to unspecified 'reforms' in different areas or unprioritised 'priorities' in a wide

Premier	Period	Coalition
Jonathan Motzfeldt	1979-1984	S
	1984-1987	S + IA
	1988-1991	S
Lars Emil Johansen	1991-1995	S + IA
	1995-1997	S + A
Jonathan Motzfeldt	1997-1999	S + A
	1999-2001	S + IA
	2001-2002	S + A
Hans Enoksen	2002-2003	S + IA
	2003	S + A
	2003-2005	S + IA
	2005-2007	S + IA + A
	2007-2009	S + A
Kuupik Kleist	2009-2013	IA + D + KP
Aleqa Hammond	2013	S + A + PI
	2013-2014	S + A
Kim Kielsen	2014	S
	2014-2016	S + A + D
	2016-2018	S + IA + N
	2018	S + N + NQ + A
	2018-2019	S + A + NQ
	2019-2020	S + NQ
	2020-2021	S + D + NQ
	2020-2021	S + NQ
Múte B. Egede	2021-2022	IA + N
	2022-	IA + S

NAALAKKERSUISUT.GL

Table 1. Greenlandic politics has been dominated by Siumut-led governments since 1979, with ever more rapidly changing coalition partners – interrupted only by two periods with IA in the lead.

range of areas. Large parts of the text can therefore be the same from one government base to the next, or as in the parties' election campaign materials. At the same time, coalition talks are concentrated on matters high on the current political agenda and on seats in Naalakkersuisut, committees and the boards of the government-owned companies. In these texts, parties may also explicitly agree to disagree if a policy area creates a conflict dimension that otherwise cannot be resolved – e.g., in relation to mining involving the extraction of uranium.

Based on international coalition research, one would expect coalitions to be formed by parties with the least possible ideological distance. However, this has far from always been the case in Greenland. Most spectacular was when in 2009 the socialist IA and liberal Demokraatit joined forces, not least to exclude Siumut. Following on from the movements of the parties on the axis of independence (see p. 252) and the differing emphasis the parties place on various economic issues, only very few party combinations have seemed impossible in the self-government era.

EMIL HELMS/RITZAU SCANPIX, 2021

Elections to Inatsisartut, on 6 April 2021. With a voter turnout of 65.8 %, it was the lowest turnout since home rule was introduced in Greenland in 1979. However, the turnout was 61.0 % in Nuuk and 71.4 % in Ilulissat, so large regional differences were observed.

Parliamentary practice

As in other parliamentary systems where government springs from the legislature, Inatsisartut balances between two roles in relation to Naalakkersuisut: On the one hand, it is necessary for Naalakkersuisut to be backed by its majority in Inatsisartut, otherwise no political decisions and legislation would be made. On the other hand, Inatsisartut must make sure that Naalakkersuisut acts according to the rule of law. Inatsisartut may eventually enact a vote of no confidence and thus remove Naalakkersuisut if it does not comply with the law.

In the first – legislative – role, members act primarily as party members and thus as either opposition or as the parliamentary base of the government. In the second – monitoring – role, the members ideally act as a unified body to control the government, disregarding party lines. In practice, the political game often plays out more complexly: Naalakkersuisut does not agree with its parliamentary base on policy. Or the Inatsisartut members identify themselves rather as party soldiers than as a control body and tend to stick up for a minister whose acts seems to be at odds with the law. No-confidence motions have periodically been a frequently used tool when the opposition, as well as at times members of government parties, have disagreed with the policy pursued. Particularly in a number of cases involving the misuse of public funds for personal consumption, a number of Naalakkersuisut members have had to resign, directly or indirectly as a result of parliamentary oversight. Early elections may be called by a majority of Inatsisartut by law. The incumbent premier of Naalakkersuisut can only call an election if a majority of Inatsisartut has passed a vote of no confidence in him or her, or by tabling a bill which must then be passed by a majority.

Both the legislative and supervising role of Inatsisartut are undertaken primarily when in session, secondarily in committees. Meetings in the chamber are traditionally held during a spring session and an autumn session, where, among other things, the Budget for the following year is adopted. Extraordinary sessions may be needed for various reasons.

CHRISTIAN KLINDT SØLBECK/RITZAU SCANPIX, 2021

A new generation of female politicians has reached key positions – even if some of them are primarily Danish-speaking. Naaja H. Nathanielsen (IA) was first Naalakkersuisoq for Housing, Infrastructure, Natural Resources, Justice and Equality and later for Finance.

According to law, Inatsisartut is required to establish a number of permanent committees: The Scrutiny of Eligibility Committee, the Legal Affairs Committee, the Finance and Fiscal Affairs Committee, the Audit Committee and the Foreign and Security Policy Committee. In addition, a number of special committees are formed to discuss issue-specific legislation, etc. on behalf of the entire Inatsisartut and any ad hoc committees, e.g., one to follow the work of the Constitutional Commission, for example. Parties are formally allotted committee memberships proportionately (again following d'Hondt's method), but in practice parties can negotiate for seats on the committees they prefer.

Legislation is negotiated in three plenary readings; each of the two first readings may refer a bill to one or more committees. Motions for resolutions other than legislative amendments are most often considered in two plenary sessions and possibly in a committee. Important or controversial matters are often negotiated by one or more members of the Naalakkersuisut directly with representatives of parliamentary party groups in between the formal proceedings. Major reforms are often prepared in commissions with representation of interest organisations and experts carrying out investigative work and/or providing recommendations, not least if political disagreement makes it opportune for the governing coalition to kick the matter to the corner.

In order to attend to important areas of its control function, Inatsisartut appoints an ombudsman to whom citizens can complain about the administration's specific proceedings and decisions. Inatsisartut's control over the expenditure of the national treasury is carried out through the Audit Committee, while the auditing of the national treasury accounts is carried out by a private firm of auditors. Generally, the Danish Public Accounts Committee does not deal with the finances of the Self-Government, other than ascertaining whether the Danish block grant has been properly transferred. Inatsisartut and the committees are given secretariat assistance by the Bureau for Inatsisartut; in addition, the national treasury funds secretariats for the party groups and also provides grants for election campaigning. There is regular criticism from the Inatsisartut Committee that information provided by Naalakkersuisut is deficient and/or delayed in relation to both the legislative work and the control functions of Inatsisartut.

A number of matters weaken the continuity of special committee work and, more seriously, popular confi-

PREMIERS OF NAALAKKERSUISUT 1979-2022

CARL CHRISTIAN RESTING-JEPPESEN/ARKTISK INSTITUT, 1996

JONATHAN MOTZFELDT (1938-2010)

Motzfeldt was born into a hunting family in the settlement of Qassimiut near Qaqortoq. He graduated as a teacher from Ilinniarfissuaq Teacher's Training College in Nuuk in 1960 and graduated as theologian from the University of Copenhagen in 1966. He then worked as a priest in South Greenland until 1979.

As co-founder of Siumut, Motzfeldt became a member of the Greenland Provincial Council from 1971 until he assumed the office as Greenland's first Premier when home rule government was established in 1979. He held this office until 1991 and again from 1997 to 2002. From 1979 to 1988, in 1997 and again from 2003 to 2008, he was also speaker of the Inatsisartut (Parliament).

Motzfeldt's stature came close to that of a 'father of the nation', and he is remembered as one of the politicians who brought the Greenland society from the colonial period to home rule and on to self-government.

KELD NAVNTOFT/RITZAU SCANPIX, 1997

LARS EMIL JOHANSEN (1946-)

Johansen was born in Illorsuit, Uummannaq, into a politically engaged family. He graduated as a teacher from Ilinniarfissuaq Teachers' Training College in Nuuk in 1970 and soon after became a full-time politician. Johansen was co-founder of Siumut, member of the Greenland Provincial Council from 1971 to 1975 and member of the Danish Parliament from 1973 to 1979 and again from 2001 to 2011. He was member of Inatsisartut from 1979 to 1997 and again from 2013 to 2021 where he served as speaker. In between he served as Naalakkersuisoq with a variety of portfolios.

From 1987 to 1997, he was party chairman of Siumut, and in parallel with this office, he served as Premier from 1991 to 1997. As Premier, Johansen had a party programme that aimed at creating a strong home rule government that would build a Greenlandic society on Greenland's own terms. In breaks from politics, he was executive vice president of the government owned Royal Greenland and engaged in private business. He returned to politics in 2001 and represented Siumut first in the Danish Parliament and since in Inatsisartut until 2018.

CARSTEN SNEJBJERG/RITZAU SCANPIX, 2008

HANS ENOKSEN (1956-)

Enoksen was born and raised in Itilleq, where he later learned commerce. In 1995, he was elected to Inatsisartut) and used his experience from local politics and his desire to bring prosperity to the outer districts when he took over the chairmanship of Siumut in 2001.

Enoksen was Premier from 2002 to 2009, but resigned as party chairman after losing the election. He left Siumut in 2014 to form Partii Naleraq as he wished to push his core policy priorities with a greater focus on the outer districts. Minister (Naalakkersuisoq) in various coalitions from 2016 to 2018.

FINN FRANDSEN/POLITIKEN/RITZAU SCANPIX, 2013

KUUPIK KLEIST (1958-)

Kleist was born in Qullissat, where he grew up with foster parents until the coal mining town closed. He finished primary school in Sisimiut and then went to Denmark where he graduated from Birkerød State School in 1978 and as a social worker from Roskilde University in 1983. In the following years, he held a number of executive positions in the Departments of Education and Foreign Affairs of the Home Rule Government and at the Greenland School of Journalism, after which he became a full-time politician from 1991 to 1995 as Minister (Naalakkersuisoq) in the coalition government between Siumut and IA. He was then elected to the Landsting (Inatsisartut) for the first time in 1995. From 2001 to 2007, Kleist was a member of the Danish Parliament, from 2007 to 2014 he was chairman of IA, and during the same period, from 2009 to 2013, he was Premier of Naalakkersuisut in coalition with the Democrats and Kattusseqatigiit Partiiat. Since leaving politics in 2014, Kleist has been a private consultant and held various public positions of trust.

JENS NØRGAARD LARSEN/RITZAU SCANPIX, 2013

ALEQA HAMMOND (b. 1965-)

Hammond was born in Narsaq into a family of hunters and grew up in Uummannaq. In her youth, she travelled around much of the world and, attended the Arctic College in Canada from 1991 to 1993 where she acquired strong language skills. Following a career in both tourism and the ICC (see p. 142), she was elected to the Landsting (Inatsisartut) for Siumut in 2005, and held different government positions.

From 2009 to 2014, Hammond was chair of Siumut, and in 2013-14 she briefly served as Greenland's first female Premier of Naalakkersuisut.

From 2016 to 2019, she was a member of the Danish Parliament, and from 2018 to 2021, she was again a member of Inatsisartut. After a few years as a member of the Nunatta Qitornai party, she returned to Siumut.

NAINA HELÉN JAMA/TT/RITZAU SCANPIX, 2019

KIM KIELSEN (1966-)

Kielsen was born in Paamiut and was trained as a sailor, after which he sailed for KNI for a number of years. He later followed in his father's footsteps and became a police officer in Nuuk in 1996. He took leave from the police to start social prevention projects for young Greenlanders, after which he was elected to the Landsting (Inatsisartut) for Siumut in 2005. He was a member of the government from 2007 to 2009, and from 2014, he served as party chairman and Premier of Naalakkersuisut in rapidly changing coalitions. Kielsen has held the position as speaker of Inatsisartut since April 2022.

MARTIN SYLVEST/RITZAU SCANPIX, 2022

MÚTE BOURUP EGEDE (1987-)

Egede was born in Nuuk, but grew up in Narsaq in South Greenland. He attended university in Nuuk from 2007 to 2013, after which he took over the management of the family feedstuff business. He became chairman of the IA youth organisation and in 2015 was called upon to take up a seat in Inatsisartut as an alternate. In 2016, as member of Naalakkersuisut, he was briefly in charge of mineral resources.

Egede was elected to Inatsisartut in 2018, after which he took over as IA chairman. The 2021 elections returned a victory, which gave the party the opportunity to lead Naalakkersuisut for only the second time since 1979.

dence in the political system. Frequent coalition reshuffles trigger reconstitutions of committees each time a member of Inatsisartut is appointed to Naalakkersuisoq. Similarly, a game of 'musical chairs' is triggered when a politician switches from one party to another. For example, in 2018, a member of Naleraq defected to Siumut just a few weeks after the election. Before the same election, Atassut had lost both of his mandates due to party defections, so the party had to collect signatures to become eligible again. The frequency of such party defections in Inatsisartut and municipal politics has led to suggestions that a mandate be locked to a party, not individual politicians. Similarly, 'dual mandates' is an ongoing theme in several parties: Is it a strength for his or her political work that an Inatsisartut member has a seat in the Folketing or a municipal council at the same time, or is it prohibitive for tending to the work with due care? The two problems become entangled when a dual mandate prevents a party defector from taking over the seat as an alternate.

The tone of the public political debate in Greenland can be fierce, both on social media and in the commentary track on online media articles, but also in regard to debates in Inatsisartut. A number of Naalakkersuisut members and party chairpersons have had to take sick leave due to stress. Several significant profiles have left politics citing the personal costs. At the same time, there is a demand for local talent which makes it easy for particularly younger and well-educated people to switch career tracks from politics.

Economy and governance of the Self-Government

The disadvantages of the insular community (see p. 234) and welfare ambitions means that Greenland has a large public sector with a number of centrally planned economic features. There are several reasons for this. The task of Self-Government is to run a welfare state, partly inherited from the Danish model, for a very small population spread over a very large territory.

To a wide extent, the disadvantages of the insular community become primarily relevant due to the ambitions of the Self-Government. Welfare, settlement pattern and self-government are conditions for the Self-Government's task solving given by the voters. Consequently, considering the size of the public sector there are two ways forward: developing the business sector and prioritising the service level in different areas so that revenue and expenditure can balance. And to organise the administration and organisation of the public sector most effectively in relation to the needs and conditions of society. Here, the political focus has often been directed at inappropriate structures that have been taken over from the Danish colonial administration.

Total public expenditure is around 60 % of the GDP and around 40 % of all employed persons are employed in the public sector, including management, education, health and social services and institutions. In total, just under 19,000 people work in the public sector (see Figure 7). If only full-time employees are included, in 2019 there were well over 11,000 full-time employees in the public sector, of which just over 4,000 are full-time employees in the Self-Government and well over 6,800 in the municipalities.

As regards the economy of the Self-Government in the strictest sense of the word, economists have identified an 'abyss' between developments in the treasury's revenues and expenditure driven by, on the one hand, uncertainty about the sources of revenue and, on the other hand, increased de-

Figure 7. Employees in the municipalities and self-government in the period 2015-2020. The number of employees has remained relatively stable in municipalities and self-government, with 12,750 and 7,089 employees, respectively, in 2015 and 13,005 and 7,444 employees, respectively, in 2020.

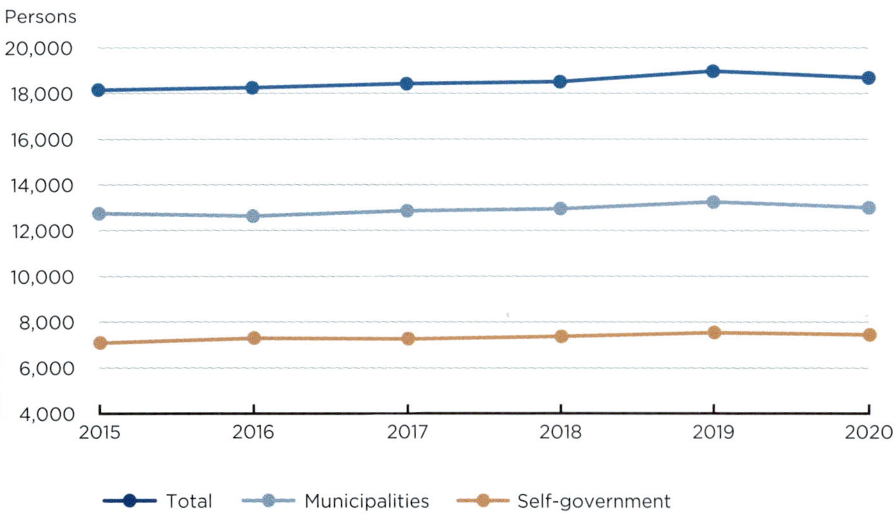

RASMUS LEANDER NIELSEN OG ULRIK PRAM GAD, 2022

mand for welfare services. According to the Greenland Economic Council, all else being equal, the expenditure will increase corresponding to 5-6 percentage points of the gross domestic product, in line with a growing number of elderly people in the population. However, there is uncertainty about whether future generations will work longer into old age by virtue of better health, as well as the proportion of elderly people who require most care who would prefer their old age in Denmark. Up to and including 2008, the retirement age was 63, in 2009 it was extended to 64, in 2010 to 65. In 2017, the retirement age was further increased to 66, and in 2021 to 67. In 2017, the retirement age was raised to 66, and in 2021 to 67.

Government revenues follow economic trends, especially in the fishing sector. The block grant of just under DKK 4 billion and other expenditure incurred directly by the Danish State means that both private and especially public revenues and consumption opportunities are greater than that generated by the domestic production of goods and services. Similarly, compared to many other countries, the Greenlandic economy is less threatened by, for example, financial crises and pandemics, as the block grant acts as a buffer against external economic shocks and as a safety net for international borrowing.

In the same way as under the Home Rule Act, the block grant is continuously regulated in accordance with price developments in Denmark, but under the Self-Government Act, increased block grants are not included when new functions are taken over from the state. Any revenue from the mineral resources area (corporate taxes, royalties, etc.) exceeding DKK 75 million is divided equally between the Self-Government and the Danish State by reducing the block grant with the Danish State's share. In addition to revenue from taxes and duties, the treasury has revenue in the form of dividends from government-owned companies, the profits of which, however, may have been generated by publicly paid service contracts or politically approved tariffs/prices.

In recent years, there has been a boom in macroeconomic activity, mainly due to favourable export fish prices representing more than 90 % of total exports. This has led to a strengthening of the general government revenue and profit on the government finances.

A comparison of the industries' share of the total production in 2020 shows that the industry of 'public and personal services', consisting of the production of public services for administration, health and education, is the largest industry and represents almost one third of the society's production. 'Fisheries, hunting and industry' (including the fishing industry's factories) together account for 20 % of the production and the 'construction industry' accounts for around 14 % of the total production and is thus the third largest industry.

As far as the public sector is concerned in the broader sense, there is a recurring desire from politicians, Greenland Business Association and the Economic Council, among others, that it should be slimmed down. A key argument is that increasing efficiency in the public sector would free up labour and thus the opportunities to develop private industries. In order to become less dependent on external transfers and fluctuations in stocks, quotas and market prices in fisheries, efforts are being made to diversify the economy, primarily by promoting mining and tourism.

Organisation of public administration

The public administration consists of the Self-Government, the municipalities (see p. 261) and the institutions of the state (see p. 241). The Self-Government executive branch is headed by Naalakkersuisut. The task of the central administration is to serve, advise and ensure that Naalakkersuisut makes decisions based on an informed basis, as well as to implement political decisions and administer within the framework of legislation adopted by Inatsisartut. Typically, a Naalakkersuisoq (minister) has several portfolios and accordingly the responsibility for several units of the central administration. Mostly, the central administration has been organised into about 10 cabinet departments, slightly fewer executive agencies, and as number of decentralized units. Over time, core government responsibilities reappear in various combinations, such as finances, fisheries, education, infrastructure, business, health, social affairs, etc. Other portfolios are established in order to secure or signal special political attention areas and may have relatively short lifespans, such as settlements and outlying districts or independence. Departments primarily support the political role of their respective naalakkersuisoq in relation to the legislative process and negotiations, while executive agencies work within a specified set of rules with specific decision-making powers delegated. Decentralized units undertake similar tasks where close citizen contact is important, as well as tasks that are of an operational character.

Spread over nine floors in what is known as the Self-Government Tower in Nuuk Centre, one or typically several departments work on each floor. The premier's department, however, remains in the old administration building on the other side of the pedestrian street in Nuuk, where the Inatsisartut chamber can also be found. Sizeable sums are spent on reorganisation, interpretation (see p. 151) and recruitment. In key ministerial posts (e.g. fisheries), major replacements and new coalition formations mean that departments move to another floor in the Self-Government Tower to gather newly combined portfolioscu under the same minister. Despite recurring demands to slim down the public administration, several functions under Naalakkersuisut appear to be understaffed compared to the ambitions of the given government coalition. The shortage of educated Greenlandic labour demanded by the departments (typically lawyers and economists) leads to the recruitment of of a steady stream of new Danes, often imported shortly after having received their diplomas. An ongoing theme in the political debate is how formal education should be weighted in relation to practical experience and Greenlandic language skills. Much locally recruited as well as imported labour seem to prefer employment in private or government-owned companies where the pay is higher and the distance to daily politics is longer. Each year, an amount in the millions is spent on external consultancy services in the departments and underlying institutions of the Self-Government. If new areas are taken over from the Danish State, it will create additional demand for qualified labour.

Operations of as well as construction works relating to public services are to a large degree organized in government-owned companies. The companies' activities include traditional public areas of activity, such as infrastructure and utilities, as well as business activities in fishing, shipping and retail, which in other countries are more widely left to the market. Overall, the government-owned companies are organised as either net managed enterprises (Asiaq, Mittarfeqarfiit and Nukissiorfiit) with public subsidy grants or in limited

EMIL HELMS/RITZAU SCANPIX, 2021

The new administration building, commonly referred to as the 'Self-Government Tower', rises above Nuuk Centre. In the background, Store Malene is shrouded in fog.

MARTIN LEHMANN/RITZAU SCANPIX, 2019

liability company form where some of the most important include Great Greenland (sealskin products, etc.), Greenland Holding (business promotion), INI (housing), Kalaallit Airports (airports), KNI (e.g. retail), Royal Arctic Line (shipping), Royal Greenland (fish and shrimp), Tusass (formerly Tele Post Greenland) (tele and postal services) and Visit Greenland (tourism). The majority of the government-owned limited liability companies are 100 % owned by the Self-Government, but in a few companies, government holds only a minority share, (e.g., Grønlandsbanken).

The US military chose Kangerlussuaq as its central airport because its continental climate provides a uniquely high level of regularity. Air Greenland's Airbus A330-200, called *Norsaq*, serves Copenhagen Airport from Kangerlussuaq – but the plan is for its replacement, an Airbus A330neo, to fly direct to Nuuk.

The tasks of the municipalities

The municipalities carry out a range of civic tasks such as the operation of day care centres, schools and nursing homes, the social area (with the exception of 24-hour institutions), local engineering and environmental management, and local business development. The Self-Government Act does not provide precise provisions on the division of tasks between the Self-Government and the municipalities, but municipal autonomy is provided for in Section 82 of the Danish Constitution. Municipal tasks are attended to on the basis of delegation and decentralisation of tasks adopted by Inatsisartut in the relevant legislation or, within the framework of current legislation, by Naalakkersuisut.

The Local Government Act establishes, among other things, that the municipal committee structures and the tasks of committees, budgetary processing, the tasks of the administration and the supervision of the municipalities as described in Inatsisartut Act No. 22 of 18 November 2010 on the Administration of Municipalities. The rules governing municipal government defines in terms of municipal law what the municipalities

100 % GOVERNMENT-OWNED LIMITED COMPANIES

Naalakkersuisut appoints the boards of the government-owned limited companies, while the relevant ministries prepare for and negotiate service contracts with each company. Naalakkersuisut and the boards communicate on an ongoing basis. Only 100 % government-owned companies are included in the following.

Air Greenland*
Great Greenland
Greenland Holding**
Greenland Oil Spill Response
Illuut
INI
Kalaallit Airports Holding
KNI
NUNAOIL
Royal Arctic Line
Royal Greenland
Sikuki Nuuk Harbour
Tusass A/S (formerly Tele Post Greenland)
Visit Greenland

* IN 2019, THE GOVERNMENT OF GREENLAND TOOK OVER THE DANISH STATE'S AND SAS' SHARES IN AIR GREENLAND AND BECAME THE SOLE SHAREHOLDER.

** GREENLAND HOLDING HAS NO TURNOVER.

SECRETARIAT OF THE BOARD OF DIRECTORS

FILIP GIELDA/VISIT GREENLAND, 2019

The Qeqqata Kommunia office in Maniitsoq houses administration, citizen services and a library.

may use their funds for in order to meet the general needs of the local community. The Self-Government negotiates block grants with the municipalities every year.

Municipalities before 2009

In 2009, a major structural reform was implemented in Greenland. The history of Greenlandic municipalities can be traced back to 1857 when governorships were created on a trial basis. In 1908, municipal councils (62) were introduced with the Act on the Government of the Colonies in Greenland of 27 May 1908, but the scheme did not come into effect until 1911. At the same time, two provincial councils were introduced and the governorships were discontinued. The municipal councils were tasked with poverty assistance, support and care for public order as well as paying expenses for the common benefit of the municipality.

In 1950, the Act on Greenland's Provincial Council and Municipality Councils was introduced. The 16 municipalities in West Greenland (Thule and East Greenland were not included until the 1960s) took over some of the tasks that had previously been under the colony administrators, including roads, bridges, refuse collection and fire services. The municipal reform of the mid-1970s involved a transfer of some state areas of responsibility to the municipalities and thus a changed distribution of duties and responsibilities, including the local treasurer function and the right to determine municipal income tax. In 1994, the white paper from the Local Government Reform Commission became the basis for a new governance regime for the municipalities and the additional transfer of tasks to the municipalities from the Home Rule Government.

The mandate of the structural committee from 2003 was to assess the advantages and disadvantages of various models for a changed distribution of duties and responsibilities in public administration. On the basis of the different models, the commission was to carry out an assessment and submit proposals on the geographical and population criteria, including what municipality sizes should form the basis for a future municipal structure. The main features of the structural reform were that the Home Rule Government and since then the Self-Government should primarily carry out legislative and supervisory tasks, and that the municipalities were given more operational and construction tasks. In order to implement the increased task volume, municipalities had to be merged into larger entities.

Municipalities after 2009

Prior to 2009, the elected assemblies of a Inatsisartut (Parliament), consisted of 18 municipal councils and 56 settlement councils with more than 380 elected officials. Following the structural reform, the elected assemblies were reduced to the Landsting (Inatsisartut) and four municipal councils; less than 100 elected officials. Small towns or local areas are assured a seat on municipal councils, and only parties that nominate candidates there, can win these mandates.

The structural reform entailed the revising of a number of executive orders, circulars and regulations in order to reduce micromanagement and to ensure more efficient case handling. The purpose of the simplified legislation was to ensure greater transparency in terms of requirements, rights and competences of the Home Rule Government, municipalities and citizens. The structural reform transferred tasks worth around DKK 1 billion from the Home Rule Government to the new regional municipalities. By decentralising the tasks, rationalisation gains of around DKK 200 million annually could be achieved. This analysis, however, is not supported by subsequent economic impact analyses.

The Structural Committee recommended that there should be at least 8,000 inhabitants in each municipality and proposed a merger of the 18 municipalities into the following four regional municipalities: Qaasuitsup Kommunia, Qeqqata Kommunia, Kommuneqarfik Sermersooq and Kommune Kujalleq. Following a referendum in 2018, a division of Qaasuitsup Kommunia into two municipalities was carried out: Avannaata Kommunia and Kommune Qeqertalik. Thus, there are currently five municipalities in Greenland (see p. 318).

The areas that were actually transferred in the connection with the structural reform were the establishment of a common citizen service portal, the centralisation of the tax and debt collection area, the transfer of planning authority to the municipalities, the transfer of pedagogical-psychological advisory services, extensive remedial instruction and the disability area and the centralisation of business support schemes.

The other areas were not transferred to the municipalities, including responsibility for negotiating service contracts on the supply of goods, shipping and airline traffic, electricity, heating and water supply, ports, the housing and construction areas. In addition, the health areas such as alcohol abuse treatment, health care and parts of the prevention area as well as a number of operational areas in the field of education were also not transferred to the municipalities. Finally, the pension area, maternity benefits, rent subsidy, child allowance, 24-hour care centres and alimony continued to be administered centrally.

A recurring theme is how a better division of competences between municipalities and the Self-Government is ensured. Decentralisation is often an electoral theme involving discussions about the future of the municipalities – possibly in new municipalities or a return to past divisions. Historically, an institutionalised cooperation has existed across municipalities through the KANUKOKA (joint municipal cooperation organisation)

ANINGAAQ R. CARLSEN/VISIT GREENLAND, 2019

Aasiaat Town Hall performs a wide range of functions. Asiaat is the main town in Kommune Qeqertalik, and 44 % of the town's jobs are in public administration and services.

AQQALU AUGUSTUSSEN/KIVITSISA, 2020

All five municipalities in the country are behind the Kivitsisa ('Let's elevate') project. The project is carried out in collaboration with the National Education Authority and the Institute of Learning. By improving internet and iPad access for all students, Kivitsisa seeks to ensure that all students have the opportunity to be active and creative with new technology, which, together with an investigative and project-oriented approach to learning, should raise the level of learning in primary schools.

which has, however, been discontinued, and a new coordinating body has not been established to replace it.

The large geographical spread and low population combined with the desire to maintain a welfare society (see p. 232) makes the public administration relatively large, also at the municipal level. At the same time, it means that the municipal room for manoeuvre and the avenues of action for the municipalities become highly dependent on key national political decisions on the placing of infrastructure, the placing of public enterprises, educational institutions, etc.

KANUKOKA was formed in 1972 as a joint municipal cooperation organisation for the purpose of attending to the common interests of the municipalities in negotiations and to provide advisory services and assistance to member municipalities. In 2016, Kommuneqarfik Sermersooq left KANUKOKA and the other three municipalities decided to move the head office from Nuuk to Maniitsoq. In 2018, the member municipalities decided to dismantle KANUKOKA and find other ways for the municipalities to work together.

The political coordination group, consisting of Naalakkersuisut and the five mayors, still cooperate. The group meets two to three times annually to deal with inter-authority issues and common public issues.

The municipalities still cooperate regularly, for example, in 2017 the five municipalities cooperated to raise the level in primary secondary schools through the Kivitsisa project (»Let's elevate«). This cooperation is unique as it is the first time that all municipalities join forces to develop primary schools. Here, they have joined forces on a project to develop the teachers, the teaching and the learning methods using e.g. technology. An iPad project is partly fund-financed (with DKK 50 million) and partly self-financed by the municipalities (with DKK 100 million). Since the start of the project, the project management has focused on upgrading the qualifications of teachers and extending of the installations in towns and settlements that will enable the use of iPads in education. The project is expected to be fully implemented by 2024.

Another example is the establishment of the national waste company, ESANI A/S, which was founded on 1 February 2019. The company's main purpose is to create solutions for the treatment of all types of waste in Greenland. The board of the company was to have members from all the municipalities in the country. Ten months after its establishment, however, Avannaata Kommunia withdrew from the company as it wanted other local waste solutions for its settlements. Thus, the joint municipal cooperation projects are challenged by the very different natural geographical and socio-economic conditions of the various municipalities.

At national level, a forward-looking plan for new airport extensions primarily in Nuuk, Ilulissat and Qaqortoq (the airport package) has been adopted, as well as hydropower plants to supply Qasigiannguit and Aasiaat, among other things. In terms of mineral resources, a stop to oil and gas exploration and a prohibition on exploitation of minerals containing radioactive elements above a certain level have been introduced. This type of large-scale national political decisions has a considerable local impact on the municipalities' room for manoeuvre, as it very much defines the business and employment opportunities of municipalities, the mobility between the municipalities as well as the revenue and expenditure of the municipalities.

Socio-economic players and social advocacy institutions

In addition to the administrative players under the Self-Government and municipalities, a number of societal institutions play prominent roles. This applies to key employee and occupational organisations, but increasingly also a quite unique construction with spokespersons for various vulnerable groups.

SIK

The labour union SIK (Sulinermik Inuussutissarsiuteqartut Kattuffiat) dates back to the Greenland Workers' Association, GAS, which was formed in 1956 but was given its current name in the late 1970s. SIK organises all unskilled workers and some skilled workers as well as various groups of students and trainees. The union has entered into a general agreement with employers, it is a negotiating party in connection with collective agreements and has historically worked for the elimination of the birthplace criterion, worked for equal pay and has been a formative party to a number of major reforms for the development of present-day Greenland. It also participated in the EC-critical ANISA in the early 1980s and was thus a factor in the withdrawal from the forerunner of the EU together with, among others, the Siumut and IA as well as KNAPK.

However, as the largest labour union, SIK has also been criticised for failing to ensure an acceptable level of pay for many of its members.

KNAPK

KNAPK (Kalaallit Nunaanni Aalisartut Piniartullu Kattuffiat) is the association for fishermen and hunters.

Greenland halibut and shrimp are Ilulissat's primary income sources, and longline fishing for Greenland halibut employs many dinghy fishermen.

MORTEN RASMUSSEN/BIOFOTO/RITZAU SCANPIX, 2015

ULRIK BANG/RITZAU SCANPIX, 2007

A major demonstration in Nuuk took to the streets to urge politicians to prioritise what they call 'better lives for children' over prestige projects and runways. According to the association, there is a shortage of at least 400 teachers and 250 educators in the country.

KNAPK was founded in 1953 as a consolidation of local associations following a meeting between the KGH (Royal Greenland Trading Department) and the office of the Governor. KNAPK is headquartered in Nuuk, but has a number of local departments and councils for various species (shrimp, Greenland halibut and snow crabs, etc.).

The association works to develop the business of Greenlandic fishermen and hunters economically, societally and culturally. Conflicts of interest sometimes arise between the Department of Fisheries and Hunting and KNAPK, as well as SQAPK, which represents some of the dinghy fishermen in the Greenland halibut districts in North Greenland.

Sulisitsisut
— Greenland Business Association
Established in 1966, Greenland Business Association (GBA) is Greenland's largest business association. The association works to safeguard the interests of Greenland's business community for both small and medium-sized enterprises to large corporations operating in international markets. GBA is organised into 10 industry committees and, as employer organisation, represents the vast majority of companies and the vast majority of private businesses. Together with SIK, it has also raised issues regarding social dumping and the challenges posed by foreign labour.

Every two years, GBA organises the Future Greenland conference, which attracts speakers and Danish top politicians and business professionals as well as international guests. However, it has also been criticised for failing to ensure that the conference adequately locally rooted in the Greenlandic community.

Social Advocacy Institutions
Up through the 2000s, changing governments have had an increased focus on social challenges in society.

Advocates have been appointed in the area of children, disability and the elderly, respectively, to personify this key area and to let the individual politically independent advocate become the face of the focus area. Although all three institutions carry the name 'advocate', these are three widely different institutions. This applies in terms of powers, tasks, organisation, secretariat and budget grants.

In order to promote the rights and interests of children in society, and as part of implementing the United Nations Convention on the Rights of the Child, which Greenland acceded to in 1993, it was agreed in 2011 to create a National Advocacy for Children Rights. Based on this, the MIO was set up in the same year, aimed at promoting children's rights and interests. The National Advocacy for Children Rights consists of the Children's Advocate, the Children's Council and a secretariat. On paper, it is a politically independent institution, but it is Naalakkersuisut that appoints the Children's Advocate and the Children's Council. The Children's Council advises the Children's Advocate, who is responsible for employing secretarial staff. MIO works to translate the UN Children's Convention into concrete initiatives that improve children's everyday lives. The Children's Advocate provides advice and guidance to children as well as to authorities, institutions, organisations and private individuals. In addition, the institution must monitor, disseminate knowledge, influence the public debate and make proposals for improving children's conditions, and report to the United Nations on how Greenland is complying with the UN Children's Convention. The Children's Advocate travels all over Greenland to talk to children about their living conditions, after which reports are submitted to authorities and the public. In recent years, the Children's Advocate has become a distinct voice in the public debate and has ensured children's conditions a more prominent place on the political agenda.

As a result of Greenland acceding to the UN Convention on the Rights of Persons with Disabilities in 2012, Inatsisartut passed a law in 2017 establishing a National Advocacy for Persons with Disabilities. Tilioq was established in the same year and is to ensure the practical and citizen-centric implementation of the UN Convention on the Rights of Persons with Disabilities. The Disability Advocate is appointed by Naalakkersuisut and aims to promote rights and interests in society for persons with disabilities. Unlike the Children's Advocate, Tilioq does not have an advisory council, but consists of the Disability Advocate and a secretariat. The Disability Advocate provides advice and guidance to people with disabilities and must also advise authorities and institutions and disseminate knowledge and influence the debate with proposals for improving the conditions for people with disabilities. Tilioq works to ensure that legislation and practices meet the UN Convention on the Rights of Persons with Disabilities. Through its work, Tilioq has since 2017 made it clear how difficult it is to live with a disability in Greenland.

In 2020, yet another advocacy institution was established, with Inatsisartut passing the law on a National Advocacy for the Elderly in 2019. Unlike the other two advocacy institutions, the National Advocacy for the Elderly is not based on a UN convention. Instead, the National Advocacy for the Elderly should be seen as part of Naalakkersuisut's policy for the elderly. Unlike the other advocacy institutions, the National Advocacy for the Elderly neither has a council nor a secretariat, and is instead be served by the secretariat of the National Board of Social Services. The duties of the Advocate for the Elderly are to ensure and promote the opportunities and interests of the elderly and to draw attention to and provide information about the conditions of the elderly in society. Although being an advocacy institution similar to MIO and Tilioq, it has been allocated no funds to ensure that it can meet its purpose to the same extent as the two other advocacy institutions.

Illustration by Ivalu Risager in connection with the publication of the population survey 'Attitudes to Disability 2020'. The survey, conducted by Tilioq in cooperation with Ilisimatusarfik, revealed the public's attitudes towards people with physical and mental disabilities. One of the objectives was to highlight and address any negative attitudes towards disabled people.

IVALU RISAGER/TILIOQ, 2020

LINDA KASTRUP/RITZAU SCANPIX, 2015

Qullissat emerged as a mining town in 1924 and became one of the country's largest towns, but when the coal mine closed in 1972, all its inhabitants were forced to leave. Today, Qullissat is visited exclusively as a holiday destination; families who lived there have resumed using the houses.

POPULATION AND DEMOGRAPHICS

The Greenlandic Inuit, the indigenous people of Greenland, commonly call themselves *Kalaallite*, Greenlanders, and the country *Kalaallit Nunaat*, the land of Greenlanders (see p. 13). The Greenlandic Inuit are one of the few indigenous peoples in the world who are the majority population in their own country.

Greenlanders were granted Danish citizenship in 1926, and therefore the terms 'born in Greenland' and 'born outside Greenland' are used in official statistics to distinguish between 'Greenlanders' and 'Danes'. Statistics Greenland's Statistics 1994 Survey of Greenlandic Living Conditions showed that the vast majority of people 'born in Greenland' perceived themselves as Greenlanders and that most people 'born outside Greenland' perceived themselves as Danes.

Size and settlement of the population – distribution by settlements

Greenland is not only the world's largest island, but it also consists of many settlements that can be characterised as insular communities consisting of 17 towns and 54 settlements, as well as sheep farms and stations (see p. 323).

The traditional way of life among the Greenlandic Inuit when the colonisation of Greenland and the mission activities began was semi-nomadic, originating in a permanent winter settlement and changing summer settlements depending on the prevalence of game and fish. Both the trade (Royal Greenland Trading Department, KGH) and the mission had an interest in the population becoming more permanent residents. The number of settlements increased, partly due to an increasing population, up to the early 1900s, but has since been declining. The population in the towns (term from 1721: 'colonies') has been increasing from the 18th century, and, in the early 1950s, it exceeded the population 'outside the towns' (see p. 143). The term 'outside the towns' is a collective term for 'settlements', which make up by far the largest part of the group as well as the sheep farms in South Greenland and stations.

The overall political objectives stated in the government plans from the 1950s and 1960s to modernise Greenland, the so-called G-50 and G-60 plans (see p. 133), were, among other things, to improve of the living standards of the Greenland population and a commercial focus on cod fisheries by, e.g., establishing fish factories in a few selected towns on the west coast. The modernisation plans built on a premise that both welfare improvements and the realisation of economies of scale in the fishing industry necessitated a population concentration in substantially fewer inhabited areas.

The concentration policy involved the depopulation and thus relocation from many settlements, including the mining town of Qullissat (1972). This policy met a great deal of local resistance and became one of the driving forces of the political movement towards Greenlandic autonomy. The conditions of the settlements (e.g., the establishment of factory plants) were therefore improved in the period following the introduction of the Home Rule. The population in the settlements stabilised in the 1980s, while migration from settlements (and a number of smaller towns) took off after the turn of the millennium, not least following a number of political initiatives such as the elimination of the same-price system for retail goods (1994) and energy and water taxes (2005). The same-price system on energy and water was reintroduced in 2018.

The population in Greenland has been increasing since the 1901 census and until 1991, when the Greenlandic population figure had reached 55,520. Since the early 1990s, the year-to-year population has changed only slightly in an upward or downward direction and was 56,421 in 2021. The highest population figure to date was 56,969 in 2005.

While the overall population has remained relatively stable over the past 30 years, there have been more significant shifts in the proportions of individual population groups, and a comparison with the development in the size and composition of the population from the beginning of the 20th century also shows substantially greater changes.

Accordingly, in the first half of the 20th century, the population had doubled to 23,647 by 1951, and doubled again by 1971 to reach 47,935 (see p. 145). There are several reasons for these significant changes: an improved state of health (including the eradication of tuberculosis), a reduction in infant mortality, an increase in fertility and immigration of mainly Danish labour.

Size of population and distribution by age groups

In the 1960s, twice as many children per year group were born compared with the year groups ten years before and ten years after. This is why the year groups born in the 1960s are called 'the large year groups«.

The number of live births reached the highest in 1966 when 1,781 children were born. In order to limit population growth, the Danish Ministry for Greenland and the Danish Health encouraged the use of intrauterine devices (IUD) in fertile Greenland women. In the period 1966-1970, 4,500 IUDs were fitted.

When the IUD campaign started, there was no legal basis to fit IUDs in underage girls without parental consent. Legislation was changed in 1970 so that doctors in Greenland could guide young girls from the age of 15 on contraception without the consent of their parents. At a UN conference in 1974, harsh criticism was voiced over family planning in previously colonised countries such as Greenland, and the UN called for everyone to be allowed to decide how many children they want and when.

In 1975, barely 10 years after the number of live births peaked and after the IUD campaign was carried out, the birth rate was 815, thus a decline of more than 50 %.

The increasing numbers in the groups 17 to 24-year-olds and 25-64-year-olds up through the 1980s are, among other things, the large year groups that have become young people and adults, just as the growing number of 0-6 year-olds in the 1980s and 1990s are children of the large year groups. They are therefore what can be called an echo effect. Citizens aged 65 and over represent the smallest age group throughout the period, but it is slightly increasing, especially in the latter part of the period. As a result of increasing life expectancy, the size of the group has tripled overall in the almost 40 years.

Figure 8. Ageing of the Greenlandic population 1977-2020. The division into groups is based on significant changes in the relationship of children, young people, adults and the elderly to education and the labour market: the 0-6-year-olds (preschool age), 7-16-year-olds (primary school), 17-24-year-olds (youth education), 25-64-year-olds (working), 65 and over (retirement age).

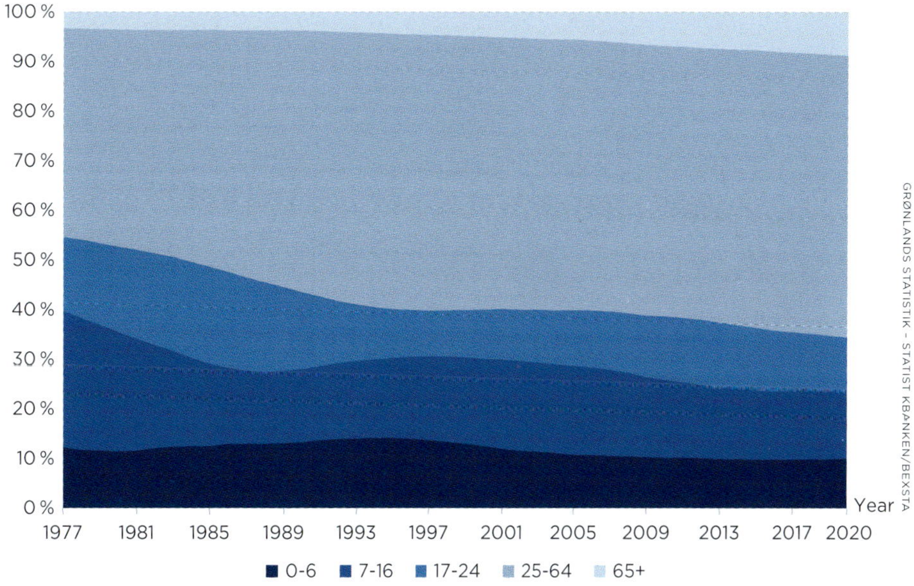

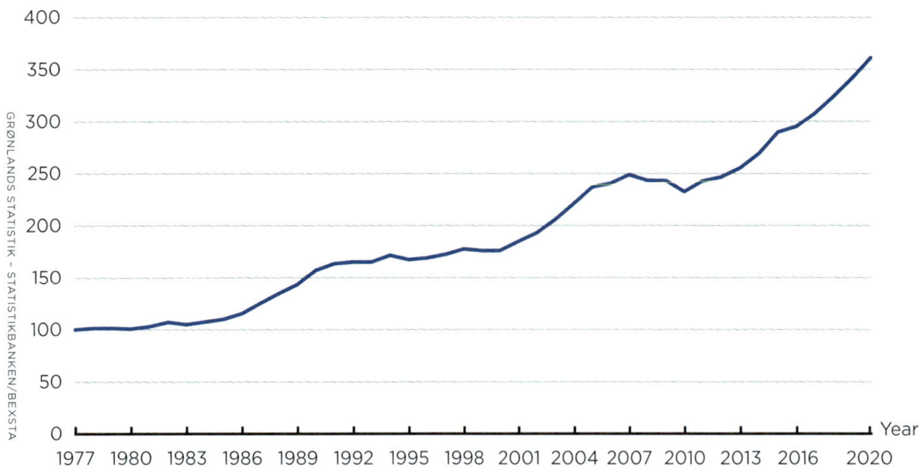

Figure 9. The development of the 60-64-year old 1977-2020 highlights a challenge for Greenland: that a more significant proportion of the population is in an age group outside or exiting the labour market, while average life expectancy is increasing. Index 1977=100.

Size and distribution of population by gender

Until 1961, women made up the largest group of the Greenland population. This was primarily due Greenlandic men dying while hunting or fishing. The preponderance of women had narrowed by 1960, partly due to fewer deaths at sea. The main reason why men have made up a majority of the population since 1961 is the large number of posted Danish workers (typically single men) and later the group of on-call workers. A majority of the workers who migrated to Greenland were and still are single. At the same time, over the past 25 years (since 1993), there has been a net emigration of women born in Greenland of around 200 annually.

Greenlandic societal development is often described as both rapid and profound, and population trends reflect this development.

The gender distribution among the working population has changed significantly, while the total number of people employed in the period 1965 to 2001 doubled. While 76 % of all employed people in 1965 were men, the proportion fell to 67 % in 1974 and 1987 and further to 54 % in 2001. Over the past decade, the distribution has been more stable and the proportion of men has been between 54 % and 56 %.

Labour market attachment, the participation rate, has been slightly higher in the period from 2010 to 2018 for men than for women, but it has remained stable for both genders. For men, it has been 76-77 % and for women 71-72 %.

Women's attachment to the labour market has not only increased, they increasingly complete educational programmes at all levels. Accordingly, in all the years since 2003, more women than men have graduated from a higher education. The level of education is generally increasing, and by 2019 a greater proportion of women had completed an education programme above primary school level.

The size and distribution of population by place of birth

The increase in the population of both persons born in and outside Greenland was exponential in the period 1951-61. For the Greenlandic part of the population this was due to e.g. improved health conditions and increased fertility. For the people from outside Greenland, typically Danes, this was due to the need

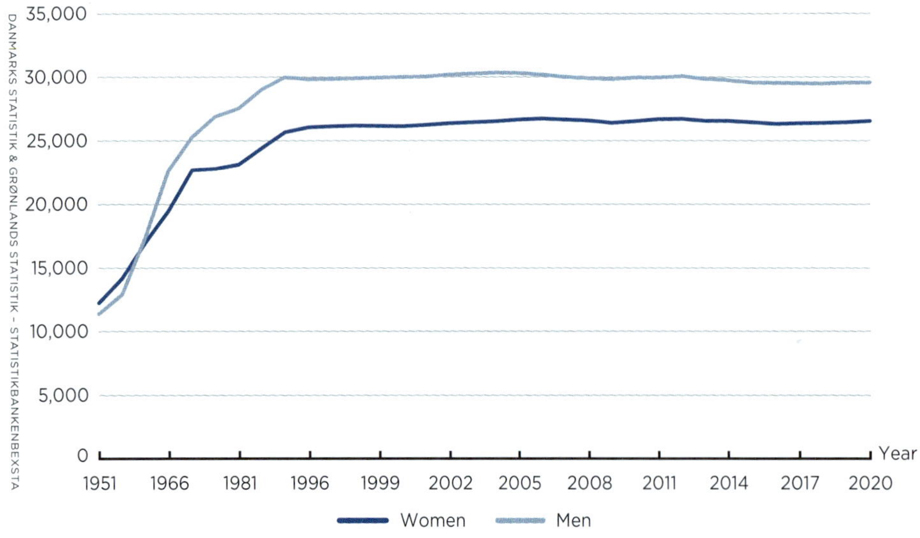

Figure 10. Population and distribution by gender 1951-2020 (selected years). The predominance of men since the 1950s is mainly due to the fact that the majority of citizens not born in Greenland are men. Also to a lesser extent, more women moving out of Greenland.

for construction workers, craftsmen, technicians and administrators. Since the basic view of Danish planners and politicians in relation to carrying out these tasks was that the need for labour would be short-lived, the Danish State skilled as well as unskilled labour to undertake the various parts of the modernisation process, while training local labour was not part of the plans.

However, the scope of external labour did not diminish, neither during the modernisation period nor after the introduction of Home Rule Government in 1979. The pace at which the Home Rule Government took over areas of responsibility from the Danish state was much faster than expected. Since it was not possible to train Greenlanders to take on jobs immediately in the areas that were taken over, it was therefore often necessary to call in labour with the necessary qualifications in order to solve the many new tasks, which increasingly included tasks in the municipalities' and the Home Rule Government's (later Self-Government) administrations, service sectors (education, health and social services) and the publicly owned companies.

The number of inhabitants born outside Greenland peaked at almost 10,000 in the late 1980s. During those years, Home Rule was in a deep economic crisis that led to a reduction in the number of called-in labour. Since then, both the proportion of the population born outside Greenland and the number of immigrants have been declining, and in 2020 it was just about 6,000.

In order to get a more differentiated picture of the population composition than the division of 'born in Greenland' and 'born outside Greenland', for the period 1977 onwards, Statistics Greenland has included the birthplace of parents and grandparents in addition to the person's own place of birth. For each combination option, a basic code is determined. Table 2 illustrates the distribution of the entire population by this basic code.

The group of people born in Greenland having two Greenland-born parents is the most frequently occurring. The group counts between 46,000 and 47,000, representing about 82 % of the

LIFE PICTURE COLLECTION/SHUTTERSTOCK/

Population statistics for Greenland do not include US military personnel. During World War II, there were about 4,000 Americans at Bluie West One in Narsarsuaq, and during parts of the Cold War, the Thule Air Base was the largest 'town' in Greenland by far – even after the thousands of construction workers had gone home.

Figure 11. Population and growth by place of birth (born in and outside of Greenland) 1901-2020 (selected years).

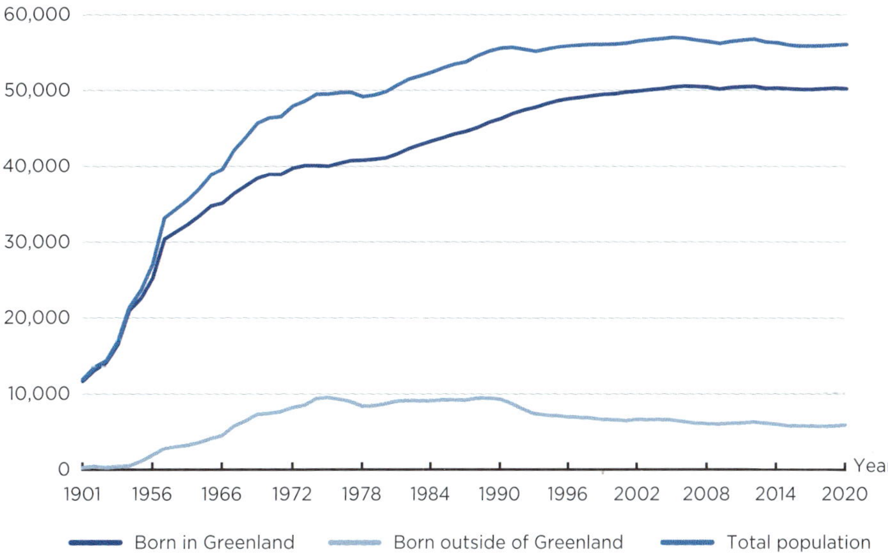

STATSMINISTERIETS GRØNLANDSAFDELING, DANMARKS STATISTIK OG GRØNLANDS STATISTIK - STATISTIKBANKEN/BEXST1

	1977	2000	2021
Born in Greenland with two Greenlandic-born parents	77 %	82 %	82 %
Foreign-born with two foreign-born parents	17 %	10 %	9 %
Greenlandic-born with two foreign-born parents	1%	1%	1%
Foreign-born with one Greenlandic-born parent	1%	1%	1%
Greenlandic-born with one foreign-born parent	4 %	6 %	6 %
Foreign-born with two Greenlandic-born parents	0 %	1%	1%

GRØNLANDS STATISTIK - STATISTIKBANKEN/BEDST9A

Table 2. Population by own and parents' place of birth 1977, 2000 and 2021.

population. The proportion of people born outside Greenland (typically in Denmark) whose parents were also born outside Greenland (also typically in Denmark) has been reduced by more than 40 %.

Population trends – the population accounts

How the overall population number develops, i.e. population growth, depends on four conditions: the number of births and deaths, and the number of immigrants and emigrants. The difference between births and deaths is referred to as excess of births, and the difference between the emigrants and immigrants is called the net immigration.

Over the past 70 years, the Greenlandic population accounts have had a positive balance, i.e. a population increase, for most years (see Figure 12).

The excess of births was highest in the 1960s when the so-called 'large year groups« were born. The years and short periods in which the total population has been in decline coincide with a positive net emigration, that is, a period in which emigration has been greater than immigration. This happened, for example, in the years before the introduction of Home Rule Government, in the period around 1990 and again in the period from 2005 to 2015. People born in Greenland typically emigrate to Denmark, and the number of Greenland-born people in Denmark was around 16,770 in 2020 (up by almost 2,400 since 2009).

There has been an excess of births in all the years since 1950, although due to declining fertility rates, it has been declining since the late 1980s. The fertility rate in 2019 was 2.08 children per woman between the ages of 15 and 49. The declining fertility rate is not least due to a high number of legal abortions. In the period 1993 to 2019, the annual abortion rate was between 740 and 1,000, while during the same period the number of births varied between 780 and 1,180. In the period since 2013, the number of legal abortions each year has exceeded the number of births.

Causes of death

The causes of death among the Greenlandic population in the period 1924 to 2019 (see Figure 13) cover several stages in the development of Greenland over the past almost 100 years: the beginning transition from hunting to fishing in a largely 'closed' Greenland (first decades in the 1900s), the modernisation period in the 1950s and 60s, the period leading up to the introduction of Home Rule Greenland, the period of Home Rule

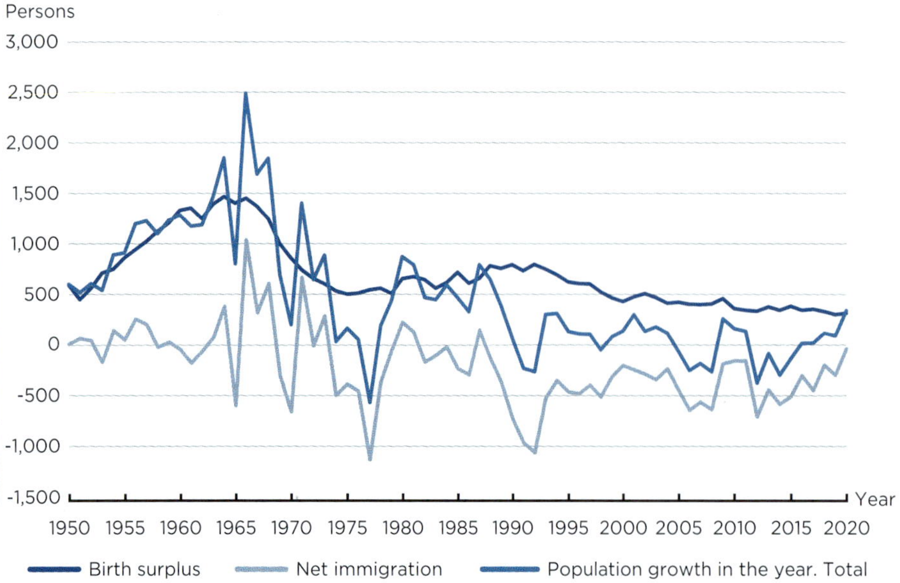

STATSMINISTERIETS GRØNLANDSAFDELING, DANMARKS STATISTIK OG GRØNLANDS STATISTIK - STATISTIKBANKEN

Figure 12. Population trends from 1950-2020. The population account is the net result of the inflow of citizens (births and immigrants) and the outflow (deaths and emigrants). The population account can also be expressed as: birth surplus + net immigration.

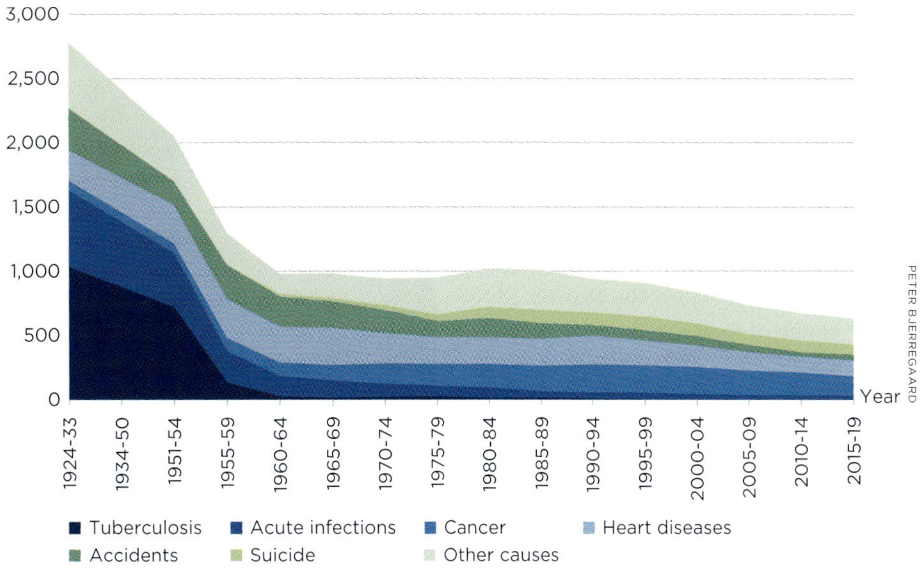

Figure 13. Causes of death in the Greenlandic population. Age adjusted. 1924-2019. The age composition of the Greenlandic population varied from 1924-2019, and diseases are typically more common in some age groups than in others. An age adjustment has therefore been made to allow comparison of causes of death over the period.

Government and the beginning of the Self-Government period.

The causes of death reflect, in various ways, the developments in technology (both in health sector and the main professions), in political objectives and in lifestyle. One of the factors that has changed significantly throughout these periods is tuberculosis which, until the early 1950s, was epidemic and was a frequent cause of death until the mid-1950s. A focused and multi-pronged health strategy that involved both the establishment of hospitals (sanatoriums) and prioritising housing to replace peat huts and other outmoded housing resulted in the near-eradication of tuberculosis (see p. 130). The development in deaths from other acute infectious diseases other than tuberculosis (e.g. pneumonia) exhibits the same pattern and is, not least, a result of improved living conditions, hygiene and improved housing conditions as well as health care initiatives.

Another cause of death that has been reduced is accidents which through the first part of the period typically involved deaths among men who worked at sea. During the period up to 1960, a gradual transition took place from hunting in kayaks to hunting and fishing in dinghies and larger boats. This also reduced the risk of fatal accidents, but accidents were still a major cause of death until the early 1990s.

At the same time, throughout the period, heart disease/disorders have been a major cause of death, while cancer was a less frequent cause of death at the beginning of the period. However since the mid-1960s death due to cancer has been among the most frequent. Some of the reasons often mentioned, is that several types of cancer are so-called lifestyle diseases.

Suicide used to be a very rarely occurring cause of death, but in the wake of modernisation and, not least, from the early 1980s, suicide rates increased – primarily among young men – and it is one of the major causes of a relatively low average life expectancy among men.

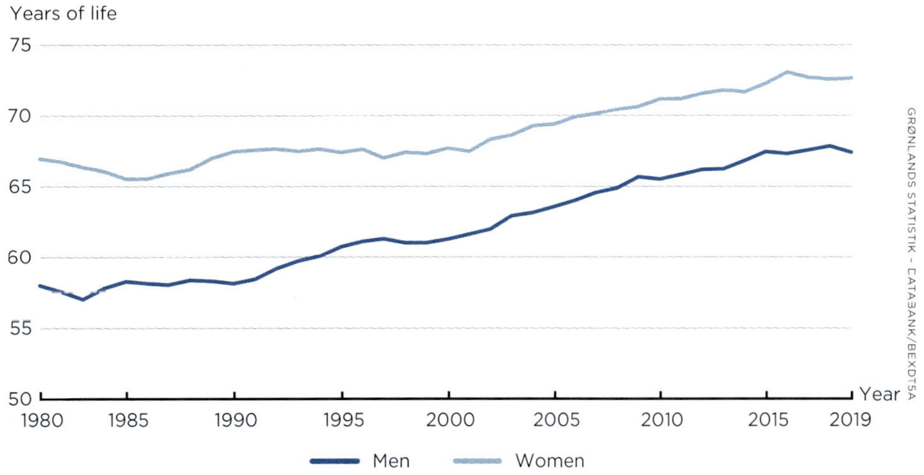

Figure 14. The average life expectancy of the population in the period 1980-2019 'Average life expectancy' is used as an overall concept for the evolution of the population's average life expectancy, calculated based on continuous 5-year periods. Age is to be understood here in the sense of remaining life. Average life expectancy is therefore life remaining at birth (age 0). The figure shows the trend in average life expectancy of newborns since 1996.

Average life expectancy of the population

In the period 1996 to 2015, women's average life expectancy increased from 69 years to just under 74 years, while the average life expectancy for men over the same period increased from just over 63 years to 69 years (see p. 273). Key reasons for the shorter life expectancy of men are that, as mentioned, suicides and accidents are frequent among younger men.

Although the average life expectancy of both men and women has increased significantly, it is considerably lower (about 10 years) than in the Nordic countries.

INFRASTRUCTURE

The Greenlandic infrastructure is significantly different in several areas than in Denmark and most other countries because Greenland's towns and settlements function as insular communities.

Due to the climate, permafrost and the wide and deep fjords, and because more homes are located on islands, establishing long road connections is complicated and very expensive. These are challenges that also exist elsewhere where a connected road network has been established to some extent. However, for Greenland, the challenges, combined with the very small population spread over a vast area of land, have meant that no inhabited places are connected by road to other inhabited places, apart from a few sheep farms in South Greenland. It is therefore not possible

In the dark, the infrastructure is visible here in one of Nuuk's newest districts, Qinngorput, located at the foot of the mountain Ukkusissaq (Store Malene). Qinngorput is located about 5 km south of Nuuk Airport.

MADS PIHL/VISIT GREENLAND, 2014

MORTEN RASMUSSEN/BIOFOTO/RITZAU SCANPIX, 2020

All installations such as water, drainage and electricity for various buildings lie freely on rocks and mountains. The water and drainage systems are heated to prevent them from freezing. Electrical installations are shown here.

to commute on a daily basis and the individual settlement is largely dependent on their own business base.

Supply infrastructure

The government-owned limited company, Nukissiorfiit, supplies electricity and water to towns and settlements, as well as district heating to some homes in most towns.

Electricity supply

For Nukissiorfiit, the great distances and the insular community mean that there is no nationwide or regional electricity grid, which is why each town and settlement have their own electricity supply. The only exception are the towns of Narsaq and Qaqortoq in South Greenland, both of which get electricity from the Qorlortorsuaq hydropower station. In addition, there are hydropower stations at Nuuk, Sisimiut, Ilulissat and Tasiilaq. Renewable energy from hydropower accounted for 67 % of Greenland's electricity consumption in 2018. The power stations in all other towns and settlements are operated with diesel generators.

District heating

In most towns, Nukissiorfiit provides district heating based on residual heat from the electric power station; in some places, the incineration plant also supplies heat. In towns supplied by hydropower, some residential areas are heated with electric heating.

The country's other homes are heated by oil-fired boilers or, especially for smaller homes, the widely used kerosene heaters. Large multi-storey housing blocks or terraced house areas often have a communal heat station.

Water supply

In the vast majority of settlements, Nukissiorfiit uses surface water from one or more nearby lakes, while river water is used in some places. The water is purified and treated in a local waterworks.

In winter, about ten settlements, primarily in North and East Greenland, get their water from tanks that are filled during summer. In addition, there are about ten towns and smaller settlements on small islands that wholly or partially use reverse osmosis (RO) for seawater desalination.

In the large towns, the majority of homes are connected to a water network, and most homes without a water connection get water delivered by a tanker to a water tank in the home. In the small towns, far from all homes are connected to a water network, and in the settlements very few households have a connected to a water system. Here, citizens have to fetch water in plastic containers from a tapping house. It is not uncommon that you need to go 500 m to a tap house and, for some settlements, the distance can be longer.

The settlements of Naajaat and Nutaarmiut in Upernavik district and Qeqertat in Qaanaaq district have no water supply. Here, citizens have to fetch water from a river or a lake, and in winter, they have to melt large clumps of ice that are collected on the sea ice.

The limited water resources and/or the absence of a supply network mean that personal water consumption is very modest in many places. Approx. 5 % of the population has an average water consumption of less than 20 litres daily and about 12 % has a consumption below 40 litres. This modest water consumption for cooking, bathing and laundry, etc. affects health and personal hygiene.

GEORG JENSEN/ARKTISK INSTITUT, 1963-65

Water supply is also important for business development and reduces opportunities for local processing of fish, including in the large Greenland halibut districts of Uummannaq and Upernavik. The Greenland halibut exported from here is therefore largely unprocessed, and the value added by processing does not benefit Greenland. Despite modest water supply in several places, the districts of Uummannaq and Upernavik, measured per capita, contribute above average to national export income.

Ice floes are loaded for drinking water for the Upernavik inhabitants 1963-65. Most of the drinking water in Greenland is produced from meltwater collected in drinking water lakes or taken directly from rivers, but ice floes are also still used to some extent. They are melted in smelters, a very expensive production method, costing about 600 DKK per m^2.

Price structure

It is cost-intensive to operate the many local power stations and waterworks where several settlements cannot be reached by boat for parts of the year. All piping needs to be kept frost-free with electric heating cables, and many settlements have a modest customer base. In order to ensure equal competition and social conditions geographically, a common price has been adopted politically across the country. This leads to significant cross-subsidisation, where towns with hydropower and large water resources help to finance the operation of the other settlements.

Wastewater

Most towns have a sewer system to some extent, but all towns have residential areas without sewers. In the larger towns, some of homes not connected to a sewer system have a sludge tank that is emptied by vacuum tankers. The remaining households use the toilet bucket (bag toilets), which are emptied by the collection of latrines. In the smaller towns and small settlements, the majority of homes are not connected to a sewer system. At national level, about a quarter of the households have a toilet bucket.

There is no treatment plant in Greenland, so wastewater is led untreated into the sea. The majority of homes with no connection to a sewer system or sludge tank lead the grey wastewater from the kitchen sink, possibly bath water and washing machine water, out on the ground.

The municipalities are in charge of wastewater management, while Nukissiorfiit is in charge of water supply. The division of responsibilities is instrumental in the limited sewer system in several towns. For example, the municipality has not provided a sewer system for Upernavik – a town with 1,100 inhabitants – because Nukissiorfiit does not provide sufficient water for the operation of a sewer system.

Waste management

Waste management is a municipal task. In all places, waste is collected and temporarily or permanently deposited at the dump (waste landfill). Some larger towns and few settlements have old incineration plants. In several locations, planned incineration of the dump is sometimes carried out. Spontaneous combustion can occur in the waste, which can result in harmful smoke. Due to strong winds and animals rooting around on the dump, scattered waste is often seen in nature and the sea.

In all locations, the intention is for environmentally hazardous waste and scrap metal to be sent to Denmark for treatment. In 2019, the inter-municipal waste company, ESANI, was established. ESANI intends to build new and larger incineration plants in Nuuk and Sisimiut to replace all existing ones, and the plan is to sail all waste suitable for incineration to Nuuk and Sisimiut. Avannaata Kommunia in the north has chosen not to be part of ESANI as it estimated that the waste transport would be too cost-intensive and complicated considering the municipality's many small settlements that cannot be reached by boat for parts of the year.

According to the Self-Government's Waste Action Plan 2020-2031 which was drawn up in 2020, landfills must be phased out and from 2024 emissions from waste incineration must comply with the EU requirements.

The dump in ilulissat. In most towns and settlements, all the waste has been dumped for decades, and open burning has been carried out regularly to reduce waste volumes. A central waste management system is being implemented, where all combustible waste must be packed in the individual settlements and sent to incineration plants in nuuk or sisimiut and all metal and chemical waste must be shipped to denmark.

ASHLEY COOPER/SHUTTERSTOCK/RITZAU SCANPIX, 2009

Freight

The government-owned Royal Arctic Line (RAL), in cooperation with the Icelandic shipping company, Eimskip, carries out virtually all shipping between Europe and Nuuk. The vast majority of cargo is shipped from and to Denmark. In Nuuk, cargo for the west coast's other settlements is transhipped to RAL's smaller ships, and cargo north of Ilulissat is transshipped again in Aasiaat. Cargo for the east coast is transshipped in Reykjavik to a RAL ship built for ice navigation. Cargo transport to the settlements is partly provided by the subsidiary, Royal Arctic Bygdeservice.

Due to sea ice, Uummannaq and Upernavik can usually only be reached by ship from May to December, and Tasiilaq and the settlements on the east coast from May to November. Ittoqqortoormiit on the east coast and Qaaaaq in the north are only called on twice a year. In winter, navigation in Qeqertarsuup Tunua (Disko Bay) is irregular or cancelled due to sea ice, and in South Greenland, the multiyear ice can periodically make navigation difficult or prevent it altogether.

The price for freight is the same across the country, but the handling charge varies and is determined by location, and a handling fee is charged for each transshipment. RAL's costs vary greatly and are especially high for settlements without a port where the cargo has to be barged in. Therefore, the price structure includes a significant degree of cross-subsidisation. Exports are indirectly subsidised by the fact that freight rates from Greenland are approximately one third of the price of freight to Greenland. To support fish exports, the price for exporting a 20-foot refrigerated container is nearly a tenth of the price of the imports.

Ship passengers

Royal Arctic Line's subsidiary, Arctic Umiaq Line, operates a passenger ship between Qaqortoq and Ilulissat, calling in nine towns and three settlements. The fare is partially subsidised.

Disko Line (owned by the Danish Topas Rejser) sails between most towns and settlements in South Greenland, Central Greenland and Qeqertarsuup Tunua with small passenger boats subject to a service contract with the Self-Government. During the winter months, Qeqertarsuup Tunua is serviced instead by Air Greenland's helicopters.

Otherwise, sailing takes place with private boats and dinghies, and in several places it is possible to charter a boat.

Air transport

Greenland's airports are operated by Mittarfeqarfiit, owned by the Self-Government. The country has two international airports, Kangerlussuaq (Søndre Strømfjord) and Narsarsuaq, with long tarmacked runways capable of receiving transatlantic aircraft. Both were built by the United States during World War II. Kangerlussuaq serves as Greenland's primary gateway and international airport. It is the hub for domestic flights, while Narsarsuaq is primarily used as a local airport for South Greenland and, in the summer months, connects to Denmark.

In the 1990s, short tarmacked runways were established for small propeller aircraft in the towns of Paamiut, Nuuk, Maniitsoq, Sisimiut, Aasiaat, Ilulissat and Upernavik. The same was the case in the settlement of Qaarsut with a helicopter connection to Uummannaq. Qaanaaq in the north has a gravel runway. On the east coast there is a US built gravel runway by the settlement of Kulusuk with a helicopter connection to Tasiilaq as well as the settlements in Ammassalik district. The Nerlerit Inaat (Constable Point) gravel runway, established by an oil exploration company, has a helicopter connection to Ittoqqortoormit. Nerlerit Inaat is only served by air from Iceland. The airports have helicopter connections to the remaining towns and most settlements. Most of these towns have heliports with a tower, runway lights and waiting room, while the remaining towns and settlements have more simple helistops. In 13 settlements, Mittarfeqarfiit has no permanent installations.

Air Greenland

Air Greenland flies throughout the year between Kangerlussuaq and Copenhagen Airport several times a week, and in summer also between Narsarsuaq and Copenhagen Airport. For domestic airports, Air Greenland uses smaller propeller aircraft, which are also used on the routes to Reykjavik from Nuuk and Ilulissat. For other destinations, helicopters of different sizes are used.

Flight frequency depends on the size of the place. Nuuk has several daily connections to many destinations, while the smaller towns and largest settlements connect a few times a week. The small settlements with helistops have a weekly connection during periods without ship or boat connection, while settlements without helistops do not have regular service.

The Self-Government subsidises the flight operations for the small towns and settlements. For other routes, Air Greenland uses internal cross-subsidisation where, primarily, the route between Kangerlussuaq and Copenhagen and to a lesser extent the route between Nuuk and Kangerlussuaq

COWI, MASANTI OG ZESO ARCHITECTS, 2018

Illustration of Kalaallit Airport's winning project in Nuuk. Cowi and sub-consultants Masanti and Zeso Architects won the tender in 2018 to build new airport buildings in Nuuk, Ilulissat and Qaqortoq. Construction work in Nuuk started in 2019.

help to finance routes with smaller passenger bases. Despite subsidies and cross-subsidisation, domestic flights over longer distances are very expensive.

Regularity

Regularity is affected by the Arctic weather and there are often long delays or cancellations. In order to make optimal use of capacity, priority is given to larger destinations. It is therefore not uncommon for smaller destinations to not be served despite good weather because the aircraft get weather-bound in one of the larger towns.

Air Iceland Connect and Norlandair

The only other companies with permanent routes to Greenland are Air Iceland Connect, which has flights from Reykjavik to Kulusuk, Narsarsuaq, Nuuk and Ilulissat, as well as Norlandair, which has flights from Reykjavik and Akureyri to Nerlerit Inaat.

New airport structure

In Nuuk and Ilulissat, Kalaallit Airports International A/S is constructing new and longer runways (2,200 m) capable of receiving transatlantic jets. The intention is that both runways should be completed by 2024. The plan is that the government-owned Kalaallit Airports Domestic A/S will establish a medium-range runway (1,500 m) in Qaqortoq for domestic and medium-sized aircraft from Iceland and North America. This runway is also expected to be completed in 2025.

Ports and roads

The Self-Government owns the 150 port facilities used for freight and landing of fish as well as passenger traffic in most settlements. Royal Arctic Line A/S operates the Self-Government ports in virtually all towns,

KÅRE HENDRIKSEN, 2019

The port and factory in the settlement of Innaarsuit, Upernavik district. Measured per capita, Innaarsuit is one of the settlements with the largest trade in Greenland halibut and the largest contribution to Greenland's export income.

while KNI Pilersuisoq operates the ports in the other towns and most settlements.

In Ittoqqortoormiit and Qaanaaq, as well as in four settlements, cargo must be barged to a barge offloading bridge. There are three settlements without port facilities where cargo is barged directly to the coast.

Road network

The municipalities are in charge of local road facilities. In the large towns, the majority of roads are paved, while the smallest towns and settlements primarily have gravel roads. A number of small settlements have no roads but only trail systems.

Telephony and Internet

Tusass (formerly Tele Post Greenland) is owned by the Self-Government and operates telephony, Internet and mobile connections to virtually all residences.

Greenland is connected internationally via a submarine cable from Qaqortoq to Iceland as well as from Nuuk to Canada. The submarine cable connects Qaqortoq and Nuuk and has been led to Aasiaat via Maniitsoq and Sisimiut. Because large icebergs in Qeqertarsuup Tunua (Disko Bay) often scrape across the seabed, it is not considered feasible to take the submarine cable further north.

All settlements between Nanortalik district in the south and Uummannaq district in the north are connected by an approximately 1,500 km long main radio link with 48 radio relay stations located on mountain peaks. The main radio link has several branches, one

THOMAS RAVNSHØJ JOHANSEN/TUSASS, 2020

Tusass has carried out successful tests using wind turbines and solar cells at radio link stations, such as this one at Meqquitsoq south of Nuuk. The station has subsequently been optimised to run primarily on renewable energy.

of which goes to Upernavik and the district's settlements in the north.

Qaanaaq, Tasiilaq and Ittoqqortoormiit are connected via satellite, and from Qaanaaq and Tasiilaq small radio links have been established to the settlements. In the areas covered by satellite, the capacity and speed is significantly lower than in towns connected to the submarine cable. The price for mobile telephony and mobile data is the same for all settlements, while fixed broadband has higher tariffs in Qaanaaq and Ittoqqortoormiit.

HOUSING

For outsiders, the epitome of Greenland housing is often the small colourful wooden houses that are beautifully tucked into a mountainside. But for most Greenlanders, today's reality is something else. A large part of the Greenlandic population lives in flats, and multi-storey housing blocks are the predominant form of housing in large towns.

The period up to approx. 1950

Until World War II, housing for the general population was first and foremost a private matter, unless you had staff accommodation. The Greenland population lived predominantly in peat houses or primitive wooden houses. Housing conditions were miserable, and around 1940, a third of all deaths were caused by tuberculosis – a direct result of the poor housing conditions.

The original peat houses were built from the materials available (see p. 193), but as the Greenlanders became more settled, they also began to build houses that intertwined the Greenlandic and Northern European building customs (see p. 194). From around 1900, some peat houses were fitted with internal wooden walls and a permanent roof made of wood, and at the same time year-round habitation became more common. The 'sealed' peat houses and wooden walls caused problems with moisture and rot, and this is presumably one of the reasons why tuberculosis became so prevalent.

Some built their own wooden houses, but this was difficult; Greenlanders were unable to obtain a mortgage, and at the same time building a wooden house required planning, importing materials and knowledge of carpentry. Greenlanders were hunters and fishermen, not craftsmen.

The period approx. 1950 to 1965

After World War II, the Danish State assumed responsibility for housing for the population. The Greenland Technical Organisation (GTO) designed and built a number of standard houses which shot up both in towns and settlements. These were small and rationally built houses with no facilities, but an improvement on the peat houses. During a relatively short period of years, the standard houses became the predominant form of housing – the small colourful wooden houses, which have since become the epitome of Greenlandic building culture.

In the late 1940s and in the 1950s, the expectation was that Greenlanders were primarily fishermen. Therefore, Danish workers were sent up to build the houses, and in the beginning Greenlanders, were not allowed to participate in the construction. However, this changed and gradually so-called self-build houses were delivered as assembly kits on favourable terms. The idea was that people should be able to build a house themselves with the help of their neighbours. The self-build houses are therefore of varying quality.

The period approx. 1965 to 1975

The Danish focus on streamlining fishing and the fishing industry and the resulting wish to centralise the population in fewer settlements led to a new paradigm for construction and planning in the 1960s. During the period, the GTO built the concrete balcony access blocks found in all large towns. In 1965, the GTO built block P in Nuuk, a 200 m long and 6-storey building that at one time accommodated 1 % of the Greenlandic population (see p. 197).

The period approx. 1980 to 2000

The introduction of home rule in 1979 heralded a new shift in housing construction. Particularly, the Aarhus-based architectural firm, Friis & Moltke, made its mark on the building style. Moltke had travelled in Greenland after the war, and around 1950, Friis worked for the Greenland Provincial Government and designed some of the first small wooden houses. So they both had a background in Greenland. Up through the 1980s they were behind a number of multi-storey housing blocks and terraced houses in e.g. Nuuk, Sisimiut and Ilulissat. A distinctive feature of these was an architectural 'Greenlandisation' of the tower block building that was clearly inspired by the colourful wooden houses – a trend that is also seen in other buildings.

In 1994, INI A/S was created as a management company for the public housing.

The period approx. 2000 to 2021

Housing construction underwent some gradual changes until the next major paradigm shift, which came after the introduction of Self-Government (2009).

The challenges included major housing shortages in several towns,

Part of the housing development in Ilulissat consists of lower terraced houses, many of which overlook the ice fjord.

HANNIBAL HANSCHKE/REUTERS/RITZAU SCANPIX, 2021

the fact that the public sector was the largest provider of rental housing as well as the fact that the private housing market was very small. Until now, public housing construction had been tax-financed, but in order to allow loan financing for the housing construction, the Self-Government set up the housing company Illuut A/S in 2009. The Self-Government wanted rational housing, mainly with free-standing multi-storey housing blocks (to reduce the scale of infrastructure), and, above all, it wanted many multi-storey housing blocks of the same type to be built in order to keep the price down.

The Self-Government issued a competition for the new housing that was won by the firm of architects, Clement & Carlsen, with Permagreen as the contractor. This partnership has constructed a number of free-standing multi-storey housing blocks, the Tuapannguit towers in Nuuk being the first. Since then, this type of housing has been built both in the Qinngorput district of Nuuk as well as in other large towns, though at varying heights depending to the need and local conditions.

From 2015 onwards, most of the new construction has been in the form of multi-storey housing blocks, and Nuuk in particular has experienced an intense construction boom towards the end of the decade. Construction is to an increasing degree being made with private clients and is of such a magnitude that it can be difficult to get contractors to submit tenders for public-sector construction projects. As a result, the price level has risen precipitously.

Thus, the Self-Government and the municipalities have difficulty in providing adequate rental housing to cover both a social need, relocation due to work or education and the need for staff accommodation.

Ownership

Previously, the share of publicly owned rental housing was two-thirds of the total housing stock, but the share is declining. For a number of years, the Self-Government has pursued a policy that more people should own their own home, including getting tenants to buy their home. However, since up to 40 % of the population relocates each year, it is limited how many have a long-term interest in owning their own home. However, the proportion of owner-occupied and cooperative housing in the large towns has increased.

The Nuussuaq district north of town centre of Nuuk was established in the late 1970s and expanded during the 1980s Nuussuaq, today Nuuk's second-largest district, is characterised by dense low-rise buildings, and many of the buildings are located in smaller enclaves around outdoor spaces.

REBECCA GUSTAFSSON/VISIT GREENLAND, 2016

Housing market and financing

Only the four largest towns have an actual housing market where it is possible for private individuals to borrow money to finance a home. In the rest of the country, you cannot get housing loans or mortgage loans. Here, buying a home and renovating privately owned homes must be financed by either consumer loans or savings.

Maintenance needs

In general, the old public-sector rental homes are in poor condition. Rent has been fixed politically and has not allowed regular maintenance. Accordingly, a great need exists for renovation and/or demolition. In 2018, the Self-Government assessed the renovation needs to amount to approx. DKK 1.6 billion primarily for homes built in the 60s and 70s. The Technical University of Denmark assesses that the need is somewhat greater, and also expects that a con-

MADS PIHL/VISIT GREENLAND, 2015

Sheep farm in Tasiusaq in South Greenland, one of almost 40 sheep farms in South Greenland.

siderable need for renovating the newer housing will arise in the years ahead.

Housing in the small towns and settlements poses a particular problem. Since there is no housing market, it can be difficult for owner-occupiers to get financing for modernisation and refurbishment. It also requires knowledge, overview and competent advice – something that is largely only found in the large towns. At the same time, a major investment requires you to be certain that you will remain in the home for a number of years or that you will be able to sell the home at a price that repays the investment. This is far from the situation in the small settlements, and it is a contributory factor in the declining housing standards.

INDUSTRY AND LABOUR MARKET

The insular community and limited population numbers mean that market economy price formation in most towns and settlements in virtually all vital areas is absent or extremely limited because there is only one or very few providers of a given product or service. At the same time, trade between settlements is modest because the size of the Greenlandic domestic market only allows home production within a few areas. Thus, only 14 % of the total freight is between settled locations in Greenland, while the remaining 86 % of cargo is comes directly from a Danish or international supplier to the individual settlement. Similarly, internal air freight is very modest. In addition, the climate-related challenges that further complicates transport (see p. 315).

The insular community with small communities also means that there are tasks that are vital to society in a number of settlements where the customer base, however, is too small and makes it is impossible for the local population to pay prices that correspond to the actual costs. Consequently, a number of services and utility tasks are carried out by the public sector or by government-owned companies, whereas the same services in countries with which Greenland is typically compared are

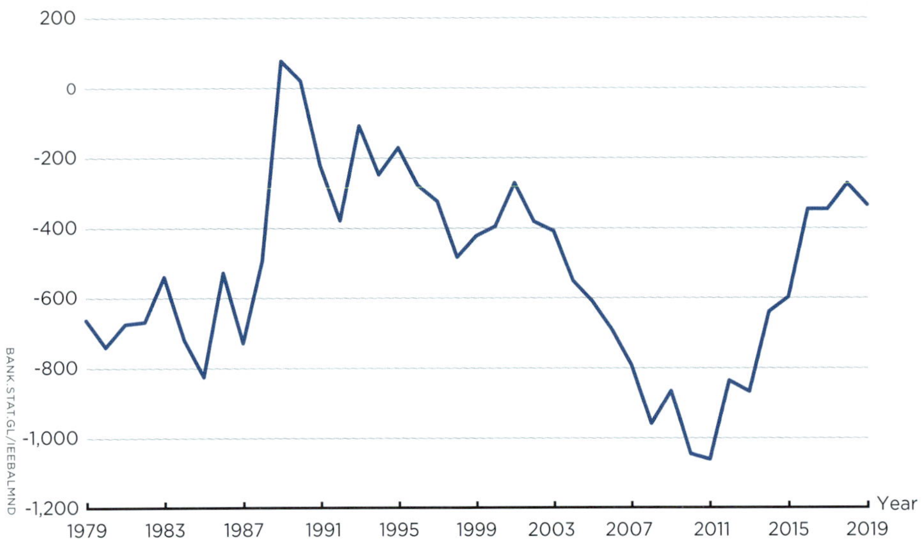

Figure 15. Trade balance from 1979 to 2019 at fixed 1979 rates. From 1979 to 2002, GDP deflator, according to Statistics Denmark, was used for conversion to fixed rates. From 2003, the GDP deflator from Statistics Denmark is extrapolated with the development of the GDP deflator from Statistics Greenland.

often provided by private companies. This contributes to a large public sector.

Since 1979, Greenland has had a trade deficit, except for 1989 (see Figure 15).

Business structure and employment

In 2019, an average of 26,994 people who had an officially registered address in Greenland were in employment per month. There were 3,987 companies, roughly half of which are made up of sole proprietorships within fishing and hunting. The value of the total value creation in 2019 was DKK 32,405 million.

The four pillars of business

Historically – and currently – Greenland has been highly economically dependent on exports of fish and shellfish, accounting for 95 % of the total exports of goods. This makes Greenland extremely vulnerable to fluctuations in stock sizes as well as fluctuations in world market prices. Since the 1980s, the objective has therefore been to increase the business and revenue base by supporting business development in other areas. Greenland's future economy is therefore based on what is known as 'the four pillars', which consist of fishing, sheep farming and hunting, mineral resources, tourism and land-based industries. Currently, mineral resources and tourism only make a modest contribution to the economy and employment.

The insular community structure means that the public sector is involved in all the pillars via government-owned limited compa-

Municipalities	Number of employees	Number of enterprises
Kommune Kujalleq	2,776	334
Kommuneqarfik Sermersooq	11,604	1,341
Qeqqata Kommunia	4,634	524
Kommune Qeqertalik	2,813	434
Avannaata Kommunia	5,109	1,242
Other / undisclosed	58	112
Total	26,994	3,987
Industry sector		
Fishing and related industry and trade	4,402	1,987
Building and construction business	2,025	260
Wholesale and retail trade	3,078	252
Transport and cargo handling	2,079	202
Public administration and service	11,238	0
Other service industries	3,900	1,214
Undisclosed	272	72
Total	26,994	3,987

BANK.STAT.GL/ARDBFB3 & ESD1A & ESD3A

Table 3. Employment and number of enterprises by municipality and industry, 2019.

Lumpfish are sold on local stalls around the country. Males are sold primarily as whole fish, while it is only the roe from females has commercial value.

EMIL HELMS/RITZAU SCANPIX, 2021

nies and government-owned companies funded by a net appropriation. Furthermore a number of business grants, including subsidies to entrepreneurs, are provided. Public administration and services include employees in the municipalities and the Self-Government, including in the education, health and social services sector and the technical area.

Fishing, sheep farming and hunting

Fishing, fishing-related industry and trade are treated as one because the industry's dominant players handle the entire chain – from fishing over processing to marketing and selling. For example, the government-owned Royal Greenland A/S, which is Greenland's largest company, has both an sea-going and a coastal fishing fleet. Royal Greenland's land-based production includes 37 facilities along the west coast of Greenland. The production units are based on local species with shrimp, Greenland halibut, cod, snow crabs and lumpsucker (roe) being the main species. In addition to its facilities in Greenland, Royal Greenland also owns factories in Canada and Germany, as well as sales offices in Canada, China, Japan and several European countries.

The industry's largest privately owned player is Polar Seafood Greenland A/S, established in 1984 as a sales and marketing firm for a number of privately owned trawlers. Polar Seafood has a Greenland halibut factory in North Greenland where local fishermen sell their catch as well as a smokehouse and a shrimp packing plant in Esbjerg, facilities in Russia and a number of European countries. The subsidiary, Polar Raajat, has factories in Nuuk and Aasiaat, which together have the world's largest production of peeled shrimp. Apart from inshore Greenland halibut, the raw products are primarily fished by the company's own trawlers.

There are other private trawler shipping companies and a large group of cutter owners as well as a significant number of dinghy fishermen who also figure as self-employed and who contribute more than a third of the value of the inshore fishing industry. In the case of Greenland halibut, dinghy fishermen account for 60 % of the total sales value (2018).

Thus, the fishing industry and, to a lesser extent hunting and sheep farming, have a total of 1,895 self-employed persons. Consequently, almost half of Greenland's companies are engaged in fishing and fish processing and accounted for 18 % of total value

Figure 16. Distribution of jobs by industry sector for people with an officially registered address in Greenland, 2020.

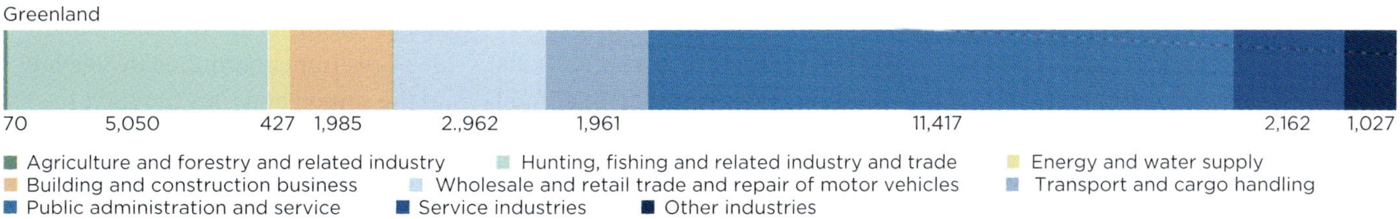

ARBEJDSPLADSERNES FORDELING PÅ BRANCHER

creation in 2019. In terms of employment and economy, this is the second largest sector after public administration and services, which do not include government-owned companies.

Greenland's fishing territory stretches 200 nautical miles from land. The fishing industry is divided into off-shore fishing (sea-going) and inshore fishing (coastal fishing up to 3 nautical miles from the baseline). Just under 60 % of the total catch value is caught offshore (2018), while inshore fishing with smaller vessels contributes a little over 40 % of the total catch value and has the greatest importance in terms of employment.

The majority of exports consist of unprocessed or more lightly processed raw products. So, for example, unpeeled shrimps constitute about 60 % of the value of shrimp exports and correspondingly about 80 % of the value of Greenland halibut exports consist of whole frozen fish. When such a large proportion of exports is not processed before export, it is partly due to the fact that some of the demand is for unprocessed raw products and partly due to the lower wage and production costs in the countries where processing is carried out. This means that a significant part of the value added in the processing is realised outside of Greenland. Both the shrimp exports and the part of the Greenland halibut exports that originates from offshore fishing are MSC certified.

In 2019, approximately 4,400 people were employed in fishing, fishing-related industries and trade (including self-employed fishermen and hunters), meaning that this industry constituted 16 % of employment.

Most of the trawlers for offshore fishing are registered in the large towns such as Nuuk, Maniitsoq, Sisimiut, Aasiaat and Ilulissat. They sell some of their catch in these towns, while a not insignificant part of the catch is exported directly without first being traded in Greenland. For many years, it has been a challenge to recruit sufficient crew and especially ship's officers for offshore fishing, partly because of the long voyages away from the family. Therefore, a not insignificant proportion of ship's officers come from the Faroe Islands and, to a lesser extent, Iceland. The Greenlandic Manning Executive Order stipulates the how many of a trawler's crew must be Greenlandic: In 2021, it is 100 % of the deck crew and 60 % of the ship's officers.

The employment and socio-economic significance of onshore fishing varies and, in some districts, it plays a modest role due to limited fishing resources and/or insufficient sales options. The shrimp trade is concentrated in the towns of Nuuk, Sisimiut, Aasiaat and Ilulissat, while the Greenland halibut is very important to the employment in the towns and settlements of Ilulissat, Uummannaq and Upernavik districts.

In addition to fishing, this pillar also includes sheep farmers and hunters. In 2019, there were 39 sheep farmers in South Greenland with a total ewe stock of 17,374 animals. Four of these farmers keep cattle and by 2019 had a total cattle herd of approximately 300. There were two caribou breeders with a total herd of about 1,600 caribou. The bulk of the sheep, lambs and cattle are slaughtered in the Neqi A/S slaughterhouse in Narsaq, which also cuts and processes products for the Greenland domestic market.

In 2019, there were just under 1,900 active hunters with commercial hunting licences. The condition for holding a commercial hunting licence is that the bulk of the income comes from hunting and fishing. Depending on the hunting season and the season of the year, many hunters also engage in fishing, and for some, fishing is actually their main occupation. The hunters' incomes vary greatly. About half earned less than DKK 150,000 per year in 2019, while one third earned more than DKK 200,000 and are thus at the same level of or above the collective wage for unskilled labour.

Extraction of mineral resources

Extraction of mineral resources in Greenland was a key focus area for the Danish State. During the colonial period and up to 1979, there were several mining projects, the most important ones being the cryolite quarry in Ivittuut, the coal quarry in Qullissat, the marble quarry and later the lead and zinc mine in Maarmorilik, and the lead and zinc mine in Mestersvig on the east coast. The cryolite quarry in Ivittuut is the one of the mining projects that until 1979 had provided the greatest socio-economic contribution over an extended period of time. Add to that a number of exploration licences for minerals and oil.

In the time before the introduction of home rule, revenue from mineral resources accrued only to the Danish State, but with the introduction of home rule in 1979, Greenland became entitled to part of the revenue. In their report, the Greenlandic-Danish Self-Government Commission of 2004 addressed three focus areas – including mineral resources – as the right to the underground was a key Greenlandic wish in the negotiations on self-government. With the introduction of the self-government in 2009, Greenland's Self-Government acquired ownership of the subsoil and thus the right to control and exploit mineral resources in the Greenlandic subsoil. As part of the self-government agreement, it was agreed that the part of the revenue that exceeds DKK 75 million annually is to be shared equally between the Self-Government and the Danish State. The Danish State's share is offset against the block grant. Administration of mineral resources was transferred in 2010.

In the first years following the introduction of self-government, initiating projects for the exploration of mineral resources were given high priority in order to provide the economic latitude for autonomy. However, despite the awarding of several exploration licences, only a few and small mining projects have so far been established which make a modest socio-economic contribution. Although, Greenland has even very large mineral deposits of a high value, extraction is challenged, in part, by world market prices and, in part, by climate conditions and the current infrastructure, which results in very high start-up and operating costs. In addition, the potentially largest project, Kuannersuit (Kvanefjeld) at Narsaq, contains uranium which makes extraction both environmentally and geopolitically problematic and thus a political topic (see p. 323).

Similarly, oil exploration in Greenlandic waters constitutes a political cross field. Cairn Energy's 2010/11 drilling programme off the west coast of Greenland resulted in investments in the range of DKK 5 billion. Since then, exploration for oil has been limited, and by mid-2020 there were seven active hydrocarbon prospecting licences and four active hydrocarbon exploration licences. In the summer of 2021, Naalakkersuisut decided to stop exploration for – and thus potential extraction of – oil and natural gas.

In 2019, 93 people were employed in extraction of mineral resources, of which 33 came from outside Greenland. Those employed in extraction of mineral resources account for 0.3 % of

Greenland Anorthosite Mining is conducting a test drilling in the Qeqertarsuatsiaat (Fiskenæsset) mountain. The mining company has located large quantities of the mineral anorthosite, which is used for producing fibreglass and aluminium, among other things.

HANNIBAL HANSCHKE/REUTERS/RITZAU SCANPIX, 2021

MADS PIHL/VISIT GREENLAND, 2016

Sevens Seas Voyager passes Qilertiki mountain near the settlement of Aappilattoq in Ikerasassuaq (Prince Christian Sound). Generally speaking, cruises are the tourist activity that contributes least to Greenland's economy in per-tourist terms because cruise ships usually carry all supplies, and tourists have relatively short stays ashore.

the total employment. The total number also includes the local employees in the towns' quarries.

A broader political recognition has emerged during the 2010s that if extraction of mineral resources is to become a supporting element for an economically independent Greenland, it will require that more and very large projects for extraction of mineral resources also produce and generate a socio-economic surplus over a long period of time.

Tourism

In order to broaden the business base, efforts have been made for decades to turn tourism into a genuinely leading business, and tourism has experienced gradual but limited growth. In 2019, Greenland received approximately 87,000 international air passengers, of which about half came for business purposes as consultants, researchers as well as short-term employees in health care, education, construction, tourism, etc. There were also 46,633 cruise tourists, some of whom had flown to and from Greenland and are therefore included in the total number of air passengers.

Currently, the main tourist destinations are Ilulissat with Isfjorden and Ammassalik district to which some one-day tourists arrive via Iceland. Most parts of Greenland have great potential for tourism, but in several places the infrastructure is a barrier to the development of tourism. Some of these locations are currently visited to a limited extent by cruise tourists.

Even though tourism is very modest compared to North European countries, it contributes to the employment of about 800 full-time equivalents in accommodation and hospitality (3 % of total employment). Of these, 227 come from outside Greenland (2019). However, many of the overnight stays relate to business tourists, and local people make up a significant part of the customer base of the restaurants.

The land-based industries

The land-based industries is a common term for Greenland's other industries and thus becomes a relatively non-homogeneous group. The following presents main industries in this group.

The building and construction industry consists mainly of private companies. The majority of building and construction activities are publicly financed and the private housing market is relatively modest. This implies that priorities for investments in construction are most made politically, where in recent decades there has been a tendency to prioritise the large towns of Nuuk, Sisimiut, Ilulissat and Qaqortoq, which have been identified as growth centres. For instance, during the period 2010 to 2019, investments in Nuuk have been approx. 15 times higher per capita compared to Upernavik. In the small insular communities, the uneven rate of investment makes it difficult to sustain private construction companies. Consequently, there is a decrease in the number of building and construction companies, which in 2019 accounted for 6.5 % of the country's companies.

In recent years, there has been an intense construction boom mainly in Nuuk and to a lesser extent in Sisimiut and Ilulissat, and this does not include the planned expansion of the airports in Nuuk and Ilulissat and the establishment of an airport in Qaqortoq. This has created increased employment in these towns and has pushed wage levels and construction costs upwards.

In 2019, 2,024 people were employed in building and construction (7.5 % of total employment), 1,954 of which in the towns and the remaining 70 in the settlements. The building and construction industry accounts for the third highest value creation and accounted for 12.6 % of the total value creation in 2019.

The wholesale and retail trade employs 3,078 people (11 % of total employment) and accounts for 7.7 % of the total value creation. Most towns and some settlements have privately owned shops, and there are some wholesale businesses in Greenland. Most of the retail trade is handled by KNI A/S, Pisiffik A/S and Kalaallit Nunaanni Brugseni.

Established in 1992 as a subsidiary of the publicly owned KNI (Kalaallit Niuerfiat), Pisiffik A/S has shops in six of the country's largest towns. In 2001, Pisiffik A/S was privatised and the controlling interest is owned by NorgesGruppen and the merchant fund KFI, which acquired ownership from the Danish company Dagrofa in 2015. Since 2001, Pisiffik has acquired and operated a number of specialty shops under their original names.

KNI A/S is a 100 % publicly owned company whose main purpose is to provide the Greenlandic people with groceries in the towns and settlements where it is not possible to run a business on commercial terms. KNI consists of the business units Polaroil, Pilersuisoq, KNI Engros, the sheep slaughterhouse Neqi A/S in Narsaq and the webshop Pisisa.gl. Polaroil is the sole importer and main distributor of oil and fuel for all of Greenland and has filling stations in virtually every settlement. KNI Engros is primarily a supplier to Pilersuisoq, which has retail shops in most towns and virtually all settlements and undertakes the supply obligation to the

A grocery shop owned by the self-government supplies the inhabitants of the settlement of Kapisillit with everything from baby food to hunting rifles. With about 52 inhabitants, the settlement is one of the many settlements in Greenland in which all grocery shopping takes place at one small shop.

PETER KLINT/POLITIKEN/RITZAU SCANPIX, 2014

smaller settlements on behalf of the Self-Government. The additional cost of operating the smaller shops is covered primarily through profits from KNI's larger shops. The price level in the Pilersuisoq areas is therefore higher than in the competing towns with several retail shops. In addition, KNI is in charge of postal services, banking, helistops and ports in a number of towns and settlements on behalf of other companies. Since some of these tasks would not be carried out under general market conditions, the company is paid by the Greenland Self-Government to undertake societal tasks subject to service contracts.

Kalaallit Nunaanni Brugseni (KNB) was formed in 1991 as one joint association of the Brugsforeninger (cooperative societies) in eight towns. There are a total of 17 Brugsen shops. The Brugsen shops in the eight towns are, like KNB, run by member-elected boards.

Transport and cargo handling employs 2,079 people (8.3 % of total employment) and accounts for 10.5 % of the total value creation. The subindustries are land transport, ship transport, air transport as well as auxiliary transport companies.

The land transport industry, in addition to municipality-owned bus companies in some of the large towns, consists predominantly of private companies. The proportion of people employed in land transport is around 1.5 %.

Ship transport represents 2 % of total employment. The main operator in this industry is Royal Arctic Line A/S that, like its subsidiary Arctic Umiaq Line, which provides passenger transport along part of the west coast, is a government-owned shipping company.

The air transport industry accounts for 2.1 %. The main operator in this industry is Air Greenland A/S, which since 2019 has been 100 % owned by the Self-Government, after having been owned by the Self-Government, the Danish State and SAS.

Employment in auxiliary services in transport is 2.7 % and includes freight forwarders, mechanics, boat yards, etc. In addition, the government-owned airport company Mittarfeqar-

Ship from Royal Arctic Line in the Container Port of Nuuk 2021. The container port Sikuki Nuuk Harbour A/S has facilities for container traffic, trawlers, cruise ships and the capacity for managing large volumes of transit cargo.

HANNIBAL HANSCHKE/REUTERS/RITZAU SCANPIX, 2021

Transport and cargo services account for 8.3 % of Greenland's total employment – of which 2.1 % is air transport. The government-owned Air Greenland, the main player in this sector, signed a purchase agreement for nine new H125 helicopters in 2022.

MADS PIHL/VISIT GREENLAND, 2015

fiit (established in 1988) operates the country's 13 airports for fixed-wing aircraft, eight heliports (primarily in towns) and 36 helistops and thus constitutes a significant part of the industry.

Other service businesses include a number of sub-industries, the largest of which are information and communication with 643 people employed (2.4 % of employment), administrative and auxiliary services with 437 people employed (1.9 % of employment) and energy and water supply with 451 employed persons (1.7 % of employment). Other service businesses accounted for 20.1 % of the total value creation. The main operator in the information and communications sub-industry is Tusass (formerly Tele Post Greenland), which is a publicly owned limited company. Tusass A/S is in charge of postal services as well as the Internet, mobile and telecommunications and employs around 400 people. Greenland's energy and water supply is provided by the government-owned Nukissiorfiit, which has about 400 full-time employees.

Public administration and service is the country's largest industry (measured by employment), employing 11,238 of wage earners in 2019 (42 % of total employment) and accounted for 28.7 % of the total value creation. The public sector consists of the five municipalities as well as the Self-Government and includes services in public administration, healthcare, education, social institutions, judicial system and defence. The judicial system and defence are undertaken by the Danish State (see p. 235). In public administration and services, social institutions have the largest share with 16 % of the total number of people employed, while educational institutions have a 9 %, general public administration has an 8 % and the health sector has a 6 % share of the total number of people employed.

Participation rate, employment and unemployment

The average monthly workforce totalled 27,141 people in 2019. Of this, 25,754 people were in employment, while the remaining 1,387 people were registered as unemployed. The working-age population (18-65 years) amounted to 36,552 people.

The remaining group of 9,411 people outside the workforce makes up a quarter of the total working-age population. Just over a quarter of people outside the workforce, or 6.5 % of the working-age population, are early retirement pensioners. This is because, for periods of time, some Greenlandic municipalities chose to transfer the long-term unemployed to early retirement because unemployment benefits were paid by the municipality, while the early retirement was financed by the Home Rule/Self-Government. Since 2015, the expenditure has been split between the Self-Government and the municipality.

People with disabilities and on maternity/paternity leave make up a fifth of the population outside the workforce, roughly evenly divided between those with disabilities and on maternity/paternity leave.

The remaining just over 3,000 people of working-age who are outside the workforce are neither working, under education nor registered as any of the aforementioned groups. Of these, 18- to 25-year-olds make up just under 2,000 people. This group is referred to as the 'youth target group' and is considered to be in a vulnerable position socially and economically. Several analyses show that there is generally a negative correla-

HUNTING AND SUBSISTENCE ECONOMY

KÅRE HENDRIKSEN, 2019

Greenland halibut fishing near the settlement of Qeqertat in the Qaanaaq district. More and more fishermen in the district supplement their income by fishing Greenland halibut. However, the fact that there is only one factory in Qaanaaq town is a challenge for a fishery. At the same time, the largest and most abundant Greenland halibut can be caught near the settlement of Qeqertat 60 km further inside the fjord.

Historically, hunting for marine and terrestrial mammals has been a prerequisite for survival, and in colonial times hunting and fishing constituted the predominant occupation, employing over 80 % of the workforce until the late 1930s. Today, hunting, fishing and related activities employ 4,400 people and 16 % of the working population.

The vast majority of fishing is part of the measurable economy and thus directly reflected in the gross domestic product (GDP), while a significant part of the catch of and proceeds of the hunt for marine and terrestrial mammals, birds, etc. are traded and consumed without being part of the directly measurable economy, meaning that no precise statements exist of the socio-economic importance of these catches.

HUNTING – A PRIMARY OCCUPATION

2,097 people hold a commercial hunting licence, of whom 991 live in the towns and settlements (2021). The condition for obtaining a commercial hunting licence is that at least 50 % of the registered income comes from hunting and fishing. The commercial hunting licence entitles the holder to hunt for selected species subject to quota, to trade game and fish and to sell them at the hunter's sales outlet 'Brættet' in the towns.

For most commercial hunters, fishing constitutes a large part or the majority of their commercial activity and income, but most supplement it with hunting. For example, there are 977 hunters holding a licence to dinghy fishing for Greenland halibut in the Qeqertarsuup Tunua (Disko Bay), Uummannaq and Upernavik districts, of these, 39 % deliver less than 10 tonnes and 22 % deliver less than 5 tonnes (2020). And fish processing plants only to a very limited extent buy other species than Greenland halibut. Of these, 39 % buy less than 10 tonnes and 22 % buy less than 5 tonnes (2020).

The Greenland Ministry of Fisheries and Hunting estimates that a dinghy fisherman must sell more than 30 tonnes of Greenland halibut annually to have a profitable fishery business, which just 16 % of these dinghy fishermen do. For the majority, the modest income from fishing does not reflect poor personal finances, but that fishing is a complementary or primary occupation. As the plant manager in Upernavik puts it: Most of the dinghy fishermen are subsistence fishermen; they fish when they need cash and otherwise make a living from hunting.

SALE OF CATCH AND PROCEEDS OF THE HUNT

Part of the catch and proceeds of the hunt is marketed through official channels such as the sales outlet 'Brættet' and through supermarket chains. This includes most of the meat from large whales subject to quotas such as the common minke whale and the fin whale, but also other marine and terrestrial mammals, fish, birds, berries, etc. Similarly, many sealskins are sold for processing by Great Greenland. Overall, it is the sales and purchases that, for each hunter, provide documentation of sufficient turnover to maintain a commercial hunting licence. However, a significant part is sold through personal networks and via the Internet. For example, it is common to be offered mattaq (narwhal skin) in Central and South Greenland on festive occasions, which comes from catches in Upernavik or another northern district. In addition, it is still customary to give meat to family members who cannot catch it themselves or, in the small towns and settlements, to other residents.

The informal sale to and through family members or similar is not usually perceived as an 'underground economy' but is rooted in culture. Exquisite goods from other districts have always been bartered, and this can be compared with farm sales from the Greenland's 39 sheep farms.

In addition to commercial hunters, 5,205 people hold a recreational hunting licence (2021), which entitles them to catch seals and, to a limited extent, hunt for caribou, muskoxen and birds.

In 2021, the catch and proceeds of the hunt totalled 64,000 seals, 3,300 harbour porpoises, 9,500 caribou, 1,800 muskoxen and birds, etc. For non-quota species such as seal, there is probably a quite significant dark figure. In addition, the catch also includes 158 common minke whales and three fin whales. Overall, a considerable amount of meat for a population of 56,000, and no formalised exports exist for any of these species.

Accordingly, the catch and proceeds of the hunt substitutes a significant amount of meat or other food that would otherwise have been imported, and about one-third of families in the Survey of Living Conditions in the Arctic, SLiCA (2004-06), reported that half or more of their meat and fish consumption comes from their own catch and proceeds.

Fin whales, common minke whales, beluga whales, narwhals, caribou, muskoxen, polar bears and walruses are subject to quotas. Quotas regulate the number of animals available for catching and hunting. All persons wishing to participate in hunting and catching must hold a valid hunting licence, and recreational hunters only have access to hunt a limited number of caribou and muskoxen, while other species are reserved for commercial hunters.

PETER LINDSTROM/VISIT GREENLAND, 2019

At Great Greenland A/S, a tailor prepares dyed seal skins. In 2021, Greenlandic hunters sold just over 32,000 seal skins to the company, which was founded in 1972 under the name Grønlandsgarveriet and which, since 1982, has been a limited liability company owned by the government of Greenland.

tion between long periods outside of employment or education and the ability to support oneself later on.

In 2019, the participation rate among 18-65 year olds was 70.5 % for the whole country. There are some differences between the municipalities with the participation rate being highest in Qeqqata Kommunia at 72.9 % and lowest in Kommune Kujalleq with a participation rate of 64.4 % (see Figure 17).

Employment has been rising since 2014. To be registered as unemployed, a citizen must go to a labour market office to look for work and must not have been registered as unemployed in the same month.

The unemployment rate in 2020 was roughly at the same level as in countries with which Greenland is often compared, e.g. the rest of the Nordic region. At the same time, the need for qualified labour is great, which is why a significant share of labour is imported. Thus, 12 % of the employed were born outside of Greenland. An unemployment rate of 5.1 % and a quarter of the working-age population outside the labour market as well as a continuing need for outside labour suggest a need for structural changes in the labour market and, not least, a higher level of education.

EDUCATION

The proportion of the population aged between 16 and 74 who only have a primary and lower secondary school-leaving certificate after 10th grade was at 60 % in 2019. The general low level of education is, among other things, a reflection of the business structure in Greenland, where fishing and hunting have been, and still are, of great importance, both economically, culturally and historically. It can also be seen in how the level of education is distributed at municipal and national level. Towns and settlements that are located far away from Nuuk have a lower level of education. Kommuneqarfik Sermersooq, where Nuuk is located has, with 51 %, the lowest proportion of the population who only has the primary and lower secondary school-leaving exam. Conversely, Avannaata Kommunia, with 71 %, has the highest proportion of the population whose longest completed education is primary and lower secondary school. The regional difference in education levels is also seen in relation to higher education. For people holding a bachelor's, master's and Ph.D. degree, Kommuneqarfik Sermersooq is at 18 % in 2019. At the other end of the education scale, Avannaata Kommunia is at 6 %.

In general, more women start on secondary and short vocational and higher education programmes, re-

Figure 17. Employed, unemployed and people outside of the labour market as a percentage of the total workforce, 2019.

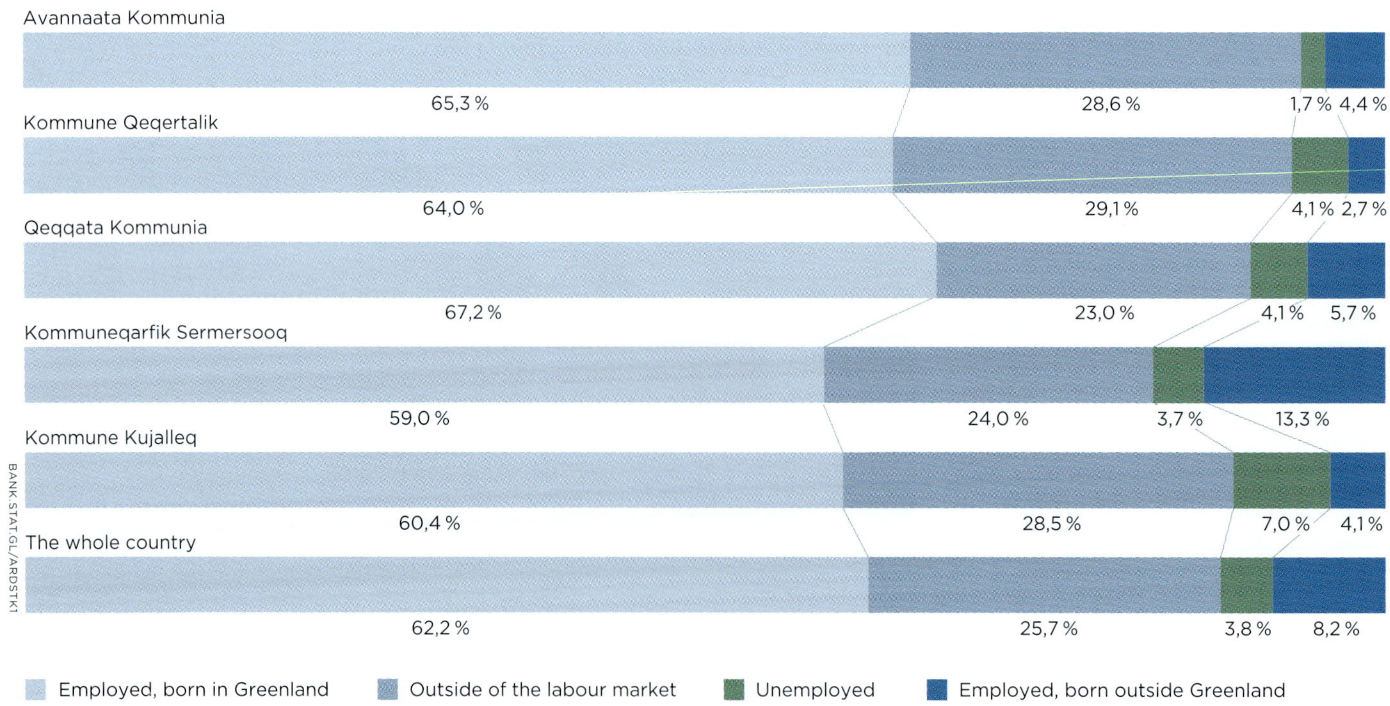

High school graduates, Nuuk, 2020. The day was marked by the graduates raising the Erfalasorput together. After primary school, many young people have to move to take up vocational training or attend one of the country's four high schools. Over half of the population lives in the four towns with high schools. However, for some people, starting secondary education means a journey of more than 1,000 km.

ANINGAAQ R. CARLSEN/VISIT GREENLAND, 2020

spectively, while more men start on the longer vocational education programmes.

Over the past ten years, the proportion of adolescents entering secondary education or vocational education programmes has been stable at around 40 % of a youth year group. This also means that around 60 % of adolescents do not continue in an education programme within the first two years of leaving primary and lower secondary school. The Self-Government's goal is to bring that proportion down to 25 % by 2024.

Primary school education

In 2021, there were a total of 75 municipal primary schools in the five municipalities, one private school (in Nuuk) and one school for the hearing impaired (in Sisimiut). Children growing up on sheep farms in South Greenland are mostly home-schooled. The 75 municipal primary schools are distributed with 23 town schools and 52 settlement schools. In general, primary schools in the settlements are organised differently than schools in towns. settlement

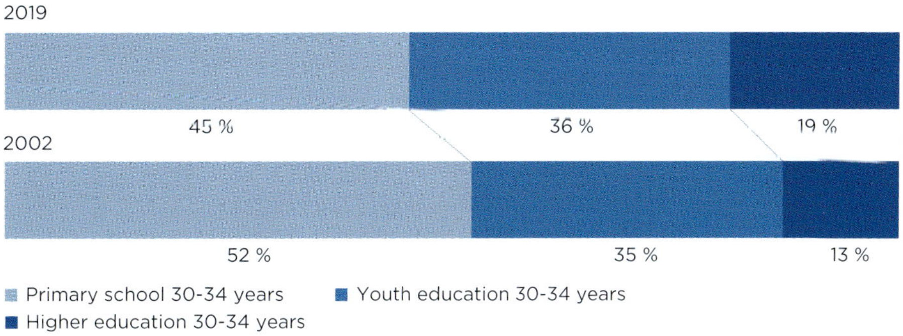

BANK.STAT.GL/UDDISCPROH

Figure 18. The highest educational attainment of 30-34-year-olds in Greenland in 2002 and 2019, respectively.

schools are characterised by having fewer pupils, and classes with different year groups are taught together. For the small settlements, teaching in the settlement school ends after the 6th to 8th grade depending on the size of the school. Students are then required to move to the nearest town to finish primary and lower secondary school. Only three settlement schools have the necessary staff qualifications to allow students to finish grade 10 in the settlement. Some settlement schools send the children to town schools already after the 6th or 7th grade. Settlement schools often only have few teachers, of which some are not trained as teachers. Both trained teachers and untrained part-time teachers teach all subjects between the 1st grade and the 7th or 8th grade. In the 23 schools in the towns, classes are divided from 1st to 10th grade.

Since the introduction of home rule in 1979, there has been an increase in the number of trained teachers from the Teachers' Training College, Ilinniarfissuaq, in the then Godthåb (now Nuuk), and out of the current almost 1,100 teacher positions in Greenland, about 90 are filled with teachers trained outside Greenland. 182 teachers work in primary and lower secondary schools without teaching qualifications and are employed as part-time teachers.

Between 2010 and 2019, there has been a 15 % decrease in the number of primary and lower secondary school pupils, in particular due to the small youth year groups, and in 2019, the overall number of pupils in primary and secondary schools was 7,499.

Over a period of more than ten years, both the National Education Authority and Statistics Greenland have documented a declining level in 10th-grade pupils' school-leaving exams in all subjects. However, Greenlandic is an exception, where the level is stable. There is no clear-cut correlation between settlement schools and urban schools and the pupil's benefits from the teaching.

In 2017, the five municipalities entered into a collaboration on the development of the teachers' didactic skills through the project Kivitsisa, meaning »Let's elevate«. The goal of the project is bilateral: All teachers and pupils must have iPads, and teachers at all schools must take a course to upgrade their didactic qualifications. The aim of Kivitsisa is for more pupils to complete primary and lower secondary school with a better result. The Kivitsisa project, partially fund-financed and partially funded by municipalities, is expected to be fully implemented in 2024.

Youth education programmes

In 1986, the Landsting (Inatsisartut) adopted the introduction of a three-year high school education in Greenland. It was to be based on the Danish higher preparatory examination, and at the same time it should be possible to incorporate features from Greenlandic cultural life and identity. In 1996, the higher preparatory examination programmes were converted into GU (high school). The Landsting (Inatsisartut) decided in autumn 1997 to work to establishment a technical high school (HTX). In 1999, a bill was passed to this effect, and the HTX programme enrolled the first students in 2000. As the last element within the field of education, the high school programmes were transferred to the Self-Government on 1 January 2009. Greenland now has four high schools, in 2021, had the following number of students: Qaqortoq in Kommune

KÅRE HENDRIKSEN, 2015

Skateboard ramp in the town of Qeqertarsuaq, the only town on the island of the same name. Qeqertarsuaq, the size of Zealand, in Qeqertarsuup Tunua (Disko Bay), is located about 100 km from the mainland. In clear weather, the island can be seen from all the towns in the bay. The town of Qeqertarsuaq is beautifully situated right on the water's edge, with the basalt mountains of Lyngmarksfjeldet and Skarvefjeldet towering behind it.

The school Atuarfik Hans Lynge is located in Nuuk, in the district of Qinngorput. Waterfalls and lakes characterise the landscape around the school, and the mountain Ukkusissaq, Store Malene, is the closest neighbour. The modern school is constructed with play and learning in mind. The school opened in 2011 and has grades 1-10 and two or three classes at each grade. In the summer of 2022, there were 489 pupils and 45 teachers.

LARS O. DEMAND-POORT, 2020

Kujalleq 139 students, Nuuk in Kommuneqarfik Sermersooq 445 students, Sisimiut in Qeqqata Kommunia 171 students and Aasiaat in Kommune Qeqertalik 262 students. In addition, it is possible to complete high school as distance learning on the eGUX programme.

The completion rate of high school educations in 2019 was 48 % with an average age of 23. The large dropout rate originates in many factors, of which distance to the home is a significant one. In addition, the students' social background, language skills and general academic skills are important.

43 % of the students who completed a high school education in 2017 commenced higher education within two years. This is a slight increase of 2 percentage points compared to the year group before. However, the proportion has been declining for a number of years, and a 32 percentage point decline can be seen compared to students from 2008 when 75 % commenced higher education within two years of graduating from high school.

18 % of the students who completed high school in 2017 continued on vocational education programme. Overall, the proportion who have started either higher education or vocational education within two years has dropped from 80 % (student year group 2008) to 61 % (year group 2017).

The decrease in the proportion continuing in higher education must be seen in the context of the number of students. The number of students completing the three-year high school programme in the period 2010-14 increased by 56 %, while the number who commenced higher education programmes was on par with previous years.

Student numbers have been increasing with 347 starting on education programmes in 2005 and 500 in 2019. In 2005, 150 completed high school; by 2019 the number had increased to 314.

Vocational education programmes

Vocational education programmes in Greenland is a practical education aimed at a particular trade. For example, this could be engineering, office work or offshore work. In Greenland, a total of 67 different vocational education programmes are offered.

The average age at completion was 30 years in 2019. What characterise these education programmes is that they are alternating programmes, i.e. alternating between theoretical school stays and practical training. A large part of the education programmes can be completed in Greenland, but there are also programmes that require school stays in Denmark.

The vocational education programmes range widely in relation to the official period of study, and the statistics are therefore divided into short vocational education programmes lasting from about six months to two years, and general vocational education programmes that last at least two years. 63 % of the students who started on a short vocational education programme in 2015 completed the programme within two years. After another two years, 65 % had completed the programme.

In the general vocational education programmes, which last at last two years, the completion rate is lower. Within four years, 36 % had completed the programme they began in 2012. After six years, 42 % had completed the programme. The number of students has been increasing with 566 commencing programmes in 2005 and 795 in 2019. In 2005, 269 finished a vocational education programme; by 2019 it had risen to 449.

The engineering programmes are gathered at Kalaallit Nunaanni Teknikimik Ilinniarfik (KTI), headquartered

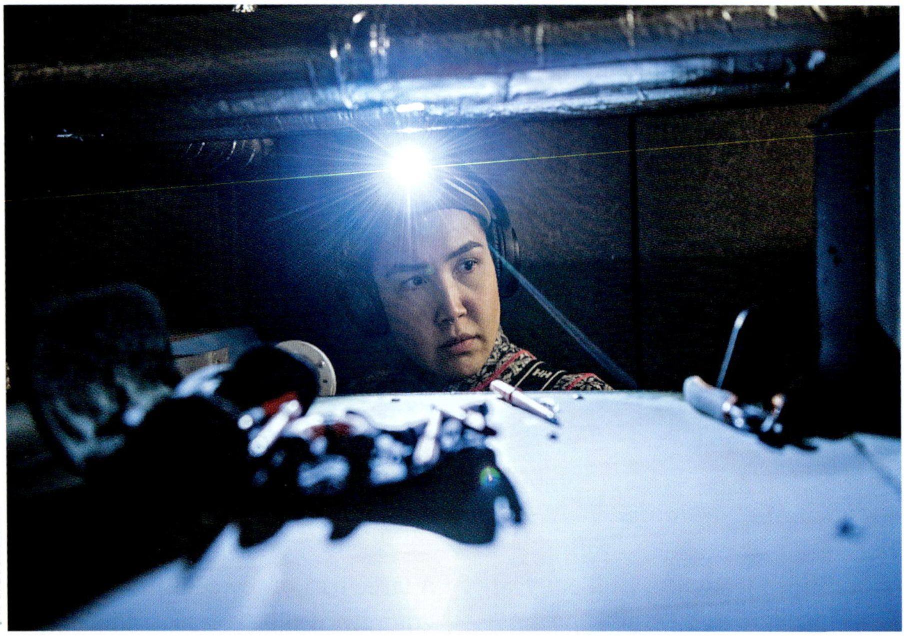

NUKISSIORFIIT, U.A.

An apprentice electrician employed by Nukissiorfiit inspects an engineering basement at the heating plant in Qassimiut. Nukissiorfiit strives to train its own workforce, and young people of both genders are encouraged to pursue technical education.

in Sisimiut, where also the building and construction programmes, Sanaartornermik Ilinniarfik, and the school of minerals & petroleum, Aatsitassarsiornermik Ilinniarfik, are housed, while Tech College Greenland, Saviminilermik Ilinniarfik, is located in Nuuk.

The business school, Niuernermik Ilinniarfik, has departments in Nuuk and Qaqortoq, and similarly, Imarsiornermik Ilinniarfik (Greenland Maritime Centre) has departments in Nuuk and Paamiut, while the food education programmes are gathered at INUILI in Narsaq.

Higher education programmes

The first education programme in Greenland besides primary and lower secondary school was the teacher/catechesis training at Ilinniarfissuaq, established in 1847 in Nuuk and in 1848 in Ilulissat. For more than 100 years, until the establishment of the Ilisimatusarfik/Inuit Institute in 1984 (renamed Ilisimatusarfik/University of Greenland in 1987), teacher training was the highest education available in Greenland.

In 1974, teacher training was joined by the training programme to become a social education worker. In 1995, the training programme, under the name Perorsaanermik Ilinniarfik/College of Social Education, PI/SPS, moved from Nuuk to Ilulissat. In addition to the education programme for social worker at bachelor level, PI/SPS provides four other vocational education programmes aimed at social work. 58 % of those enrolled complete social worker programme.

Ilisimatusarfik/University of Greenland provided the first master's degree programmes in 1987. The new university park, Ilimmarfik, which has the Greenland Institute of Natural Resources as its nearest neighbour, was inaugurated in 2008. Today there are four institutes at Ilisimatusarfik: The Institute of Society, Economics & Journalism, the Institute of Culture, Language and History, the Institute of Health & Nature and the Institute of Learning. Each institute has a number of departments, and together the four institutes cover over ten departments. In addition, a PhD school has been established.

A planned bachelor programme in Arctic biology in 2021 has not yet obtained funding.

The number of enrolled students at Ilisimatusarfik has increased over the past ten years; from around 500 students in 2009 to about 700 in 2019.

In addition to Ilisimatusarfik, it is possible to take higher education in Sisimiut where the Technical University of Denmark (DTU), in cooperation with Greenland's Self-Government, offers a four-year BSc programme at the Centre of Arctic Technology (Artek), which accepts students from both Greenland and Denmark. The first 18 months takes place in Sisimiut, and a six-month internship and the six-month final project usually takes place in Greenland or alternatively somewhere else in the Arctic. The remaining 18 months takes place in Denmark. Since the start-up of Artek in 2001, 135 bachelors of Arctic engineering have graduated from the programme, including 56 who have completed high school in Greenland. 68 graduates worked in Greenland in 2020, including 50 who have completed high school in Greenland.

Pinngortitaleriffik, the Greenland Institute of Natural Resources, primarily a research institution, has been providing master's programmes for biology students through the Arctic Science Study Programme, ASSP, since 2015. Since its start-up in 2015 to autumn 2021,

158 students have completed the programme. Graduate students at ASSP primarily come from universities outside of Greenland, as a BSc degree in natural sciences had not yet been established in 2021.

HEALTH AND CARE

In Greenland, a newborn boy can expect to live for 68.3 years, while a newborn girl can expect to live for 73.0 years. However, the average life expectancy for men in settlements is two years less than for men living in towns, while there is no difference between the average life expectancy of women in towns and settlements. Public health in Greenland is monitored through repeated population surveys carried out since 1992 by the National Institute of Public Health on behalf of the Self-Government. Most recently, the survey was conducted in 2018, and here 60 % of people aged 15 and above rated their own health as 'good' or 'really good' and 89 % stated that they have a good quality of life.

The 2007 United Nations Declaration on the Rights of Indigenous Peoples (see p. 141) has led to important debates about basic health rights among indigenous peoples, including the people of Greenland. In continuation of this, there is an increased awareness of the importance of health research in the Arctic that contributes knowledge that helps promote well-being and health, rather than solely pointing out health problems. Healthcare research and practice in Greenland have only, to a limited extent, included knowledge of fundamental values and health perceptions in Greenlandic culture. However, such social and cultural determinants of

Campus Ilimmarfik, Ilisimatusarfik, University of Greenland, was completed and ready for the first student cohort in 2008. The university park in Nuuk has a beautiful view of the Sermitsiaq mountain. In addition to Ilisimatusarfik, it also houses the National Library Groenlandica and the Greenland National Archives. Next to Ilmarfik is the Greenland Institute of Natural Resources, which was finalised in 1998. Ilimmarfik and the Greenland Institute of Natural Resources are decorated by Aka Høegh and Anne Birthe Hove.

ULR K BANG/ILISIMATUSARFIK, 2016

health are now increasingly being recognised and included.

The population of older people is increasing, and although the majority of older people want to stay in their own homes as long as possible, some, due to the lack of nursing homes and homes for the elderly in the settlements, have to leave their native place to receive proper care.

Half marathon, 18 August 2019 in Kulusuk, Ammassalik district. Greenland has five health regions, each with a single regional hospital. The national health service has staff at around 70 locations, and a total of 55 healthcare stations are distributed around the country. The settlements' access to public health services is limited compared to the towns.

The health sector

The Greenland health sector has staff at around 70 locations. There are currently approximately 120 medical positions, of which about 60 are permanent staff, and there are about 300 nurse positions, of which about 200 are permanent staff. There are approximately 30 dentists, 15 of whom are permanent staff.

The health area has been a Greenlandic area of responsibility since 1992, when the area was taken over from Denmark. In 2011, the health sector became regionalised, meaning that 16 health districts with their own district medical officer and hospital were centralised into a total of five health regions with one regional hospital in each region. The five towns with regional hospitals are Ilulissat (Region Avannaa), Aasiaat (Region Disko), Sisimiut (Region Qeqqa), Qaqortoq (Region Kujataa), and Nuuk has Queen Ingrid's Hospital, which is both a regional hospital for Region Sermersooq and national hospital for all of Greenland. The regional hospitals have doctors have a broad education and expertise, including surgical experience. They are able to deal with certain emergency situations that require surgical intervention such as Caesarean section or an appendectomy. About 60 % of the population lives in one of the five largest towns and thus belongs to a regional hospital.

In the other 11 towns with a hospital, the hospitals now serve as either large or small health centres with limited medical care. However, the

JONATHAN NACKSTRAND/AFP/RITZAU SCANPIX, 2019

Doctor Heidi Noasen (right) at work on the medical hotline at Queen Ingrid's Hospital in Nuuk, together with her colleague, doctor Kristina Roed. Public health in Greenland is monitored through population surveys carried out since 1992 by the National Institute of Public Health on behalf of the Self-Government. The latest survey was conducted in 2018.

CHRISTIAN SØLBECK, 2020

hospital in Tasiilaq on the east coast also serves as a regional hospital since the population on the east coast is often prevented from reaching the regional hospital in Nuuk due to extreme weather conditions and long transport time.

All disciplines and health regions in the country use telemedicine, but the extent to which it is used varies. However, telephone and fax machine is still needed in those places where the Internet is too unstable and the connection too bad. Online meetings and videoconferencing are also widely used as part of the telemedicine solutions.

Births

With the health reform in 2011, it was decided that births should take place in the regional hospitals and not locally in the nearest hospital. Being a place of birth requires special level of preparedness, and based on the perinatal guidelines, there are currently major challenges associated with retaining births, even in regional hospitals outside of Nuuk. Many citizens are sad that they cannot give birth in their hometown, since births are an event that brings together the family and brings joy in the local community. The women have free choice of hospital, and can therefore choose to give birth at Queen Ingrid's Hospital in Nuuk instead of the regional hospital in the health region where she lives. It is not possible to give birth in the settlements and in towns without a regional hospital. Women who come from places other than the five towns where there are regional hospitals must, therefore, travel to and stay in the town where they are due to give birth a few weeks before the birth.

Access to health services and medical care in settlements

Access to health services and medical care is very limited in the settlements compared to the towns. The settlements have healthcare stations staffed by one or more settlement healthcare workers, depending on the number of inhabitants. The aim is for the consultation to be staffed by a trained settlement healthcare worker, but this is not always possible, meaning that there are quite a few unskilled healthcare workers. In the largest settlements, one nurse is often the standard. There are a total of 55 healthcare stations distributed around the country. Virtually all of these are equipped with a telemedical device, which allows telemedical consultation to be accessed by the larger hospitals or specialists.

Recruitment of healthcare workers

Recruitment of healthcare workers is a general challenge in many places, and this is very much the case in Greenland. The challenge is particularly related to three circumstances: increased medical specialisation, lack of trained healthcare workers and competition for staff resources. The growing specialisation of both medical and care staff makes it difficult for small communities such as the ones in Greenland where people with broad competences are needed more given that the population base is limited. Furthermore, the Greenland health sector has a very high staff turnover. There are about 1,400 salaried employees in the health sector, and approximately 950 of these are permanent employees. Despite this, around 1,400 recruitments were made in 2019.

Every year, 14 nursing students are admitted to the University of Greenland, just over half of whom com-

Average lifespan	
Men	68,3
Women	73
Health and well-being, self-assessed (Population survey in Greenland 2018)	
Good or really good (%)	59,6
Good quality of life (%)	89,3
Poor mental health (%)	23,3
Severe overweight (BMI >30) (%)	27,0
Health habits (Population survey in Greenland 2018)	
Daily smoking (%)	52,3
Never consumes alcohol (%)	19,0
Non-harmful alcohol consumption (incl. abstinence) (%)	58,0
Drink juice or soda every day (%)	41,3
Social and health challenges (Population survey in Greenland 2018)	
Food insecurity: Have experienced food shortages without the possibility to buy food (%)	8,8
Proportion of 18-29-year-olds who have been sexually assaulted before the age of 18 (%)	26,0
18-29-year-olds with suicide attempts in the past year (%)	7,4
Parents with children under the age of 18 with potentially harmful alcohol consumption (%)	38,6

HTTPS://WWW.SDU.DK/DA/SIF/RAPPORTER/2019/BEFOLKNINGSUNDERSOEGELSEN_I_GROENLAND

Table 4. Public health has been monitored through regular population surveys conducted by the National Institute of Public Health on behalf of the Self-Government since 1992. The latest survey was conducted in 2018.

plete the education programme. By 2020, a total of 182 nurses had been trained in Greenland. However, this is far from enough to meet the need. To become a doctor in Greenland, you have to travel to Denmark or another country to get a medical education. Experience from the Greenlandic Houses in Denmark shows that only about half of the medical students from Greenland complete their medical studies, and of them only about half move back to Greenland once they graduate.

Citizens' perspective on the health sector

If you ask the citizens, they do not always feel well enough informed about the possibilities and limitations of the services provided where they live. The expectations of citizens and the service that is offered locally do not always match. Citizens say that they want to be able to speak Greenlandic when in contact the health sector. This is challenged by the fact that the majority of doctors and nurses in the Greenlandic health sector have either a Danish or other background and do not speak Greenlandic. Therefore, an interpreter is often needed in the meeting between citizens and the health sector, which involves a wide range of common challenges, e.g. misunderstandings when dispensing medicine and providing information about examination results. The linguistic barriers also cover significant cultural differences between the health workers and the citizens, which may lead to misunderstandings and an awkward dialogue.

Social and health challenges

Over the past 100 years, Greenland has undergone an epidemiological transition in the health area, which means that the threat of infectious diseases has gradually been taken over by an increase in lifestyle diseases. Today, most people die from cancer and cardiovascular disease, whereas the third most frequent cause of death is suicide. The increase in lifestyle diseases is attributed to changes linked to marked changes in lifestyle towards more sedentary lifestyle and a dietary pattern that has changed from primarily comprising of marine mammals and fish to comprising much more of imported meat and carbohydrates. In addition, there is an increasing intake of sugary drinks as well as fast food.

Overweight, smoking and a harmful consumption of alcohol and cannabis constitute the three biggest challenges to public health. While overweight and smoking are well known factors (diet, smoking, alcohol and exercise) with consequences for the individual's health, the public health consequences of alcohol and cannabis are predominantly social and the reason why some children have a difficult childhood.

Since 2007, Greenland has been working to improve public health

PEQQINEQ - HEALTH AND BALANCE

The Greenlandic word for health, peqqineq, refers to the fact that something is in its proper place and that there is equilibrium. Many have a holistic understanding of health where values such as restraint, honour, strength, respect and accepting man's powerlessness to those things that are greater, e.g. Sila (meaning both weather, knowledge and consciousness) and Pinngortitaq (nature), play a central role.

Genetics, biology, living conditions, education, housing conditions and income are well-known factors that are obviously also important to health in Greenland. However, it is equally important to look at health from a broader perspective to ensure a culturally relevant view of citizen health. *Peqqineq* also refers to communities, family, environment and experiences. The concept of health is thus different from the typical Western view of health where we talk more about 'health' versus 'disease'.

If the healthy choice is to make sense to the individual, it is crucial that the understanding of health reflects the values and strengths of citizens. In the Greenlandic view of health, emphasis is placed on being well as a whole human being, meaning that balance must be ensured in **body, soul and spirit**. In recent years, health research in Greenland has become more aware of the link between physical and mental health, while spiritual health is not being addressed to the same extent. Spiritual health is described by older citizens as being part of their faith in God, but also the spirit as part of an act, a way of being or thinking that can characterise or describe persons or communities.

Pinngortitaq (nature) is described by many as a kind of epicentre of their lives. Nature is where you get your food through hunting, and where you prepare your catch. Nature is also where many find strength, peace and renewed energy. Summer is the season for collecting stocks for the long winter. Love, acceptance and respect of nature is considered one of the healthiest traits a human being can have.

Culture and language are essential to health, and speaking in one's own mother tongue is of great importance to man's well-being and ability to formulate both symptoms of illness and well-being. Language is decisive to the way people in Greenland perceive health and disease.

Family, relationships and community are central values of health; close family ties can be an emotional safeguard against depression, and the family as well as close relationships constitute a strength in society and a resource for the individual citizen. It is therefore also essential to consider the whole family when developing preventive and health-promoting measures.

Local values and strengths can differ from town to town, from town to settlement and from settlement to settlement. Local values can be linked to recurring events in a town or a settlement that foster a sense of community and a common spirit among the citizens. Values and strengths can also be about the surrounding nature of a place and the local hunting and activity opportunities.

Kalaalimerngit (Greenlandic diet) is, if you ask the citizens, more than energy and vitamins. Nature and the Greenlandic diet are closely linked. Eating *kalaalimerngit* is considered by many to be a healthy choice, despite warnings about the content of pollutants among certain marine mammals. The Greenlandic diet is associated with many emotions and contributes to an experience of common identity because it often gives pleasure to enjoy and share the food with others.

The spiritual side of her identity as a Greenlander is strong for art photographer Jukke Rosing. With the photo, she seeks to show that Greenlanders are at one with nature yet also float between heaven and earth without losing their connection to the earth,

JUKKE ROSING, 2016

through the national public health programmes Inuuneritta I, II and III. In the first two public health programmes, particular focus was placed on risk factors for lifestyle diseases. With the current public health programme, which runs from 2020 to 2030, focus is on the good life for children and on developing health promoting initiatives in close partnerships with municipalities and in dialogue with citizens.

Since 2007, Greenland has been working to improve public health through the national public health programmes Inuuneritta I, II and III. The current public health programme, Inuuneritta III, which runs from 2020-2030, focuses on the good life for children and on developing health-promoting initiatives in close partnerships with municipalities.

Since the mid-1990s, population surveys in Greenland have shown that especially generations born in the 1970s and 1980s have been burdened throughout their upbringing by alcoholism and violence in their home as well as by a high incidence of sexual abuse. Of these generations, as many as upwards of 70 % have grown up with alcohol problems in their home. However, over the past 20 years, there has been some improvement, and for the generations born in 1995 and later, 37 % have grown up with alcohol problems in their childhood.

For the generations born in the 1970s and 1980s, almost 40 % have suffered sexual abuse before the age of 18. In this area too, there has been an improvement, although it remains a serious social and health challenge. For the generations born in 1995 and later, the incidence of sexual abuse before the age of 18 has fallen to 20 %. In 2018, Naalakkersuisut launched a strategy against sexual abuse for the period 2018-2022, Killiliisa – Let's set boundaries. The overall objective of the strategy is to reduce the number of cases of sexual abuse over the next five years so that it does not occur at all among 15 to 29-year olds, and from 2022 onwards, the intention is to achieve year groups who do not experience abuse at all.

Despite the positive development, numbers remain high, and the consequences of growing up with physical abuse and alcohol problems in the home have an impact well into adolescent and adult life. The high incidence of alcohol problems, violence and abuse during childhood is a major cause of distress and suicide among young people in Greenland.

Suicide

The suicide rate in Greenland is one of the highest in the world. Over the past 40 years, it has remained unchanged with an average of 100 suicides per 100,000 people. Although the overall incidence remains unchanged, a development can be seen where more young people under the age of 18 die by suicide. Overall, it is especially young men aged 20-24 who die by their own hand, with an average suicide rate of more than 400 suicides per 100,000 person years. Compared with the 1970s, a demographic shift has also taken place. The highest suicide rates are now found in East Greenland and Northwest Greenland, while a decrease has occurred in the towns on the west coast and especially in Nuuk, where back in the 1970s the first significant increases in the incidence of suicides were seen. The high incidence means that the majority of citizens have lost close relations and family members to suicide.

Social services and support

A total of 4,642 people over the age of 18 received public benefits in 2019. This corresponds to an annual expense of more than DKK 74 million. The 4,642 people who received public benefits make up just over 10 % of the total

KHAN TARIQ MIKKEL/RITZAU SCANPIX, 2007

population aged over 18. In 2019, Statistics Greenland described 4 % of citizens as economically disadvantaged, measured as having a disposable income below 50 % of the median income (excluding students and residents at social institutions). The figure varies from town to settlement with an average of 3 % economically disadvantaged citizens in the towns and 8 % in the settlements. There is also a large variation between the towns: In Tasiilaq, the proportion of economically disadvantaged citizens is at 11 %, while Qeqertarsuaq has the lowest proportion of economically disadvantaged citizens at 1 %.

Children in day-care facilities

In 2020, 3,672 children were enrolled in public preschool day-care facility, and they were distributed among 165

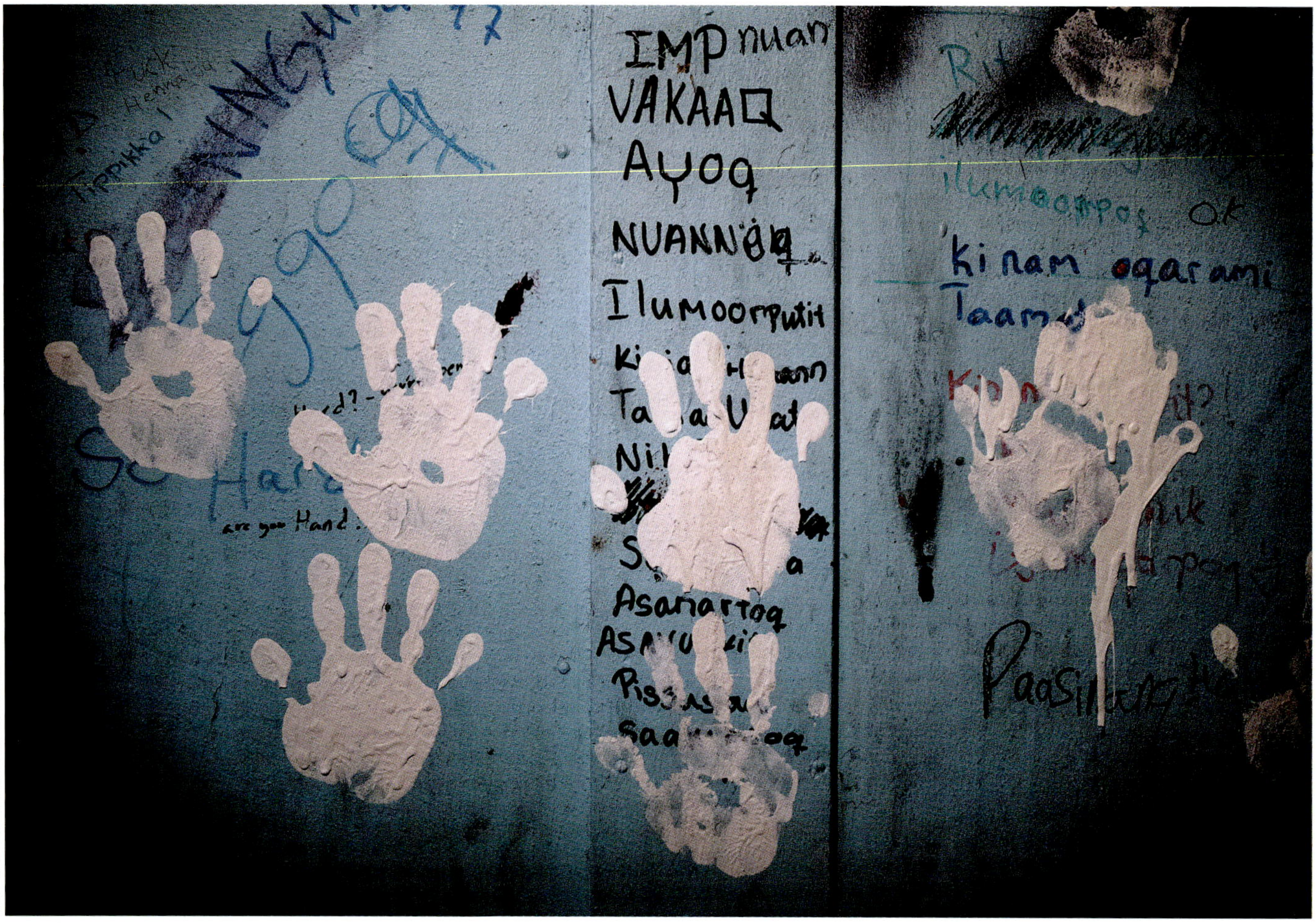

MARTIN LEHMANN/RITZAU SCANPIX, 2019

Tags on housing block in Nuuk. In Nuuk, there was a 33 % over-representation of reported offences in the period 2014-2018, and the crime rate was consequently 33 % higher than would be expected, given the size of the population. Kommuneqarfik Sermersooq and the Greenland Police are working together on early intervention to prevent patterns of criminal behaviour among children and young people from escalating.

municipal day-care institutions nationwide. This corresponds to 75.2 % of all children in the age group 0-5 years in Greenland being enrolled in a preschool day-care facility. The proportion of enrolled children varies from town to town. Thus, 90 % of 0-5 year olds in Paamiut and Maniitsoq are enrolled in a preschool day-care facility, while less than 50 % in Upernavik and almost 60 % in Qaanaaq and Tasiilaq are enrolled. A daycare facility is available in all towns, while not all settlements have municipal preschool day-care facilities.

Children placed outside the home

Based on statistics from the five municipalities, in 2019, just over 300 children and young people under the age of 18 had been placed in foster families. In addition, close to 300 children and young people under the age of 18 were placed in one of the country's 22, 24-hour institutions. Overall, around 600 children and young people, corresponding to 4.5 %, were placed with people other than their parents.

The elderly population

The population of elderly people is increasing and will present a significant challenge in the health and social services area in the coming years. The trend shows that, compared with the past, the elderly in the municipal senior citizens' centres are to a greater degree being medicated, undergo rehabilitation and have other extensive needs for help which help drive up the healthcare costs. Among the elderly over the age of 55, in 2018, 77 % could be categorized as chronically ill (includes diabetes, obesity, high blood pressure, etc.).

The majority of elderly people want to stay in their own home for as long as possible, but the lack of senior citizens' centres and homes for the elderly in the settlements means that some elderly people have to leave family, their native area and local environment to move to a home possibly in another town or settlement that is better suited to their needs.

Crime

The amount of dangerous crime committed against people is generally high. The number of reports of violence and sexual assault increased significantly in 2020 and, in both cases, is well above the level of the rest of the Unity of the Realm measured per capita. The Greenland Police has noted that the large increase in 2020 may be related to the campaign 'Report sexual assaults – we will help you' launched in early 2020.

In 2020, there were 89.6 reports of violations of the Criminal Code per 1,000 inhabitants. This figure is twice as high as corresponding reports in Denmark and ten times as high as corresponding reports in the Faroe Islands. It would be better to compare with other Arctic Inuit communities, but it has not been possible to find comparable figures. The figure has risen in recent years and was at 62 reports per 1,000 inhabitants in 2018. In particular the number of homicide and violent crimes as well as sex offences have increased steadily over the past five years. 748 sex offences were reported in 2020, which corresponds to 13.1 per 1,000 inhabitants. That is 13 times higher than the corresponding figure both in Denmark and the Faroe Islands.

The scale of offences against property, which includes shoplifting and burglary, has conversely declined and almost halved since 2016. In 2020, there were 8.1 burglary reports per 1,000 inhabitants. In comparison, the figure in the same period was 6.6 in Denmark and 0.5 in the Faroe Islands. Youth crime poses a particular problem since the level of crime among 15- to 17-year-olds is steadily increasing.

PLANS

After the social-engineering drive of the Danish state in the 50s, 60s and 70s, and the accelerated takeover and establishment of Greenlandic institutions during the first period of home rule, Greenland politics and societal development has switched gears since the millennium. On the one hand, big ambitions for an autonomous welfare state, a self-sustaining economy and a people capable of shouldering both continue to be the driving force. On the other hand, the next steps on the road are monumental and heavy, and the politics is characterised by disagreements on priorities and the balancing of different interests.

The introduction of Self-Government in 2009 sparked pride and a can-do attitude, but after just a little over a decade, the mood is more subdued. With the solid revenues from fishing, the national economy and the national treasury are healthy, and Greenlanders are taking up more positions in management and positions requiring specialised qualifications. However, transfer of responsibilities has come to a standstill (see p. 246) and neither mineral resources nor tourism have become key economic industries (see p. 282).

Technocratic attempts to formulate the basis for overall societal priorities, such as the 2000 Structural Policy Action Plan and various attempts

The mainstream of Greenlandic politics points towards independence, but a counter-current embraces the Unity of the Realm. In 2017, Foreningen Rigsfællesskabet (the Unity of the Realm Association) held a small torchlight procession at the palace square of Christiansborg.

OLAFUR STEINAR RYE GESTSSON/RITZAU SCANPIX, 2017

CHRISTIAN KLINDT SØLBECK/RITZAU SCANPIX, 2022

A new era in relations between Greenland and Denmark may be in sight; Greenland is stronger in foreign policy than ever, and Danish Prime Minister Mette Frederiksen (S) seems more willing to apologise for the darker chapters of history. Here she is seen with the Premier of the naalakkersuisut, Múte B. Egede (IA), with students from GUX Nuuk. The two governments agreed to set up a commission to look into Denmark's historical relationship with Greenland. After many difficulties, the Greenland Constitutional Commission is about to produce a draft constitution for an independent Greenland.

to formulate a comprehensive regional policy have, as a result of political disagreement, been nipped in the bud or have only haphazardly been translated into concrete reforms. A recurring theme of both technocratic plans and the political debate has been the size of the public sector. The public administration is criticised for being too large for such a small society – and it clashes with the diseconomies of scale that result from the ambition that Greenland should manage all areas on an equal footing with larger nation states. The desire to let the private sector take over more areas clashes with the diseconomies of scale coming with the insular community consisting of 71 separate communities that are so small that many tasks would not constitute a market for more than one player – and often not even one (see p. 234).

Instead, political attention has been on a number of sectoral reforms designed to finetune economic efficiency, to develop human resources or lift infrastructure to a new level. Part of the reform plans (e.g. in the fisheries sector) end up as minor adjustments after a tug-of-war between well-known interests (the contribution made by large efficient companies to exports and the national treasury; maintenance of the culture and settlements of coastal fisheries; employment in land-based production plants). Others remain ambitious objectives without the necessary resources being available (e.g. English as a first language, reforms to the education sector). It is in this light that one needs to look at the very unique character that corporate social responsibility (CSR) has acquired in Greenland. In other countries, CSR is carried by private companies that try to gain legitimacy by not just selling certain products and services demanded by the market, but also by taking into account societal values demanded by their stakeholders. In Greenland, CSR is largely carried by publicly owned companies. In addition to solving the tasks given to them by their owners, the companies try to compensate for their owners' lack of priorities or possibilities of implementing policy in other areas via various actions under the heading CSR.

Great visions for development lose momentum when the driving projects fail to materialise (e.g. Alcoa's aluminium melter furnace and numerous mining projects based on foreign investment). In some reform areas, regional ambitions have been driven into a corner, either directly because regional growth centres are afraid that Nuuk will take it all, or indirectly because an investment package, in order to make everyone happy, ends up being prohibitively expensive and operationally questionable. Along the way, a number of pipe dreams have cost dearly. Despite stacks of consultancy reports, countless citizens' hearings and endless casework in departments and government-owned limited companies, it is ultimately a political decision to what extent one dares to believe that an airport or mine will bring growth or destruction.

Similarly, it will be a political decision whether the benefits in the form of self-determination and self-respect would outweigh the disadvantages in the form of uncertainty and more direct pressure from the great powers when the Constitutional Commission makes a proposal for a constitution. Initially, the Commission was tasked with writing two constitutions: one to be used in conjunction with the Danish Constitution, and another to form the framework for an independent Greenland. After a messy process that was difficult to get going, the much-delayed Commission has only opted to work with the latter. The expectation is that the Constitution will recommend that an independent Greenland can enter into *'free association'* with, for example, Denmark. The next step would then be a new round of negotiations with Denmark. Opinion polls point to it being crucial whether a new scheme will affect the welfare in Greenland.

The ambitions of self-sustainability and autonomy have been driving real structural changes, especially during the Home Rule years, when it took place under slogans about solutions based on Greenlandic conditions. Under these slogans, subsequent generations have looked for international inspiration for the next step: The demands for home rule, rights, self-government and autonomy can only be understood in the light of global decolonisation, of the international movement of indigenous peoples and of the experiences gained by the other peoples of the North Atlantic who have loosened ties with Copenhagen. Many of the major plans for economic self-sustainability can only be understood as part of international trends – and several of the concrete steps have been inspired by other peoples and countries whose challenges are more similar to those of Greenland than Denmark. In this light, the home rule and self-government years can only be seen as a successful effort to reduce dependence on Denmark and to find Greenlandic solutions to Greenlandic problems. Not just by looking inward, but by looking outward.

EMIL HELMS/RITZAU SCANPIX, 2021

On Tuesday, 6 April 2021, there were elections for Inatsisartut (Parliament) – here, the moment is captured in the Inussivik hall in Nuuk. In Denmark, it is possible to vote by post in Inatsisartut elections at the places where postal votes are usually cast.

MUNICIPALITIES AND TOWNS

ANINGAAG R. CARLSEN/VISIT GREENLAND, 2017

Winter scenery from Tasiilaq, the largest town on Greenland's east coast. The signpost speaks very clearly about the great distances between the towns in Greenland. Only settlements in the Tasiilaq district are within a distance of 100 km.

Greenland is the largest island in the world. Much of the country is made up of ice or tundra, and the Arctic climate is not always well disposed to settlement. There are only a few and relatively small towns and settlements scattered like stamps along the coasts. With a population of 56,421 (1 January 2021) and a total area of 2,486,000 km², most of the country is uninhabited, with population density being the world's lowest. The population density is 0.14 per km², and if the ice sheet is included, then the figure is as low as 0.025 inhabitants per km².

In a historical perspective, the Inuit settled where the hunting opportunities were – depending on the time of year – and where there was a natural harbour and protection from the harsh climate. In areas of solid ice during winter, access to hunting on the ice was essential. The Inuit returned to the best hunting grounds, and they often became of a more permanent nature. Many of Greenland's towns and settlements are founded on the former hunting grounds of the Inuit, some at whale stations set up by foreign whalers, and others where missionaries or government planners saw fit. Finally, it may be a combination of the above.

Photo pages 312-313. Twilight over Qaqortoq (Myggedalen) looking towards Nuup Kangerlua in Nuuk. The development is a nice mix of colourful GTO standard houses and modern single-family houses. In the background is the newer district of Qernertunnguit and just above that, the town's landmark Sermitsiaq.

REBECCA GUSTAFSSON/VISIT GREENLAND, 2016

Here, the main focus is on the towns where most of the population lives. Only 12-13 % of the population live in settlements and sheep farms and a small proportion in stations in open land. The latter typically has research, technical infrastructure or sovereignty enforcement as its purpose.

On or along the ice sheet, there are a handful of research stations, e.g. Summit at the top of the ice sheet. It is run by American researchers in collaboration with Greenlandic, Danish and international researchers. Another type of station is the Sirius Dog Sled Patrol's headquarters in Daneborg, Northeast Greenland, whose main purpose is to enforce the sovereignty of the Unity of the Realm in the National Park. Danmarkshavn, originally founded by the Denmark Expedition in 1906, in the middle of the National Park in Northeast Greenland, is a station with a gripping history, and the station is still permanently staffed by six people and used by scientists, technicians and occasionally by the Sirius Dog Sled Patrol. Since 1948, it has served as a weather station, and the long data time series are invaluable for international aviation and for research – including in climate change.

Tussas (formerly Tele Post Greenland) operates numerous of radio chain stations which form the backbone of communications infrastructure. Today, the majority are unmanned.

The individual towns and settlements – and virtually all other settlements – are to be seen as isolated insular communities with a defined infrastructure and supply lines and dictated by geography and climate, especially the long, dark and cold winter.

Over the winter, many of the northern towns and settlements have solid sea ice. When the ice is safe, transportation often takes place by dog sled, snowmobile and even cars across the ice. The purpose is hunting and fishing from the ice or transporting goods and persons between the settled locations. Longer journeys and all imports and exports of goods are either paused or by air and helicopter during the period when the solid ice does not allow navigation. The further north, the longer the period when the settled locations are effectively shut down from the outside world. However, the period of solid ice which is safe for fishermen, hunters and people in general to travel on getting shorter due to climate change. That does not mean that the period of closure will be shortened.

	Population (2021)	Area	Of this, built-up area
Location		km²	km²
Nuuk	18,800	11,55	5,20
Sisimiut	5,622	12,82	2,44
Ilulissat	4,710	9,18	1,82
Qaqortoq	3,037	7,2	1,23
Aasiaat	3,035	3,66	1,33
Maniitsoq	2,486	3,63	0,98
Tasiilaq	1,916	4,03	0,88
Uummannaq	1,428	1,51	0,79
Narsaq	1,392	6,02	1,04
Paamiut	1,247	2,39	0,79
Nanortalik	1,169	1,88	0,78
Upernavik	1,076	1,23	0,62
Qasigiannguit	1,071	2,31	0,64
Qeqertarsuaq	838	3,67	0,65
Qaanaaq	619	4,88	0,45
Kangaatsiaq	534	0,96	0,30
Ittoqqortoormiit	361	0,84	0,39
Towns	49,341	78	20
Villages	7,080	85	7
All of Greenland	56,421	2,486,000	28

Tabel 5. Greenland's towns, population and area as of 1st January 2021.

BEFOLKNINGSTAL: GRØNLANDS STATISTIK. AREALER: ASIAQ GRUNDKORT: LANDAREALER SAMT BYGNINGER, VEJE MV.

ANDREAS FAESSLER, 2013

Danmarkshavn in Northeast Greenland. The station is permanently manned and used by the Sirius Dog Sled Patrol and researchers. Since 1948, Danmarkshavn has served as a weather station operated by DMI and Tusass (formerly Tele Post Greenland).

Often the inhabitants of the towns and settlements instead have a long period of unsafe ice, which makes navigation and other transport difficult or impossible.

Sled dogs and dog sleds are associated with the solid ice in the winter and are found only north of the Arctic Circle as well as in the Ammassalik district on the east coast. This is simply for practical reasons, but is also secured by legislation to protect the particularly suitable and adapted sled dog from mixing with less reliable dog breeds.

Although Sisimiut is an open-water town, the town has an active sled dog culture with a large catchment area and sled tracks all the way to Kangerlussuaq. Nowadays, dogs are often harnessed to the sled to satisfy tourists and citizens who simply want the experience. Further north, the development is the same, but fishermen and hunters still use the sled as a means of transport to and from the hunting grounds. For long periods, the Ammassalik district is isolated behind the multiyear ice – 'storis' (great ice), but strong ocean currents often mean that the ice is not passable. Much of the dog sledding between Tasiilaq and some of the settlements in the districts therefore takes place over land.

A number of towns and settlements in southern Greenland do not have solid ice (see p. 128). The heat from the Gulf Stream is the primary factor and ensures that the towns on the west coast from Nanortalik in the south to Sisimiut in the north are open-water towns throughout the year. Thus, the towns and associated settlements can be reached by boat year-round – in principle. The storis, which originates in the Arctic Sea and over summer drifts down the east coast, around Uummannarsuaq (Cape Farewell) and northwards on the west coast, gives rise to major problems with maritime accessibility, especially in South Greenland, but during periods also all the way up to Paamiut. Despite the storis (great ice), sea navigation does take place, but in certain seasons it is more the rule than the exception that delays occur, or that the storis simply packs up so much that for days the waters completely close to navigation.

The ice and climate are terms that Inuit live with and adapt to. The same applies to infrastructural challenges related to each settlement being an insular community. There are no roads and railways between the settled locations, and people and goods are transported by air or sea – by dinghy, boat or ship – in winter in North Greenland over the ice.

Each town or settlement must be self-sufficient with electricity, water, heating and telecommunications, waste and waste water management. If the power plant – or supply line from the hydroelectric plant – suffers a failure, there is nowhere else to get electricity from. For this reason, power outages are common. The corresponding situation applies to the supply of water, heat and telecommunications. There is no backup in the grid – because it does not exist.

Greenland has 17 towns and 56 settlements, two of which in the Uummannaq district have been closed since 2017 due to a tsunami triggered by a large mountain slide. The size of the towns varies from barely 400 inhabitants in the smallest, Ittoqqortoormiit, to about 19,000 in the capital, Nuuk. On an international scale, they are all small towns. Similarly, the population of the settlements varies from 475 in Kangerlussuaq to 14 in Kangerluk (as of 1 January 2022). This means that some settlements have more inhabitants than the smallest town, and the concept of a town is merely used for administrative purposes. The population figures used here are all from 2021, unless otherwise mentioned.

Until a local government reform in 2009, there were 18 municipalities in Greenland, and the settlements that were home to a municipal office were defined as a town, while the other settlements of the municipality were settlements. The municipalities took their name after the town, except for Ammassalik Municipality, where the town changed its name to Tasiilaq along the way. The old municipality towns have maintained their town status after the 2009 reform, and the former municipalities are today referred to as districts. The exception is Ivittuut Municipality, where the town of Ivittuut and later Kangilinnguit (the Grønnedal naval station) were vacated and the settlement of Arsuk came under Paamiut district.

THE FIVE REGIONAL MUNICIPALITIES

After the municipal reform in 2009 and an adjustment in 2018, when the northern municipality was divided into two, Greenland is administratively divided into five regional municipalities. As mentioned, there were 18 municipalities until 2009, some of which had less than 1,000 inhabitants. It provided proximity to citizens, at least in the towns, but also resulted in major challenges in relation to tax bases, public and private service levels and building and maintaining knowledge and competences, particularly in the social sector.

TOWN NAMES IN GREENLAND

The Greenland Place Names Committee was established in 1989, and during the 1990s the committee introduced official Greenlandic place names. The Greenland Self-Government Act made Greenlandic is the main language. In this context, the Greenlandic place name is always used, often supplemented by the Danish name in parentheses. The descriptions reflect the practice that has long been in place (though to a lesser extent now) of giving towns, settlements, neighbourhoods, roads, and other place names two names. For each town, the two names, if they exist, will be explained, e.g. for the town of Aasiaat (Egedesminde).

EMPLOYMENT IN DISTRICTS AND TOWNS

The town descriptions include a figure showing the percentage distribution of employed and self-employed persons within nine selected occupational groups together with the percentage distribution of occupations at national level. The percentage distribution of occupations in the settlements as a whole is also shown for the 11 districts. For the following districts, figures for the occupational distribution of the settlements are not shown because the population is too small to ensure the required level of anonymity: Qaqortoq, Paamiut, Aasiaat, Qasigiannguit, Qeqertarsuaq and Ittoqqortoormiit.

The figures are based on a special data run by Statistics Greenland using fewer than the 16 industries usually used by Statistics Greenland. The reduction in the number of industries has been chosen in part to provide a better graphical overview and in part because some of the 16 industries in Greenland have so few employees that a percentage cannot be given. Therefore, some closely related industries have been grouped together and some industries have been moved to the category 'Other industries'.

In many cases, the figure will not include all nine selected categories because the number of persons employed within a given category is too small. In this case, the category concerned has been moved to 'Other industries'. This particularly applies to the settlements.

The category 'Fishing and related industry and trade' includes all commercial hunters – including the 130 or so commercial hunters who do not have a reported any sales. In order to obtain a commercial hunting licence, income from hunting and fishing must constitute at least half of the previous year's income. This implies that commercial hunters without reported sales contribute primarily to the subsistence economy of the household.

KÅRE HENDRIKSEN, 2015

En route from Ikerasaarsuk to Kangaatsiaq with freshly caught cod. Trading capacity in Ikerasaarsuk is limited, so fishermen often have to sail the 20 km to Kangaatsiaq to trade. Part of the sailing takes place offshore.

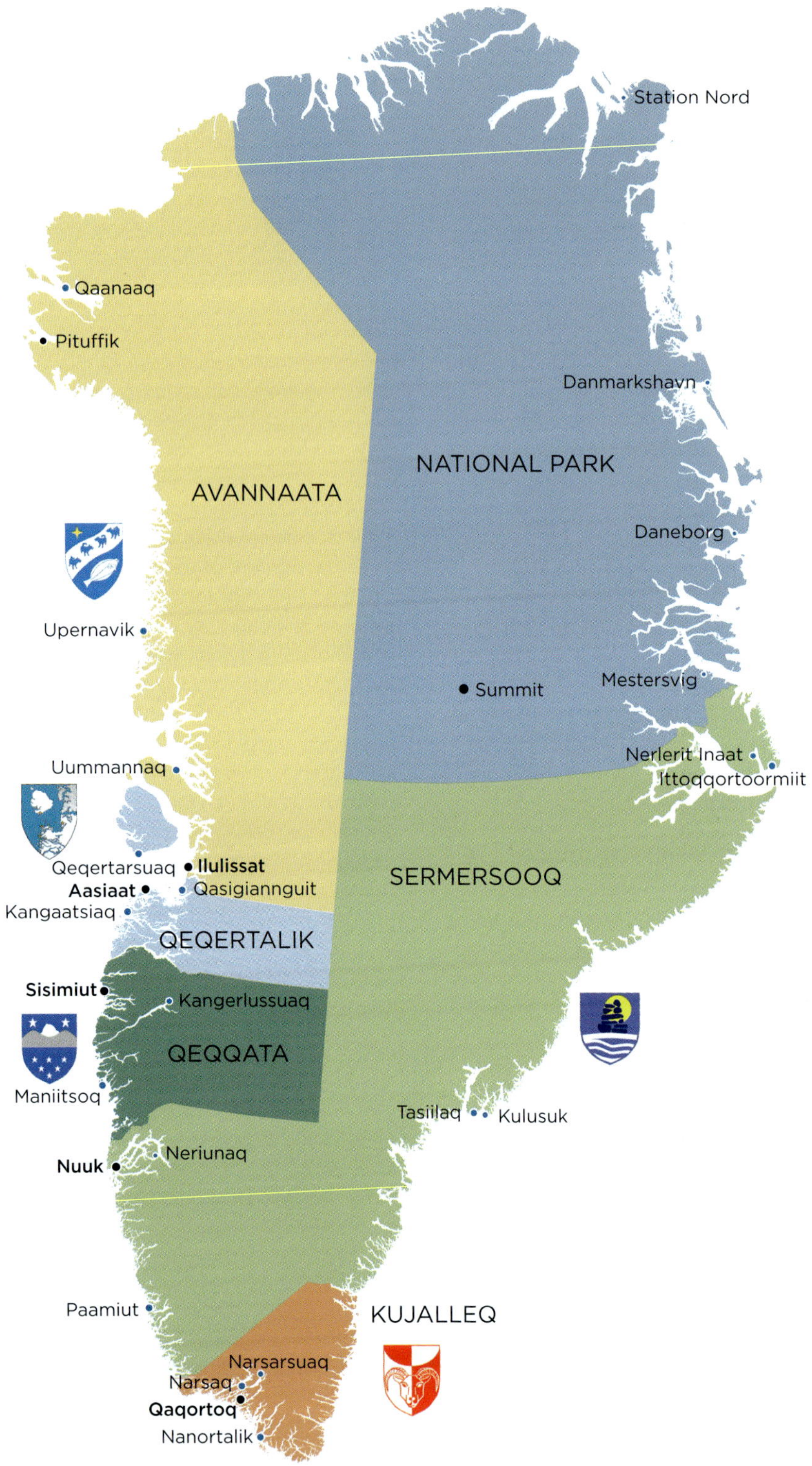

Greenland's municipalities with towns and airports, and the National Park with stations.

ASIAQ, 2022

In connection with the local government reform, significant responsibilities were transferred to the regional municipalities from the Self-Government, not least in the social sector. A struggle is still seen to make ends meet financially and legislatively. The distance from towns that previously had their own town hall and their own administration, and the settlements not to forget, to the municipality's main town and town hall is substantial both geographically, mentally and perhaps especially in a democratic sense.

Local democracy is a struggle in settlements and towns which report to the main town of the municipality. Residents of Qaanaaq may find that they are ignored by the town hall in Ilulissat, more than 1,000 miles to the south. In Sermersooq, the main town of Nuuk is also the country's capital, so the problem is reinforced for the municipality's two towns and associated settlements in East Greenland. A physical journey from Ittoqqortoormiit to Nuuk is very costly and, although it takes place by plane and helicopter, it can easily take several days each way. Naalakkersuisut and the five major municipalities make great efforts to create equal social, economic and healthcare cohesion throughout the country.

The challenge did not diminish when the national association of Greenland's municipalities, Kanukoka, was decommissioned in 2018, as a delayed response to the local government reform. Since then, Naalakkersuisut and the regional municipalities, typically represented by the mayor, have tried to gain common ground at political coordination meetings. The municipalities have a lot in common, but there are also major cultural, climatic, economic and geographical differences around the negotiating table, and it is often difficult to reach agreement. A Nordic model welfare society is not a simple thing to achieve in a country with a small population that lives scattered over an enormous area of great climatic and geographical variation.

Municipalities are large measured on their geographical extent. On the other hand, they are very small in terms of population figures. A third of the country's population lives in Nuuk, but the town is, with its barely 20,000 inhabitants, one of the world's smallest capitals.

The area of the municipalities includes the ice sheet, and this explains why the areas of Avannaata and Sermersooq are very large in comparison with the other municipalities, but also in a global perspective. Sermersooq is not only the largest municipality in the Unity of the Realm, but also one of the largest municipalities in the world.

More than 60 % of Greenland's population live in the open-water towns of Nanortalik, Qaqortoq, Narsaq, Paamiut, Nuuk, Maniitsoq and Sisimiut on the southern west coast. The National Park in Northeast Greenland is outside the municipal subdivision and is administered by the Self-Government.

This chapter describes the towns from south to north and the municipalities in the following order: Kommune Kujalleq; Kommuneqarfik Sermersooq; Qeqqata Kommunia; Kommune Qeqertalik and Avannaata Kommunia. Under each municipality, the towns are also described from south to north. The two towns on the east coast are described as the first in the section on Kommuneqarfik Sermersooq – followed by the towns on the west coast. But first, the chapter gives a general introduction to settled locations, which are not towns: settlements and settled locations with agriculture and sheep farming.

TOWNS AND SETTLEMENTS

Since the Danish modernisation of Greenland in continuation of World War II (see p. 128), settlements, their importance and development have stirred debate, both culturally, economically and politically. The debate is often emotive and characterised by cultural markers that cannot be easily rationalised or made up in economic models.

The Danish State's major development plans G-50 (see p. 127) and G-60 (see p. 128) aimed, on the one hand, to improve health and housing conditions and thus the living conditions of the population, and, on the other hand, to develop a self-sustaining economy based primarily on fishing. In order to reduce the cost of housing construction, health and education, etc., as well as to ensure manpower for fishing and fish processing, part of the G-50 and G-60 was about decommissioning settlements and bringing together the population in open-water towns.

During the first generation after World War II, the Greenland population doubled (see p. 145) while the number of settlements halved, and most of the increase in population was seen in the large towns. However, right up to the turn of the millennium, the total population in the settlements remained roughly stable at around 10,000 people, because some moved to the large settlements that gradually came to be included in the fisheries. It was largely the excess of births in the settlements that contributed to overall increase in population because the urban birth rate gradually declined and is below two children per woman today.

Up through the 1960s and 1970s, popular opposition to the G-50 and G-60 and the associated centralisation emerged and that became an important argument for the desire for home rule (see p. 144). After the introduction of the Home Rule in 1979, the Naalakkersuisut (the Government) and Inatsisartut (the Parliament) took several steps to support the settlement population.

Livelihood of the settlements

Since the turn of the millennium, the population of the settlements has decreased significantly – from 10,219 in 2002 to 7,010 in 2021 – a decrease of 30 %. However, where the population decreases in most settlements, it increases in others, while some have a stable population. The explanation is

Municipality	Area, km²	Population	Share in %
Kujalleq	32,000	6,454	11,4 %
Sermersooq	531,900	23,460	41,6 %
Qeqqata	115,500	9,305	16,5 %
Qeqertalik	62,400	6,312	11,2 %
Avannaata	522,700	10,820	19,2 %
The National Park	972,000		
Outside towns and villages		70	0,1 %

GRØNLAND I TAL 2021, GRØNLANDS STATISTIK; WIKIPEDIA

Table 6. Geographical size and population of the Greenland municipalities, 2021.

KÅRE HENDRIKSEN, 2015

On the way back to Qaanaaq after a hunting trip west of Siorapaluk. An area with currents has made the sea ice unsafe, so traffic takes on the narrow ice foot between the mountain and the ice pack.

primarily to be found in the overall livelihood of each settlement and the population's abilities and opportunities to exploit the natural resource base.

For the vast majority of the settlements, the livelihood is fishing combined with hunting. Often the catch and proceeds of the hunt are not, or only to a limited extent, part of the measurable economy of the settlement. However, the catch and proceeds of the hunt contribute significantly to the local food base, and particularly the small settlements maintain the old traditions of giving meat as gifts to people who do not themselves have the opportunity to go hunting, while in the larger settlements, the catch and proceeds of the hunt are distributed primarily within the family structures.

At the same time, the catch and proceeds of the hunt contributes to the overall income of each hunting family, either through exchange for other catch, berries or similar products from other districts, or it is sold by informal channels across most of the country. This makes much of the hunting and part of the fishing in the settlements a non-measurable subsistence economy. Part of the catch is included in the measurable economy, including when it is traded at fish factories in some of the settlements. This applies, for example, to parts of the whale meat, mattak (whale skin), caribou and muskox caught by the settlement population.

Part of the explanation for the overall decreasing population of the settlements should thus be found in changes in the livelihood of each settlement. When cod disappeared from the southwest Greenland coast due to international overfishing in the North Atlantic from the late 1980s, most of the livelihood of the settlements in this area disappeared, and this is where the largest and fastest depopulation of settlements is seen. During the same period, fishing for Greenland halibut started gaining ground in Uummannaq and Upernavik districts, which is why the population of some of the settlements in the districts continued to increase. These settlements are among the towns and settlements that have the greatest contribution to export income measured per capita.

In terms of time, the cod fishery collapse partly coincides with Greenpeace's campaign against sealskins and the imposition of international restrictions on the sale of walrus and narwhal tusks, polar bear skins, etc., that challenged the economy of the settlements that had few alternative sources of income.

Another important factor in the development of each settlement is whether the necessary infrastructure to exploit the resource base is maintained. Unless the settlement has an trading facility or factory within a reasonable radius, it will not be possible to exploit the local resource base, this is a challenge for a number of settlements. In a number of settlements, the trading facility has insufficient freezing capacity, meaning that trading needs to stop during the period of sea ice because the warehouse cannot be emptied.

Other settlements are challenged by insufficient or lack of fresh water supply for fish production, insufficient electricity capacity, or by the lack of sufficient draught in the harbour. At the same time, settlements are challenged by changing pricing structures, meaning that certain goods and services are much more expensive in the settlements than in towns. In particular, young people's desire for increased opportunities for education and jobs, better public and private services, for example in the field of healthcare, as well as actual cultural and leisure time activities.

Maintaining the settlements

For decades, it has been debated whether the decentralised settlement pattern of 17 towns and 56 settlements is too expensive, and in particular whether maintaining the settlements is too costly. This is a complex and nuanced issue. There is a lot of difference between settlements, just as between towns. A number of towns and settlements have lost their natural livelihood and therefore barely contribute to exports or the national food supply, and the primary occupation is to maintain the town or settlement. A challenge for insular communities where commuting is not an option.

Measured per capita, public spending is at roughly the same level for towns and settlements, but it reflects very large disparities in service levels that are far better in the large towns. If the service levels of the settlements were to be up to the level of larger towns, the settlements would become significantly more expensive per capita. On the other hand, in terms of the working age group of the unskilled, fishermen, hunters and people outside the labour market, public spending per capita is significantly less in the settlements than in towns. This is because the settlement population is far more self-sufficient.

The realisation that town versus settlement is a false contradiction is gaining a foothold in the Self-Government and municipalities, where there is an increasing focus on the overall livelihood of each settlement and the extent to which society supports the opportunities for the people to exploit the natural resource base through infrastructure, pricing structures, etc.

In addition, a number of settlements have great untapped tourism potential and some settlements have gradually changed their status to 'summer settlements' with leisure huts for the nearby town, where the settlement's permanent inhabitants contribute to running the summer settlement.

KÅRE HENDRIKSEN, 2011

In addition to hunting and fishing, the sea ice is used for transport and as a sports and playground area. The school in the settlement of Kullorsuaq, Upernavik district, organised a sports day.

KÅRE HENDRIKSEN, 2022

GTO standard house in the settlement of Tiniteqilaaq, located on the beautiful Sermilik ice fjord in the Ammassalik district, the second most visited district in Greenland. With its 75 inhabitants, Tiniteqilaaq is visited by several tour operators for day trips or longer stays.

ASIAQ, 2023

Map of South Greenland, stations, agricultural and sheep/domestic reindeer farming as well as towns and settlements. For Neriunaq, see p. 318

The pattern of settlement in Greenland has always been dynamic, and there will also be settlements that grow and thrive, and settlements that depopulate and close.

The two settlements of Kangerlussuaq and Narsarsuaq were originally founded by the Americans as airports during World War II (see p. 125) and today serve as the primary entrances for passengers to and from Greenland. Therefore, they are described separately in the following sections on the towns – under their respective municipality.

LOCATIONS WITH AGRICULTURE AND SHEEP FARMING

The first known farms with livestock in Greenland are attributed to the Norse who settled around the year 985, after Erik the Red had visited and named the country (see p. 95). In more recent times, the first full-time farm was founded in 1924 in the settlement of Qassiarsuk, associated with the original farm of Erik the Red.

Livestock farming was already an additional livelihood in the settlement of Igaliku (see p. 118) from 1783, where dairy cattle and goats grazed in the place where the ruins of Garðar, the diocese of the Norse, are located. Livestock farming with cattle and goats thus gradually became widespread in South Greenland up through the 19th century.

Agriculture and sheep farming, with few exceptions, are concentrated in South Greenland, where the subarctic climate deep within the fjords is sufficiently warm and winter is generally milder and shorter than farther north.

The predominant part of Greenland agriculture is directly linked to sheep farming. The sheep graze out throughout the summer, and the fields are used for growing winter fodder, which is supplemented by imported concentrates.

The number of farms has gone from 54 in 2005 to 38 in 2019, while the number of large farms with more than 500 ewes has remained stable at 14-18 farms over the same period.

MADS PIHL, VISIT GREENLAND, 2015

Since the turn of the millennium, sheep farming has evolved into intensive production from small to larger farms. The average size of a farm is 28.5 ha (2015) and most of the country's farms, about 68 %, are found in the Narsaq district (2015). Farms are mostly run by individual families.

Life as a sheep farmer in Greenland is very special, not only because of the Sub-and Low-Arctic climate, but because most sheep farmers are cut off even more than the residents of the towns and settlements from the outside world, from infrastructure, public and private services, cultural offerings and something as obvious as schooling for their children. They have to manage most of it themselves. As well as raising sheep and growing crops, sheep farmers solve tasks that others would ask a carpenter, electrician, plumber or IT expert to do. And then they often teach their children themselves. It is not without reason that the sheep farmers' children have a reputation for becoming some of the country's most outstanding adults. They have been taught to solve the problems themselves in any given situation, any time of the year. The profession constitutes a strong culture with proud and hard-working farmers.

Attempts to raise cattle have started within the last ten years, primarily near or in Narsaq. In addition, there are two farms producing meat from domestic reindeer in Isortoq caribou Station in the Qaqortoq district and Tuttutooq in the Narsaq district. Isortoq, the larger of the two, is authorised to export to the EU and produces 800-1,000 domestic reindeer annually at its own slaughterhouse from a fixed breeding population of about 3,000 animals. But sheep remain the most common animal in the landscape.

Over the years, substantial investments have been made in developing the infrastructure of the sheep farming district, and the Self-Government provides grants for trading and establishment support to sheep farmers, and since 2004 for the operation of

Tasiusaq sheep farm north of the settlement of Qassiarsuk. Greenland's 38 sheep farms contribute to the country's self-sufficiency in lamb supply and, to a limited extent, vegetables and potatoes. The socio-economic importance of sheep farming is relatively modest. However, the development of self-sufficiency is a political priority.

slaughterhouses. Modern agricultural buildings have been erected, and to ensure supply through hard winters, production of fodder has increased and reliance on imported concentrates is decreasing.

The vision of the Kommune Kujalleq and the Self-Government is to make South Greenland 'Greenland's pantry' with intensive agriculture, supplying the whole country with potatoes and other root vegetables, lettuce, vegetables and meat from cattle, pigs, sheep and caribou. Seen from an agricultural perspective it is possible; the challenges are to make it profitable and thus competitive in terms of price, and to ensure a stable capacity, supply and quality compared to the massive imports from abroad of the very type of products that are often subsidised, e.g. through the EU agriculture subsidies.

To support agriculture and sheep farming in the efforts to achieve the goals described, the Self-Government operates the Upernaviarsuk experimental station, located a few kilometres from Qaqortoq.

At the experimental station, various crops for consumers and types of winter feed are tested in Arctic conditions, just as the station operates an agricultural school offering education in agriculture, horticulture and sheep farming (see p. 323). Furthermore, small-scale forestry experiments are carried out.

Agriculture forms a small part of the national economy and most of the production is sold domestically. Exports represent a modest 0.1% of the country's total exports (2015) and are almost exclusively products of lamb or caribou meat.

Farmers are constantly looking to develop and optimise the farm and make it more profitable. Solar cells and hydropower have found their way to farms over the past few decades, saving many thousands of gallons of fuel for the benefit of both the environment and economy. The future of agriculture lies in secondary income. Many offer hostel rooms, raise cattle or grow vegetables and potatoes, which can be sold in town.

POPULATION TRENDS

Until the Danish Constitution of 1953, when Greenland became a Danish county, most of today's towns were colonies. The other areas were either remote trading posts with some form of trade and commerce or settlements. The majority of the population lived in remote trading posts and settlements and thus outside the colonies. The modernisation to offer better health and housing conditions was accompanied by an almost explosive population growth – from just over 20,000 residents at the end of World War II to over 40,000 20 years later. A population growth that, because of the Danish centralisation policy, almost exclusively took place in the towns.

The description of the individual towns includes a figure showing population trends from 1886 to 2021 in which only periodic data are available up to 1977. During the first period from 1886 to 1947, only the part of the colony's (town's) population born in Greenland is included, but there were relatively few non-Greenlanders living in the country during the colonial period.

During the same period, 1886-1947, the residents of East Greenland and the then Thule (Qaanaaq) are not included because Ammassalik was not established until 1894 and Scoresbysund (Ittoqqortoormiit) was not established until 1921, and the two districts had a special status in the Danish Greenland administration, while Thule only became a formal part of Denmark in 1938.

Numbers for Nanortalik, Narsaq and Kangaatsiaq are only available from 1960 because they did not have colony status, and the first two only achieved town status around 1950 and Kangaatsiaq only as late as 1986.

MADS PIHL, VISIT GREENLAND, 2015

Chinese cabbage is harvested at the Upernaviarsuk experimental station near Qaqortoq. Vegetables and potatoes are sold locally and further up the West Coast, including Nuuk.

ANINGAAQ R. CARLSEN/VISIT GREENLAND, 2018

KOMMUNE KUJALLEQ

With an area of about 32,000 km^2, Kommune Kujalleq, the southernmost municipality of Greenland located between 59° and 62° north, is the smallest of the five municipalities with just under 6,500 inhabitants. The southern location and a branch of the Gulf Stream provide a relatively mild climate and hence opportunities for agriculture and sheep farming, which is why South Greenland is often referred to as 'Greenland's pantry'. The municipality includes the towns of Nanortalik, Qaqortoq and Narsaq, as well as 11 settlements. The administrative centre is Qaqortoq.

Kommune Kujalleq is the only place in the country with a road connection between settlements, as several sheep farms are linked by dirt roads and connected to the nearby settlements. The Qorlortorsuaq hydropower plant supplies the towns of Narsaq and Qaqortoq with electricity. Narsarsuaq is home to one of the two international airports on the west coast.

The municipality is also home to several historic settlements from the Middle Ages, Norse ruins lie side by side with modern farms. Some of the archaeological sites and nearby sheep farms are part of the UNESCO World Heritage Area of Kujataa (see p. 118).

The region is of geological interest and rich in mineral resources with several active or potential mining projects, including the gold mine in Nalunaq (Kirkespirdalen) northeast of Nanortalik. To this should be added projects such as Killavaat Alannguat (Kringlerne) near Qaqortoq and Kuannersuit (Kvanefjeld) 7 km from Narsaq, both containing rare earth elements. Kuannersuit also has significant amounts of the radioactive elements uranium and thorium.

A possible mine at Kuannersuit has triggered concerns about radioactive contamination, including deteriorating or impossibility of business opportunities such as tourism, sheep farming and fishing, and undermining the potential of the UNESCO World Heritage Area of Kujataa. Whether it should be allowed to extract radioactive elements is a question that over decades, locally and nationally, has led to political turmoil, demonstrations and debate. The turmoil has been fuelled by opposition to nuclear power around the world. After the 2021 election, Inatsisartut (Parliament) passed an act that does not immediately allow the extraction of radioactive elements. In the April 2021 municipal elections, IA, who are declared opponents of uranium, won the office of mayor.

The municipality's business strategy is based on four pillars: fisheries, food production, tourism and extraction of mineral resources, and the strategy is closely linked to the national political decision to establish an airport in Qaqortoq.

The municipality is challenged by relocation from the region, in social fields and by lack of employment. There is a desire to generate economic growth, new jobs and revenue from extraction of mineral resources, but not at the expense of developing agriculture, fishing and tourism.

The gravel road from Qassiarsuk to the sheep farm Tasiusaq. The only places in Greenland with road connections between inhabited areas are the gravel roads between sheep farms and the nearby settlements in South Greenland.

VALDEMAR REN/RITZAU SCANPIX, 2021

The large mountain plateau Kuannersuit (Kvanefjeld) northeast of Narsaq. The approximately 2.5 m^2 large area rises 500 -670 m above sea level.

Nanortalik

Nanortalik is located about 100 km northwest of Uummannarsuaq (Cape Farewell) and is Greenland's southernmost town. Nanortalik means 'the place with polar bears' after the bears that come with the multiyear ice – 'storis' (great ice) – down the east coast and around Uummannarsuaq. In the colonial period, Nanortalik was a trading station under the colony of Julianehåb (today: Qaqortoq) and has never had a Danish name. The town has 1,169 inhabitants, with 474 living in the five settlements of the district. The district and especially the southern part are regularly isolated by the storis (great ice) in spring, summer and early autumn.

Nanortalik, street map. 1) Hotel Kap Farvel. 2) Health Centre. 3) Nanortalik Museum. 4) Primary school. 5) Fish factory. 6) Nanortalik Church. 7) Tourist office. 8) settlement hall. 9) Sports centre.

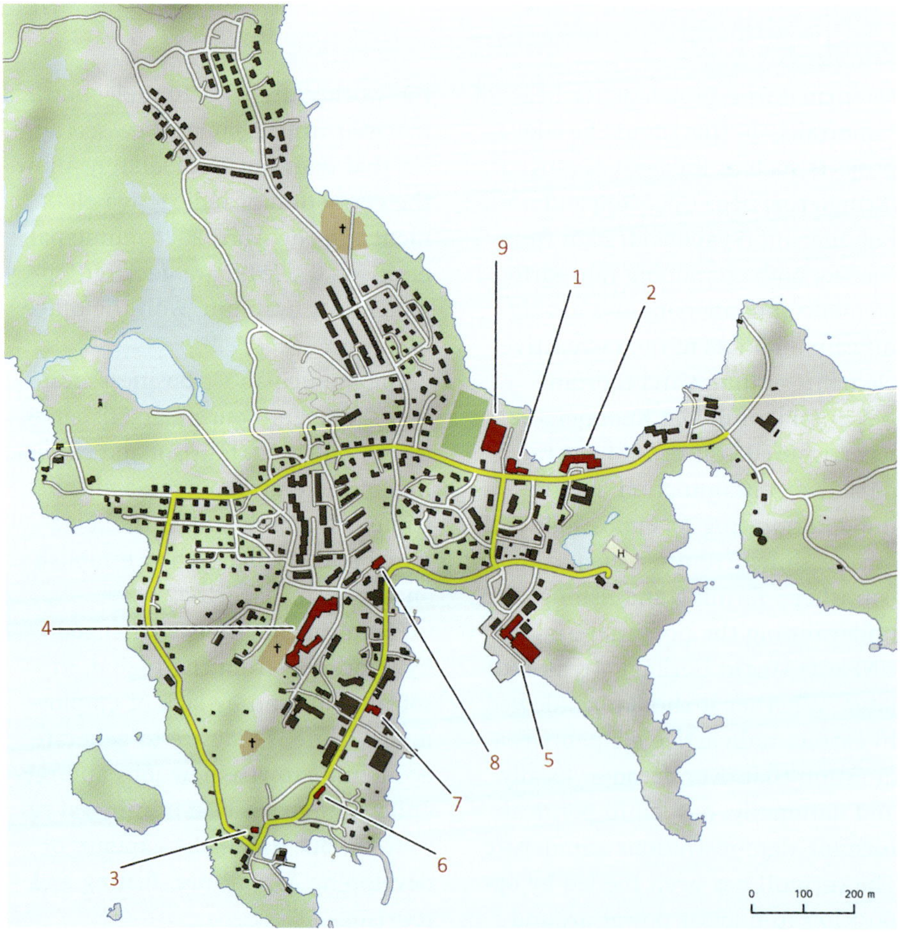

ASIAQ, 2022

History of Nanortalik

The storis provides good hunting opportunities and the district has been relatively densely inhabited with many settlements for both Inuit and Norse. In 1797, a trading post was established to increase the trading in Greenlandic products.

Beginning in the late 19th century, the basis of hunting was undermined by the massive seal hunting by Canadians on the banks at Newfoundland and the intensive seal hunting by Norwegians in the storis along the East Coast. It brought about starvation and poverty in the early 1900s, especially in the southernmost part of the district.

To counter it, the Greenland Provincial Council under the Danish state supported a gradual conversion from hunting to cod fishing from the early 1900s, and it gradually lifted the district out of poverty. However, the storis made regular fishing impossible in the southernmost part of the district and brought about a partial relocation to the north, primarily to Nanortalik.

With post-war decolonisation, Nanortalik became an autonomous municipality in 1950, but the problems with the storis meant less investment in trading and infrastructure than in the open-water towns.

Since the collapse of the cod fishery around 1990, Nanortalik and the settlements have been characterised by a lack of business base, unemployment and social problems. The settlement population went to Nanortalik, growing to 1,564 inhabitants in 1995. Since then, the town has been gradually depopulated, a process that accelerated with the municipal amalgamation in 2009, when a number of administrative tasks were moved to Qaqortoq.

Figure 19. Population trends in Nanortalik 1960-2021.

GRØNLANDS STATISTIK

MADS PIHL/VISIT GREENLAND, 2016

The wooden church of Nanortalik was constructed in 1914-1916 and listed in 2004. The church rests on a boulder foundation, and the very high-pitched roof is covered with swarf.

Geography and infrastructure of Nanortalik

The area around the town is characterised by high, steep mountains and green valleys, while the district further north gradually transitions to lower and ice-scoured mountains. Near some of the district's settlements, there are sheep farms, and not far from Nanortalik is the country's only natural forest with several meters high willow and birch, and coniferous forest has also been planted (see p. 74).

Low pressure caused by the encounter between the cold Irminger Current (Polar Current) and the warm Gulf Stream at Uummannarsuaq provides a climate with much rainfall and powerful storms where the average temperature is somewhat lower than in the fjord systems at Qaqortoq and Narsaq.

Situated on a flat peninsula, the town consists of single-family houses, terraced houses and, to a lesser extent, blocks of flats. The water supply network covers virtually the entire town, but the quality of the water from the lake in the town is challenged during certain periods, especially after spring thaw and heavy rain. Central parts of the town have sewer systems, while smaller parts of the town in the south, east and west use toilet buckets and collection of latrines.

The Royal Arctic Line has weekly freight services to the town and Disko Line has weekly passenger services to the settlements and Qaqortoq. Cancellations due to storis (great ice) are common. Air Greenland has helicopter service to Qaqortoq three or four times a week and to Narsarsuaq a few times a week.

There are accommodation options, including Hotel Kap Farvel and a hostel, and the town has a supermarket, some smaller shops and a small hospital. Primary school is the town's only school. Youth and higher education are taken in other towns.

Trade and industry in Nanortalik

Just under half of those employed work in public administration and services, including education and health, and one-sixth in fisheries and

MADS PIHL/VISIT GREENLAND, 2016

Nanortalik lies on a flat plain surrounded by steep alpine mountains. The town is named after the polar bears that follow the Storis (great ice) down the east coast in search of seals and other sustenance, sometimes coming ashore near the town before beginning the long trek back over the ice sheet or around Uummannarsuaq (Cape Farewell) and up the east coast.

Figure 20. Distribution of jobs by industry sector for people working in the town of Nanortalik, the settlements in Nanortalik district and Greenland as a whole, 2020. As illustrated in the figure, there are no entries in the category 'Agriculture, forestry and related industries' in the settlements since the sheep farms are located outside the administrative boundaries of the settlement. In contrast, some settlements with schools, shops, health facilities, etc. act as service centres for the surrounding sheep farms.

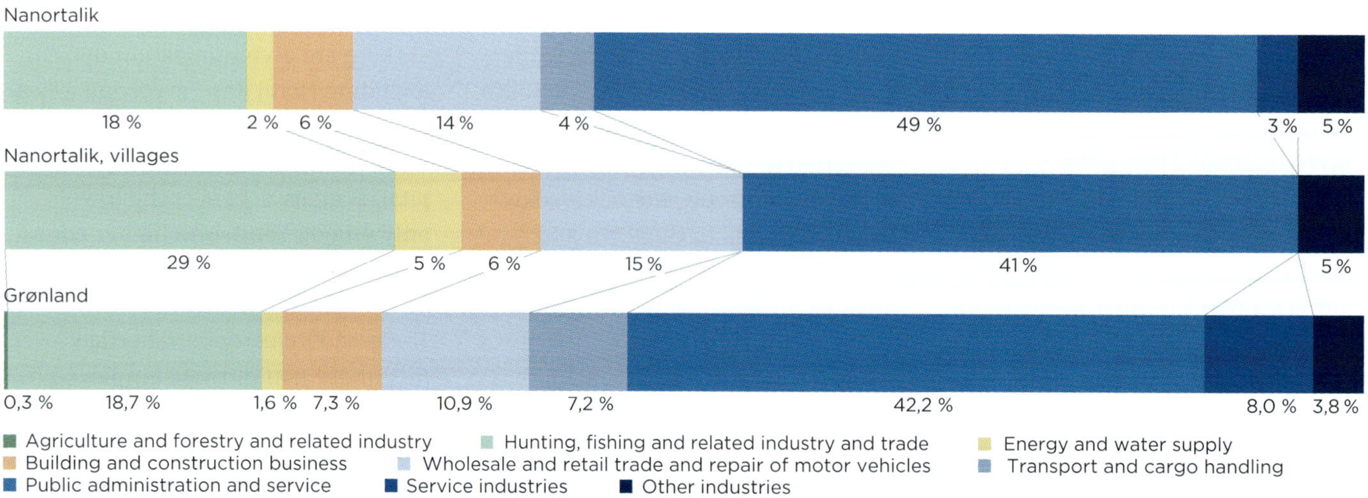

fish processing. The Arctic Prime Fisheries' plant has up to 15 employees, trading mainly cod, the majority of which come from offshore fishing, and other species such as Greenland halibut, lumpfish, redfish and Atlantic wolffish. The gold mine in Nalunaq (Kirkespirdalen), when in operation, contributes jobs to the town. There are jobs in commerce and a number of smaller industries, such as building and construction. The hotel and restaurant sector had approximately 30 employees in 2019.

There is great potential for tourism in the district with the steep mountains, green valleys, sheep farms, storis (great ice) and icebergs, and a lot of climbing and kayaking tourists are already seen.

Qaqortoq

Qaqortoq means 'white' in Greenlandic. One explanation is that the river with the vast amounts of trout was well attended by gulls that made up an impressive white spectacle when people approached, and the many gulls took off. Another is that the old settlement was located on very light rocks (at the current reserve power plant).

History and geography of Qaqortoq

The area has in all likelihood been inhabited by the Norse, but the Danish name Julianehåb is in honour of widowed queen Juliane Marie (widow of King Frederik V) and was given by Norwegian merchant Anders Olsen at the founding of the colony in 1775 (see p. 118).

From the beginning, the Danish mission was in competition with the Moravians (see p. 115) to Christianise the population of the region. The Moravians settled in Imerissoq in 1774 and named their mission field Lichtenau.

Julianehåb soon became an important hub throughout the region for trade, mission and mineral extraction. Throughout the 19th century, the geology was mapped and small mines were started. It stopped in the early 1900s, but interest still exists. The colony established the test stations Nordprøven (now Narsaq) in 1830 and the Sydprøven (now Alluitsup Paa) in 1850.

Over the years, Tasersuaq (Storesøen) northwest of the town has been a favourite place of residence, with berries along the shore and trout in the lake. In the middle of the lake lies Allagarsivit (Kongestenen), where Anders Olsen engraved his name in 1777. Several royal monograms adorn the stone, albeit they have hardly ever been there. Tasersuaq is, then as now, the town's primary drinking water supply.

In 1929, a sheep slaughterhouse was built for the autumn slaughtering; for the rest of the year, it served as a fish factory. The slaughterhouse introduced canned food to the local population, which was a major advance for food storage, which had otherwise traditionally been done by drying fish and meat. Every household received a tin opener, and the slaughterhouse distributed meat and fish recipes to the locals. The slaughterhouse soon became too small, moving to Narsaq in 1955.

Qaqortoq was and is characterised by the short river dividing the town in two. In colonial times, the Greenland houses were typically located on the western side of the river, the houses of the colony on the eastern side.

The town is located in a hollow facing the fjord south of the town. The location inside the fjord and the high mountains around the town provides good protection against foehn, northerly and westerly winds. The climate is mild, with an average of 7.5°C in summer and -0.1°C in winter. There is some precipitation (972 mm/year), most in autumn and least in spring.

Development and infrastructure of Qaqortoq

In the post-war years, the population of Qaqortoq increased. People from the settlements moved to the town to accommodate the need for labour in the factories and in the administra-

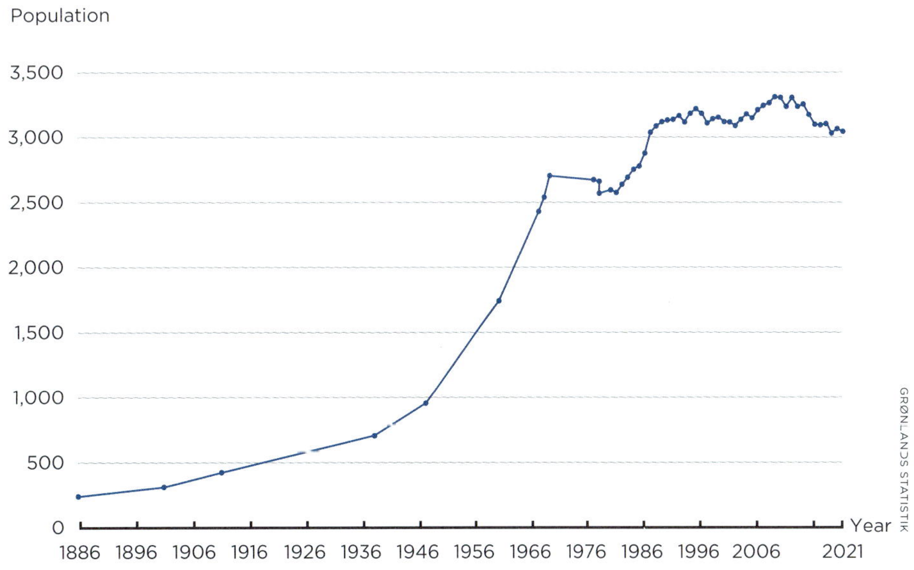

Figure 21. Population trends in Qaqortoq 1886-2021.

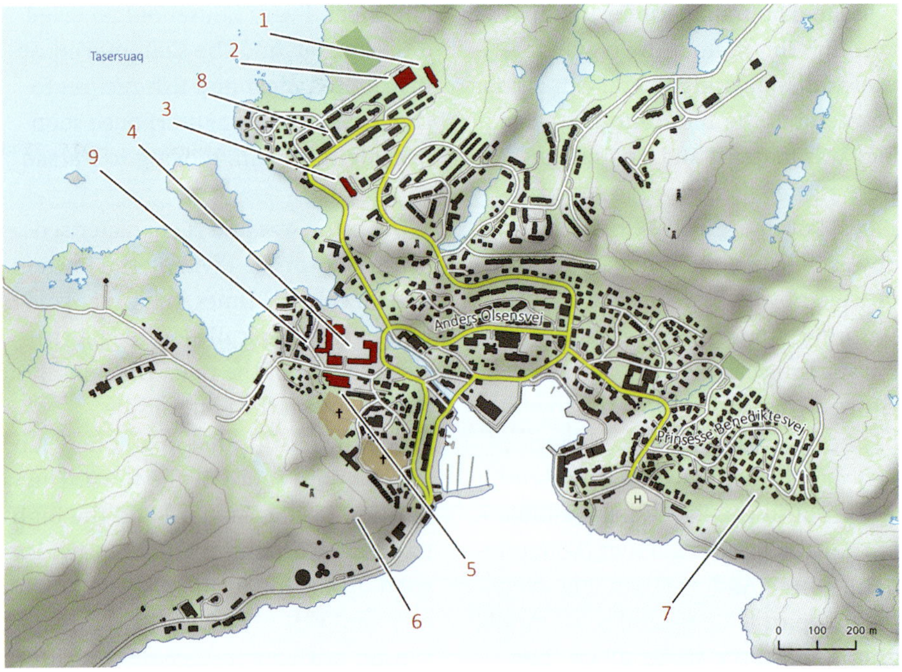

ASIAQ, 2022

Qaqortoq, key map. 1) College. 2) Sports centre. 3) Regional Hospital. 4) Primary school. 5) High School & Business College. 6) Radiofjeldet. 7) Alleriisanngivit (district). 8) Illoqarfittaaq (district). 9) 'The school neighbourhood'.

tion, just as the amendment of the Danish Constitutional Act in 1953 (see p. 133) brought about a significant increase in Danish clerks who were to start a municipal administration. The town grew rapidly, primarily to the north and east. In 1938, there were 708 inhabitants, by 1960 the population had increased to 1,521.

To the north lies the district of Illoqarfittaaq ('the new town'), originally dominated by classic blocks of flats built by the Greenland Technical Organisation (GTO) during the modernisation (see p. 195). Several multi-storey housing block and terraced houses came into being in the period leading up to the 1980s. Two run-down GTO housing blocks were redeveloped in 2021. The district also includes the sports arena from 1973 and the Sulisartut Højskoliat (folk high school) from 1977. The college took over the GTO's barracks, which had been used by workmen from Denmark.

To the east are residential neighbourhoods with private houses between retirement homes, shops and administration buildings. In the 1960s, the housing development reached the hill towards Alleriisanngivit, and the town gradually advanced up the hill. In 1964, the Danish Princess Benedikte came to Qaqortoq and the road up the hill was named 'Prinsesse Benediktesvej'. In 2011, the old gravel football field was renovated and Greenland's first artificial turf field, inaugurated by FIFA's then president, Sepp Blatter, was a reality.

The port area has a container terminal, a fish factory, a shipyard as well as the country's only tannery, Great Greenland. At the beginning, the tannery was in private hands, but is owned today by the Government of Greenland. The primary focus is the procurement and tanning of sealskins, while the sewing room has been reduced to about 10 % of the size of previous times. The fish factory has had various owners in recent decades and has alternately processed cod, shrimp and pelagic species. To survive, the town's yard invited several functions under the same roof and today also houses a contracting company.

Figure 22. Distribution of jobs by industry sector for people working in the town of Qaqortoq and Greenland as a whole, 2020.

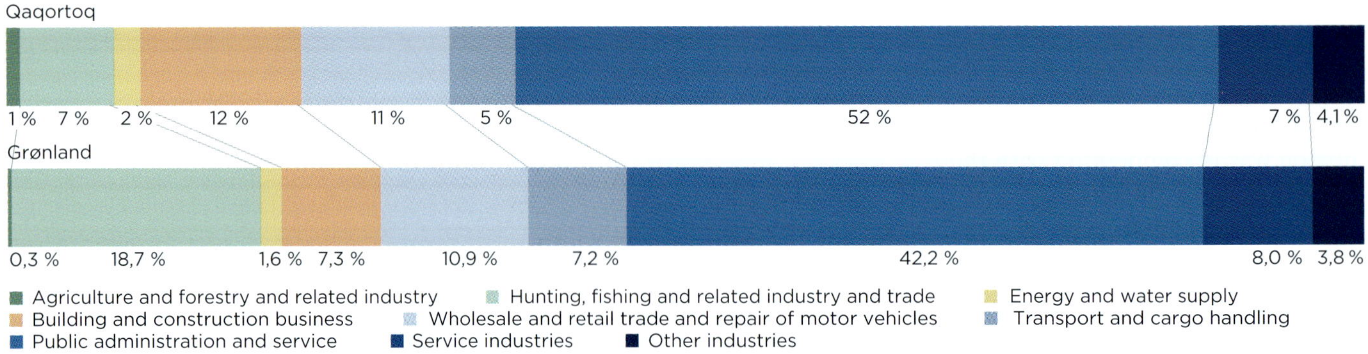

GRØNLANDS STATISTIK

MADS PIHL/VISIT GREENLAND, 2015

Qaqortoq is the main town in Kommune Kujalleq, and an important educational town with a high school and business college.

The old port warehouse has served, among other things, as a trade office, post office, travel agency and now houses Innovation South Greenland (business development) and the office of UNESCO Kujataa.

Anders Olsensvej is the town's main street with two supermarkets, town hall, self-government administration, bank, post office, hotel, settlement hall, church, restaurant and bar – and the old retirement home from 1953.

The regional hospital in Qaqortoq has approximately 60 employees in two departments and accommodates 18 patients. The hospital has responsibility for all settlements in the municipality and health centres in Narsaq and Nanortalik.

Since 2008, Qaqortoq has been supplied with electricity from the Qorlortorsuaq hydropower plant about 45 km east of the town. The heat comes from a combination of waste incineration and a conventional cogeneration plant, supplemented by electric heating from Qorlortorsuaq. A waterworks located at the mouth of Tasersuaq Lake supplies the town's drinking water. Electricity, water and heat are provided by Nukissiorfiit. The entire town is fully sewered with a few exceptions in small industrial areas on the outskirts of the town. Qaqortoq was one of the first towns to have a submarine cable, allowing Tussas (formerly Tele Post Greenland) to provide high-speed internet and telephony to the town.

Disko Line has daily connections with fast small boats to Narsarsuaq and Narsaq, and Air Greenland has several weekly connections to Narsarsuaq and Narsaq by helicopter and to Nanortalik a few times a week. Arctic Umiaq Line's coastal passenger ship sails north each week. Air transport to the rest of the country is via Narsarsuaq, with Air Greenland typically having two-to-three weekly connections to Nuuk and Kangerlussuaq and

GTO, ASIAQ ARKIV

Qaqortoq town centre and port, 1958. In the middle, behind the port, is the town square with its many historic buildings. Much of the port was fairly new at the time when the photo was taken. For example, the town's Atlantic wharf was built in 1952, while the first heliport was built 12 years later, in 1964.

Central Qaqortoq. 1) Great Greenland. 2) Fish factory. 3) The old port warehouse. (NW corner). 4) Town hall 5) Hotel Qaqortoq. 6) The Board. 7) Gertrud Rask's Kirke. 8) Town square. 9) Qaqortoq Museum and the old governor's building. 10) Police station. 11) The Old Brewery (original). 12) KGH Bakery (original). 13) The Old Shop. 14) settlement hall. 15) Church of Our Saviour.

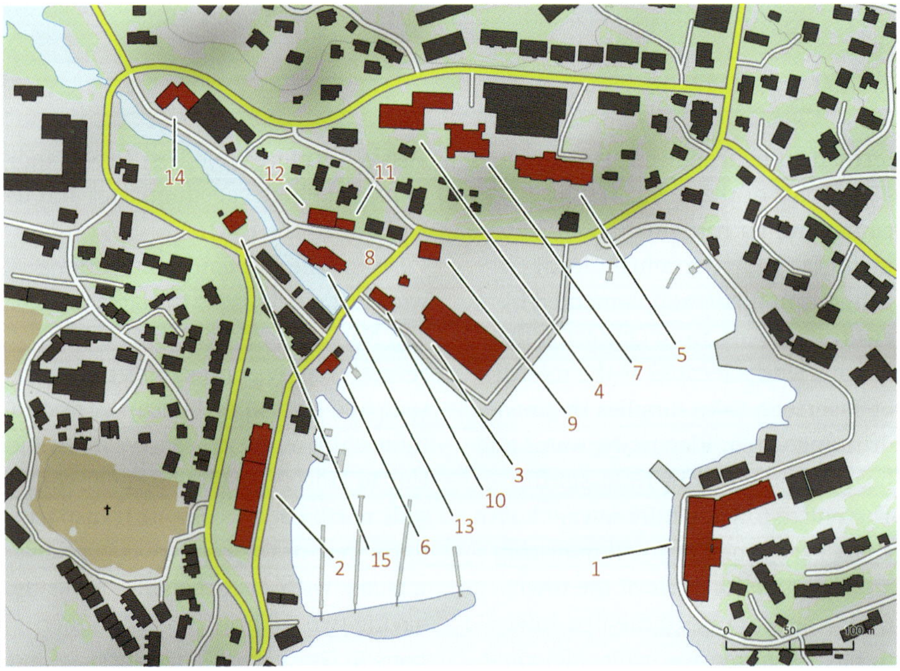

ASIAQ, 2022

in summer a weekly service to Copenhagen, just as Air Iceland operates Narsarsuaq in summer. Outside of the summer season, all passenger transport from Qaqortoq to abroad takes place across Kangerlussuaq.

The Government of Greenland has decided to build a domestic airport in Qaqortoq with capacity for close overseas destinations, such as Iceland and Iqaluit in Canada. Funding for the airport finally fell into place in Inatsisartut (Parliament) towards the end of 2021.

Royal Arctic Line arrives at Qaqortoq once a week, and from here there is connection by a small ship to the municipality's settlements.

Distinctive character of Qaqortoq

The square in the historic part of Qaqortoq is the natural centre of the town, and the fountain in the middle of the square, established by master carpenter Pavia Høegh in 1930, is the town's hallmark.

At the square is the old colonial manager's building from 1804, where the colonial manager lived on the ground floor, while the Royal Greenland Trading Department Handel (KGH) for almost 100 years had its shop on the first floor. In the attic was KGH's office until Qaqortoq Museum took over the building (see p. 167).

The small red building next to the museum, Qaqortup Katersugaasivia, is the old governant house from 1856. The governorships, the first step towards democracy in Greenland, were elected Greenlandic men who were allowed to discuss the future of the colony and the region. When home rule came in 1979, Greenland Premier Jonathan Motzfeldt decided to mark the event in the governor's building.

The old brewhouse from 1871, which is now home of *Kujataamiu Nutaaq*, is also located at the square, and the major extension was KGH's bakery from 1948. Today, a private company producing skin creams based on Greenland ingredients has taken over the building.

Beneath the former bakery is the old shop dating from 1900. The first building became too small for the needs, so already in 1901 came an extension that doubled the area of the shop. The colony shop moved in here – after nearly 100 years in the attic of the colonial manager's residence.

On the east side of the river are the settlement hall from 1937 and the public bath from 1933 with facing walls clad with Igaliku sandstone.

The 1920s marked a period of change, with the building style moving from peat to wood, and self-build houses were erected in place of the 30 or so original peat houses.

Qaqortoq is the educational centre of the region, and in the 'school neighbourhood', west of the river are the Majoriaq educational consultancy, the high school, the primary school, the business school and the library of the old primary school from 1909. Between the school neighbourhood and Qapiarfiusaaq (Radiofjeldet) to the south is the town cemetery.

The Church of Our Savior along the west side of the river was erected in 1832 (see p. 190) as assembly sets produced in Drammen in Norway. On Anders Olsensvej is the town's new church, Gertrud Rask's Kirke, erected in white concrete and consecrated in 1973.

Facing the coast is the old saltery from 1933, today a shop with fishing and hunting equipment. On a chute from the river to the saltery, hunters and fishermen could clean the catch of the day and sell it to the saltery. For the sale they received 'annotations' – the predecessor of money. For these annotations, one could, for example, go to the sales stall next-door and, carefully meted out from a board divided by lines, buy fish or meat. Therefore, sales points for freshly caught produce in Greenland are called 'Brættet' (the board) (see p. 161).

In 1993-1994, the local artist Aka Høegh made the art project Stone and Man. Here she invited artists from across the Nordic countries to design sculptures in rocks and free-standing stones around the town. It resulted in about 30 sculptures which give Qaqortoq a unique cultural expression. Aka Høegh has also decorated the chimney of the power plant, the old yard canteen and the high school aula.

For over 40 years, Qaqortoq has had solstice celebrations. Musicians, street entertainers, magicians and theatre actors have visited the town every year in the week leading up to Greenland's National Day, coinciding with the longest day of the year.

Narsaq

Narsaq, meaning 'the plain', is situated on a large, flat grassy peninsula surrounded by the fjords Tunulliarfik (Eriksfjord) and Ikersuaq (Bredefjord). The town has 1,392 inhabitants. The strait west of the town often has vast amounts of icebergs and to the northwest, the 685 m high mountain Qaqqarsuaq.

History of Narsaq

The area has been inhabited by early Inuit immigrations, but when the Norse settled from the year 985 (see p. 94), it had been depopulated over several hundred years. Up to the 1400s, the Norse developed an agricultural culture with a relatively large population in the fjord systems. Towards the end of the Norse period, Inuit began to come far south, and a cultural encounter ensued. After the Norse left Greenland around the second half of the 1400s, Inuit lived in the resource-rich area with good hunting opportunities that allowed relative population density. In the following centuries, a new cultural encounter emerged, first sporadically with European whalers and later more systematic with colonisation.

Up through the 19th century, livestock farming gradually became a complementary livelihood, primarily among the descendants of Igaliku's founders, the Norwegian merchant Anders Olsen and his Greenlandic wife, Tuperna (see p. 118).

In 1830, the Royal Greenland Trading Department (KGH) established the remote trading post Nordprøven under the colony of Julianehåb (today: Qaqortoq). The name was quickly replaced by the Greenlandic Narsaq, also in the colonial administration. Around 1880, a place of call and a trading post with a shop were established in the bay at the outlet of the Kuukkasik river. In 1919, Narsaq was Greenland's largest remote trading post with 162 inhabitants.

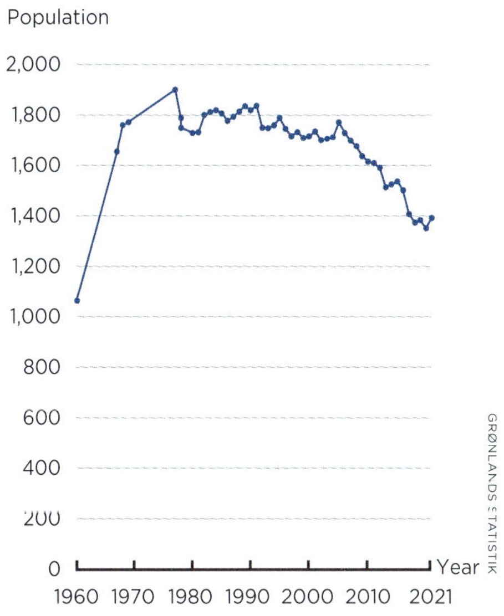

Figure 23. Population trends in Narsaq 1960-2021.

VALDEMAR REN/RITZAU SCANPIX, 2021

Narsaq used to be an important fishing town with a shrimp factory. Today, it serves as an agricultural food processing centre with the Neqi slaughterhouse and the Inuili Food College.

Narsaq, street map. 1) Slaughterhouse 2) Fish factory 3) Microbrewery 4) Inuili Food College 5) Municipal office 6) Primary school 7) Narsaq Museum 8) Health centre 9) Sports hall 10) settlement hall

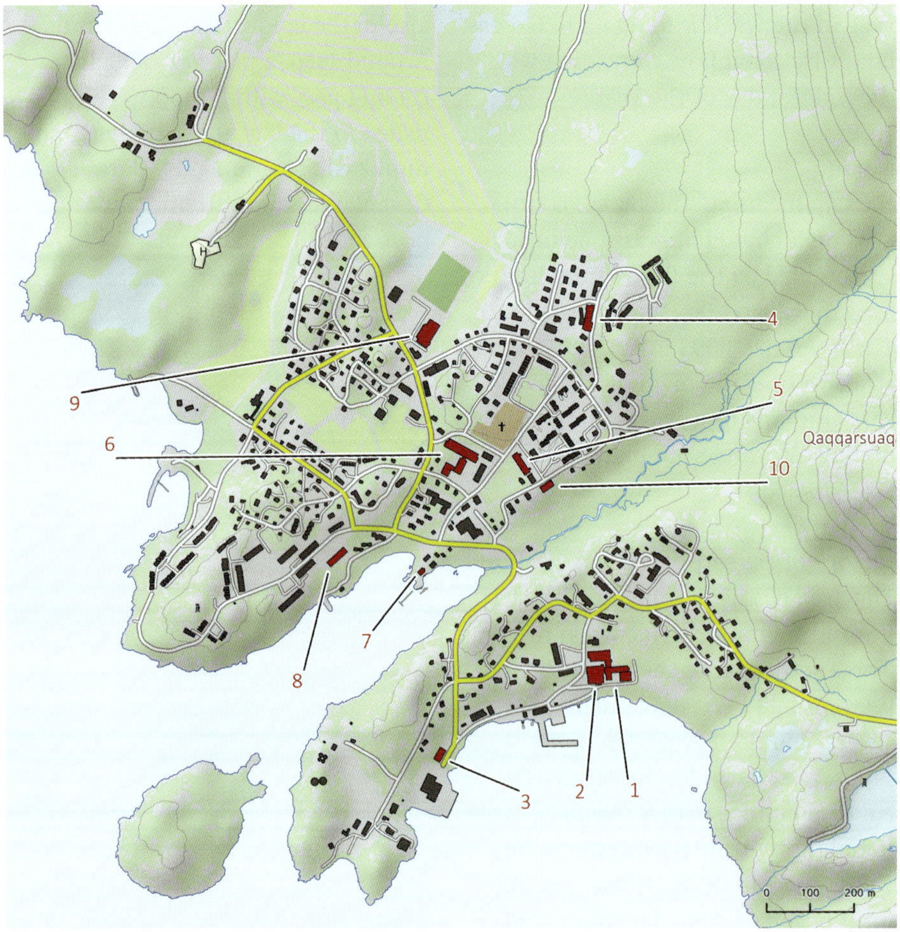

ASIAQ, 2022

From the late 1920s, full-time sheep farming became more prevalent in Narsaq and in southern Greenland. By 1930, the population had grown to 299.

During World War II, the United States established the Bluie West One airport (see p. 125) at Narsarsuaq at the head of the Tunulliarfik fjord as well as a weather and telegraph station, Bluie West Three, at Simiutaq. Narsaq is located right in the middle, and an interaction with the American soldiers occurred.

After World War II, during modernisation, Narsaq was prioritised because there were good seafaring conditions, while Qaqortoq was more often isolated by the storis (great ice). Moreover, the fjords at Narsaq had favourable conditions for cod and shrimp fishing, just as the town was the centre of sheep farming.

By 1947, the population had grown to 434. In 1950, Narsaq Municipality was created and Narsaq was granted the status of town in 1959.

KGH established a shrimp cannery and a sheep slaughterhouse in 1951, and 40 homes were built for employees. On the same occasion, the slaughterhouse in Qaqortoq was closed. In the late 1940s, fox farms were established and from 1964 to 1969 mink farms – using fish waste from the factory as feed.

To ensure supplies, a wharf was built for small freighters in 1952, and in 1959 the fish factory was extended to include freezing of shrimp and cod fillets. All of South Greenland's fish processing activities were brought

View of the town, Narsaq 1958. A boom period for the city, with population growth and business development.

GTO, ASIAQ ARKIV

together in Narsaq, and fish and shrimp traded in the Nanortalik and Qaqortoq districts were sailed to the town. From 1961, there was a decline in shrimp trading, and towards the end of the 1980s the cod disappeared from southern Greenland waters.

During the 1960s and the following decades, there was massive construction activity with 16 private contractors and 95 employees (in 1969), and the town had the heliport in 1966.

Narsaq grew to 1,835 inhabitants in 1991, but since then the population has declined, particularly as a result of the closure of the shrimp factory, boosted by the municipal amalgamation in 2009, when administrative jobs moved to Qaqortoq.

Uranium in Kuannersuit

In 1956, uranium was discovered in Kuannersuit (Kvanefjeld) about 8 km north of Narsaq, and several test drillings were carried out. Kuannersuit was part of the Danish plans to develop nuclear power, boosted by the 1973 oil crisis, as well as of the construction of the Risø experimental plant at Roskilde Fjord. The discovery of uranium was instrumental in the Danish authorities developing Narsaq, and the expectation was that the employees of a future uranium mine would live in the town. During the period, the majority of townspeople had high expectations for the project.

International opposition to nuclear weapons and nuclear power gradually put an end to the intentions of Danish nuclear power, and popular opposition was reinforced by the reactor meltdown on the Three Mile Island in the United States in 1979. In 1984, the Folketing (the Danish Parliament) agreed that nuclear power should not be part of the Danish energy supply, and the exploration at Kuannersuit was discontinued. Inatsisartut (Parliament) adopted a zero tolerance for uranium mining in 1988, which implied that ore must contain no more uranium than natural background radiation.

In 2007, Australian company Greenland Minerals and Energy was granted a licence to search for rare earths in Kuannersuit – with uranium as a by-product. It sparked a long-running debate about environment versus jobs and revenue to fund Greenlandic independence. First, zero tolerance was lifted in 2013, and after the 2021 elections, with a clear majority in Inatsisartut opposed to the Kuannersuite project, a new and slightly less restrictive zero tolerance was adopted. The new law makes mining at Kuannersuit unlikely, and a legal aftermath remains between the mining company and the Self-Government.

The years of discussions for and against uranium and jobs versus the protection of nature and the environment and the development of agriculture, tourism and fishing have been gruelling for the people of Narsaq.

Infrastructure of Narsaq

The Greenland Technical Organisation's (GTO) focus on Narsaq has meant good infrastructure; most of the roads are paved, the water supply network is good and the town has virtually a full sewer system. The electricity comes from the Qorlortorsuaq hydropower plant, which also supplies Qaqortoq, but the plant does not yet have capacity for electricity-based heating of homes in Narsaq.

The Royal Arctic Line arrives at the town once a week with supplies. Disko Line has daily passenger services with smaller ships to Narsarsuaq and Qaqortoq and a few times weekly to the nearby settlements – however, capacity is a challenge. Air Greenland has a helicopter connection to Narsarsuaq and Qaqortoq a few times a week.

Trade and industry in Narsaq

Today, Narsaq is the centre of food processing in southern Greenland.

GREENLAND MINERALS LTD/HANDOUT VIA REUTERS/RITZAU SCANPIX, 2020

Narsaq valley with sheep farm and Kuannersuit (Kvanefjeld) in the background Kuannersuit contains large quantities of rare earth elements, so there is a dilemma between protecting nature versus extracting mineral resources.

Figure 24. Distribution of jobs by industry sector for people working in the town of Narsaq, the settlements in Narsaq district and Greenland as a whole, 2020. As seen from the figure, there are no entries registered in the category 'Agriculture, forestry and related industries' in the settlements, although Narsaq is the district with the most sheep farmers. This is due to the sheep farms being located outside the administrative boundary of the settlement, while the settlement of Qassiarsuk and Igaliku, with schools, shops, health facilities etc., act as service centres for the surrounding sheep farms. Qassiarsuk and Igaliku have a combined population of 108, while the airport settlement of Narsarsuaq has 138 inhabitants or 56 % of the district's total settlement population (2021). The predominant employment in Narsarsuaq is related to the airport's operation, so the 'Transport and cargo handling' industry will be disproportionately large for the district's settlements.

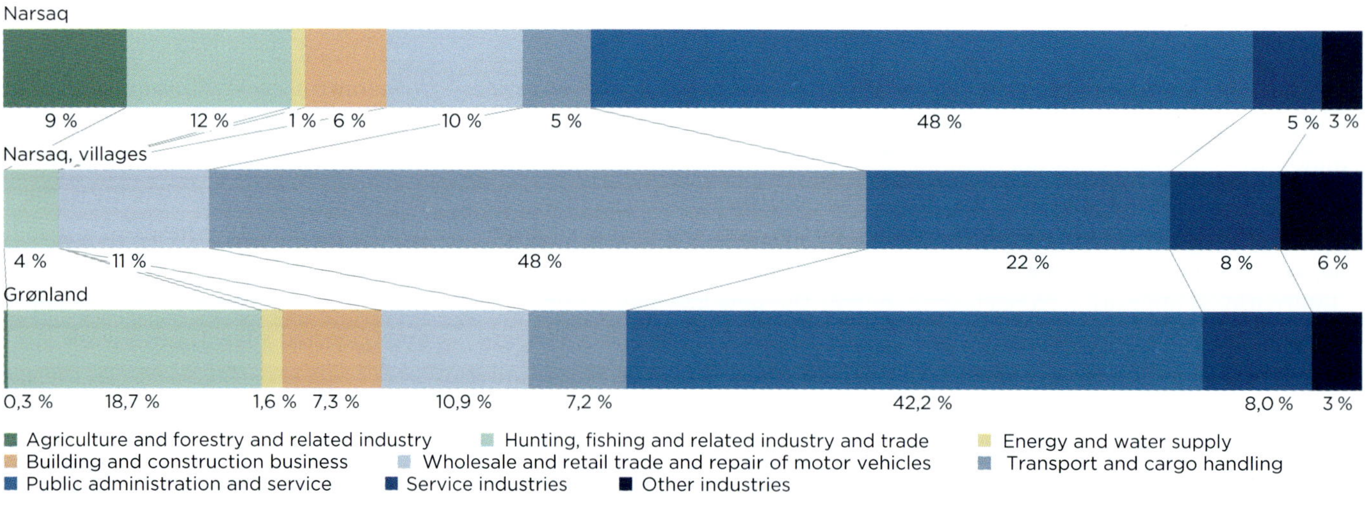

GRØNLANDS STATISTIK

The slaughterhouse Neqi A/S has approximately 80 employees during the three months of the slaughtering season and about 25 for the rest of the year. In addition to sheep farming, Narsaq is home to Greenland's largest cattle farm, beginning in 2004 with three cattle of the Irish Dexter breed in Narsap Ilua. Out of about 250 cattle, the farmer sent 58 animals to the slaughterhouse in 2020.

Local business initiatives include the Greenland Brewhouse, established in 2008 and now taken over by Qajaq Beer, the Inuili Food College Greenland with about 25 employees, and the Royal Greenland's factory, which mainly trades in cod and roe, has up to 10 employees. In addition, the primary employment in Narsaq is in public administration and service.

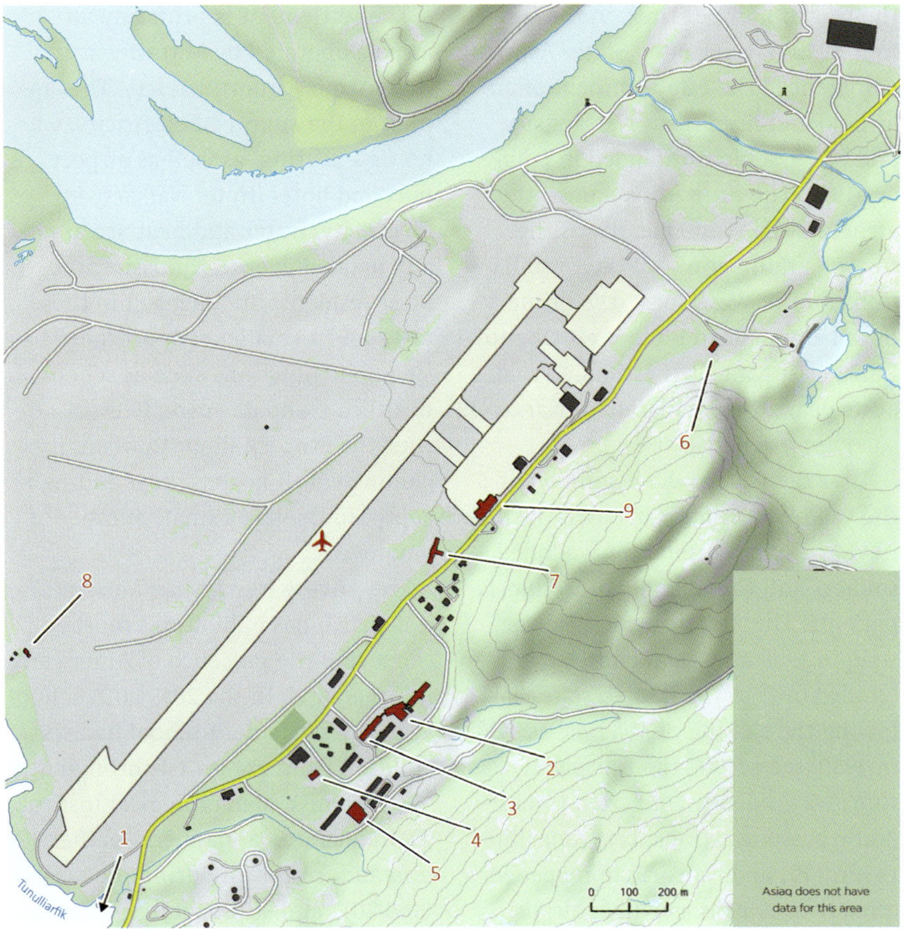

ASIAQ, 2022

Narsarsuaq, street map. 1) Towards the port. 2) Hotel Narsarsuaq. 3) Health care station. 4) School. 5) Grocery shop. 6) Hostel. 7) Narsarsuaq Museum. 8) Radiosonde station. 9) Terminal building.

Narsarsuaq

Narsarsuaq means 'great plain' and refers to the moraine plain between the Tunulliarfik fjord and the Kiattuut Sermiat glacier, where the settlement and airport Narsarsuaq are located.

The population has been below 200 since the mid-1970s, varying from 116 in 2016 to 191 in the year 2000. Today there are 138 inhabitants. In the past, there have been considerably more (see p. 271).

History of Narsarsuaq

When the United States entered World War II, Greenland became an important link in the North Atlantic air bridge: Aircraft could touch down and refuel en route from the US to

MADS PIHL/VISIT GREENLAND, 2015

Narsarsuaq at sunset. The airport is surrounded by rich vegetation with natural trees and shrubs up to seven metres high. Additionally, there are a large number of planted trees, which gradually grow into a forest.

the war in Europe. Thus, in the summer of 1941, the United States Army commenced the establishment of an airfield and hospital at Narsarsuaq, and the first aircraft landed in January 1942. The Americans named the airfield Bluie West One. It was the first of a number of military installations on the West Coast, the primary ones being named Bluie West One to Bluie West Nine. It is estimated that, at its peak, the population of Narsarsuaq approached 2,000.

After the war, at the beginning of the Cold War, Americans used the airfield as a base hospital for wounded soldiers and Greenlandic tuberculosis patients. The Americans left Narsarsuaq in 1958 and the facilities were handed over to the Danish state. Following the sinking of *M/S Hans Hedtoft* in 1959 (see p. 190), the Danish authorities gradually converted transatlantic passenger transport from ship to aircraft, and Narsarsuaq reopened already in 1959 as a civilian airport.

Infrastructure of Narsarsuaq

Since its reopening in 1959, Narsarsuaq has served as South Greenland's hub for air, helicopter and ship traffic to destinations such as Narsaq, Qaqortoq, Nanortalik, settlements in the region as well as to Kangerlussuaq – later also Nuuk. With changing frequency and occupancy, typically during the summer months, there have been connections to Copenhagen and Iceland. The Royal Arctic Line arrives at the settlement with freight once or twice a month, and Disko Line arrives with passengers from the region in fast boats, alternatively by helicopter when the storis (great ice) makes navigation impossible. Air Greenland has helicopter services to Narsaq, Qaqortoq, Nanortalik and some of the southern settlement. The frequency is adapted to air travel.

The settlement is primarily an airport operated by Mittarfeqarfiit (Greenland Airport Service). This applies to the supply of electricity, water and heating, as well as airport, port and hotel (Hotel Narsarsuaq). Electricity, water and heat are gradually handed over to Nukissiorfiit. Most residents are engaged in these tasks and a smaller proportion in basic public functions such as school, healthcare and day care. Pilersuisoq runs the grocery shop. Tourism, including youth hostels, boat and helicopter tours, has a small number of employees.

The private Narsarsuaq Museum (see p. 173) has exhibitions on the Norse and the presence of Americans in Greenland. In the 2019/2020 school year, the settlement school had six children from grades 1st through 8th.

The Danish Meteorological Institute (DMI) operated a manned weather station and ice centre during the period 1959-2017 and still operates, in cooperation with Tussas (formerly Tele Post Greenland), a radiosonde station in Narsarsuaq.

There is no bridge over the river, but there are dirt roads to the sheep farms to the north and west and to the settlement of Qassiarsuk. Road links could potentially be continued to Narsaq. A possible road connection to the south to Igaliku is hindered by steep mountains and the Qooqqut Fjord.

The Greenland Arboretum in and around Narsarsuaq is one of the most extensive forest border arboreta in the world and includes over 100,000 transplanted trees (see p. 76).

With Inatsisartut's (Parliament's) decision to construct an airport in Qaqortoq, Narsarsuaq is to be downgraded to heliport and the future of the settlement is intensely debated at the turn of 2021/22.

KOMMUNEQARFIK SERMERSOOQ

The metropolitan municipality extends from 60°N and up to 71°N north and includes the towns of Paamiut, Nuuk and three settlements on the west coast, as well as the towns of Tasiilaq and Ittoqqortoormiit and five settlements on the east coast. Sermersooq has an area of approximately 531,900 km^2, and it is the country's most populous with just under 23,500 inhabitants, of whom more than 19,000 live in the capital Nuuk.

As the only municipality, Sermersooq stretches across the ice sheet to the east coast, where Tasiilaq is the largest, and Ittoqqortoormiit the country's smallest and most isolated town. In terms of navigation, the east coast is challenged by the storis (great ice). Tasiilaq and the district's five settlements have supply ships five months a year, and Ittoqqortoormiit only two annual ships, which obviously challenges food and goods supply. East Coast telephony and internet are based on satellite connectivity and thus significantly slower and more limited than in the rest of the municipality.

The municipality is geologically interesting and has several major mineral exploration projects.

In East Greenland there are sled dogs, while they are not allowed on the west coast of the municipality.

Nuuk is the government town with parliament building for Inatsisartut (Parliament), the headquarters of Naalakkersuisut and most of the central administration, as well as most government-owned infrastructure companies and several research and education institutions. The contrast with the municipality's other towns is large in terms of physical geography and socio-economy. The business potentials exploited on the east coast

FLORIAN LEDOUX/SHUTTERSTOCK/RITZAU SCANPIX, 2016

Royal Arctic Line docks in Tasiilaq. The first ship of the year arrives in June, and the town receives ships until the beginning of November. By then, stocks should be filled with supplies for the next six months.

and south of Nuuk are limited and the municipality is characterised by economic inequality.

Since the municipal amalgamation in 2009, the municipality has been led by the IA and the party regained the office of mayor in April 2021. The municipality's planning strategies focus on core values such as sustainability, quality of life as well as democracy and service and focus on coastal fisheries, the construction industry and the elderly. Over time, the UN Sustainable Development Goals have gained greater attention.

For a number of years, the municipality has experienced a sharp but unevenly distributed population growth that is assumed to continue. Part of the capital strategy is linked to Naalakkersuisut's plans to expand the existing airport at Nuuk into an international airport. The municipality expects that it will attract more tourists, more business and provide more jobs to the capital.

In the 2009 local government reform, the towns on the east coast wanted to form their own municipality, but were rejected, citing that the population is too low. The reason for merging the two east coast towns with Nuuk was and is that the capital provides greater economic latitude. However, municipal cohesion is challenged by unevenly distributed growth and by an experience that decisions are taken very far away.

Tasiilaq

Located on the shore of a fjord that looks like a lake, the town got the name, Tasiilaq, which means 'the place that looks like a lake'.

Tasiilaq has a population of 1,916, which has remained roughly stable

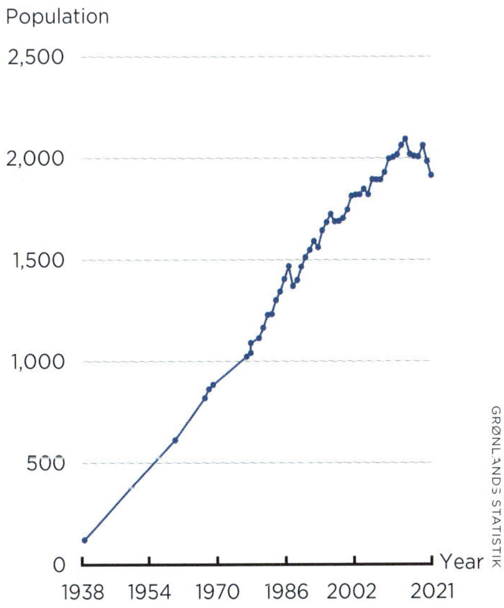

Figure 25. Population trends in Tasiilaq 1938-2021.

KÅRE HENDRIKSEN, 2022

East Greenland's largest town Tasiilaq with 1,916 inhabitants viewed from Qaqqartivakajik (Sømandsfjeldet). In the background is the distinctive Polhem Fjeld at 1,030 m. The town is named after the fjord, which looks like a lake and has a narrow mouth.

for the past 10 years (2,062 in 2014) after a prolonged period of growth.

Tasiilaq is located on the southeast of the island of Ammassalik, which, with an area of about 730 km², is slightly larger than Bornholm. There are five settlements in Ammassalik district, with the largest being Kuummiut, Sermiligaaq and Kulusuk, all of which have 200-250 inhabitants, and the smallest Tiilerilaaq and Isertoq.

History of Tasiilaq

Most of the east coast is characterised by being isolated in behind the storis (great ice) that drifts with the Irminger Current (Polar Current) from the North Pole down the east coast and around Uummannarsuaq (Cape Farewell), where it meets the Gulf Stream, dissipates and disintegrates.

The storis makes it very difficult to reach the coast, not least in the in the sailing ship era, and the population of the district first made contact with Europeans in 1884 in connection with the Umiak Expedition led by Gustav Holm. At that time, the vast district had about 400 people living in a highly developed Stone Age community, but hunger and malnutrition prevailed because the Norwegians hunted seals out in the storis, and the hunting opportunities within the storis were therefore modest.

The town was founded as a Danish trading station and mission in 1894 under the name Angmagssalik, which is derived from ammassat (the small capelin) and means 'the place where ammassat is found'. Later, the spelling was changed to Ammassalik and it is maintained as the name for the district, while the other districts of the country are named after their main town.

The location of the colony was chosen because there was good shelter for sailing ships, and the storis did not enter the fjord to a significant extent, but the location was poor when it came to hunting opportunities. The colony therefore grew slowly and in

Tasiilaq, street map. 1) Tourist office. 2) Youth centre. 3) Tasiilaq Church. 4) Municipal office. 5) Primary school. 6) Ammassalik Museum. 7) Health centre. 8) Sports centre. 9) settlement hall. 10) Hotel Angmagssalik. 11) Hotel The Red House.

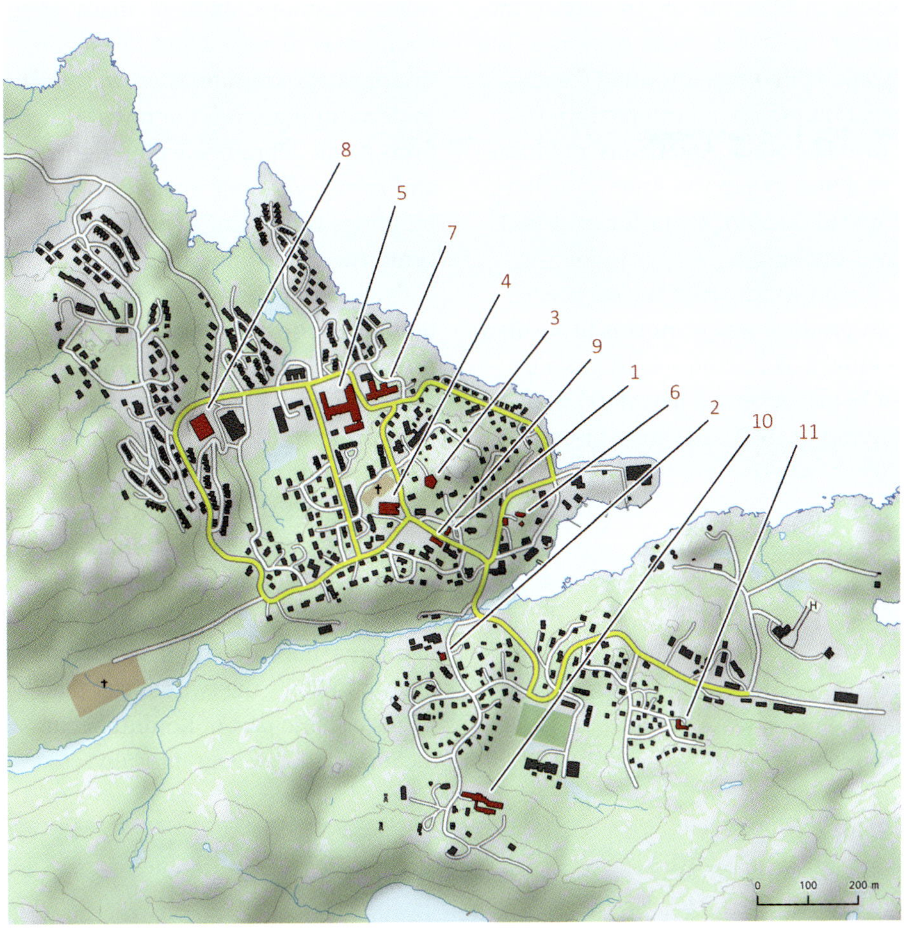

ASIAQ, 2022

1939 had 122 Greenland-born inhabitants, while in the district, 750 lived in smaller settlements with better hunting opportunities, where the population primarily lived off hunting that had become easier with rifles and implements made of iron.

In 1942, the US Air Force established the Bluie East Two airport at the settlement of Ikateq as part of the air bridge to Europe, and an interaction with the inhabitants of the settlement occurred. The base closed in 1947.

In the post-war years, the Danish State considered that the business base was extremely modest, which is why the main focus was on the 'survival' of the population, while investments in infrastructure and business development were limited. From the late 1950s, however, the cod fishery was developed, and for a period up through the 1960s there was some cod fishing at Kuummiut and in the mouth of Ammassalik Fjord.

In conjunction with the DEW Line flight warning system across the ice

UKENDT, 1970

In 1970, Tasiilaq was hit by a particularly severe Piteraq (see p. 22) of up to 90 m/s, with over half of the houses being damaged and many destroyed, and the school blew away. Some of the children were sent to Denmark while the school was being rebuilt. Since then, a lot of work has been done to ensure better building structures to withstand the strong winds, and warning lights and sirens have been set up in the town to warn people of a forthcoming Piteraq.

sheet, in 1956, the US Air Force established the DYE-4 gravel runway at Kulusuk, which was also used for civil aviation. Again, an interaction occurred between Americans and local residents. When the Americans left Kulusuk in 1991, Mittarfeqarfiit took over operations.

Foehn winds from the ice sheet can produce severe storms in Tasiilaq – known as Piteraq – with gusts up to 70 m/s. After a particularly fierce Piteraq in 1970 with gusts of up to 90 m/s, where more than half of the houses sustained damage, many were completely damaged and the school disappeared, consideration was given to decommissioning the town. However, it was rebuilt, warning lights and sirens were put up, and the buildings were reinforced to withstand the fierce wind loads.

The natural and other business potentials of the district are not comprehensively mapped, and only modest investments have been made in infrastructure that supports business development.

Tasiilaq was the main town of Ammassalik Municipality and shares fate with the only other town on the east coast, Ittoqqortoormiit, barely 850 km to the north: At the municipal amalgamation in 2009, the two former main towns became part of Kommuneqarfik Sermersooq. Geographically, culturally and linguistically, they are far from the decision-makers in Nuuk on the other side of the ice sheet. At the time of the amalgamation, some administrative jobs disappeared, causing a drain of skills. Coupled with the small population base, a challenge is seen in maintaining a level of welfare, especially in the social and healthcare sectors. The municipality, the Self-Government and the Danish State are jointly trying to lift the town out of the social problems with various measures. In 2021, a youth centre opened in the town to provide a safe environment for a meaningful leisure life for young people.

Geography and infrastructure of Tasiilaq

Kulusuk Airport is the district's primary approach route and typically

Next to Ilulissat, the Ammassalik district is Greenland's largest tourist destination. From day-trippers flying to Kulusuk and hotel guests staying in Tasiilaq to hikers, kayakers, skiers, climbers and dog-sledders.

ANINGAAQ R. CARLSEN/VISIT GREENLAND, 2017

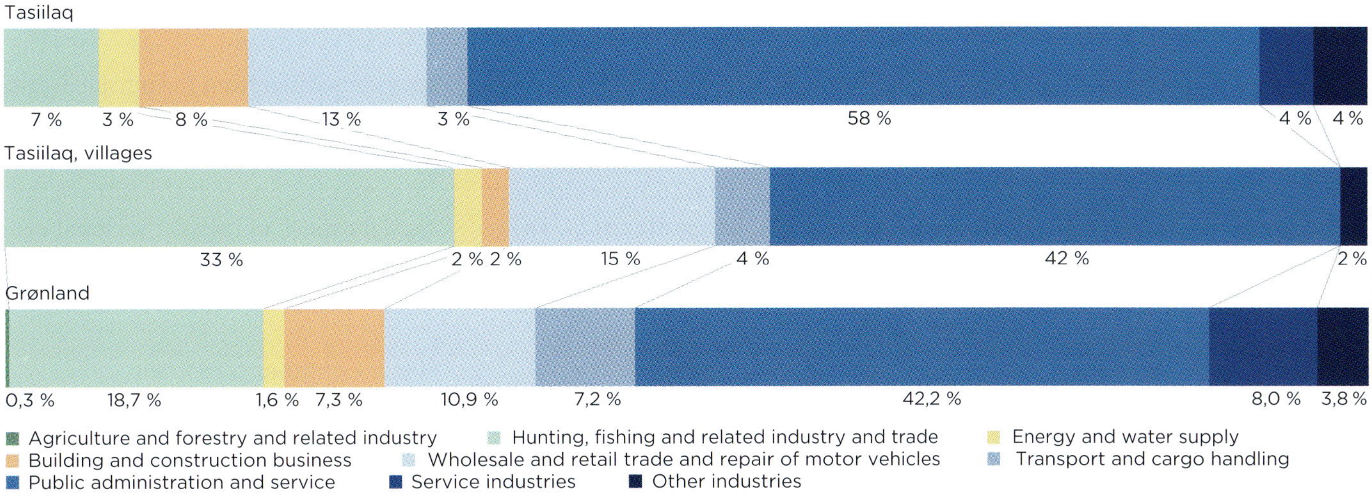

Figure 26. Distribution of jobs by industry sector for people working in the town of Tasiilaq, the settlements in Tasiilaq district and Greenland as a whole, 2020.

has two weekly flights to Nuuk and two to three weekly flights to Iceland. From Kulusuk, there is a helicopter service to Tasiilaq and the district settlements. Between Tasiilaq and some of the settlements in district, transportation by dog sled or snowmobile is possible in winter.

Royal Arctic Line's ice-reinforced container ships call at Tasiilaq and the settlements every fortnight from June to early November.

The settlement in Tasiilaq is predominantly single-family houses and terraced house. Since 2004, the electricity supply has come from a small hydropower plant a few kilometres northwest of the town, and the old diesel power plant serves as an emergency power plant. In addition to electricity, Nukissiorfiit also supplies water to the town. Some dwellings are not connected to the supply network, so citizens have to get water from tapping houses or have it delivered. There is a municipal sewer network, but most of the houses still have toilet buckets and collection of latrines.

Telephony and internet are via satellite, making it is slow and costly for citizens, businesses and for the supplier, Tusass (formerly Tele Post Greenland).

The town's topography with significant elevation differences means that some roads are very steep, and it puts demands on the brakes and traction of vehicles, especially when the roads are covered in snow and ice. Off-road ATVs are therefore very popular in the town.

Tasiilaq has a small health centre with medical staff. Children can finish primary school in the town, and in the oldest grades, the school also teaches children from the settlements who live in the town's boarding house for school children.

Tourism, trade and industry in Tasiilaq

With its location amidst the harsh nature with the characteristic high steep mountains, the storis (great ice) and access to the ice fjord Sermilik with icebergs from several glaciers as well as air connection to Iceland via Kulusuk, Tasiilaq – and Ammassalik district – is Greenland's second most visited tourist destination. Guests via Iceland contribute the majority of the 19,271 annual overnight stays on the east coast, incl. Ittoqqortoormiit (2019).

There are two hotels with restaurants, several cafes, guest houses and private accommodation, as well as a dozen tour operators in Tasiilaq, offering many kinds of experiences such as sailing and hiking, dog sledding, kayaking, etc. The rich culture and history of the district also attracts quite a few elderly tourists.

The Ammassalik Museum (see p. 171) has been housed in the town's old church above the port since 1990 and has a significant collection that illustrates the East Greenlandic cultural heritage. In addition to the church building, the museum consists of a chapel, a photo museum, a potato house and a peat house.

Tourism is growing, but fishing and hunting remain the main industries of the district, with seal and Greenland halibut as the primary species.

In 1989, Tussas (formerly Tele Post Greenland) opened Filateli which sells Greenland stamps to collectors worldwide.

Most people employed are in the public sector in education, administration, healthcare, municipal management as well as in the government-owned infrastructure companies, such as Nukisiorfiit, Mittarfeqarfiit, Air Greenland and Tusass (formerly Tele Post Greenland). In addition, some are employed in retail and construction and tourism also provides a lot of jobs.

Ittoqqortoormiit

On the border with the National Park in Northeast Greenland and at the mouth of the world's largest fjord system, Kangertittivaq (Scoresby Sound), lies the northernmost town on the east coast, Ittoqqortoormiit. The East Greenlandic name means 'the place with the big houses' and refers somewhat unusually to buildings and not nature. The name suggests that the town is isolated from the outside world – an isolation enhanced by the fact that for most of the year, the fjord is closed by winter ice, and that the entry to the fjord for much of the year is closed by the broad belt of storis (great ice) from the Arctic Ocean in the north.

History of Ittoqqortoormiit

The town's location provides good hunting opportunities with seal, narwhal, walrus, polar bear and muskox being the main ones. The hunting opportunities and a political need to mark Danish sovereignty in Northeast Greenland were the reason why the Scoresby Sound Committee, on the initiative of the polar explorer Ejnar Mikkelsen, among others, founded the town as the colony of Scoresby Sound. In 1925, 70 people from hunting families in Ammassalik as well as 15 people from the west coast and Denmark, to be included in the administration and operation of the colony, moved to the present-day Ittoqqortoormiit. Initially the plan was for the town to be located elsewhere, but drift ice that destroyed the ship's rudder determined the location.

The first houses were established on the west side of the river mouth, and since then the town has grown to the opposite side of the river. The

MARTIN N. JOHANSEN/BIOFOTO/RITZAU SCANPIX, 2013

The church in Ittoqqortoormiit was completed in 1929. Several uses were envisaged at the time, and the building was used as a school and an office for a while. Today it is used as a priest's office and store. The warmer climate had resulted in subsidence of the church at the western end. The church was closed in 2017 and reopened in 2019 after the foundation had been renovated.

terrain is gently rising and rocky with very little natural stand.

Like in Ammassalik district, the population speaks the East Greenlandic dialect which differs significantly from West Greenlandic.

Climate change has caused the land-fast ice on the fjord to settle about two months later than just a few years ago, and it breaks up earlier in the summer. The period allowing transport across the ice by dog sled or motorised vehicle is thus noticeably shortened, and it has, among other things, the unfortunate consequence that polar bears come near the town more often because their hunting opportunities on the ice are limited. Citizens and guests of the town need to pay extra attention and make sure to have a hunting rifle nearby and be able to use it if a polar bear with threatening behaviour shows up.

With 361 inhabitants, Ittoqqortoormiit is the country's smallest town. The population peaked at 539 residents in 2006, but has since stabilised just below the 400.

At the abandoned settlement of Uunarteq (Cape Tobin) about 8 km south of Ittoqqortoormiit are Greenland's warmest springs at just under 62°C, and within a 14 km radius of the town are three to five other sources with the potential for sustainable building heating and geothermal power generation.

Infrastructure of Ittoqqortoormiit

The district airport, Nerlerit Inaat (Constable Point), about 37 km northwest of Ittoqqortoormiit, was established in connection with oil exploration on Jameson Land in the early 1980s. In winter, it is often not possible to operate the helicopter service to the town due to drifting snow and darkness. Therefore, there has long

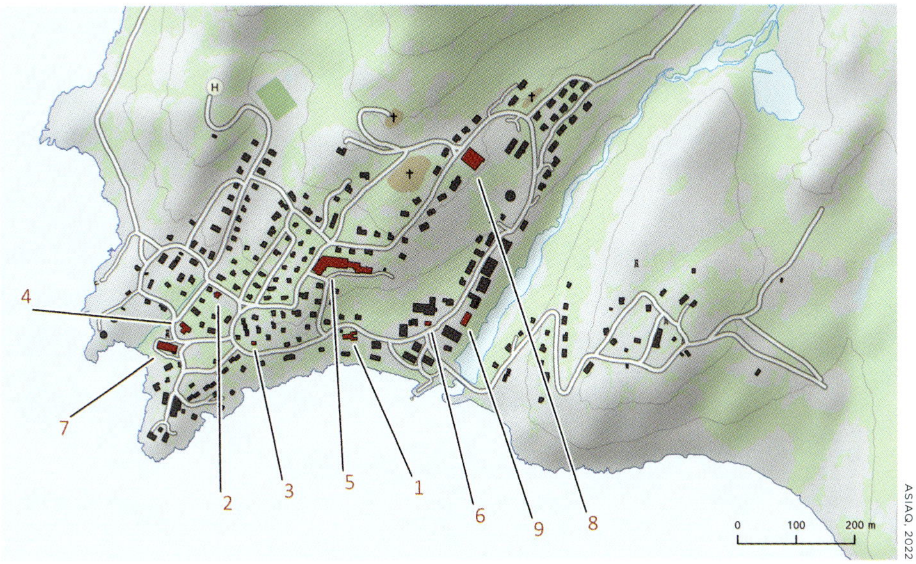

Ittoqqortoormiit, street map. 1) Ittoqqortoormiit Church. 2) Police station. 3) Fire station. 4) Municipal office. 5) Primary school. 6) Ittoqqortoormiit Museum. 7) Health centre. 8) Sports centre. 9) settlement hall.

been a local desire to establish an airport at Ittoqqortoormiit.

Air Greenland no longer serves Nerlerit Inaat, and Icelandair's former connection between Kulusuk and Nerlerit Inaat has also stopped. At the Self-Government's latest bidding round, the licence went to Icelandic Nordlandair, which has a weekly flight to Reykjavik and a weekly flight to Akureyri. This means that a trip to the rest of Greenland will be more expensive and that the traveller needs an overnight stay in Iceland – each way.

Royal Arctic Line is calling Ittoqqortoormiit with container ship twice annually, in July and September.

Communication with the outside world is via satellite and connections are substandard in terms of uptimes and bandwidth. For citizens and for business, it is a basic condition which means that connections via the Internet are slow and unreliable and, moreover, a costly affair for Tussas (formerly Tele Post Greenland).

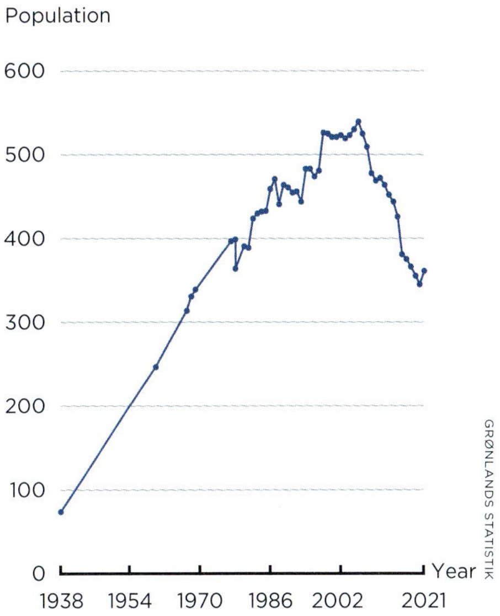

Figure 27. Population trends in Ittoqqortoormiit 1938-2021. See highlighted text

The town is supplied with electricity from a diesel plant. As the first step towards greener energy, Nukissiorfiit established solar cells on one of their buildings in 2021, and they are expected to reduce diesel consumption by up to 22,000 litres per year. The town

HELENA LARSSON/TT/RITZAU SCANPIX, 2016

Ittoqqortoormiit viewed from the sea. Far north, behind the mountains, is Northeast Greenland. The drift ice typically comes from the Greenland National Park, which the Ittoqqortoormiit district borders. However, the town does not act as a gateway to the national park.

Figure 28. Distribution of jobs by industry sector for people working in the town of Ittoqqortoormiit and Greenland as a whole, 2020.

is supplied with water from a nearby lake as well as tank for storing winter water. The water network is modest, but supplies essential municipal buildings and the hospital, as well as individual dwellings. The same buildings have water closet and septic tank. The other citizens get water from one of the two tapping houses or have it delivered. Grey water is discharged to the ground, and the town has no sewer system, so most people use toilet buckets and collection of latrines.

Most of the road network is gravelled, but as of 2017, a project began with gradual establishment of concrete paving on primary roads.

The town has all the most basic functions sized to the needs of the town – that is, small. The fire and police station as well as the municipal office are all about the size of a sin-

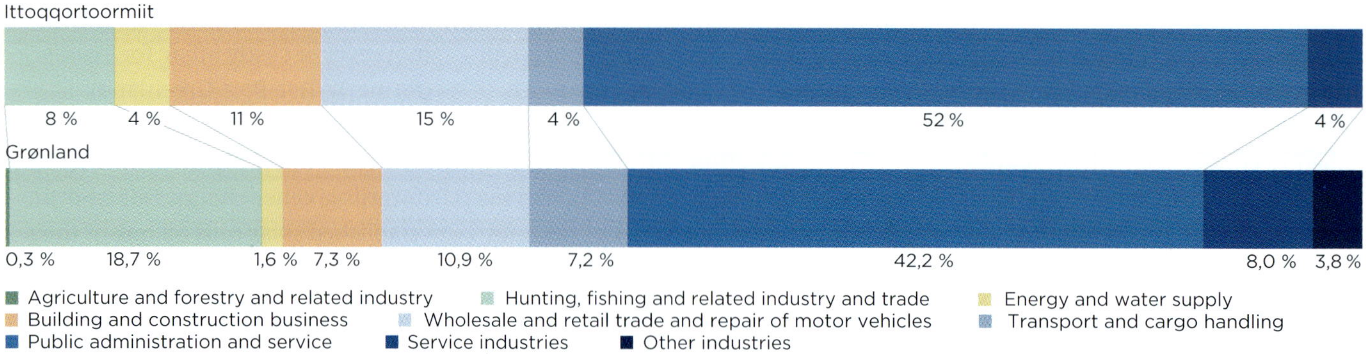

GRØNLANDS STATISTIK

gle-family house, and the sports hall is effectively a gym hall. The health centre has four to five employees, including two nurses who report to the regional management. There is no permanent doctor in town.

There is a very popular artificial turf football field, a small ski lift and a cross-country ski run close to the town.

There is a road to Qinngaajiva (Walrus Bay) to the northwest of the town, where the wide sandy beach with mountain and fjord views is frequently used. During World War II, the US Air Force had a weather station here.

Trade and industry in Ittoqqortoormiit

The local population has traditionally lived off hunting of marine and terrestrial mammals, as well as birds supplemented by fishing. Game and fish are directly included in the maintenance of hunting families, and there is a significant degree of subsistence economy. Trading options are limited, just as the ice cover limits the potential for commercial exploitation of the large amounts of shrimp and Greenland halibut in the sea outside the town. Kommuneqarfik Sermersooq is working to reopen a muskox slaughterhouse with purchase and sales to the rest of the country.

Tourism is a developing industry and the district is open to nature and hunting tourism. The harsh and pristine nature just outside the town and in the vast hinterland holds potentials, but tourism development is challenged by flying conditions.

Paamiut

Paamiut's location at the mouth of the fjord Kuannersooq (Kvanefjord) has given the town its Greenlandic name meaning 'those who reside by the mouth'. Founded as a colony by merchant Jacob Severin in 1742, the town was given its Danish name, Frederikshåb, after then-Crown Prince Frederik – later King Frederik V.

History of Paamiut

When the colony was founded in 1742, the Inuit were already settled on the site as hunting opportunities were good, just as a natural harbour on the lee side is always a benefit to a settlement. During colonial times, whale and seal hunting were the primary livelihood of the town. Paamiut and the settlement of Arsuk formed an independent municipality until the municipal amalgamation in 2009, when Paamiut became part of Kommuneqarfik Sermersooq.

The Gulf Stream ensures that Paamiut is an open-water town that can be reached by boat all year round. During some periods, boat traffic can be impeded by the storis (great ice) coming from the south, from the east coast of Greenland.

On the entrance to Paamiut lies an old shipwreck, the trawler Greenland Star, which ran aground in 1984. A contributing cause of the grounding was the fog for which the town is known. The frequent fog is, among other things, caused by Sioraq (Frederikshåb Isblink) north of the town. It is one of the few glaciers in West Greenland that extends all the way to the coast, where it spreads out in about 25 km width.

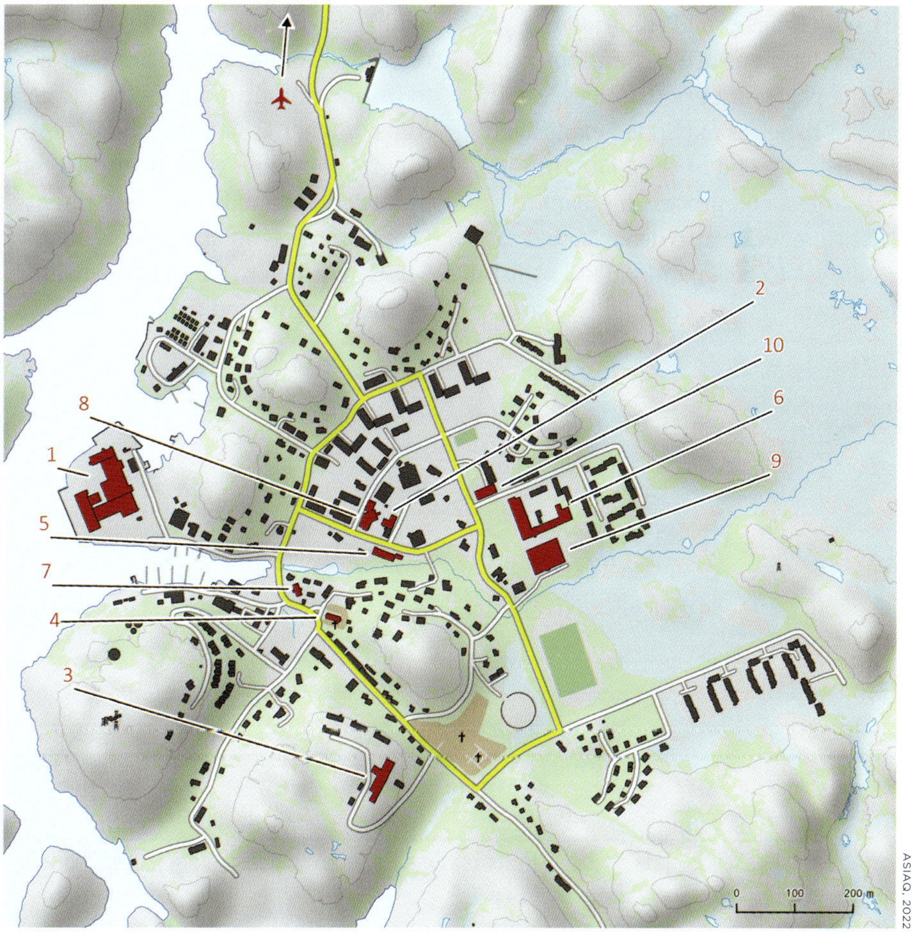

Paamiut, street map. 1) Fish factory. 2) Hotel Paamiut. 3) Greenland Maritime Centre. 4) Fredens Kirke 5) Municipal office. 6) Primary school. 7) Paamiut Museum. 8) Health centre. 9) Sports centre. 10) settlement hall.

ANINGAAQ R. CARLSEN/VISIT GREENLAND, 2019

46 registered commercial hunters in Paamiut (2021) trade in crab, cod and roe with Royal Greenland. A glimpse of their facilities can be seen at the bottom left.

White-tailed eagles are frequent in the area and have lent their name to the town's sports association Nattoralik which has proud traditions.

Development of Paamiut took off in the post-war period, and especially as cod fisheries expanded explosively towards the end of the 1950s. In 1967, the largest cod processing plant at that time was constructed here; Paamiut was to be Greenland's No 1 fishing town.

Meanwhile, the Greenland Technical Organisation (GTO) planned the development of Greenland from Copenhagen, including the urban development of Paamiut. The town was to be one of four main towns in the country where the population was to congregate rather than live scattered in settlements and small settlements. Based on the G-50 and later G-60 development plans (see p. 133) as well as the booming cod fishery, the GTO planned and executed a major expansion of Paamiut. The objective was a town of up to 10,000 inhabitants, Greenland's second largest. Thus, Paamiut was the first town in Greenland to have a full sewer system, and the basic infrastructure with electricity, water and heat supply, roads, housing, school, sports hall and other institutions was developed with this objective in mind. The sports hall in Paamiut was Greenland's largest until Inussivik in Nuuk was inaugurated in 2002. Much of the urban development was on former plains and wetlands, which the GTO was adept at draining and developing. The GTO's plans to move people from settlements to selected towns and into large blocks of flats have since met much resistance in Greenland – and in Paamiut.

When the cod disappeared from Greenland waters, it hit hard. The settlements of Narsalik and Avigait, south and north of Paamiut, respectively, closed and the town's development stagnated. Its population peaked in 1988 at 2,433 inhabitants, which was just under 1,000 more than the 1,247 today and thus far short of the 10,000 that was the Ministry of Greenland's target. Today, the existing building stock and infrastructure are maintained, while worn-out and vacated blocks of flats are demolished.

Infrastructure, trade and industry in Paamiut

The port is the focal point of the town. This is where supplies come in and out, and this is where the fish

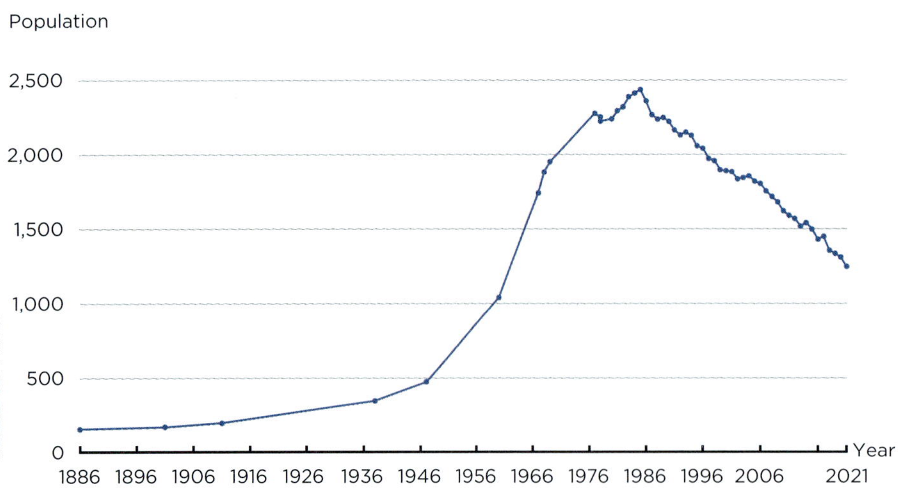

GRØNLANDS STATISTIK

Figure 29. Population trends in Paamiut 1886-2021

Paamiut viewed from the northwest. In the middle of the photo is the school, Atuarfik Tuiisaq, built in 1972, which goes up to the 10th grade. To the left of the school is the yellow sports hall, often the venue for the town's cultural events.

ANINNGAAQ R CARLSEN, VISIT GREENLAND, 2019

factory is located. The port, which is large and has good depth, handles cargo, fishing vessels and passengers by boat or ship and also disposes of a marina with several pontoon bridges. The coastal ship calls at Paamiut twice a week in northbound and southbound directions, respectively. Royal Arctic Line calls at the town weekly with freight and food.

Until 2007, when Paamiut got its own airport north of the town, it was served by helicopters via the heliport in the middle of the town. The airport is primarily served by Air Greenland with Dash 8 aircraft and fixed routes to Nuuk in the north and Narsarsuaq in the south.

The fish factory has survived times of hardship by making continuous adaptations, and is operated by Royal Greenland in 2021. The factory mainly processes crab, cod and lumpfish roe, all seasonal productions. In order to secure the 65 or so jobs throughout the year, Royal Greenland also started to produce Greenland halibut fillets in 2020.

In addition to the public workplaces, shops, hairdresser and fish factory, there are some small trade and contractor firms and a small hotel. The townspeople, the municipality and the Self-Government are working purposefully on commercial initiatives that can benefit and develop the town. The port and the airport play a key role. Most promising is the export of clean drinking water from the hinterland, while the idea of using the town's port as a transit port for international navigation and transhipment of cargo is probably some way off. With better traffic links and marketing, tourism can be developed and expanded.

The primary school in Paamiut is called Atuarfik Tuiisaq and the school's current buildings were inaugurated in 1972. A few years later, there were nearly 800 pupils – a contrast to the barely 200 pupils in 2021. The town's public library is found at the school.

The town houses a branch of the Imarsiornermik Ilinniarfik (Greenland Maritime Center), which provides basic training and courses in the maritime and fishing field. School uptake is declining.

Figure 30. Distribution of jobs by industry sector for people working in the town of Paamiut and Greenland as a whole, 2020.

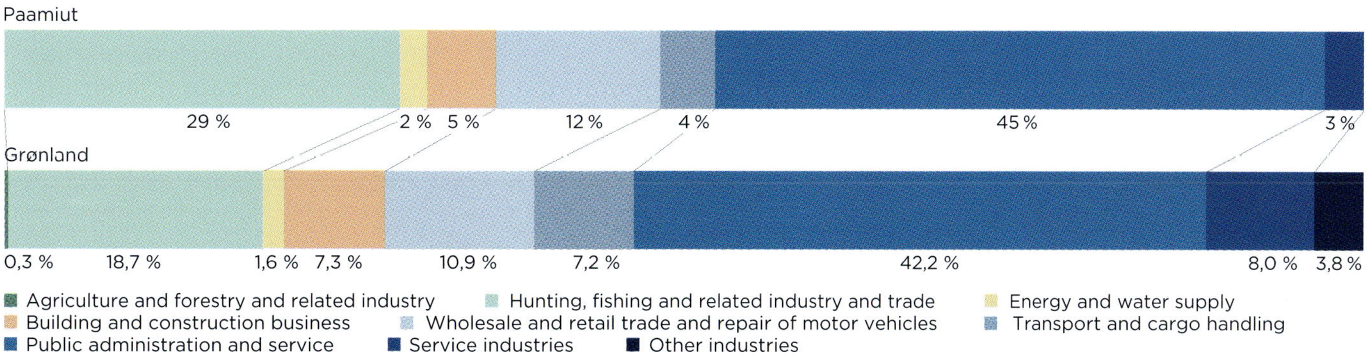

GRØNLANDS STATISTIK

NUUK

Nuuk means 'headland' in Greenlandic and describes the place where the town was founded in 1728. The town is located at the mouth of the extensive fjord system Nuup Kangerlua (Godthåbsfjorden) on the southwest coast, about 250 km south of the Arctic Circle. ILL: 53+54+55

Nuuk is the main town in Kommuneqarfik Sermersooq, Greenland's capital and seat of the government, where Inatsisartut and Naalakkersuisut (parliament and government) work. With a population of 19,261, one-third of the Greenland population lives in the town (2022).

There are two settlements in Nuuk district: Kapisillit further into Nuup Kangerlua and Qeqertarsuatsiaat about 130 km to the south. They have a total population of 223. Opposite Kapisillit lies the sheep farm Neriunaq in an area with many leisure huts. The abandoned settlement of Qoornoq midway between Nuuk and Kapisillit is a popular sojourn, especially during the recreational sailing season.

Nuuk overview map- 1) Herrnhut Mission House 2) High School (GUX) 3) Port Administration 4) Godthåbhallen 5) Emergency services 6) Nuuk Art Museum 7) University and Greenland Institute of Natural Resources 8) Ukaliusaq School 9) Samuel Kleinschmidt School 10) Business College 11) National Theatre of Greenland 13) Malik swimming baths15) Queen Ingrid's Hospital 16) Institution for convicts 17) Greenland Court 18) Hans Egede Church.

ASIAQ, 2022

HISTORY OF NUUK

Long before the town was founded, the mild summer climate at the head of the fjord system attracted the Norse, who have left multiple traces of settlements that are collectively referred to as Vesterbygden. Similarly, with changing settlements, the Inuit have for centuries enjoyed the optimum hunting opportunities in the fjord system.

The missionary Hans Egede arrived in the area in 1721 and settled on Kangeq (Håbets Ø) at the entrance to Nuup Kangerlua (Godthåbsfjorden). This was not an optimum location and was therefore abandoned in favour of the Godthåb colony in 1728. Håbets Ø and Godthåb are named after the ship *Haabet*, which carried Hans Egede and his entourage (see p. 113). 2021 marked the 300 anniversary of Greenland being colonised and the introduction of Christianity, which sprung from Godthåb.

The Godthåb colony was established in Nuutoqaq (Kolonihavnen) at a natural harbour on the western side of the headland, Nuuk. Some of the

Morning in Qaqortoq/Myggedalen. The area consists mainly of tightly positioned GTO standard houses decorated in different shades and with gables facing the water.

REBECCA GUSTAFSSON/VISIT GREENLAND, 2016

Central Nuuk. 1) Hans Egedes House. 2) Cooper's workshop (historical). 3) National Museum of Greenland. 4) Teachers' Training College (historical; now GUX-adm.). 5) Church of Our Saviour. 6) Inspector's residence (historical). 7) Qivittoq house. 8) Doctor's residence (historical). 9) The American Consulate (historical). 10) Nuuk Local Museum. 11) The Board. 12) Hospital (historical). 13) Nuuk Centre. 14) Naalakkersuisut administration. 15) The Landsting. 16) Inussivik sports hall. 17) Greenland Maritime Education Centre. 18) Town hall. 19) The School of Iron and Metal (KTI). 20) Niels Lynge's House 21) Katuaq Cultural Centre.

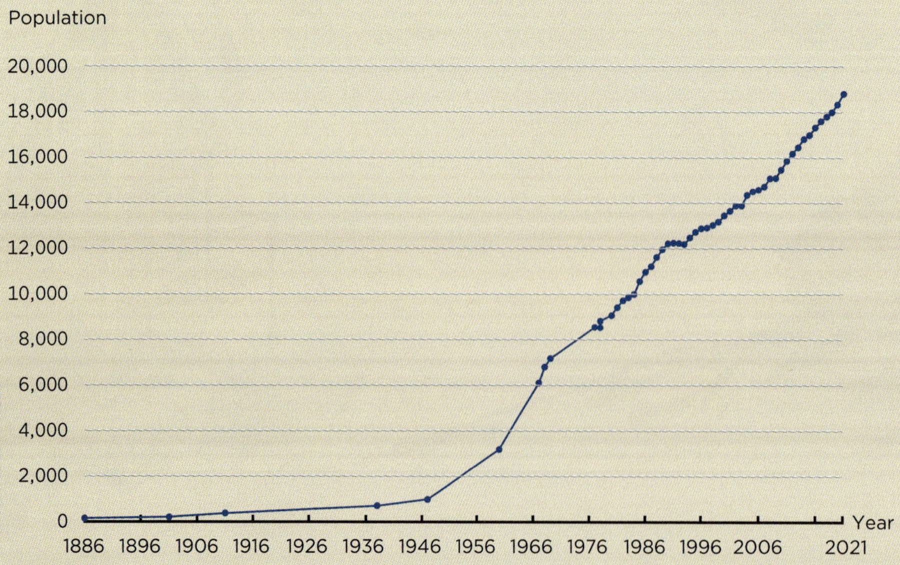

Figure 31. Population development in Nuuk 1886-2021.

buildings and facilities built by the mission and trade in the 1700s and up to the early 1900s still exist today.

The oldest buildings in Nuuk

Hans Egede's stone house from 1728 is Greenland's oldest preserved building and now serves as the reception rooms of the Government of Greenland. Next to the house lie the old goat barn and carpentry workshop (built in the 1890s). On the other side of the path lies the original cooper's workshop built in 1758, later a bakery and painting workshop and today a workshop for the Greenland National Museum (see p. 166), which in close vicinity and since 1978 has been housed in the original warehouses from 1929-30 that was expanded in 1992.

At a suitable distance from the colony, the Moravian mission built the mission house, which was completed already in 1747 and is now used by the Greenland Ombudsman Institution. It was also here that Ilisimatusarfik (University of Greenland) started in 1984.

Ilinniarfissuaq (the Teachers' Training College) was founded in 1847 and is the oldest educational institution in Greenland. The college building on the outskirts of Nuutoqaq is from 1907 and its Nordic-inspired architecture was included in the former coat of arms of Nuuk Municipality. The building now serves as an administration building for the secondary education programme GUX. In 2008, the teachers' training college became an institute under – and moved to – Ilisimatusarfik, University of Greenland.

Annaassisitta Oqaluffia (Church of Our Saviour) is Greenland's cathedral. The church was built in 1848-49,

KARSTEN SCHNACK/BIOFOTO/RITZAU SCANPIX, 2004

In the foreground, the old teachers' training college building, which now serves as the administration of GUX Nuuk. In the background is the cathedral and the statue of Hans Egede.

and the church tower was added in 1928. Next to church lies the old rural deanery from 1847.

In Nuutoqaq, there are a number of buildings from the 1800s, e.g. the inspector's house and the qivittoq house (both from 1831), the provisions stall (1850), another cooper's workshop (1887) and a crew house from 1848. Further south lies the current official residence of the Premier, originally built in 1906 and used as a doctor's residence. That is why it is yellow, which is the colour of the health care system.

The government seat of Nuuk

In 1911, Greenland had two elected provincial councils, one for North Greenland seated in Qeqertarsuaq (Godhavn) and one for South Greenland seated in Nuuk (Godthåb), which served as advisers to the Danish State. During World War II, the administration was gathered in Nuuk under the leadership of the chief administrative officer for North Greenland, Eske Brun, while the chief administrative officer for South Greenland, Aksel Svane, was in charge of Greenland's interests in Washington, and in 1940, the United States established a consulate in Nuuk, which was active until 1953.

In connection with the Greenland Commission's major modernisation plan for Greenland, G-50, it was considered to move the capital to another open-water town because the commission found that the commercial opportunities in Nuuk were poor. However, when the two provincial councils were formally merged into one in 1950, it was seated in Nuuk, which was given the status as capital at the same time. The merged Provincial Council served until the introduction of Home Rule in 1979 and played a significant role in the modernisation of the country.

Naalakkersuisut has many employees in the central administration at addresses around the town, but primarily in two building complexes close to Nuutoqaq (Kolonihavnen):

The tower in Nuuk Centre from 2012 and the original administration building built in stages in the period 1966-86 next to it. The latter also houses the chamber of the Inatsisartut (Parliament of Greenland) which has 31 members.

The Greenland office of the organisation for indigenous peoples of the Arctic, Inuit Circumpolar Council (ICCs), is located in Nuuk.

URBAN DEVELOPMENT AND INFRASTRUCTURE

Nuuk developed slowly from the time of Hans Egede until around 1950; in 1947, the colony had a population of 789.

At that time, the town consisted of Nuutoqaq (Kolonihavnen) with the Teachers' Training College and some small houses as well as a road built in the 1930s to the new port on the opposite side of the headland. The port administration was housed in the now listed building at the far end of the quay. Buildings were gradually built for administrative, commercial purposes as well as housing along Aqqusinersuaq (Skibshavnsvejen), but the housing remained very scattered.

1950 to 1980

In an effort to gather the population in fewer settlements, the district of Qaqortoq (Myggedalen) was established by the Greenland Technical Organization (GTO) during the 1950s, among other things to attract hunters from the smaller settlements so that they could be part of the fishing or work in the town. The area consists of single and semi-detached houses, part of which have been updated to modern standards.

The heliport at the current football field and Inussivik was established in the early 1960s and was in use until the town got its own airport in 1979.

The Qorsussuaq school from 1952 was supplemented by two new primary schools, Atuarfik Samuel Kleinschmidt (1967) and Ukaliusaq (1969). Qorsussuaq closed in 2011 and now serves as Greenland's Maritime Education Centre (Imarsiornermik Ilinniarfik).

The first sports hall with a full-size handball court, Godthåbhallen, opened to Nuuk's citizens in 1969. The population of Nuuk reached 7,166 in the same year.

On the other side of the Quassunnguaq ridge, which bounds the town to the east, the 400-rtalik road was established, initially for the town's new power station (built in 1975) and a commercial area, both in Sarfaannguit.

The tunnel under Nuussuaq, the large headland, was built at the same time to provide access to the town's then waste landfill on the other side where the incineration plant is located today. The tunnel is, then as now, one-lane, although the town's road users have long needed two lanes.

The blocks Q, R, S and T on Tuapannguit, 1-10 and P on Narsarsuaq (Sletten) and 11-16 on Qapiarfiusaaq (Radiofjeldet) were built by GTO during an intense construction period from 1966 to 1974. Block 17 completed the construction of terraced houses and housing blocks on Qapiarfiusaaq some years later.

In the period up to 1979, the town on the headland had gradually become fully developed with schools, housing and a new urban centre around the town's new town hall from 1980.

In 1979, Nuuk Airport opened in the area along Quassussuaq (Lille Malene), which at that time was very far away from the town itself. The long road to the airport opened up for new urban development, which began in Nuussuaq, where the first residents moved into Ole Brandt-ip Aqqutaa in 1979.

In 1980, the population of Nuuk was 9,077 – almost 20 % of Greenland's total population.

1980 to today

The districts of Eqalugalinnguit and Paarnat followed in the late 1980s, Siaqinneq and Kangillinnguit in the 1990s and Qernertunnguit in the 00s, and most recently Issortarfimmut and Akunnerit. Nuussuaq and the mentioned areas were fully developed around 2010. Urban development in Nuuk was – and still is – primarily driven by the lack of housing.

As a precursor to the new university buildings, Pinngortitaleriffik (Greenland Institute of Natural Resources) moved into their new building in 1998, built in two sections connected by a glass canopy, which, seen from the inside, looks like a glacier crevasse. Ilisimatusarfik (University of Greenland) was added in 2007. Both are still under expansion.

Around the turn of the millennium, a road had been established and prepared for urban development in Qinngorput at the foot of the mountain Ukkusissaq (Store Malene). The area appears fully developed with supermarkets, marina, school, children's institutions, restaurants and housing.

The next phase of urban development in Nuuk will take place south of Qinngorput on the headland below Ukkusissaq (Store Malene). The area is named Siorasiorfik and is expected to accommodate housing, public services and small businesses for 2,000-3,000 citizens.

The old town centre is also undergoing massive urban development. Old and worn-out buildings, such as Block P, Q, R, S and Block 1, homes on

HANNIBAL HANSCHKE/REUTERS/RITZAU SCANPIX, 2021

Port of Nuuk. In the background are some of the multi-storey housing blocks that have been built in recent years as part of the urban development. The multi-storey housing blocks have partly replaced some of the block construction of the 1960s.

Ceresvej and worn-out industrial and construction buildings are being renovated and make room for newbuilding. Areas are included on mountain ridges and along the abandoned water reservoir. New homes are being built, typically in the form of towers, but large and small companies are also building new domiciles. The country's largest primary school is expected to open in 2024 on the site where Block P, once Greenland's largest housing block (see p. 196), used to be.

The port in Nuuk is the largest in Greenland and was expanded in 2017 with a separate container port at Qeqertaq Avalleq (Fyrø) on the eastern side of the harbour basin.

A major expansion of Nuuk Airport, including to allow international flights, has been initiated and is expected to be commissioned in 2024. Hotels and especially conference facilities and competences take up a lot of space in the capital, and the industry expects and prepares for increased activity when the expanded airport opens in 2024.

Nukissiorfiit supplies electricity, water and heat to the capital. Since 1991, Nuuk has received its power supply from the hydropower plant at the head of Utoqqarmiut Kangerluarsunnguat (Buksefjorden) about 50 km southeast of the town, and the diesel-powered power station at Sarfaannguit (Nordhavnen) today functions as an emergency power station. With the steadily increasing population, both hydropower plants and emergency supply need to be continuously expanded to ensure power to the town's citizens and businesses.

Nuuk gets its drinking water from Qallusuaq (Circussøen) behind Quassussuaq (Lille Malene). The water lake was last dammed in 2010 to increase the volume and to be able to meet the need.

Virtually the entire town is connected to a sewer system, which handles all kinds of wastewater that is

PETER LINDSTROM/VISIT GREENLAND, 2019

The new coastal path, started in 2018, marks the interaction between town and nature and offers the opportunity for beautiful hikes.

discharged untreated into the fjord. Some households have sludge tanks and even fewer toilet buckets.

The town's heat supply is a combination of district heating, electric heating as well as decentralised or individual oil-fired boilers and heating boilers.

The town's communication lines to the outside world were significantly improved when the submarine cable connecting to Qaqortoq, Canada and Iceland opened in 2008. The same submarine cable opened up for new business opportunities in the IT industry. In Nuuk, there are 2-3 large IT companies with national and international customers and an undergrowth of small companies.

Nuup Bussii is Greenland's largest urban bus company and has 4-5 routes that daily take passengers around the town from early morning to midnight. Nuuk also has two large taxi companies.

NATURE AND GEOGRAPHY

With few exceptions, room has only been made for nature in areas with steep mountains, lakes and rivers. A cross-country skiing piste is available and a coastal trail was started in 2018. The trail is planned to go all the way around the town's coastline, which marks the town's link to the fjord system.

The Nuup Kangerlua (Godthåbsfjorden) fjord system is one of the largest in the world at about 9,000 km², or about the combined size of Zealand, Møn, Lolland and Falster in Denmark. It offers ample opportunities for hunting and fishing for both vocational hunters and spare time hunters as well as magnificent nature experiences with waterways between the large islands, countless promontory, high and steep mountains and the ice fjord at its head. This largely explains why there are well over a thousand recreational boats and two marinas in Nuuk.

Nuup Kangerlua also offers cultural experiences: remnants from Vesterbygden of the Norse, abandoned settlements that serve as school camps or holiday settlements, and a geology that offers a view so far back in time that it contributes to the understanding of the earliest history of Earth.

The town's landmark, the Sermitsiaq (Sadelø) mountain, is located near the edge of the fjord system and at 1,210 m, it can be seen from most of the town.

POPULATION TRENDS, CULTURE AND HEALTH

The town's population growth is expected to continue, but the municipality's target of 30,000 residents by 2030 is hardly realistic. However, the new airport is very likely to attract both tourists and businesses, and thus create more jobs in the town. The trend of people moving from settlements and small towns to the capital doesn't seem to be changing anytime soon.

Education

Nuuk is a large educational centre with students from all over the Greenland who come to get an education at GUX Nuuk (Central Greenland Upper Secondary School), a commercial education at Niuernermik Ilinniarfik, Commercial College, or a vocational training at Saviminilerinermik Ilinniarfik, Tech College Greenland, where you can train as a mechanic, electrician and blacksmith or in transport and logistics etc.

Imarsiornermik Ilinniarfik (Greenland Maritime Centre) offers training programmes in shipping, fishing and the hunting industries.

Peqqissaanermik Ilinniarfik, Centre for Health Education, offers training as a healthcare helper, healthcare or dental assistant, porter, dental hygienist and other specialisms.

At Ilisimatusarfik (University of Greenland) you can study Cultural & Social History, Social Sciences or Language, Literature & Media. The university also offers programmes in journalism, teaching, social services, nursing and theology. The number of PhD students and research is increasing.

The acting school, together with Nunatta Isiginnaartitsisarfia, National Theatre of Greenland (see p. 210), was started in 2011.

There are many dorms around the town, but rarely enough to meet the high demand.

There are five municipal primary schools and an international private independent school (NIF) in Nuuk.

Cultural institutions

In addition to the Greenland National Museum (see p. 166), the town has: Nuuk Art Museum, Nuuk Local Museum and Niels Lynges House, all run by Kommuneqarfik Sermersooq.

The art museum was established by Svend Junge in 2005 with the purpose of organising exhibitions and showcasing his large art collection to the public. The museum handed over to the municipality in 2007. The building itself is interesting since it has previously served as a church, clinic and many other functions.

Nuuk Local Museum (see p. 172) is located in Nuutoqaq (Kolonihavnen) in the town's old boatyard. The permanent exhibition focuses on telling about the period from the 1950s and onwards when Nuuk became Greenland's capital. The museum also offers changing exhibitions and events.

Niels Lynge's House (see p. 174) is a typical self-builder house from 1950 and belonged to the priest Niels Lynge, who was also a well-known painter and poet. The house stands as when Niels Lynge and his family lived in it, and it offers a wonderful look into a private home from the 1950s.

Children playing on a hilltop in Nuuk.

PETER LINDSTRØM/VISIT GREENLAND, 2019

One of Niels' sons was Hans Lynge, a highly acclaimed writer, sculptor and painter. He made the 13 tapestry friezes that decorate the council hall in Nuuk (1982-87) and has given his name to Atuarfik Hans Lynge, the primary school in Qinngorput.

In 1997, the Katuaq Cultural Centre opened in the capital, which organises exhibitions, concerts and theatre performances as well as conferences and other cultural events. For example, since 2017, Nuuk International Film Festival has held an annual festival that shows films from Arctic filmmakers, typically in Katuaq's great hall, which is a modern first-run cinema.

Nunatta Isiginnaartitsisarfia is an independent public institution. The theatre produces three or five performances each year.

Sports Associations

Nuuk has many sports associations and offers many opportunities to play sports both indoors and outdoors during summer and winter. There are cross-country skiing pistes, a ski lift on Quassussuaq (Lille Malene), a 9-hole golf course, a riding course with Icelandic horses, countless artificial turf pitches as well as skate parks and BMX tracks. The multi-sports hall, Inussivik, was built in connection with the Arctic Winter Games in 2002. The hall is Greenland's largest and was reused for the Arctic Winter Games in 2016 (see p. 176).

Health and healthcare

The national hospital, Queen Ingrid's Hospital, is located in Nuuk. During the tuberculosis epidemic in Greenland, a national tuberculosis sanatorium was established where the new national hospital is located today. The sanatorium was popularly called Sana – a name citizens of Nuuk continues to use for the hospital. The Greenland Diabetes Center is an integral of the hospital. A donation from Novo Nordisk has now secured an independent building for treatment and research in diabetes as well as consultancy in healthy lifestyles. Construction is expected to begin in 2023 where the

PETER LINDSTROM/VISIT GREENLAND, 2019

Nuuk's cultural centre Katuaq is inspired by undulating architecture and designed by Schmidt Hammer Lassen. Katuaq is the venue for cultural events, conferences, etc., and also serves as a café. In the background is the tall hall of residence belonging to the School of Iron and Metal.

The Malik swimming baths which opened in 2003, was built by KHR Architecture. The 3,000 m² large facility is beautifully located at the entrance to one of Nuuk's marinas in the district of Nuussuaq.

REBECCA GUSTAFSSON/VISIT GREENLAND, 2016

fire station is located today – not far from the hospital.

Nuuk has two retirement homes with a total of 76 places.

BUSINESS

The capital of Greenland, Nuuk, has many public-sector employees since it is home to the administration of the Government of Greenland and the municipality. For the same reason, many of the country's large companies, private as well as public, are headquartered in Nuuk.

Within transportation and handling of goods and passengers, we have Air Greenland, Royal Arctic Line, Arctic Umiaq Line (coastal passenger ships) and Sikuki Harbour (port operations in Nuuk), Visit Greenland (tourism strategy and marketing), Mittarfeqarfiit (airport operation) and Kalaallit Airports a/s (new airports etc.).

The important utility companies, Nukissiorfiit and Tusass (formerly Tele Post Greenland), also have their headquarters and development departments in Nuuk.

Mapping, research and monitoring of the physical environment throughout the country is handled and coordinated from Nuuk by Pinngortitaleriffik (Greenland Institute of Natural Resources) and Asiaq – Greenland Survey.

The country's two banks and many insurance companies, lawyers and accounting firms are headquartered in the Nuuk.

Nuuk Imeq, which supplies beer and water to all towns and settlements subject to the Greenlandic deposit return system, has run its business from Nuuk since the start of production in 1989.

Royal Greenland and Polar Seafood are both headquartered and have fish factories in connection with the port in Nuuk. In total, the fishing industry

Figure 32. Distribution of jobs by industry sector for people working in the town of Nuuk, the settlements in Nuuk district and Greenland as a whole, 2020.

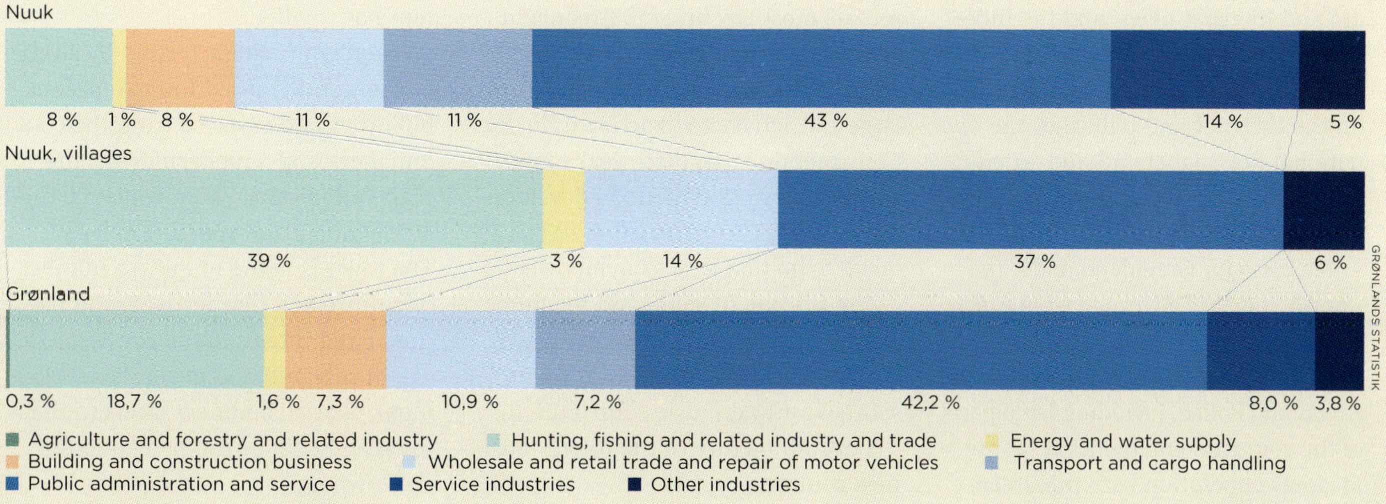

EMIL HELMS/RITZAU SCANPIX, 2021

Mural in Nuuk. There are also criticisms of the ongoing urban development and the rapid pace of construction, dominated by tower blocks, which are seen as a contrast to a romanticised picture of the former Nuuk.

employs more than 700 people in the town and has a turnover of many millions of Danish kroner.

A total of 65 people work with extraction of mineral resources, most of whom work in quarries that supply the construction industry, e.g. for the expansion of Nuuk Airport. Near the settlement of Qeqertarsuatsiaat south of Nuuk is a ruby mine run by Greenland Ruby A/S, which employs 10-20 local people.

Greenland has few nationwide media, and they all operate from Nuuk. This applies to the public service station KNR, which broadcasts TV, radio and Internet news, and the independent media house Sermitsiaq/AG, which produces and broadcasts news on the Internet and publishes the two national newspapers *AG* and *Sermitsiaq* (see p. 229) as well as the local paper *Nuuk Ugeavis*.

Key areas for Greenland under the Danish State are governed from Nuuk. This applies, for example, to the Joint Arctic Command (see p. 243) and the Greenland Police, including large parts of the country's emergency response in close cooperation with the Greenlandic authorities. Furthermore, the Danish Prison and Probation Service operates an open and a closed facility in Nuuk. The latter opened in 2019 in a newly built and top-secured building, which means that Greenland for the first time is able to provide secure detention for detainees, who previously would be detained in Denmark.

The Courts of Greenland are part of the Courts of Denmark and consist of four district courts, the Court of Greenland and the High Court of Greenland. Both the High Court, the Court of Greenland and the Sermersooq District Court are based in Nuuk. For the same reason, most law firms in Greenland are headquartered in Nuuk.

Since the 1950s, Nuuk has had a large and active construction industry. Construction takes place year round, see the description of urban development, and the industry often has so much to do that it lacks manpower.

A multitude of small and medium-sized enterprises and entrepreneurs in horticulture, fashion, boats, cars, sportswear and music contribute to the overall picture of the town's business community.

NUUK'S DIVERSITY

The town of Nuuk, its citizens and climate are characterised by their diversity. The Nuuk's citizens (called nuummiut) consist of Inuit, Danes (and a mix of the two), permanent newcomers especially from the Nordic countries and from Asian countries such as the Philippines and Thailand, but also from countries such as Germany, the United Kingdom, the United States, etc. In practice, the many different nationalities mean that English is the town's third main language, after Greenlandic and Danish.

The town's climate ranges from moderate gale and precipitation to clear skies and no wind on the same day. In February, the coldest month, the average temperature is -8.3°C and in July, the warmest month, it is 7.0 °C. Tourists are starting to discover the unique experiences that the small Arctic capital with its large fjord system has to offer.

The town is characterised by a mix of thousand-year-old Inuit traditions, buildings and customs from the colonial era and a modern urban life with restaurants, cafes, concerts, theatre, cinema and sporting events. The town is home to quite a number of culture and film festivals and national and international conferences, e.g. in research and business development, that all add new facets to the town and change the perception of the town.

QEQQATA KOMMUNIA

Qeqqata Kommunia is located between 64°N and 67°N, with the main town of Sisimiut located slightly north of the Arctic Circle. The area of the municipality is 115.500 km². From 66° and north on the west coast, sled dog keeping is allowed.

The municipality has a total of just under 10,000 inhabitants with Maniitsoq as the second town and a total of six settlements, of which Kangerlussuaq (Søndre Strømfjord) is the largest.

Kangerlussuaq is Greenland's most used international airport. It is located close to the ice sheet, 130 km east of Sisimiut.

Roughly in the middle of the north side of the Kangerlussuaq Fjord is an anorthosite mine operated by Hudson Greenland A/S. The mine was officially inaugurated at the end of 2019, but due to COVID-19, production has not really started. During production, approximately 60 employees are expected.

Since its establishment in 2009, the municipality has been led by Siumut, and in the most recent elections in 2021, the incumbent mayor was re-elected. Strengthening of social policy, promoting business development, tourism, health and prevention, and in particular education with a focus on primary education and young people outside the labour market are the key priorities of the municipality.

A large UNESCO World Heritage Site has been designated, Aasivissuit – Nipisat, covering a 20 km wide area of land, which extends 220 km from the ice sheet north of Kangerlussuaq Airport and beyond Sisimiut, but does not include the town itself. There are popular hiking routes between Sisimiut and Kangerlussuaq, and there are ongoing discussions about connecting the two with a road. Recently, ATV trails were established, which will soon be ready for use.

A road link would extend the existing road network around Kangerlussuaq that was established by the Americans, and in 1999, it was extended all the way up to the ice sheet by Volkswagen (VW) in connection with test drives on the ice sheet. That would be the first a long road link to be established in Greenland. It will link the country's second-largest town with an international airport, bringing with it major development potentials in terms of tourism and exports of superfresh fish.

A boat heads out of the long winding Kangerlussuatsiaq, the Eternity Fjord, with its steep mountain sides and many glacier outlets. The mouth of Kangerlussuatsiaq is at the settlement of Kangaamiut in the northern Maniitsoq district.

ANINGAAQ R. CARLSEN/VISIT GREENLAND, 2021

Maniitsoq

Maniitsoq, with 2,486 inhabitants, is located on an island in an archipelago area, and parts of the town are cut through by small canals where islands are connected by bridges or by rocks and earth. The three settlements of the district have 570 inhabitants.

Maniitsoq translates to 'the uneven' and refers to the topography of the town. One mountain hillock follows the next, so the town is characterised by many long stairways and walkways. In the hinterland, the town is bounded by high mountains. The town's former Danish name, Sukkertoppen, is derived from the name given by Dutch whalers to the colony's original location.

History of Maniitsoq

The Sukkertoppen trading post and colony was originally founded in 1755 by the Norwegian merchant Anders Olsen (see p. 118) in the place, where the settlement of Kangaamiut is located today, but it was moved in 1782 to its current location on Maniitsoq Island because it had better whale hunting opportunities. Incidentally, it was the same Anders Olsen who founded the trading posts Fiskenæsset and Julianehåb (today: Qaqortoq). Merchant Jens Larsen Smidt proposed the relocation in 1779, thus founding Maniitsoq in the place, where the town is located today.

In connection with the Danish modernisation of Greenland after World War II (see p. 128), Maniitsoq (Sukkertoppen) was designated as one of the prioritised four open-water towns. And because the town was well located relative to the important fishing grounds, including the Store Hellefiskebanke, the Greenland Commission considered in its major G-50 plan to relocate the administrative centre of the country from Godthåb (Nuuk) to Maniitsoq. However, the idea of relocation was abandoned. Yet, in the subsequent G-60 plan, Maniitsoq was one of the key priorities of the Greenland Technical Organisation (GTO). A vigorous expansion of the port and associated fishing industry, as well as the housing to house the growing population, was carried out.

ANINGAAQ R. CARLSEN/VISIT GREENLAND, 2021

Maniitsoq. On top of the mountain you can see a radio link station that connects the west coast. Today, Maniitsoq also has submarine cable.

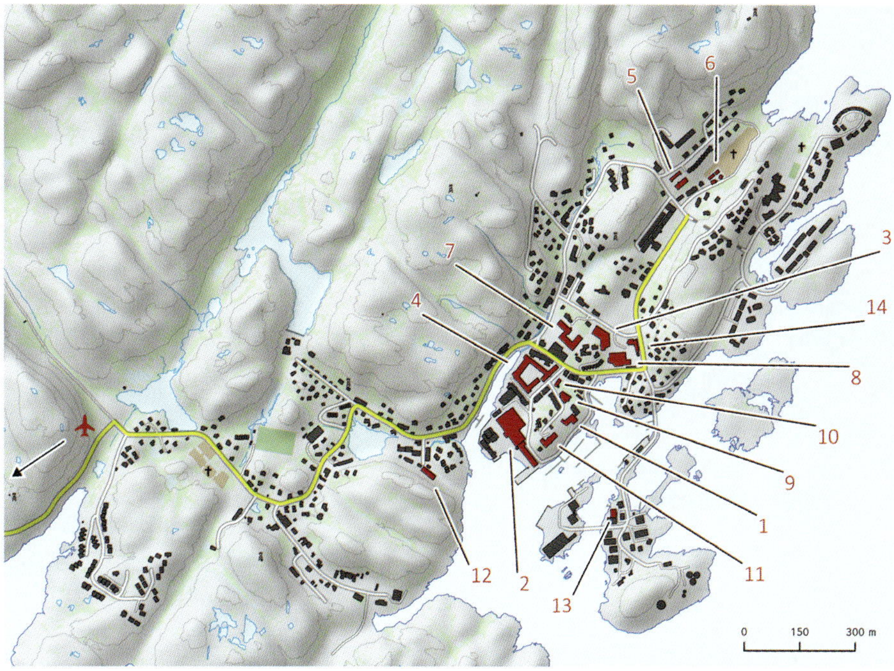

Maniitsoq, street map. 1) Maniitsoq Church. 2) Fish factory 1. 3) Municipal office. 4) Primary school. 5) Continuation school. 6) Maniitsoq Museum. 7) Health centre. 8) Sports centre. 9) settlement hall. 10) Drop-in centre (old church). 11) Hotel Maniitsoq. 12) Shipyard. 13) Fish factory 2. 14) Hotel Heilmann Lyberth.

Good fishing of Atlantic wolffish, cod and salmon contributed to the economic development of the town, but from the late 1980s the cod disappeared and it started an economic downturn that led to gradual migration away from the area. From 1989 to 2010, the Aalisakkanik Tunisassiornermut Ilinniarfik (ATI) fish processing school was located in Maniitsoq.

Infrastructure, culture and health in Maniitsoq

Due to the hilly terrain, the blocks of flats from the 1960s and 1970s are more scattered than in other Greenlandic towns, and the town is characterised by a relatively high proportion of multi-storey housing block and terraced houses.

Maniitsoq Airport replaced the old heliport in 1999. There are several weekly flight connections to Nuuk, Sisimiut and Kangerlussuaq and from there to the rest of the country and abroad.

The town has good connections by coastal passenger ships and private boat companies that call at the settlements, Nuuk and Sisimiut. As an open-water town, Maniitsoq is characterised by strong fishing and hunting traditions and a large and active marina that provides easy access to sailing trips in the fjords and to the many old summer settlements in the area.

Nukissiorfiit's diesel-based power plant supplies district heating to parts of the town, and drinking water is reclaimed from a contained lake on the northern edge of the town.

The western and eastern part of town as well as the town centre are fully sewered, but the area around the football field is not connected to the sewer network.

Maniitsoq has a health centre which falls under the Sisimiut Regional Hospital. The town's primary school seats approximately 440 pupils from 1st through 10th grades. In the oldest grades, the school has pupils from the municipality's four smallest settlements – they live in the boarding house for school children. Pupils in Kangaamiut and Kangerlussuaq finish school there.

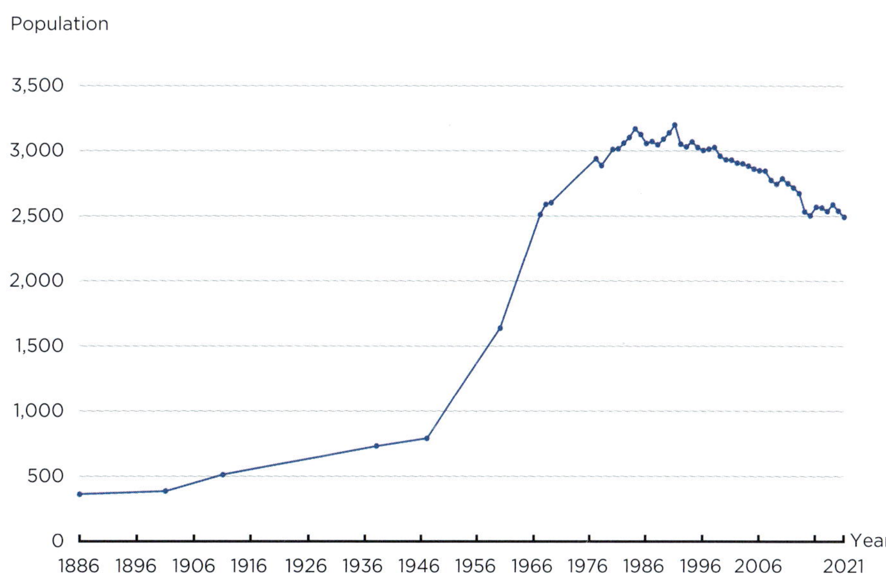

Figure 33. Population trends in Maniitsoq 1886-2021.

FILIP GIELDA/VISIT GREENLAND, 2019

The Maniitsoq Health Centre has an X-ray department, a laboratory and a pharmacy. All births and patients with severe cases are cared for in Nuuk.

Fishing and fish processing is an important industry in Maniitsoq, contributing 24 % of the employed. Royal Greenland imports and processes a variety of species and produces cod and Greenland halibut fillets, roe and dry fish. The capacity of the factory is 80 tonnes per day.

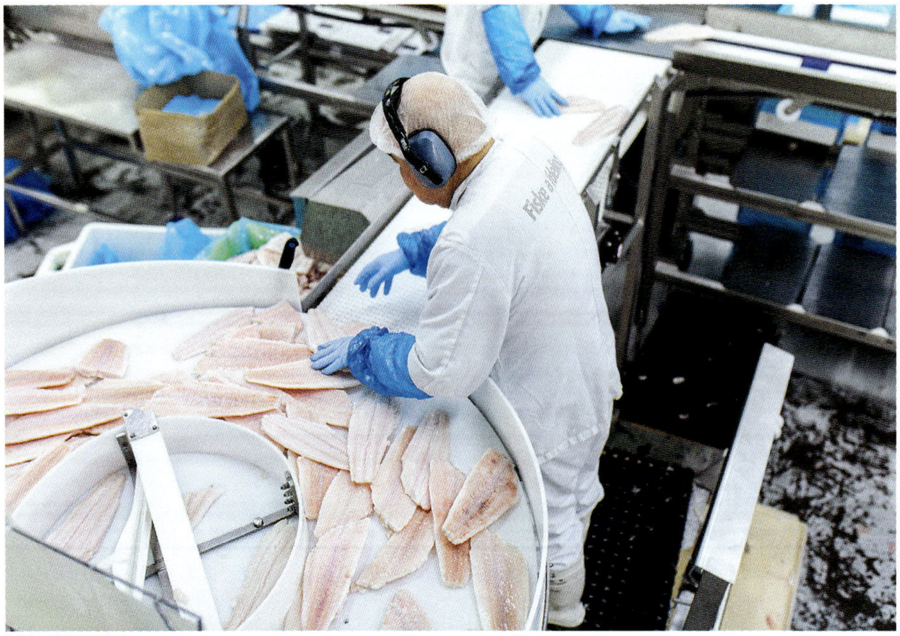

FILIP GIELDA/VISIT GREENLAND, 2019

A continuation school with an associated hall of residence was established in 2011 in the former ATI school. The continuation school can take up to 70 pupils from all over Greenland.

There are two hotels in Maniitsoq, both centrally located in the town. The largest, Hotel Maniitsoq, overlooks the harbour and strait with the many whales, one of Maniitsoq's tourist attractions.

The old church in the middle of the town centre dates back to 1864. It is erected in stone and has served as a municipal meeting place since the new church was finished in 1981.

The Maniitsoq Museum (see p. 168) is housed in some of the town's oldest buildings dating back to the mid-19th century. They were rebuilt on the site in the period 1970-1978 and listed in 1982. Originally, they served as a workshop building (erected in stone in about 1874), provisions stall (board-clad timber-framed erected in about 1850), assistant's residence (two-and-a-half-storey house built in 1844) and finally cooper's workshop (stone house bricked with boulders erected in about 1840-1950).

Population trends, trade and industry in Maniitsoq

In the inter-war period, Maniitsoq was one of the country's largest towns, and from 1950 to 2009 the town was the main town of Maniitsoq Municipality. In 2009, the town became part of Qeqqata Kommunia, and much of its administration has subsequently moved to the main town, Sisimiut. Population peaked in 1991 at 3,197. Since then it has declined steadily – in 2002 it was 2,905, and in 2010 the population was 2,784.

Northeast of Maniitsoq lie some of the country's greatest hydropower potentials, including Tasersiaq and Tarsartuup Tasersua. To exploit the energy potential, there was a plan in the 2000s to establish a larger aluminium smelter at Maniitsoq, but the plan was shelved in 2015. Other business opportunities are now being pursued – primarily in fisheries, tourism and minerals. The Maniitsoq area is rich in nickel, copper, rare earth elements, niobium, uranium, phosphate, gold and platinum. From 2005 to 2010, Minelco operated an ol-

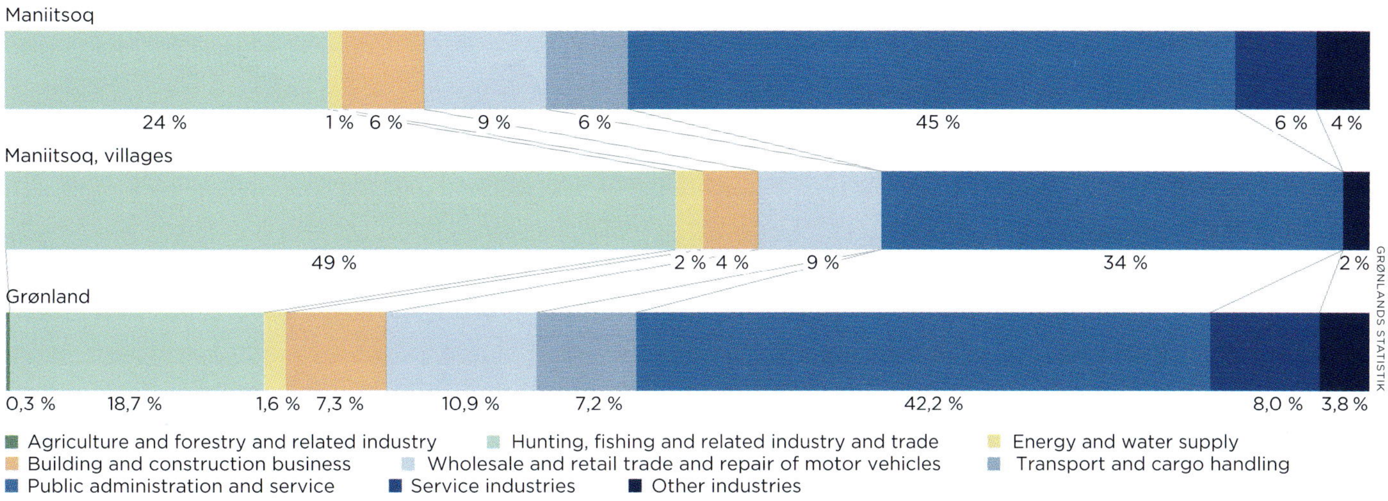

Figure 34. Distribution of jobs by industry sector for people working in the town of Maniitsoq, the settlements in Maniitsoq district and Greenland as a whole, 2020.

ivine mine in Niaqunngunaq (Fiskefjord). The mouth of the fjord is at the settlement of Atammik south of Maniitsoq.

East of Maniitsoq Island, in the vast ice-free land area with several deep fjords towards the ice sheet, there are ample opportunities for hiking and mountain walks in the exciting landscape originally shaped by glaciers. In winter, the area's ice caps, high mountains and deep valleys provide unique opportunities for cross-country skiing as well as alpine skiing. Heliskiing attracts ski enthusiasts from all over the world, but is mostly for those who do not worry about the price. Greenland's first and best summer ski resort is located on the Apussuit ice cap barely 30 km east of the town.

The two fish factories in the town has between 35 and 100 employees depending on the season. Royal Greenland produces cod, Greenland halibut, roe and dried fish, and Maniitsoq Fish ApS, owned by Polar Seafood & Sermersooq Seafood, produces cod, Greenland halibut, Atlantic wolffish, trout and roe. Local fishermen obtain higher minimum prices due to the competition of the factory.

In recent years, fishermen in the area from Sisimiut to Paamiut have fished Nutaaq cod, meaning 'new cod'. The fish is caught with pound nets, especially along the coast inside the fjords, and it is then moved to so-called net cages. The fish stays in the cage until their stomach contents disappear, typically after 10-14 days. With this, a white meat product of very fine quality is obtained. Subsequently, the fish are picked up by a well-boat and moved to the seawater pool outside the fish factory, where they calm down before being transported live into the factory for production. In 2020, Royal Greenland produced about 4,500 tonnes of Nutaaq cod, and Maniitsoq is the only place in the world where it is produced.

Sisimiut

Sisimiut is Greenland's second largest town and the northernmost, which is ice-free throughout the year, as well as the country's main fishing port. The town is located 42 km north of the Arctic Circle at the small bay of Kangerluarsunnguaq (Ulkebugten), and the Nasaasaaq mountain (Kællingehætten) is the landmark of the town. Sisimiut means 'the people at the foxholes'.

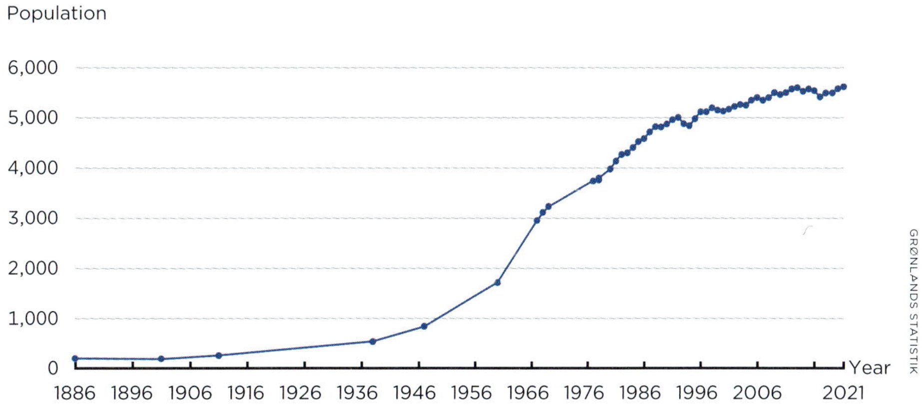

Figure 35. Population trends in Sisimiut 1886-2021.

ANINGAAQ R. CARLSEN/VISIT GREENLAND, 2020

Sisimiut with Nasaasaaq (Kællingehætten) in the background. Sisimiut is the main town in Greenland for fish and shrimp handling and processing, and it is also an essential educational town with, among other things, a high school and a construction training college, both of which are part of Kalaallit Nunaanni Teknikimik Ilinniarfik (KTI), as well as the Arctic Engineering School.

Its Danish name was Holsteinsborg named after Count Holstein, who was leader of the mission college which supported Hans Egede's mission and colonisation work in Greenland.

Sisimiut is the main town of Qeqqata Kommunia. Two smaller fishing settlements belongs to the Sisimiut district as well as Kangerlussuaq, the country's primary international airport. The three settlements of the district have 627 inhabitants and Sisimiut 5,622.

History of Sisimiut

Due to the living resources of the sea and the good hunting areas for terrestrial mammals along fjords and lakes all the way into the ice sheet, people have lived in the area from the earliest immigrations. During the period about 2,400-500 BC, the area was populated by the Saqqaq culture (see p. 90). After this and up to about year 0, the Dorset people lived there (see p. 92). Subsequently, the area was uninhabited until the Inuit came to the area around the 1300s (see p. 103).

At the time of Danish colonisation, there was great European whale hunting activity in the area. The first Danish whale hunting station to be constructed on the island of Nipisat was burned down by Dutch whalers. In 1738 and 1739, this led to armed conflicts between the Danish Navy and Dutch whalers. In 1756, the colony of Holsteinsborg was established on the island of Ukiivik (Sydbay) north of present-day Sisimiut, but due to poor hunting and trading opportunities, the settlement was moved to the current location of the town in 1764.

In connection with the establishment of the colony's port, a gradual industrialisation began. In 1914, the Royal Greenland Trading Department (KGH) opened a saltery for cod, which in 1924 was extended to the preservation of halibut. During the period 1935 to 1949, production was converted to shrimp, after which the shrimp disappeared from the nearby banks. In 1957, a new and larger salt fish plant was constructed, which in 1964 was redirected to frozen cod fillets supplemented by shrimp during periods with limited cod. And in 1962, a private shrimp factory was established, which also traded various fish as well as whale and caribou meat. In 1969, KGH founded another fish factory. In 1970, an actual Atlantic wharf was constructed in the town.

To support the growing fishing industry, Greenland's first shipyard was established in 1932 to repair large vessels and build new smaller vessels. The yard has since been expanded several times.

In 1941, the US Air Force established the airport Sondrestrom Air Base on the large plain at the head of the 180-km long fjord Kangerlussuaq (Søndre Strømfjord), which ends about 100 km south of Sisimiut. Since the mid-1950s, the airport has also been used for civilian traffic and has served as the primary international airport. A heliport was established in 1964 in Sisimiut to be able to get access to the airport at Kangerlussuaq.

In 1938, Sisimiut had a population of 541, and by 1947 it had increased to 841. The rapid population growth during the modernisation,

KÅRE HENDRIKSEN, 2013

Kalaallit Nunaanni Teknikimik Ilinniarfik (KTI), also called Tech College Greenland, with the School of Natural Resources in the background. The KTI campuses in Sisimiut and Nuuk have a total of about 650 students.

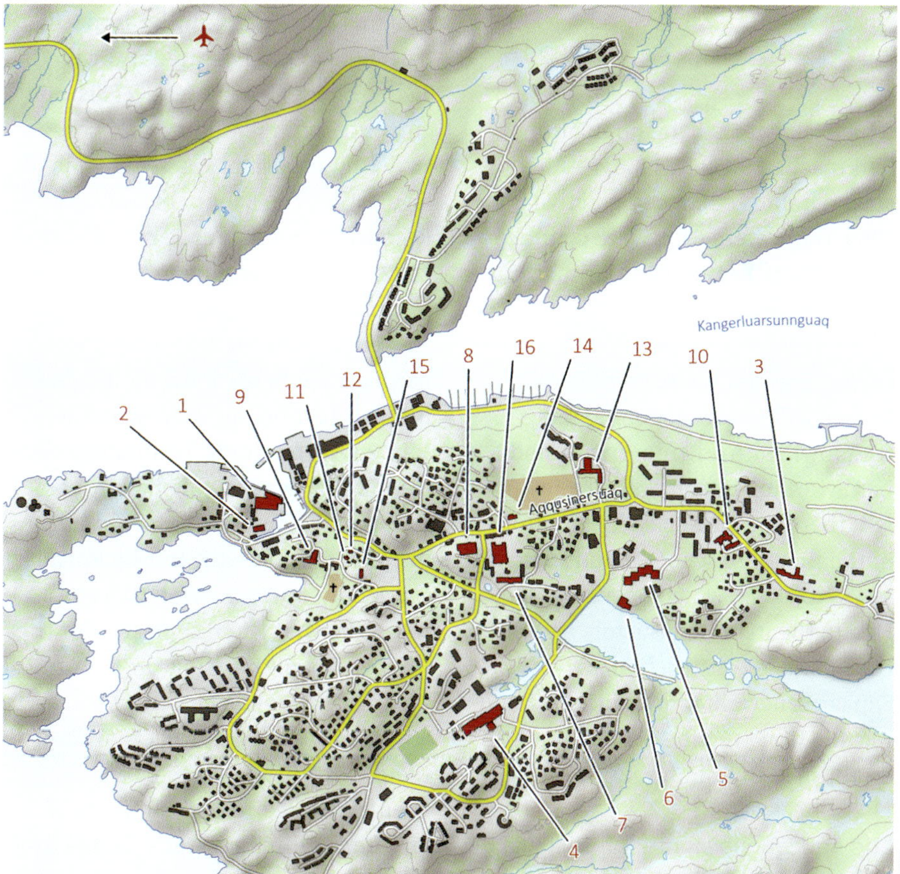

ASIAQ, 2022

Sisimiut, street map. 1) Fish factory. 2) Shipyard. 3) Knud Rasmussens Højskole. 4) KTI (Tech College). 5) and 16) Primary school. 6) Taseralik Culture Centre. 7) Town hall 8) Sports centre. 9) Hotel Søma. 10) Hotel Sisimiut. 11) Sisimiut Museum. 12) Old church (Museum). 13) Hospital 14) settlement hall. 15) Sisimiut Church.

ANINGAAQ R. CARLSEN/VISIT GREENLAND, 2020

The Taseralik Cultural Centre reflected in Nalunnguarfik (Spejdersøen). The building contains a theatre and cinema as well as a café, etc.

which meant that the population had reached 3,113 in 1968, resulted in considerable construction activity. In 1969, there were 31 private contractors with 140 employees, primarily from Denmark.

The strong focus on fishing and processing meant that Sisimiut became known as a well-functioning industrial town with an enterprising and hard-working population. The town was therefore chosen as the location of the first high school, Knud Rasmussen's Højskole, which opened in 1962. Due to the many contractors and the considerable construction activity, the Sanaartornermik Ilinniarfik, the Building and Construction School, was also established in Sisimiut in 1977.

The educational centre of Sisimiut

The establishment of Sanaartornermik Ilinniarfik was the beginning of Sisimiut becoming one of the country's most important educational centres. Since then, Sanaartornermik Ilinniarfik has expanded several times and has become the Kalaallit Nunaanni Teknikimik Ilinniarfik (KTI) educational centre, also called Tech College Greenland. KTI-Sisimiut currently offers building and construction programmes, technical college (GUX), the School of Natural Resources and language courses. In 2022, a marine engineer and technology management programme will be started in collaboration with the Copenhagen School of Marine Engineering and Technology Management.

In 2001, the Arctic Technology Centre, ARTEK, was established by the Greenland Home Rule Government in collaboration with the Technical University of Denmark (DTU) and KTI. Here, the first three semesters of the Arctic civil engineering programme are provided, and the centre serves as a base for research projects around the country. The centre has since been expanded to include parts of an international master's degree programme and a civil engineering programme in fisheries technology.

Sisimiut also has two primary schools.

The many education programmes leave their mark on the town in terms of businesses and culturally, and the town has a total of 400 dorm rooms.

Development and infrastructure of Sisimiut

Architecturally, Sisimiut is a mix of single-family houses, terraced houses and housing blocks erected as part of the Danish centralisation policy in the 1960s. The town is under development and on the other side of the narrow Kangerluarsunnguaq (Ulkebugten), by the road going to the airport, multi-storey housing blocks as well as terraced houses and single-family houses have been built over the past decade. Virtually all of the town's housing is connected to the water supply, and there is plenty of good drinking water in the town's catchment area. About 80 % of the town has a sewer system, while the remaining inhabitants use sludge tanks that are emptied by vacuum trucks, or toilet buckets and collection of latrines. In 2010, a hydropower plant opened at Sisimiut, and the former diesel-based power plant now operates as an emergency power plant.

Establishment of the domestic airport in 1998 on the north side of Kangerluarsunnguaq led to a significant improvement in passenger and, to a lesser extent, freight transport.

Sisimiut's main road, Aqqusinersuaq, goes from the harbour in the west and continues out to the open landscape to the east, where most of the town's sled dogs are located. Several of the town's shops, the two main supermarkets, hotel and folk high school are located along the road. Most of the town's road network is paved and paved trails have been established along several of the central roads and there is road lighting in most places.

In 2008, Greenland opened its second cultural centre, the Taseralik Cultural Centre, containing both a cinema and theatre hall.

Since 2002, the municipality has been working to establish a road link between Sisimiut and Kangerlussuaq in order to optimise transport to the international airport and to exploit the potential for tourism in the vast ice-free land area between Sisimiut and the ice sheet. External consultancy reports have documented the positive economic impact of the road despite heavy construction costs, and Qeqqata Kommunia has applied for permission from the Self-Government to establish the road, but the process has been protracted. As a first step, the municipality is currently establishing a trail for ATVs (small off-road vehicles) hoping that the trail can since be extended to an actual road.

WORLD HERITAGE: AASIVISSUIT - NIPISAT

The large land area between Sisimiut and Kangerlussuaq is the reason why Aasivissuit - Nipisat was entered on the UNESCO's World Heritage List in 2018. The 20-km wide land area, stretching 235 km from the ice sheet in the east to Davis Strait in the west, has been entered as a unique cultural landscape where the Inuit have hunted for 4,500 years.

Aasivissuit, located 20 km northwest of Kangerlussuaq, was the summer hunting settlement to where people from winter settlements in and around Nipisat 15 km south of Sisimiut went to hunt caribou. Dried caribou and trout were brought back as winter supplies to supplement the staples like seal, whale, fish and seabirds.

The area is entered on the UNESCO's World Heritage List both because of the archaeological findings and because of the existing hunter culture. This culture is important for the hunting profession, but also for ordinary Greenlanders who sail into the fjords in late summer and autumn to hunt caribou and catch trout (see p. 161).

Royal Arctic Line's container ships call at Sisimiut weekly throughout the year. The Arctic Umiaq Line make weekly calls with its coastal passenger ship going north and south, respectively, with Sisimiut being the northernmost town on the route during winter and only southbound connections. On behalf of Disko Line, the local Sirius Greenland company makes weekly passenger trips to the district's two fishing settlements, and Sirius Greenland offers package trips.

In 2013, Qeqqata Kommunia established an 86-m long port in Sisimiut, including a 10,000 m² storage and working area west of the Royal Greenland fish factory. The new port allows two large ships to be berthed, meaning that a large trawler or cruise ship can call at the same time as Royal Arctic Line's freighter is at the quay. Sisimiut is an important trawler port used primarily by Greenland trawlers and to a lesser extent by trawlers from other nations fishing in the Davis Strait.

In summer, Sisimiut receives about 40 cruise ships. Many disembark and embark passengers in Kangerlussuaq, after which Sisimiut is the first stop up the west coast, with some continuing to the East Canadian waters or all the way through the Northwest Passage to Alaska. Cruise ships need supplies and oil, which is why it is paramount that they can get to a port. Many studies show that tourists are more likely to disembark, go on tours of the town and buy souvenirs if their cruise ship get to the port – rather than if they have to be sailed ashore in small boats.

Trade and industry in Sisimiut

The town's main source of income is the fishing industry. The country's largest factory, owned by Royal Greenland, lands and processes cod and crab and employs about 100 people during the off-season and 150 during the peak season. Furthermore, Royal Greenland has a small factory for processing roe, which employs eight during the roe season. Fishing generates a number of related industries such as seine making and shipyards.

The total workforce in Sisimiut is over 2,700 people (2021). In terms of employment, the largest industry is public administration and service, including education and health care, accounting for 40 % of the town's jobs. Trawler fishing and coastal fishery as well as processing of fish and related industries contribute around 16 % of the jobs. Trade and auto repair account for 13 %, the transport sector 5 %, construction 11 % and other service industries 9 %.

Sisimiut has several private companies within production, construction and service. The government-owned and nationwide companies such as the housing company INI A/S and the utility company KNI A/S are both headquartered in the town. KNI has the following subsidiaries: the coun-

Figure 36. Distribution of jobs by industry sector for people working in the town of Sisimiut, the settlements in Sisimiut district and Greenland as a whole, 2020. Itilleq and Sarfannguit have a combined population of 164, while the airport settlement of Kangerlussuaq is home to 466 or 74 % of the district's settlements. The predominant employment in Kangerlussuaq is related to the airport's operation, so the industries 'Wholesale and retail trade and repair of motor vehicles' and 'Transport and cargo handling' will be disproportionately large for the district's settlements.

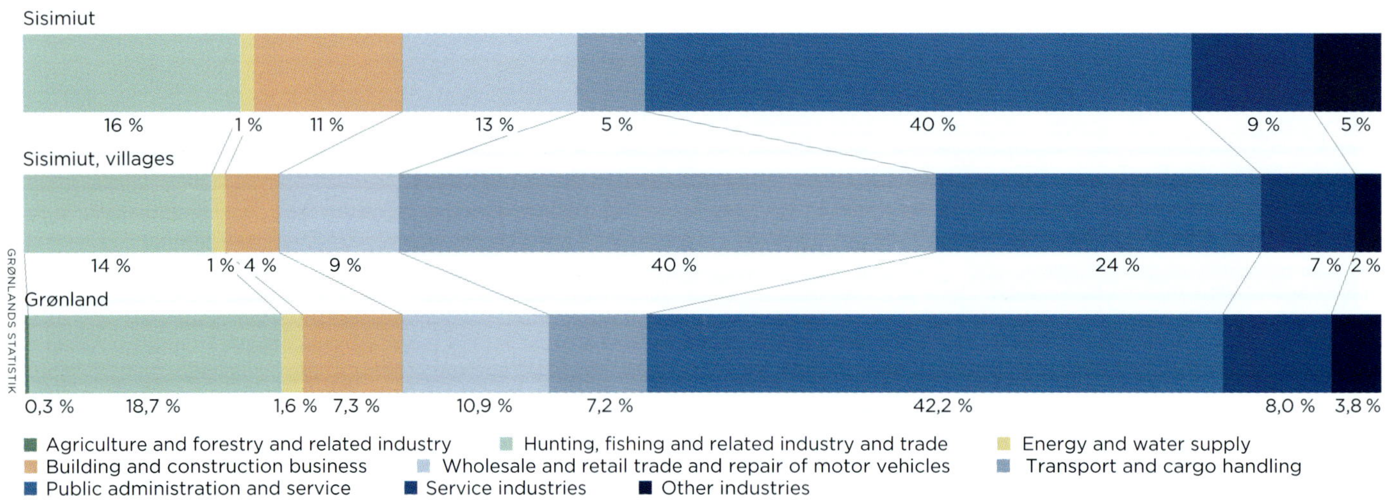

MADS PHIL/VISIT GREENLAND, 2015

Sisimiut is the southernmost town on the West Coast where sled dogs are permitted. There is an active dog sledding community with many recreational sledders and hunters who use dog sleds for hunting caribou and muskox. The opportunity to do a multi-day sledding adventure attracts a lot of tourists. Here we see people heading out to watch the Greenland Dog Sledding Championship, held in Sisimiut in 2015.

try's largest retail chain, Pilersuisoq, Polar Oil, the webshop pissiffik.gl, KNI Engros and the slaughterhouse in Narsaq, Neqi A/S. All of the subsidiaries, except Neqi, are headquartered in Sisimiut.

The large trawler fishing industry that regularly disembark and embark its crew uses the well-functioning Hotel Sisimiut Sømandshjem (Hotel SØMA) with 32 rooms, which is currently being extended, and the 38-room Hotel Sisimiut, which also offers accommodation in seven hotel flats as well as a restaurant. This is supplemented by private accommodations via Airbnb as well as a handful of cafes and restaurants.

Tourism has become increasingly important to the town. In addition to the enterprising town, its main attraction is the large accessible ice-free land area that goes all the way to the ice sheet. During winter, you can go on dog sled rides for a few hours up to three days if you take the popular ride between Sisimiut and Kangerlussuaq. Other major attractions are cross-country skiing, snowmobiling and most recently *fat biking*. The annual cross-country skiing event Arctic Circle Race, where participants cover 160 km in three days and sleep in a tent camp for two nights, has up to 200 participants and just as many volunteers (see p. 177).

During summer, tourists go hiking in the local area, take day trips to the abandoned settlement of Assagutaq and the town's nearby mountains. The 160-km long Arctic Circle Trail hiking route between Sisimiut and Kangerlussuaq is particularly popular. Municipal cabins along the route provide accommodation and safety. During summer, sports fishing in the trout rivers around Sisimiut is very popular.

MADS PIHL/VISIT GREENLAND, 2016

Kangerlussuaq viewed from south to northwest. All the buildings in the foreground were built by the Americans as part of the military base that was handed over to Greenland in 1992. The civilian part has always been on the opposite side of the runway.

Kangerlussuaq

For more than 4,000 years, the Kangerlussuaq area has been used by humans as hunting preserve for caribou hunters and trout fishermen, and you will be able to find traces of settlements, hunting activities and well-preserved ancient monuments.

Like Narsarsuaq, the Kangerlussuaq airfield was built during World War II as part of the North Atlantic air bridge that connected the United States with Europe. The United States military established the airfield under the name Bluie West 8 in 1941. After the war, the base changed its name to Sondrestrom Air Base. The Greenlandic name means 'the big fjord' and refers to the 180 km long fjord leading into the airfield. The Danish name of the fjord and the airfield is Søndre Strømfjord.

Kangerlussuaq's history

The military base remained in operation during the Cold War, but now focused on serving the many radar stations that the Americans built around Greenland. Civilian flights began in the mid-1950s, and the civilian area of the airport was built in 1960, turning Kangerlussuaq into the gateway to Greenland. Already in 1954, Scandinavian Airlines (SAS) started flying to the western United States via Kangerlussuaq, and in 1967, SAS established the Copenhagen-Kangerlussuaq regular service. The United States Air Force left the base in 1992, primarily as a result of the end of the Cold War, and the base was given its original Greenlandic name Kangerlussuaq and was handed over to the then Home Rule Government. Mittarfeqarfiit (Greenland Airports) operated and maintained the airport, hotel, supplies and infrastructure until Kangerlussuaq acquired settlement status under Qeqqata Kommunia.

The Kangerlussuaq fjord only allows boat traffic for about six months of the year because it freezes over.

Kangerlussuaq, street map. 1) Kangerlussuaq Church. 2) Municipal office. 3) Primary school. 4) Kangerlussuaq Museum. 5) Sports centre. 6) Swimming baths. 7) Terminal & Hotel Kangerlussuaq. 8-10) Hostel. 11) Joint Arctic Command.

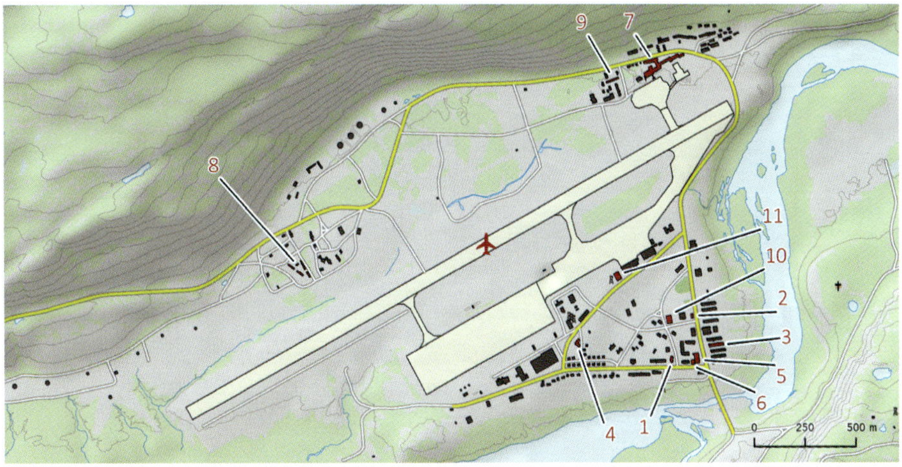

ASIAQ, 2022

Therefore, in the 1960s, a road link was considered between Søndre Strømfjord Airport and Sisimiut so that Sisimiut could be used as a transit port during the winter months.

Kangerlussuaq has been crucial as an international emergency airport for transatlantic routes, a function which the International Civil Aviation Organisation (ICAO) continues to fund.

The landscape clearly shows traces of its proximity to the ice edge, and loose sediment deposits from meltwater rivers and moraine deposits can be found throughout the area. The ice sheet to the east and Maniitsoq Ice Cap to the southwest protect Kangerlussuaq and help reduce precipitation. The wind-protected location at the head of the fjord provides optimum flying conditions in the otherwise harsh Arctic climate with winter temperatures down to and below -40 °C.

Caribou and muskoxen are widespread and breed in the area. The same applies to peregrine falcons and the greater white-fronted goose. The muskoxen were introduced in the 1960s from East Greenland and currently numbers approx. 10,000 animals. By taking a suitable vehicle, both guests and permanent residents can easily get close to the rich wildlife.

Kangerlussuaq port lies about 16 km west of the airport area, and in the hinterland to the north lies the Kellyville research station where international scientists study e.g. the atmosphere.

Some cruise ships use Kangerlussuaq to disembark and embark passengers who are sailed ashore and to the ship by small boats because it is not possible to dock in the existing port, which is very shallow. Plans are under way for a new port on deeper waters further out along the north side of the fjord.

Kangerlussuaq covers a vast area and, thanks mainly to the Americans, has one of the longest road networks in the country. From the airport, it is possible to drive towards the fjord to the port and on to Kellyville where there are a number of leisure cabins. There are roads into the ice sheet, and to the south, there are roads up to and behind the Black Ridge mountain and below it there is a road to Lake Ferguson. There is at least 75 km of road in Kangerlussuaq.

MARTIN LEHMANN/RITZAU SCANPIX, 2019

Air Greenland and the Joint Arctic Command are located side by side at Kangerlussuaq Airport. The majority of Defence aircraft in the Arctic take off and land in Kangerlussuaq.

OY-LKI
OY-OCM
www.AirZafari.gl

In Kangerlussuaq, it is possible to take a flight over the ice sheet, whose active glaciers are best seen from the air. It is also possible to go up on to the ice sheet at Kangerlussuaq, where guided tours are available.

MADS PIHL/VISIT GREENLAND, 2013

In 2022, an approx. 130-km ATV trail will open between Kangerlussuaq and Sisimiut. It is unusual and is seen as the first step toward an actual road and new development opportunities for trade and industry, fish exports and tourism. Along the trail you can experience the UNESCO World Heritage Area Aasivissuit – Nipisat (see p. 371).

A diesel generator plant supplies the settlement with electricity and district heating, supplemented by an oil-fired heating station. Drinking water is retrieved from Lake Ferguson south of the settlement. Most of the buildings are connected to a sewer system, however, some have a sludge tank. Tusass (formerly Tele Post Greenland) provides communication lines via the Greenland radio link.

Kangerlussuaq has a health care station that, in consultation with the regional hospital in Sisimiut, treats patients. The settlement has an integrated day care facility for approx. 60 children, a primary school for about 80 children and a combined sports and swimming hall. The town has no elderly care or a retirement home.

Kangerlussuaq's population trends, trade and industry and culture

During World War II, the base had more than 1,000 residents, all U.S. military personnel.

In 1978, there were 729 residents, and in 1993, the year after the military base closed, its population reached a low point at 291. Since then, the number has remained stable at 400-500 residents, and as of 1 January 2021, there 463 people lived in Kangerlussuaq.

Pilersuisoq runs a convenience store just outside the terminal building and next to a Tele-Post service centre. Most businesses are linked to the airport, including Hotel Kangerlussuaq and 3-4 hostels as well as KISS – Kangerlussuaq International Science Support, which supports the increasing number of researchers. There are some small and medium-sized tourist operators selling handicraft and offering excursions on land, on the water, in the air or by dog sled as well as trophy hunting.

The Kangerlussuaq Museum (see p. 168), which opened in 1997, includes a permanent exhibition which focuses on the military history of the base, the civilian airport and the transition into a research centre.

A drive to the ice sheet, with an 18-hole golf course at the halfway point, is spectacular. Access to the ice sheet was instrumental in allowing the international Extreme E motor race with electric rally cars in 2021 to be held near the Russell Glacier under great media attention.

Kangerlussuaq remains Greenland's main international gateway. At some point, all travellers to Greenland meet in Kangerlussuaq. The area has great tourism potential and can be developed as a hub for research and air transport between the United States and Europe. However, the outlook is uncertain with Atlantic airports in both Ilulissat and Nuuk expected to open to international traffic in 2024. On the other hand, Danish Defence has announced that they wish to maintain the airport, including as the primary base for the Joint Arctic Command's airborne operations in Greenland.

KÅRE HENDRIKSEN, 2015

Fishermen and dinghy anglers are at sea as soon as it is clear of ice. When it is not possible to get to the quay, dinghies are pulled onto the ice as far inland as possible. Dinghies are moored at the edge of the bay that forms the port of Qeqertarsuaq, where fishing is a major part of the population's livelihood and accounts for 26 % of employment.

KOMMUNE QEQERTALIK

Kommune Qeqertalik is located between 67° and 70° N and covers an area of 62,400 km². he municipality has approx. 6,500 residents. In addition to the mayoral town of Aasiaat, the municipality includes the towns of Kangaatsiaq, Qasigiannguit and Qeqertarsuaq as well as eight settlements. The municipality was established in 2018 in connection with the division of Qaasuitsup Kommunia and has since had an Inuit Ataqatigiit mayor who was re-elected in 2021.

The municipality's towns and settlements traditionally use dog sleds for hunting and study visiting. Now dog sleds are also widely used for leisure and tourism.

Interest in the geology and mineral resource potential of the area has been significant, leading to extensive oil and gas exploration in the sea and on Qeqertarsuaq (Disko Island). Global interest in oil exploration has been declining over the past ten years in step with a drop in oil prices and increased focus on the climate impact of CO_2 emissions. In 2021, Naalakkersuisut decided to stop oil exploration in Greenland. Some small-scale mineral exploration projects remain in the region, particularly on Qeqertarsuaq, but none of the projects are at an advanced stage. Nationally, politicians plan to establish a hydropower plant at Qasigiannguit, which, if established, will also provide power to Aasiaat.

The municipality intends to spread out administrative units to the four towns. The wish is to create a more holistic form of administration, and the municipality is therefore prioritising an innovative and more agile organisational structure adapted to local identity, culture and cultural heritage. Moreover, emphasis is placed on business development and fisheries, education, health and prevention, infrastructure as well as development of tourism throughout the municipality.

Kangaatsiaq

Kangaatsiaq, which means ‘the small promontory’, has never been a colony and has no Danish name. As the only district besides Upernavik, most of its residents live in the four settlements, whose population total 621 people. The town has a population of just 534 people. Kangaatsiaq has always been a sort of forgotten outer district, despite its central location between Aasiaat and Sisimiut.

History of Kangaatsiaq

The district has long branched fjords with several areas with very strong currents that ensure ice-free water all year, and the large relatively flat ice-free land area towards the ice sheet has historically been important to the Inuit because it was a good place to hunt for marine mammals, birds, caribou and other terrestrial mammals. The area was therefore an important part of income base for the colonial districts of Egedesminde (Aasiaat) and Holsteinsborg (Sisimiut). The two colonies had permanent remote trading posts in the district from the early 19th century, and by the early 20th century there were 17 permanent settlements. Kangaatsiaq district is an excellent example of the original high degree of mobility and flexibility of the Greenlandic population. People moved to where the hunting was best, and the partial transition from hunting to fishing led to the establishment of new settlements.

Until 1950, Kangaatsiaq and the district’s settlements were part of the Egedesminde colonial district, but was then given special municipal status. Most of the administrative tasks of the new Kangaatsiaq Municipality remained in the hands of Egedesminde Municipality, and Kangaatsiaq belonged under the Egedesminde school district, and it was the town treasurer of Egedesminde who was the secretary of the municipal council of Kangaatsiaq. Similarly, Kangaatsiaq was under the Egedesminde medical, commercial and police district.

In 1986, Kangaatsiaq was granted the status as autonomous municipality. Unlike the country’s other municipalities, Kangaatsiaq did not have a hospital and the schoolchildren of its settlements continued to be sent to Aasiaat to finish primary school, while the children in Kangaatsiaq was able to finish primary school in the town. In many respects, Aasiaat

Figure 37. Population trends in Kangaatsiaq 1960-2021.

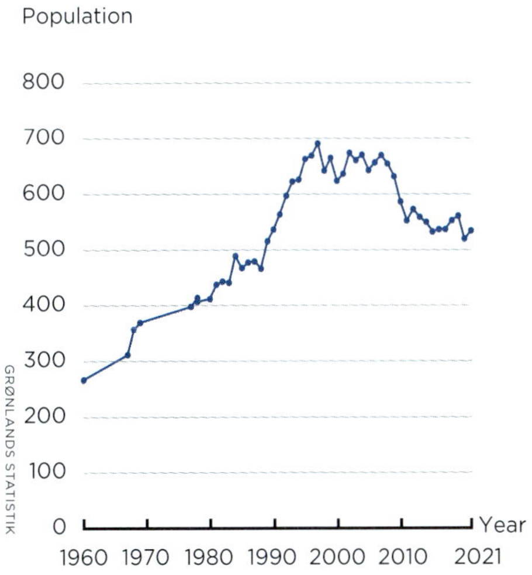

Kangaatsiaq, street map. 1) Fish factory. 2) Municipal office. 3) Primary school (old). 4) Primary school (new). 5) Health centre. 6) Sports centre. 7) Kangaatsiaq Church.

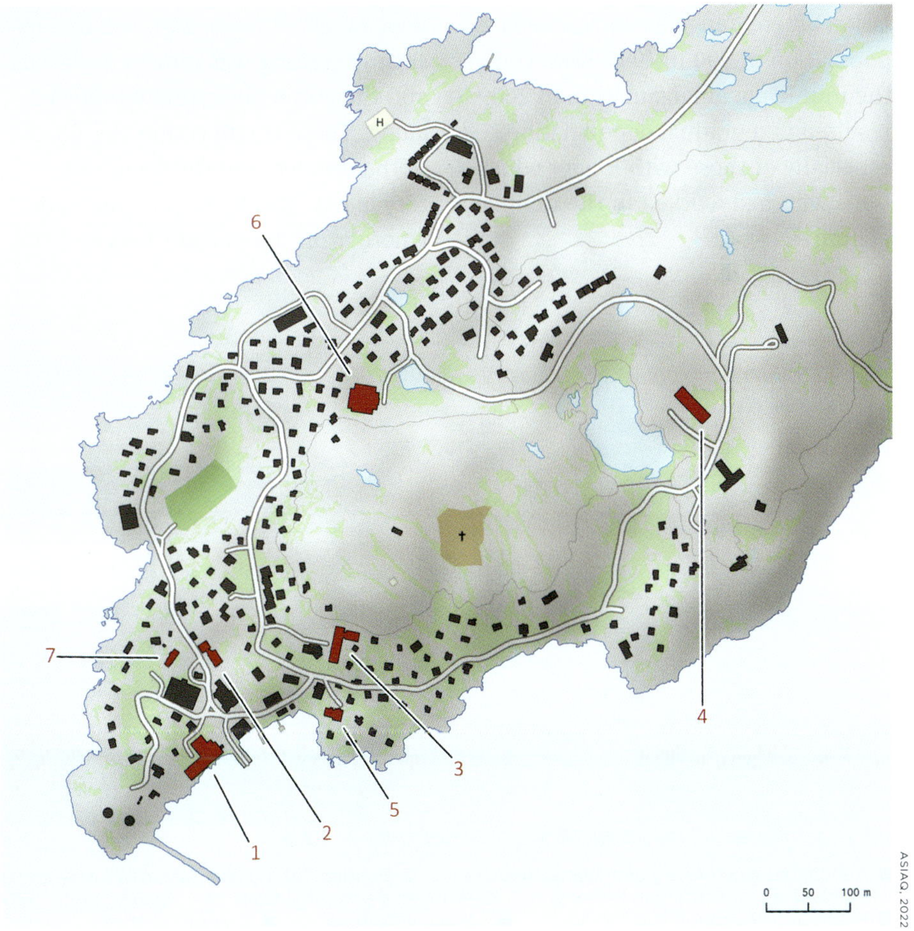

NIKOLAJ JEREMIASSEN, 2016

Kangaatsiaq is Greenland's second smallest town and the last to be granted town status. In the district, there are several long fjords towards the ice sheet, and in some places, the fjords are narrow, so the tide creates strong currents and significant water renewal. Combined with meltwater from the ice sheet, it provides oxygenated and nutrient-rich waters that support a large fish population and many seals. At the same time, the large ice-free land towards the ice sheet provides good opportunities for caribou hunting.

effectively remained the administrative centre, which is why Kangaatsiaq was and still is the town with the fewest people not born in Greenland compared with other towns.

The period as an autonomous municipality led to an increase in population where Kangaatsiaq reached a population of 690 in 1997, however, the municipal amalgamation in 2009 resulted in a gradual depopulation, especially of those with longer educations who had worked in the municipal office.

During the cod fishing period, Kangaatsiaq district had periods of good trading compared to its population size. With the disappearance of the cod at the end of the 1980s, the Kangaatsiaq factory switched to landing shrimp in 1994. The local cutter fleet switched to shrimp, but in 2000 shrimp trading was moved to Aasiaat and the cutters eventually followed along. There is still cod in the fjords of the district, and the town's factory is now run by Royal Greenland where it produces cod fillets and cod in frozen blocks as well as roe and to a lesser extent freezing of other fish species. The factory employs up to 25 people during the season.

Geography, infrastructure and trade and industry of Kangaatsiaq

Kangaatsiaq is located on a relatively low peninsula with the central functions such as port, factory, filling stations, shop, health station and the old school located on the headland to the west. A new school is being built in the east end of the town and is ex-

Figure 38. Distribution of jobs by industry sector for people working in the town of Kangaatsiaq, the settlements in Kangaatsiaq district and Greenland as a whole, 2020.

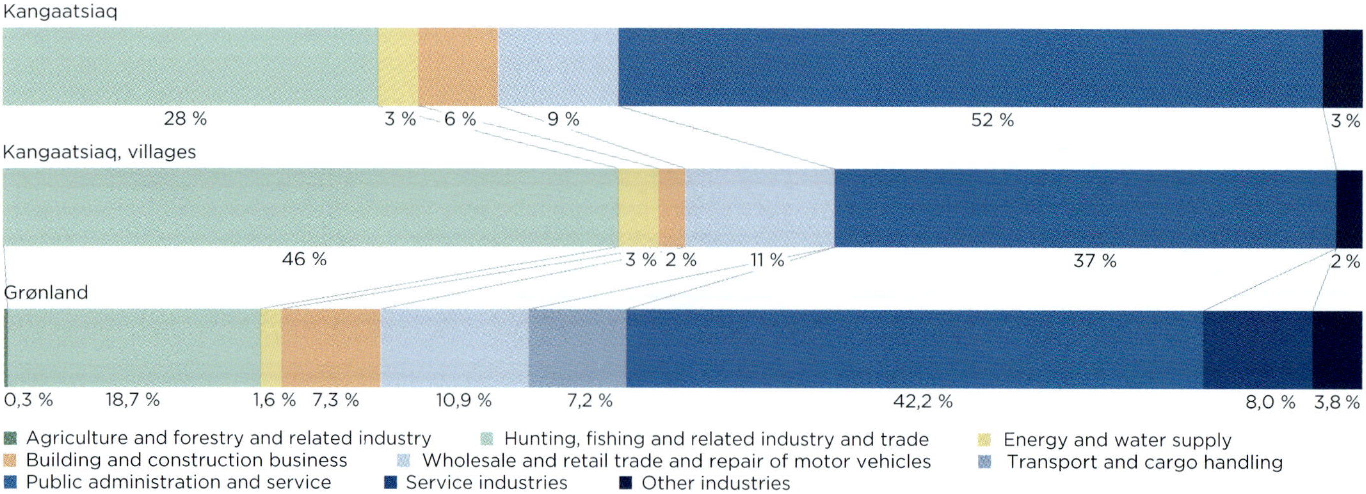

GRØNLANDS STATISTIK

pected to be complete in 2022. Most homes are single-family homes supplemented by single terraced houses. The town has a water network, however, a few homes are not yet connected, meaning that the residents have to get their own water. The town has no sewer system, meaning that toilet buckets with biobags are used, and the bags are collected and emptied by the collection of latrines.

Royal Arctic Line calls weekly at Kangaatsiaq with cargo, and the Disko Line has one weekly passenger trip with relatively small boats during the summer season, while Air Greenland serves the town once a week by helicopter during the winter months. Most passengers travel via Aasiaat.

The hunting profession has been maintained together with fishing and fish processing, and the local economy consists of a mix of subsistence economy from hunting combined with fishing as well as the revenue from administration, education and services, etc.

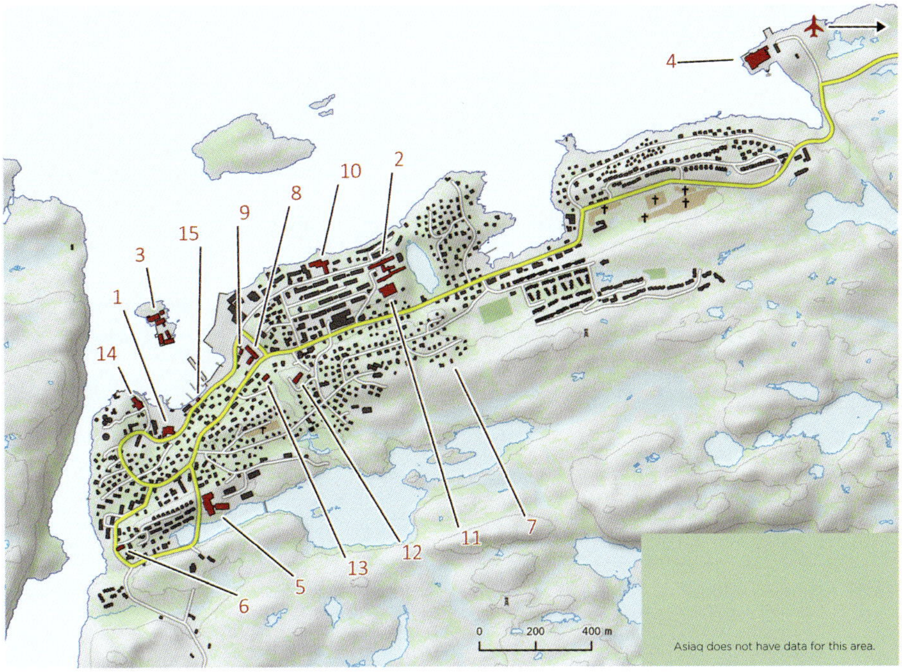

ASIAQ, 2022

Aasiaat, street map. 1) The shipyard. 2) Primary school.3). Closed fish factory. 4) Fish factory. 5) GUX Aasiaat. 6) Ado Lynge School. 7) Coastal radio station (Aasiaat Radio). 8) Town hall 9) Aasiaat Museum. 10) Regional Hospital. 11) Sports centre. 12) settlement hall. 13) Hotel Nattoralik. 14) Seamen's home. 15) The Board.

Aasiaat

Aasiaat is located on a roughly 7-km long and 3-km wide and relatively low archipelago island with very modest vegetation. The town shares the place as the country's fourth largest town with Qaqortoq. The district's two settlements have 124 residents and the town has 3,035 residents.

Aasiaat island is located in a large archipelago called 'the land of the thousand islands'. There are two different interpretations of the name: either 'the place with spiders' or the 'meeting place' where people gathered for a short period of time during the nomadic journeys in summer (see p. 105).

The town's founders included Niels Egede (see p. 113), who named the place after his father, Hans Egede.

Aasiaat's history

The Egedesminde trading post was established in 1759 by Hans Egede's son Niels Egede at Eqalussuit north of Nordre Strømfjord, located south of Kangaatsiaq. In 1763, the colony was moved to its present location because there were many Inuit summer and winter settlements in the vast archipelago and thus great trading opportunities.

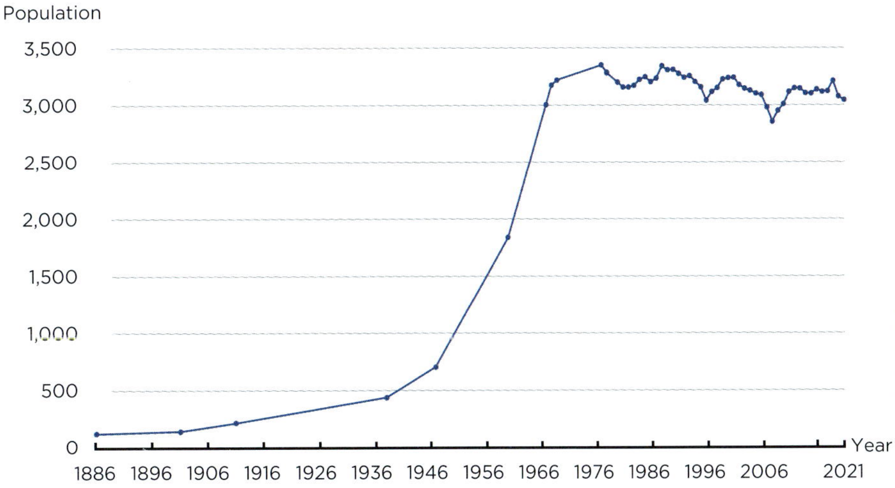

GRØNLANDS STATISTIK

Figure 39. Population trends in Aasiaat 1986-2021.

GTO, ASIAQ ARKIV

The port of Aasiaat, 1958. At that time, the town had barely 1,800 inhabitants. A few years earlier, Aasiaat (then Egedesminde) had been given one of Greenland's first ports with a draught for Atlantic ships, and the town served as a transit port for North Greenland. Aasiaat Radio was the centre for sea rescue and air communications in Northwest Greenland.

The large archipelago with the many straits provided ample opportunities for hunting from kayak and seal hunting from the sea ice in winter, and there are archaeological finds from the first immigrations and since around the year 1150 from the present Inuit.

The waters also provided a vantage point for European whalers, which had several permanent anchor and flensing sites in the district, which led to interaction and tusk trade between the Inuit and the whalers. The relocation of the Egedesminde trading post was thus part of the effort to get the whaling trade into Danish hands, and the good natural harbour, which the Dutch often used and called 'Jesse Bay, was chosen for the relocation. In the following years, this led to minor clashes between the Danish colonists and the Dutch whalers.

In 1774, the Royal Greenland Trading Department (KGH) took over the trading post. The catchment area, which by the standards of those days was densely populated, and the location at the entry to Qeqertarsuup Tunua (Disko Bay), made Aasiaat one of the most important trading posts which contributed positively to the economy of the KGH. Aasiaat therefore became the trade and administration centre for the southern part of Qeqertarsuup Tunua.

Aasiaat gradually became an educational centre where, in addition to the local children, the school received students from the colonial district's settlements, and in 1928, Greenland's first continuation school for boys was established, followed by a continuation school for girls in 1932.

A cod fishing industry gradually developed during the 1930s and contributed with increased income for

FILIP GIELDA/VISIT GREENLAND, 2019

Aasiaat is located in a beautiful archipelago in the southern mouth of Qeqertarsuup Tunua (Disko Bay) on a relatively small and flat island without much vegetation. Although the many single-family houses dominate the urban landscape, more than half of the dwellings are single-storey or terraced houses.

the population. During World War II, Aasiaat served as transit port for the American provision of supplies to Northwest Greenland, and in 1942, the Americans established the Bluie West 5 weather station. This created jobs and led to interaction between locals and American soldiers.

The transit function continued after the war and became extremely important for employment. 1948 saw the establishment of a publicly owned boat builder's yard with slipway capacity for cutters and trawlers as well as the vessels that undertook transit operations for Qeqertarsuup Tunua's other towns and settlements. Aasiaat built one of the first ports in Greenland in 1951 with capacity for the Atlantic ships of the time.

The transition from hunting to fishing gained momentum during the post-war years, and in 1949, KGH established a fish facility for the production of dried cod on Transit Island off the town, which was expanded in 1966 to a fillet factory. At that time, it was common for the sled dogs to walk freely, so the factory was placed on the island to keep the dogs away from the production during the summer fishing season.

Shrimp fishing gradually became more important. The private frozen shrimp factory, established in 1951, was acquired by KGH in 1953.

At the same time as the power station and other service facilities, Aasiaat Radio was established in the early 1950s as a sea rescue service and aircraft communication centre for all of northwest Greenland. In 2017, the telecommunications submarine cable was led to Aasiaat, and since then, Aasiaat Radio has expanded to also serve as a relay station for telephony and Internet for towns and settlements that have yet to be connected to a submarine cable. Finally, Aasiaat Radio issues weather forecasts for air and ship traffic in cooperation with the Joint Arctic Command.

Post-war modernisation led to a massive boom in construction where Aasiaat, with the vast district that also included Kangaatsiaq and the settlements, became home to Danish contractors who took on assignments across much of the country. In 1969, the town had 27 contractors which employed both skilled workers and apprentices.

ANINGAAQ R. CARLSEN/VISIT GREENLAND, 2020

Royal Arctic Line uses Aasiaat as a transit point for containers and general cargo shipment to destinations in North Greenland, which provides numerous jobs.

A secondary school was established in 1964, and the town's status as an educational centre was strengthened.

Aasiaat developed into one of Greenland's cultural centres, and from the late 1960s, the town, inspired by the youth rebellion in Europe, became central to the budding rebellion against Danish domination. In 1976, the left wing group Inuit Ataqatigiit (IA), which in 1978 turned into an actual party, was founded in Aasiaat.

In 1974, the port was deepened and expanded to allow Royal Arctic Line's new Atlantic ships to dock, and facilities were established for container transit. Since then, Aasiaat has continued to serve as a transit port for most of North Greenland.

Avannaani Ilinniarnertuunngorniarfik (North Greenland High School) was established in 1987. This meant jobs and an increased influx of young people from all over North Greenland. The high school has approximately 400 students and 28 staff and has 350 dorm rooms at its disposal. This means a small demographic over-representation of young people, who contribute to the town's active cultural life. Furthermore, the nationwide special school for children and young people with mild intellectual disabilities and learning disabilities, Ado Lyngep Atuarfia, was established in 1982.

Large-scale shrimp processing was challenged by the fact that there is relatively little water on Aasiaat Island, so most of the landing of shrimp and the processing took place in Qasigiannguit. Despite the challenge to get water, a larger shrimp factory was built in 2000 where attempts were made to secure the water supply by connecting virtually all the lakes to the supply – however, this was not an optimum solution. In 2008, the factory started using purified, but not desalinated, seawater in the production. The factory is currently owned by Polar Seafood.

In connection with the 2009 municipal amalgamation, Aasiaat Municipality became part of Qaasuitsup Kommunia, covering a geographical area from Kangaatsiaq to Qaanaaq. Later, in 2018, Qaasuitsup Kommunia was split into Avannaata Kommunia and Kommune Qeqertalik, and Aasiaat became the main town.

Since 2011, Aasiaat has been home to the Qeqertarsuup Tunua regional hospital which, besides Aasiaat, covers the districts of Kangaatsiaq, Qasigiannguit and Qeqertarsuaq.

Demography and infrastructure of Aasiaat

434 people lived in Aasiaat in 1938, and the population increased to 701 in 1947. Modernisation accelerated population growth, meaning that in 1977, 3,347 people lived in Aasiaat, later followed by a gradual decline, so that the population was down to 2,853 at the time of the municipal amalgamation (as of 1 January 2008). Since then, there has been small population growth to today's level.

Aasiaat has approximately 1,250 dwellings spread over 43 % single-family homes, 12 % semi-detached terraced houses and 45 % blocks of flats. Approx. 80 % of the town is connected to a sewer system, while the remaining citizens use toilet buckets with biobag and collection of latrines or private sludge tanks.

Electricity supply in Aasiaat comes from a diesel-powered power station at Aasiaat Port. Surplus heat supplies two district heating networks in the central parts of the town. In addition, one area also gets heat from a waste incineration plant. The latter is replaced with waste oil incineration as waste incineration is shut down. Homes without district heating are heated with oil-fired boilers.

Figure 40. Distribution of jobs by industry sector for people working in the town of Aasiaat and Greenland as a whole, 2020.

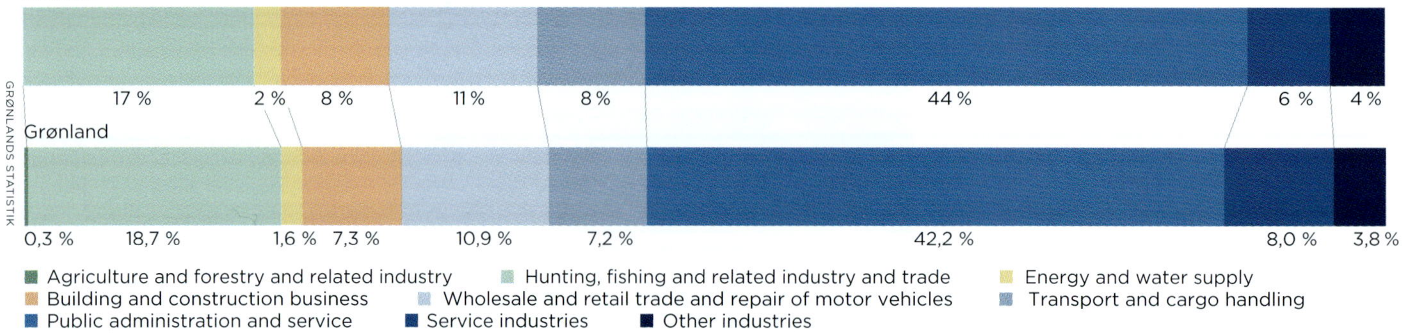

FILIP GIELDA/VISIT GREENLAND, 2019

Seamen's home in Aasiaat in winter. The inner harbour is frozen over with ice, while the town's large container port rarely freezes over.

The town has an incineration plant for domestic waste and other combustible waste that provides district heating to a small residential area. From 2024, all waste must be packed and shipped for incineration in Sisimiut, after which the incineration of waste in Aasiaat will stop and waste oil will be incinerated instead. Non-combustible waste is deposited at the town dump (waste landfill) and hazardous waste is shipped for treatment outside Greenland.

The plan is to construct a hydropower plant at Qasigiannguit in 2027 to take over the electricity supply in Aasiaat. Most of the town's roads are paved and it has a drinking water network that covers about 80 % of homes.

In 1998, an airport was established in Aasiaat to replace the former heliport, and the town has more or less daily flight routes to Ilulissat, Kangerlussuaq, Sisimiut and Nuuk. Since it functions as a transit port, Royal Arctic Lines' ships call into the port several times a week throughout the year to the extent allowed by the sea ice. During the summer months, Disko Line provides passenger service to Aasiaat with small ships several times a week, and Arctic Umiaq Line's coastal passenger ship calls twice a week going north and south, respectively.

Trade and industry in Aasiaat

Public administration and services, including education and health care, account for 44 % of the town's jobs. Besides that, the town's main industries are trawler fishing, primarily shrimp, and to a lesser extent fishing of crab and other species as well as fish and shellfish processing, which, along with hunting, account for about 17 % of jobs. Trade etc. accounts for 8 %, the transport sector 8 %, construction 8 % and other service industries 6 %.

Over the past few years, tourism has gained increasing importance and there are currently five official providers of overnight accommodation, including a hotel and a seamen's hostel as well as restaurants and cafes, etc.

Qasigiannguit

Qasigiannguit with 1,071 residents (as of 1 January 2021) is located at a small bay in the northeast corner of the Sydøstbugten in Qeqertarsuup Tunua (Disko Bay). The nearest settlement is Ikamiut with 78 residents.

Qasigiannguit means 'the small common seals' and is probably named after the breeding and shedding ground of the common seal on the sandbank at the head of the bay.

History of Qasigiannguit

The colony was established in a small southward cove in 1734 by the trading company of Jacob Severin and was named Christianshaab after the current King Christian VI. The establishment of the colony was an attempt to keep Dutch whale hunting and trade with the population under control and also to spread the mission to North Greenland.

The early colonists named the cove Bryghusbugten, a name that is still used to this day. The bay called Vierrebay by the Dutch was marked as an anchorage on Dutch nautical charts of the Davis Strait from the early 18th century when the Dutch engaged in extensive whale hunting.

In the period 1734 to 1740, Hans Egede's son Poul Egede stayed in the colony for a total of four years, during which period he laid the foundation of a Christian congregation.

The location in Bryghusbugten was abandoned after a few years. The colony was moved to the opposite side. A new farmhouse was erected, but the old farmhouse was later moved to the same location. The old houses with their exquisite Nordic building style became the centre of the small colony, which only expanded slowly

The inner part of the bay, to which the Christianshåb colony was moved a few decades after its foundation. Around the colonial buildings, a small urban community developed up the slopes to the north, and family houses were built at the head of the bay. The old harbour became a bustling fishing port in the mid-1900s around the boom in shrimp production. Today the colonial buildings are listed and house an active local museum, and the wider territory provides a setting for open-air sailing and a bustling dinghy fishing port.

GTO, ASIAQ ARKIV

in the first two centuries. The population was less than 100 people all the way up to the early 20th century.

The families provided for themselves by hunting and fishing, supplemented by collecting eggs and small secondary income from selling lingonberries that are common in the area. Up until World War II, development in trade and industry was slow, despite random new initiatives involving e.g. hunting and fishing using nets.

Development, trade and industry and infrastructure of Qasigiannguit

The localisation of major shrimp fields in Qeqertarsuup Tunua in the late 1940s set the development in motion in the following decades. Despite production being at a standstill during months of ice formation, the Royal Greenland Trading Department (KGH) set up a modern shrimp production facility in the town. Local families invested in shrimp cutters that often also had harpoon guns for supplementary whale hunting, and fishing families from outside the town and seasonal workers moved to the town over time.

The town rapidly became a modern town with a significant population growth, where most people worked in shrimp fishing and shrimp production and related industries. The factory meant that the town had water supply and electricity early on.

The residential neighbourhoods with the distinctive GTO housing benefit houses started shooting up in the Quikoq neighbourhood to the south and Illoqarfittaaq to the north, and multi-family homes, blocks of flats and terraced houses were built for newcomers. The road network was expanded in a south- to north-bound road network in three levels connected by the main road Poul Hansensvej.

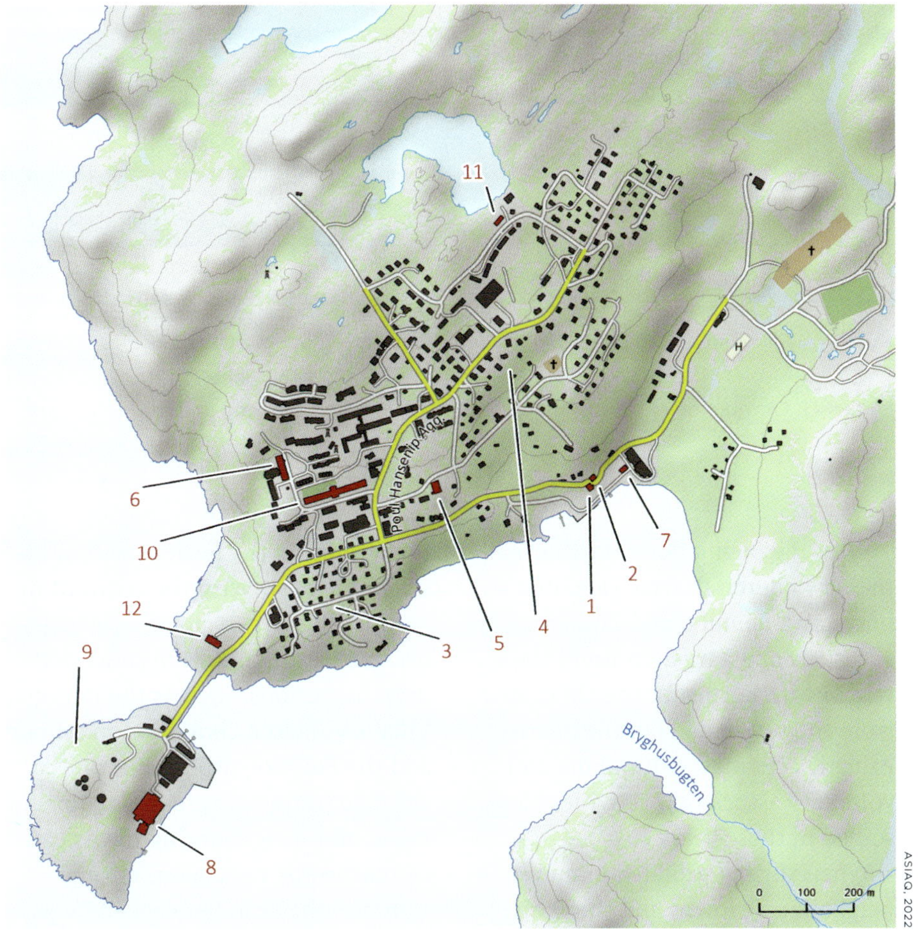

Qasigiannguit, street map. 1) New Farmhouse. 2) Old farmhouse. 3) Quikoq (district). 4) Illoqarfittaaq (district). 5) Pavia Church. 6) Health centre. 7) Boat workshop. 8) Fish factory. 9) Quilik (island). 10) Continuation school. 11) Orphanage. 12) Hotel Qasigiannguit.

Figure 41. Population trends in Qasigiannguit 1886-2021.

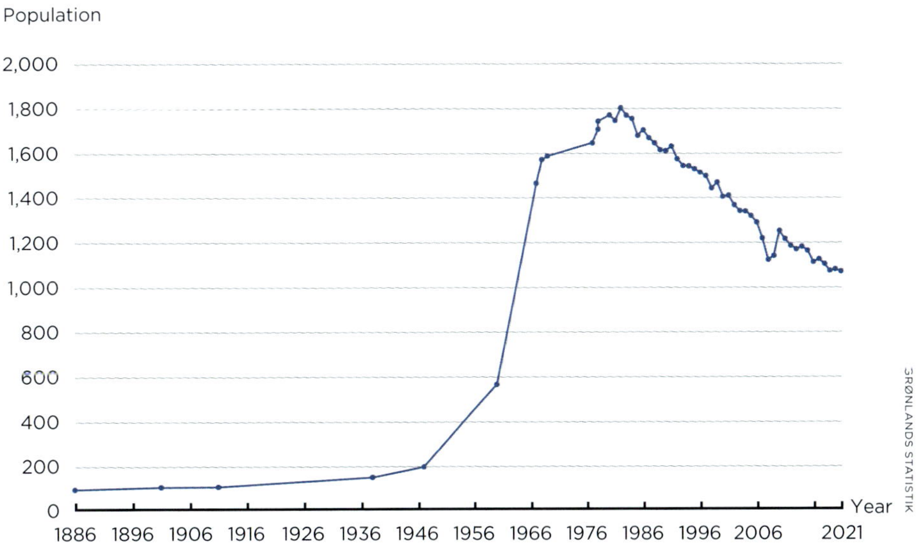

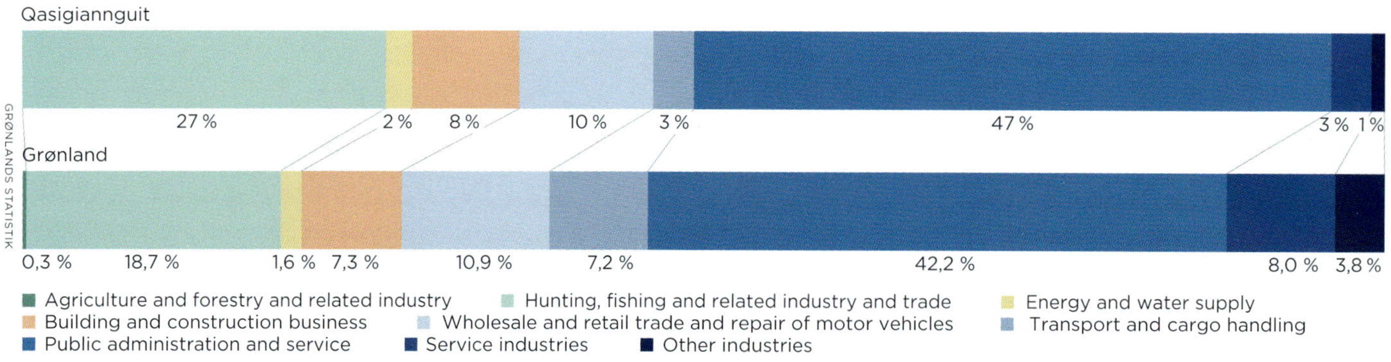

Figure 42. Distribution of jobs by industry sector for people working in the town of Qasigiannguit and Greenland as a whole, 2020.

With the population growth came the new church Paaviap Oqaluffia, a hospital, a dental clinic, children's institutions and a modern multi-track school with dorms for the children from the settlements in the district. Athletic activities, cultural life and voluntary association activities blossomed in those years.

When shrimp production moved to neighbouring towns and the Qasigiannguit factory was converted to process Greenland halibut and other fish species shortly before the 2000, the town changed character and a systematic reorientation has since been underway. The production facility in the inner harbour was demolished and the area is now used by dinghy fishermen and recreational sailors. A dinghy repair workshop has been built on the old waterfront.

The town's activity centres are now located in the town centre, Royal Greenland's production facility and the Royal Arctic Line's port are located on the island of Quilik. The change in production lead to a drop in the need for labour, leading to a noticeable outward migration from the town. The population of 1,071 is nearly half that of the blooming population of the 1980s where it peaked at 1,803 in 1982 and is considerably higher during the peak season of the shrimp fishery. With the outward migration, the largest housing blocks became vacant and were converted into institutions, such as the national Villads Villadsen continuation school and the Puilasoq orphanage.

Public transport is seasonal; during winter, neighbouring towns can be reached by helicopter, during summer, they can be reached by passenger ship, which departs daily during the peak season. In recent years, travel options have been supplemented by small private vessels.

Most people work in public administration and service, in the service sector and in private businesses, primarily construction, many work at Royal Greenland's production plant and as fishermen and hunters. The tourist industry is growing: The quiet small-town atmosphere is an attraction and also offers great excursions both on land and water.

Qasigiannguit mainly receives supplies from Royal Arctic Line's ships that call on the town once a week from April to December. Grocery trade is mainly undertaken by Pilersuisoq (KNI), which operates in Greenland's small towns and settlements. Taking into account the size of the town, it even has a strong tradition of a small private businesses. Electricity is supplied by Nukissiorfiit through a combined heat and power plant. The plan is to establish a hydropower plant in the district in the coming years. Telephony and Internet are provided by Tusass (formerly Tele Post Greenland) via a high-speed radio link connection to Aasiaat.

Most single-family homes have their own oil burner. The town's water supply comes from surface water from water lakes east of the town, and institutions as well as a large part of the town's housing have running water, while most other housing either have a water tank or get the water themselves. The town has a sewer system, but it has not been extended to cover all private homes, so many still have toilet buckets with biobags or a sludge tank. Waste management is carried out from unsorted waste bins and the town still has an open dump (waste landfill) with some degree of sorting under the inter-authority waste company, ESANI a/s.

The waters consisting of short fjords and many small islands are calm, and many people enjoy fishing in summer, hunting and berry-picking as well as camping in tents or staying in summerhouses. In particular, the abandoned settlement of Akulliit has many visitors, and during the high season it almost becomes repopulated. Fishing is mainly for capelin, trout and cod, and birds, seals and small whales are hunted for own consumption.

Historically, Qasigiannguit has a relatively dry climate. Summers are warm with an average temperature for July of 9 °C, spring and autumn are short, and the cold winter transitions into the special North Greenland sunny and frosty months from March to May. The prevailing wind direction is from NNE. However, Qasigiannguit is known for the strong Saqqarsarneq foehn from SE, which at short notice in both summer and winter roars down from the Qaqqarsuaq mountain, causing temperatures to rise by several degrees in just a few hours. In recent years, weather has been characterised by more humidity and more frequent storms, and winter ice forms less often.

The town lies beneath the Qaqqarsuaq massif with direct access to the hinterland's rounded mountains and wide stretches of lush valley and lake systems that are perfect for hiking. In winter, it provides the perfect conditions for dog sledding or snowmobiling along the traditional winter roads. The sled trails lead south into Sydøstbugten and north to Ilimanaq and the southern branch of the ice fjord, Tasiusaq, and they are used as soon as it is possible for hunting and fishing as well as outdoor activities and tourism.

Qeqertarsuaq

Qeqertarsuaq on the south side of the island of Qeqertarsuaq (Disko) is named after the island, which means 'the big island'. Qeqertarsuaq is also the name used for Qeqertarsuup Tunua (Disko Bay). The town's Danish name is Godhavn (the Good Harbour) – the only early Danish colony in Greenland where the Danish name describes the place; the town located on a bay that provided shelter and good protection for the sailing ships of the time. The island of Qeqertarsuaq is slightly larger than the Danish island of Zealand, and is mostly made up of basalt. There are several hot springs close to the town that contribute to an unusually high and strong vegetation. Qeqertarsuaq has 838 residents, and the district's settlement has 11 residents.

History of Qeqertarsuaq

The waters around Qeqertarsuaq are abundant in whales, seals and other marine mammals, and there are several archaeological finds dating back to the first migrations to Greenland. In the 17th and 18th centuries, Qeqertarsuup Tunua played a key role in European whale hunting operations, and the Dutch had a permanent whale hunting station west of the town of Qeqertarsuaq at what is known as Hollænderhavnen (the Dutch Harbour).

In 1773, a Danish whale hunting lodge was established in Qeqertarsuaq. The aim was to strengthen Danish whale hunting and to limit other countries' whale hunting and trade with the Inuit. Qeqertarsuaq, which was given colony status, was an

Qeqertarsuaq, street map. 1) Arctic Station. 2) and 13) Fish factory. 3) Schooner bridge. 4) Health centre. 5) Primary school. 6) Hotel Disko. 7) Hostel. 8) Municipal office. 9) Qeqertarsuaq Church. 10) Qeqertarsuaq Museum. 11) Sports centre. 12) settlement hall.

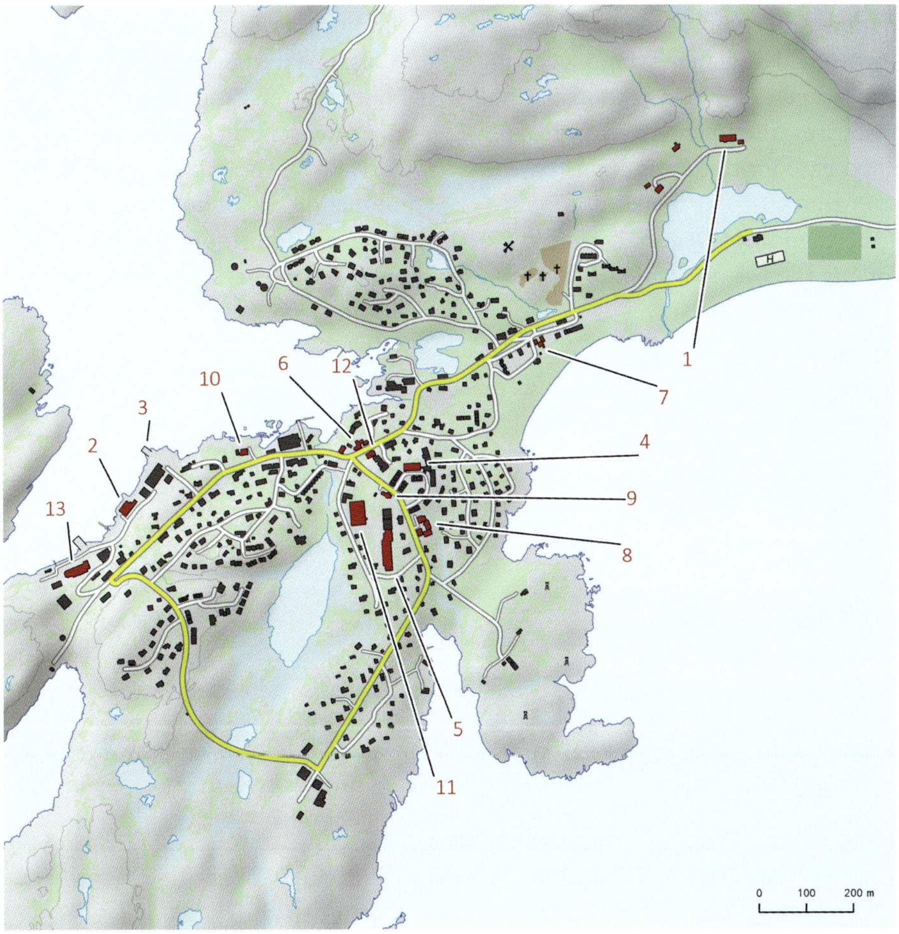

ASIAQ, 2022

important centre for the catching of bowhead whales.

The homothermic springs and the vegetation as well as the geographical conditions meant an increase in scientific expeditions from the late 19th century, and in 1906, the botanist Morten P. Porsild established a natural science station that has been the centre of multifaceted research. The station is currently run by the University of Copenhagen under the name Arctic Station.

The North Greenland Printing Company was established in the town in 1913, and from 1925 to 1950, Qeqertarsuaq served as the headquarters of the administration of North Greenland. This was despite the fact that, in 1947, the colony with its 341 residents was the second smallest town in North Greenland. As an administrative capital, Qeqertarsuaq was prioritised throughout the colonial period and, compared with other towns, had a relatively high level of education. Combined with its natural resources, the town generally had a good base for trade and industry.

During post-war modernisation, fishermen from Qeqertarsuaq were among the first to switch to cutter fishery for shrimp in Qeqertarsuup Tunua, and in 1963, a shrimp factory opened in Qeqertarsuaq, and by 1981, Godhavn Shrimp A/S produced 200 tons of shrimp and had 20-25 employees. Furthermore, four to five people worked at the Disko Laks factory, which traded salmon, seal, whale and Greenland halibut.

The shrimp fishing from cutters helped develop the town's economy and contributed to significant population growth, and in 1994, the number of residents peaked at 1,125 and the town was generally considered to be well-functioning. Shrimp fishing became centralised from the early 1990s with quotas being distributed among fewer and larger vessels. The cutter fishery for shrimp disappeared from Qeqertarsuaq, and in 2001, the shrimp trading moved to Aasiaat. With the closing of the town's shrimp factory, a large number of jobs disappeared resulting in a gradual outward migration.

As a replacement for the shrimp factory, Royal Greenland established a production of halibut frills in 1999. Frill production stopped in 2008 and about 30 employees lost their jobs.

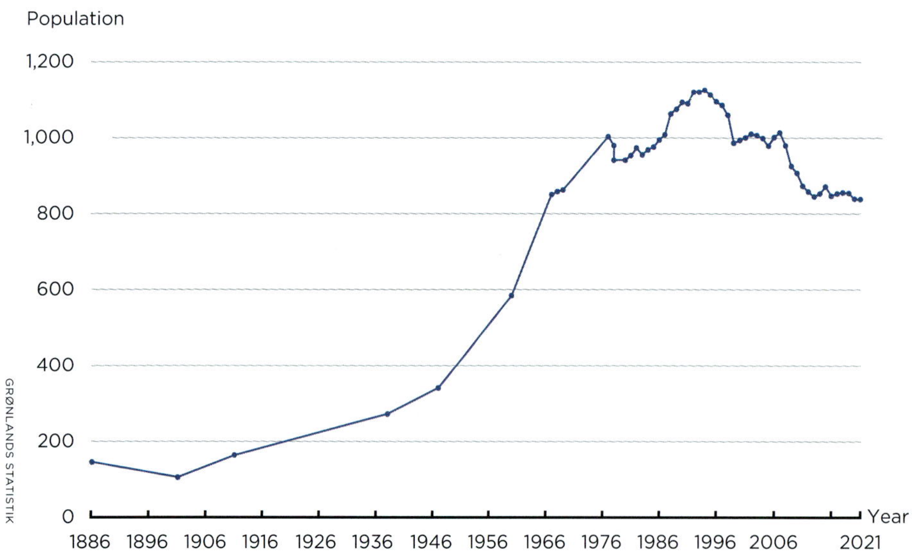

Figure 43. Population development in Qeqertarsuaq 1886-2021.

The factory was sold to Greenland Spring Water ApS and was converted into tapping spring water from the springs close to the town. Water production was unsuccessful and the factory building is now being leased by Royal Greenland, which use it for seasonal production of lumpsucker roe.

Fishermen increasingly began fishing for other species. Especially crabs gained relatively high importance in the first years, just as Greenland halibut or lumpsucker have been economically important, but never achieved the same importance as the former shrimp trade.

In continuation of the 2009 municipal amalgamation, which saw Qeqertarsuaq become part of Qaasuitsup Kommunia, a number of administrative tasks were moved to Ilulissat, which reinforced the outward migration. Qeqertarsuaq became part of the new Kommune Qeqertalik in 2018.

Development and geography of Qeqertarsuaq

Qeqertarsuaq is located on a relatively level peninsula with the high red basalt mountains as a backdrop, and there are several springs and a large river close to the town. The historic colonial buildings near the port are centrally located in the town. This is also where the church, called Vorherres Blækhus (Our Lord's Inkwell), lies. Most homes are single-family homes and, to a lesser extent, terraced houses. The town continues east along the beach, and towards the edge of the town lies the Arctic Station, a geophysical observatory, and the former ionosphere station, which now serves as a hostel. On the beach, called Sortsand (Black Sand), lies the town's football field with artificial turf as well as the heliport.

During summer, Disko Line operates passenger services to Qeqertar-

KÅRE HENDRIKSEN, 2020

suaq by relatively small boats, and during winter, the town is served by Air Greenland by helicopter. On several occasions, plans have been made to establish a runway on Sortsand (Black Sand), and the beach has occasionally been used as a runway for small aircraft.

The many springs mean that Qeqertarsuaq has plenty of good drinking water. Still, only part of the town is connected to a water supply network – the other households have a water tank and get water delivered or pick up water themselves in containers at one of the town's tapping houses.

Until 1950, Godhavn, now Qeqertarsuaq, was the administrative headquarters of North Greenland. The town's fishermen were among the first cutter fishermen to catch prawns in the 1960s, and a thriving economy and rapidly growing population emerged. Later, the town was impacted by the centralisation of shrimp fishing and processing, which led to population decline.

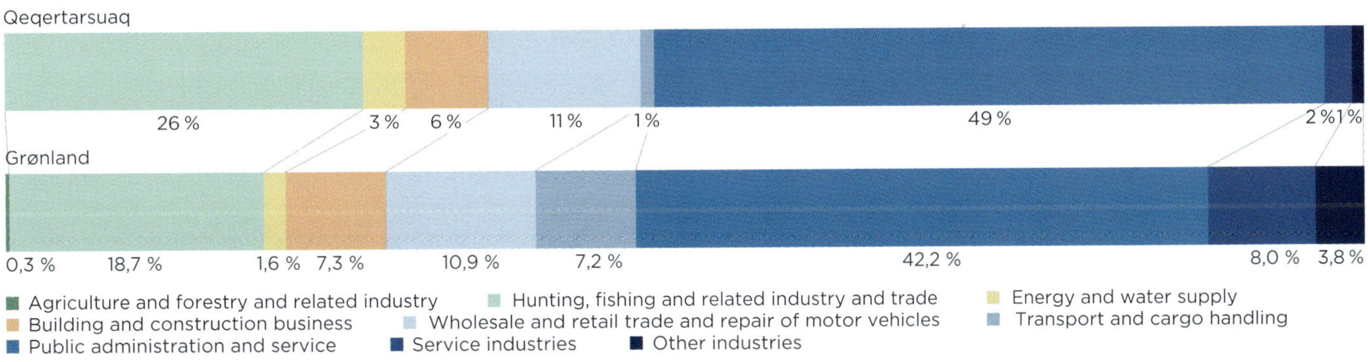

Figure 44. Distribution of jobs by industry sector for people working in the town of Qeqertarsuaq and Greenland as a whole, 2020.

Only a small part of the town has a sewer system, and most households use toilet buckets and collection of latrines. The town has a diesel-based power station. The central roads are paved while other roads are dirt roads.

Qeqertarsuaq has a small hospital that is occasionally manned with doctors and a primary school. However, people must seek elsewhere for secondary education and higher education.

The town has a supermarket and some small shops.

Trade and industry in Qeqertarsuaq

Approximately half of those in a job work in public administration, service and education, while about a quarter work in fisheries and related areas, and just over a tenth work in retail as well as in a number of small areas, including construction and tourism (2019).

Since the 1980s, several initiatives have been taken to develop Qeqertarsuaq's huge potential for tourism, which has been somewhat successful. Tourists have several accommodation options such as Hotel Disko, and there are several providers of tourist activities such as hiking and sailing trips as well as dog sled trips in both winter and summer that go out to Lyngmarksbræen glacier. Qeqertarsuaq used to have relatively many tourists, however, in recent years tourism has been somewhat challenged by the limited transport infrastructure to the town.

Uummannaq in autumn evening light where fjord ice is forming. In winter, the fjord is covered with thick sea ice, so transport to some settlements is possible by car and truck.

ANINGAAQ R. CARLSEN/VISIT GREENLAND, 2021

AVANNAATA KOMMUNIA

Stretching between 68° and 81° N, Avannaata is the northernmost municipality in Greenland. With an area of about 522,700 km^2, it is the same size as Spain and the second largest municipality in Greenland. The municipality has just under 11,000 residents who live spread over its vast coastline in the four towns of Ilulissat, Uummannaq, Upernavik and Qaanaaq as well as in 23 settlements, including the settlements in Uummannaq that in 2017 were hit by a tsunami and therefore have been evacuated. The municipality was established in 2018 in connection with the division of Qaasuitsup Kommunia.

The shortest distance from Ilulissat to Qaanaaq is 1,070 km, or almost the same distance from Copenhagen to Venice, which causes some major logistical challenges in the municipality. Its northern location and the formation of winter ice mean that the districts of Uummannaq and Upernavik can only be served by ships for six months of the year and that the district of Qaanaaq only receives two supply ships each year. This presents some challenges in terms of food and goods supply and requires large frozen stores for the export of the Greenland halibut, which is the primary source of income of the districts. Furthermore, Internet speed and availability are limited in the northern parts of the municipality because they rely on satellite connection. Water supply is very limited in some towns and settlements. Living conditions therefore differ considerably from the towns and settlements further south. Avannaata is particularly affected by climate change because changes in the stability and distribution of sea ice affect hunting and fishing as well as local transport during winter. In the past, it was possible to drive a car across the sea ice for extended periods and between several settlements, however, that period has become shorter and traffic is subject to greater uncertainty.

Despite these challenges, employment in the municipality is relatively high as are incomes because of a good fishing industry and tourism. Together with the UNESCO area Sermeq Kujalleq (Jakobshavn Isbræ) (see p. 83), Ilulissat is Greenland's main tourist destination. Throughout the municipality, dog sledding continues to play a major role in hunting and fishing and increasingly so in relation to adventure tourism.

Just like in Kommune Qeqertalik, there has been extensive offshore oil and gas exploration, which in the past raised hopes for new business opportunities and jobs but also fears about pollution of the important fishing areas.

Some mineral resource projects are underway, including at the abandoned settlement of Moriusaq in the Qaanaaq district where the companies Dundas Titanium A/S and Greenroc Mining want to extract ilmenite sand. Concerns have been expressed by e.g. the Inuit Circumpolar Council (ICC) Greenland (see p. 142) about the possible impact of increased activity on game in the Northern Water Polynya – a permanently ice-free open water area of vital importance to the marine ecology (see p. 62).

Since the 2009 municipal amalgamation, the municipality has been criticised by the northern towns and settlements for largely prioritising Ilulissat when it comes to public investment and initiatives. Thus, the wish for an autonomous northern municipality consisting of the districts of Uummannaq, Upernavik and Qaanaaq has regularly been mooted.

Since its establishment, Avannaata Kommunia, like the former Qaasuitsup Kommunia, has been run by Siumut, and the party won the April 2021 municipal elections. After a prolonged process of dividing the offices among those elected, the former mayor from Siumut chose to become an independent before being able to continue as mayor with the support of other parties.

The municipality prioritises long-term planning with focus on sustainability and consideration for the needs of the future as well as on finding new solutions adapted to local needs and challenges. At the same time, the intention is to strengthen the level of professionalism and service in the municipality.

Ilulissat

Ilulissat is Greenland's third largest town and is located at the head of Qeqertarsuup Tunua (Disko Bay) just north of the Kangia estuary (Ilulissat Icefjord). Ilulissat means 'icebergs'. The large Sermeq Kujalleq glacier at the head of Kangia is the world's most productive, moving at 40 m a day. The glacier calves around 46 km³ of ice a year, creating various sized ice floes and icebergs, some of which are extremely large. The largest icebergs become stranded on a bank about 5 km off the estuary of the fjord. At times, this causes Kangia to pack tightly until the stranded icebergs calve (capsize and become smaller) and are pushed past the bank so the fjord suddenly 'shoots out'.

The town's Danish name is Jakobshavn, named after the merchant Jakob Severin, who founded the area's first trading post in 1741.

Ilulissat is the main town of Avannaata Kommunia. 349 people live in the district's four settlements and 4,710 live in the main town.

Ilulissat's history

The ice that releases oxygen and nutrients contributes to an extraordinarily abundant wildlife in the fjord, so the estuary of the fjord has been home to several and, by the standards of those days, large settlements. The Saqqaq people (see p. 90) was, after a short gap, followed by the Dorset people (see p. 92), after which followed a long period without settlements until the Inuit settled around year 1150.

One of the country's largest known Inuit settlements is Sermermiut (place of the glacier people) at the Kangia estuary on the north side of the fjord a few kilometres from present-day Ilulissat. From the 17th century, the population increasingly came into contact with European whalers and started trading tusks.

The establishment of the trading post in 1741 provided more stable trading opportunities for the Inuit, who traded e.g. blubber, skins and ivory for firearms, equipment and beads. Slowly, the population at Sermermiut began to move the few kilometres to the colony with its trading post that in 1805 had 86 residents.

A seminary for Catechists was established in 1848, which led to in increased cultural awareness.

The last residents moved from Sermermiut in 1850, and the colony's population reached 262. Already 30 years later, the population had grown to 418. With the many people and the great considerable trade in blubber and skins, the catchment area became too small and there were periods of famine and hardship. New trading posts were established in the colonial district in order to spread the population. Ilulissat was hit by a severe tuberculosis outbreak in 1867 where a third of the population died.

As kerosene and other crude oil products became widespread in Europe, the demand for blubber dropped significantly. As a substitute for earnings from seal hunting, long-line fishing of Greenland shark became the primary source of income from the 1890s, and the shark's liver was exported due to its high oil content. Shark fishing was largely supplemented by longline fishing for Greenland halibut from the fjord ice for own consumption and for feeding dogs, which gradually led to exports to Denmark with 22 tons of Greenland halibut being traded in 1899. Thus, the transition from hunting to fishing began earlier in the Jakobshavn area than in the rest of the country. This meant that the population to a greater extent transitioned to money economy because it was more difficult to become self-sufficient as a fisher than as a seal hunter.

With fishing as the new main trade, the settlement policy changed, and during the interwar years, the Danish authorities wanted to gather more of the district's population in Ilulissat. The population in 1938 was at 528.

The decolonisation and modernisation after World War II strengthened Denmark's desire to gather the population, e.g. to be able to intensify fishing. At the same time, large quantities of cod came to Qeqertarsuup Tunua (Disko Bay), and cod fishing further boosted the economic and population growth while making the Greenland halibut less important. However, the cod adventure did not last for long, and the cod already started disappearing from Qeqertarsuup Tunua in the 1960s. At the same

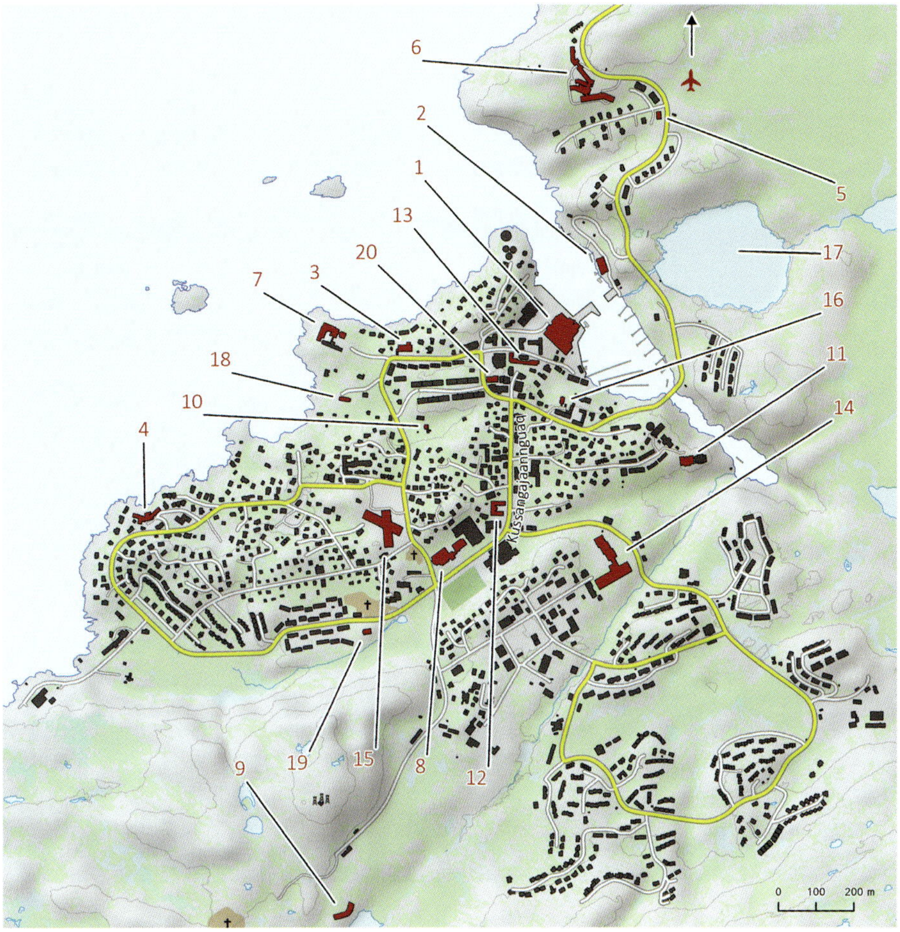

Ilulissat, street map. 1-2) Fish factory. 3) Hotel Hvide Falk. 4) Hotel Icefiord. 5) Hotel Avannaa. 6) Hotel Arctic. 7) Hospital. 8) Sports centre. 9) Icefjord Centre. 10) Knud Rasmussen's birthplace (Museum). 11) College of social education. 12) Vocational school. 13) Town hall 14-15) Primary school. 16) Ilulissat Museum. 17) Water reservoir. 18) Zion Church. 19) Qullisat Church. 20) settlement hall.

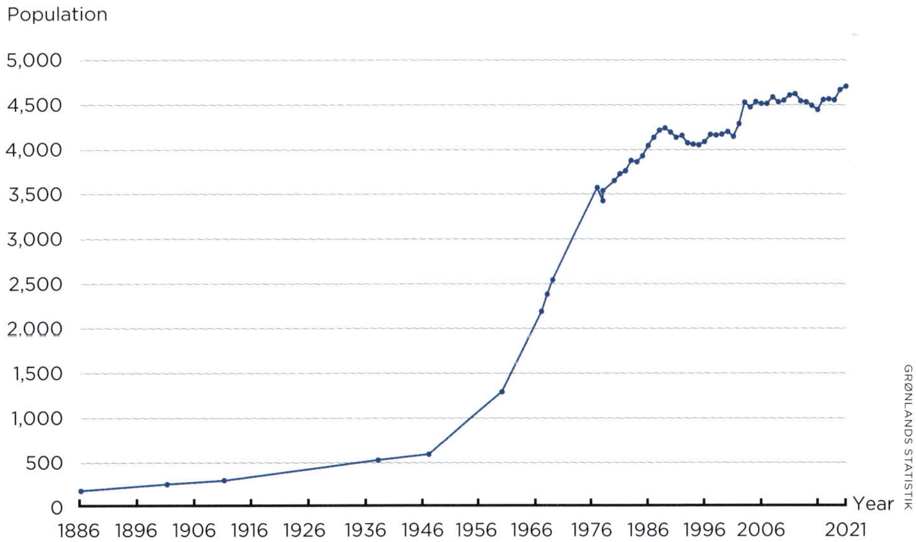

Figure 45. Population development in Ilulissat 1886-2021.

Aerial photo, Ilulissat. In the background is Kangia (Ilulissat Icefjord) and the many large icebergs it releases into Qeqertarsuup Tunua (Disko Bay).

time, the fishing of shrimp from cutters, which had much greater economic importance than cod, started gaining momentum. Activities meant that an Atlantic wharf was built in 1960, and in 1961, a dam for the water reservoir was established to ensure water supply. In 1962, the Royal Greenland Trading Department (KGH) established an actual fish factory that was expanded in 1969 to increase shrimp production. In addition to shrimp, frozen cod, Atlantic catfish, Greenland halibut, etc., were produced and the factory employed 200 employees during the season. In 1967, a private fish factory producing shrimp and frozen Greenland halibut fillets was established.

The development led to considerable construction activity, and by 1969 a total of 20 contractors employed a total of 113 people, most of whom came from Denmark. During an intense period, much of the town's central infrastructure was established such as the heliport (1965), Hotel Hvide Falk (1970), the town's two large primary schools (1974 and 1977), a sports hall (1976) as well as several large blocks of flats. Water supply and a sewer system was established in parts of the town and the hospital was expanded (1980).

In connection with the forced closure of the coal mining town of Qullissat by the Danish State in 1972, many of the locals were moved to Ilulissat where most moved in to blocks of flats in a neighbourhood that by many has since been called Qullissaaqqat (Little Qullissat).

In 1969, the population reached 2,382, and by 1977 the town's population had grown to 3,572. From the mid-1980s, fishing for Greenland halibut started again, and in 1998, Royal Greenland established a Greenland halibut factory. In 2010, a group of local fishermen founded the Halibut Greenland factory as a competitor to Royal Greenland.

Development of tourism in Ilulissat

South Greenland use to be the main tourist destination, but up through the 1980s tourists shifted their focus, and Ilulissat with Kangia (the Icefjord) became the country's most visited tourist destination. A development that was gradually supported by tourism-related infrastructure. In 1984, Ilulissat domestic airport was established approx. 3 km north of the town, and Air Greenland opened Hotel Arctic next to the road to the airport. In the coming years, several tour operators started working from the town. The increased activity led to greater demand for accommodation, which is why the accommodation for the contractors during the construction of the airport was turned into a hostel. Over the past few decades, Hotel Icefiord and Hotel Avannaa have been opened, and all hotels in the town have been expanded. Capacity was increased again in 2021 with the opening of Hotel Ilulissat in the town centre. Today, Ilulissat has five hotels, four guesthouses/hostels as well as several bed & breakfasts.

Interest in visiting Ilulissat only increased when The World Heritage Committee in 2004 included Kangia (Ilulissat Icefjord) on the UNESCO World Heritage List. The Ilulissat Isfjordscenter opened in 2021 in a partnership between the Greenland Self-Government, Avannaata Kommunia and Realdania. The centre has created much international and national attention for its architectural qualities and as an information centre for Kangia (the Icefjord) as well as the culture, nature and history of the area.

Another attraction is the museum area which features several of the old colonial houses, including Knud Rasmussen's birth home (see p. 167).

Geography and infrastructure of Ilulissat

Ilulissat and the surrounding landscape are characterised by low and soft glaciated mountains and vast plains. The town is relatively level without major elevation differences, except where the plateau descends relatively steeply towards the port. Architecturally, the town is a mix of single-family houses, terraced houses, new blocks of flats as well as housing blocks from the modernisation period under the Danish centralisation policy. Over the past few decades, the town has been extended north along the road to the airport.

Most of the town's housing is connected to the water network and there is plenty of water in the catchment area to supply the town, however, during spring thaw, the water quality may be reduced. About three-quarters of the town is connected to the sewer system, while the rest of the town has sludge tanks or toilet buckets and collection of latrines. In areas without a sewer system, primarily in the western part of the town, the grey wastewater is discharged from washing and bathing directly to the terrain.

The majority of the town's roads are paved, and the central roads have street lighting. The Kussangajaannguaq road, which goes from the square above the port in a southerly direction towards Kangia and the UNESCO area, is a popular hiking for the many tourists. Most of the 20 tourist agencies and tour operators as well as a large part of the town's shops are therefore located along the road.

Ilulissat is supplied with electricity from the hydropower station at Paakitsoq about 45 km northeast of the town. The town uses less than 60 % of the station's capacity, and the station will therefore be able to support a major expansion of the town.

MADS PIHL/VISIT GREENLAND, 2014

Cruise ship by Hotel Arctic in Ilulissat. The town's big attraction is Kangia (Ilulissat Icefjord), where the vast masses of ice from the world's largest glacier, Sermeq Kujalleq, are pushed towards the sea. 5 km off the mouth of the fjord is Iluliarsuit (Icefjeldsbank), where the largest icebergs strand, so the fjord sometimes packs tightly until the icebergs calve and are pushed over the bank.

ADAM MØRK/REALDANIA, 2021

ADAM MØRK/REALDANIA, 2021

Ilulissat Icefjord Centre opened in 2021 to share knowledge about Kangia (Ilulissat Icefjord) and the UNESCO World Heritage Site. The Icefjord Centre was designed by architect Dorte Mandrup-Poulsen, and the building's shape is inspired by the wingspan of a snowy owl. It is possible to walk on the roof and look out over Ilulissat and the ice fjord.

Ilulissat Hospital serves as the district hospital for Avannaata Kommunia. In the graduating classes, the two primary schools also have students from the district's four settlements. Perorsaanermik Ilinniarfik (College of Social Education) is located in Ilulissat and is the town's only higher education institution.

Royal Arctic Line calls at Ilulissat weekly with a container ship, which may be cancelled in winter due to ice. During summer, the town is the northernmost destination for the Arctic Umiaq Line's coastal passenger ship, and the Disko Line calls on the other towns and settlements in Qeqertarsuup Tunua (Disko Bay) with smaller ships. Domestic flights depart for Kangerlussuaq and most large towns on all weekdays, and in summer, there are also flights to Reykjavik. During winter, there is a helicopter connection to the settlements in the district and those of the towns in Qeqertarsuup Tunua that do not have a runway.

An international airport is currently being built at Ilulissat, which is expected to be ready in 2024. The airport is to have flights to Denmark in summer and it is also planned to establish summer connections to other European and US destinations. In 2022, the municipality will begin expanding the existing airport road to two lanes all the way to the airport.

In connection with the coming airport, the municipality plans a road east of the existing airport as well as areas for a new commercial port north of the airport because the physical framework makes it difficult to expand the existing port and limits the possibilities of fish factories. Nordre Næs and the area around the drinking water lake have been designated as future residential, shopping mall and commercial areas, and construction has begun at the drinking water lake.

The UNESCO area prevents urban development to the south, but the primary obstacle is the drinking water lakes and surrounding water barrier zones northeast of the town. It is therefore planned to use other lakes and thus move the water barrier zone further into the hinterland.

Trade and industry in Ilulissat

Fishing for Greenland halibut and shrimp is Ilulissat's absolute biggest source of income, followed by tourism.

The Greenland halibut is primarily caught by longline from open dinghies near the bank off Kangia (Ilulissat Icefjord). In winter, longline fishing takes place from the ice on Kangia, which is reached by dog sled. Shrimp, however, is caught from trawlers. Royal Greenland primarily produces frozen shelled shrimp and, to a somewhat lesser extent, Greenland halibut and has up to 120 employees. Halibut Greenland, which

There are almost 800 children in Ilulissat's two primary schools, Atuarfik Mathias Storch with around 310 pupils and Atuarfik Jørgen Brønlund with around 305 pupils.

SERGIO PITAMITZ/AP/RITZAU SCANPIX, 2015

has up to 40 employees, primarily processes Greenland halibut. In 2020, a total of 21,749 tonnes of fish and shellfish were traded in Ilulissat and the nearby settlements. Shrimp accounted for 63 % and Greenland halibut 33 % of the total quantity. The trading value of Greenland halibut is higher and amounted to DKK 154.7 million, whereas it was DKK 132.3 million for shrimp.

The total workforce of Ilulissat is 2,425 people. In terms of employment, the largest industry is public administration and service, including education and health care, accounting for 40 % of the town's jobs. Trawler fishing and coastal fishing as well as fish processing and related industries account for about 25 % of jobs, while trade, etc. account for 10 %, transport 6 %, construction 9 %. Total employment in the tourism sector including accommodation and hospitality is around 7 % during the low season and around 14 % during the peak season.

There are three large supermarkets and several smaller grocery shops in the town, a number of clothing, electronics and souvenir shops as well as restaurants and small cafes. There are also some bars, mainly in the hotels that, together with the venue Naleraq, make up the town's nightlife.

Before COVID-19 caused a slowdown in international tourism, there were 31,103 overnight guests in Ilulissat in 2019, of whom 11,844 were from Greenland. In the same year, there were 14,762 cruise tourists.

Figure 46. Distribution of jobs by the industry for people working in the town of Ilulissat, the settlements in Ilulissat district and Greenland as a whole, 2020.

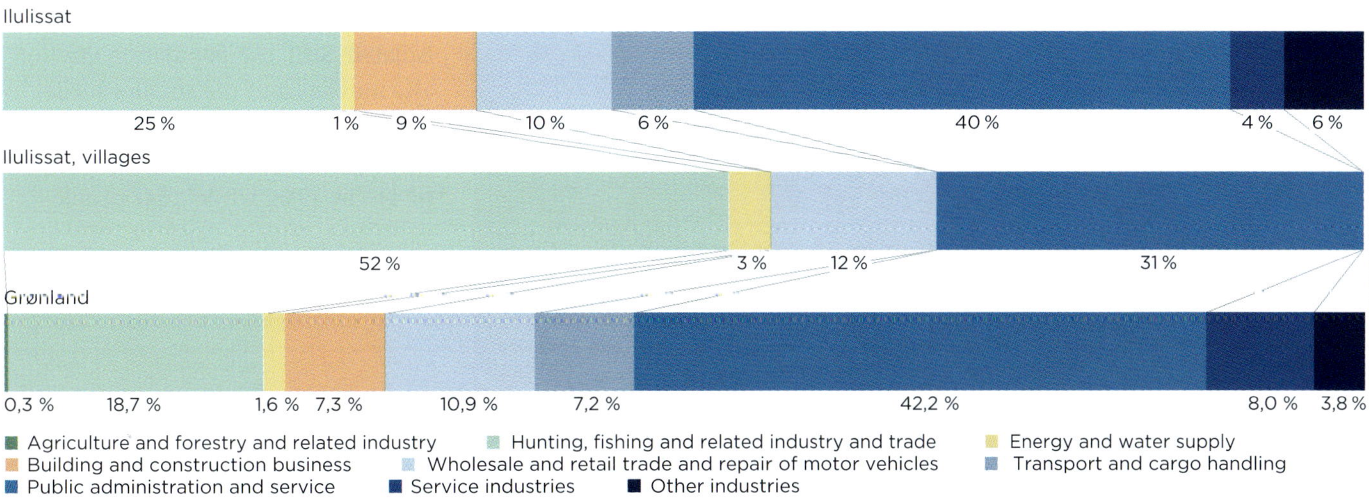

GRØNLANDS STATISTIK

Uummannaq

Uummannaq means 'invigorating' or 'heart-shaped' and is named after the town landmark, the 1,170 m high mountain north of the town. Unlike most other colonies, Uummannaq was not given a Danish name, but was formerly spelled Umanak. The town with 1,428 residents (2021) is located on a roughly 12-km^2 large island in a sprawling and beautiful fjord system with high steep mountains and at the head of the fjords you will find active and productive glaciers. The inner part of the fjord system has a mainland climate and thus often high pressure and limited precipitation. In winter, there is meter-thick ice in large parts of the fjord, so that you can drive by car to many of the settlements. Uummannaq is known for its many hours of sunshine (2,000 h/year).

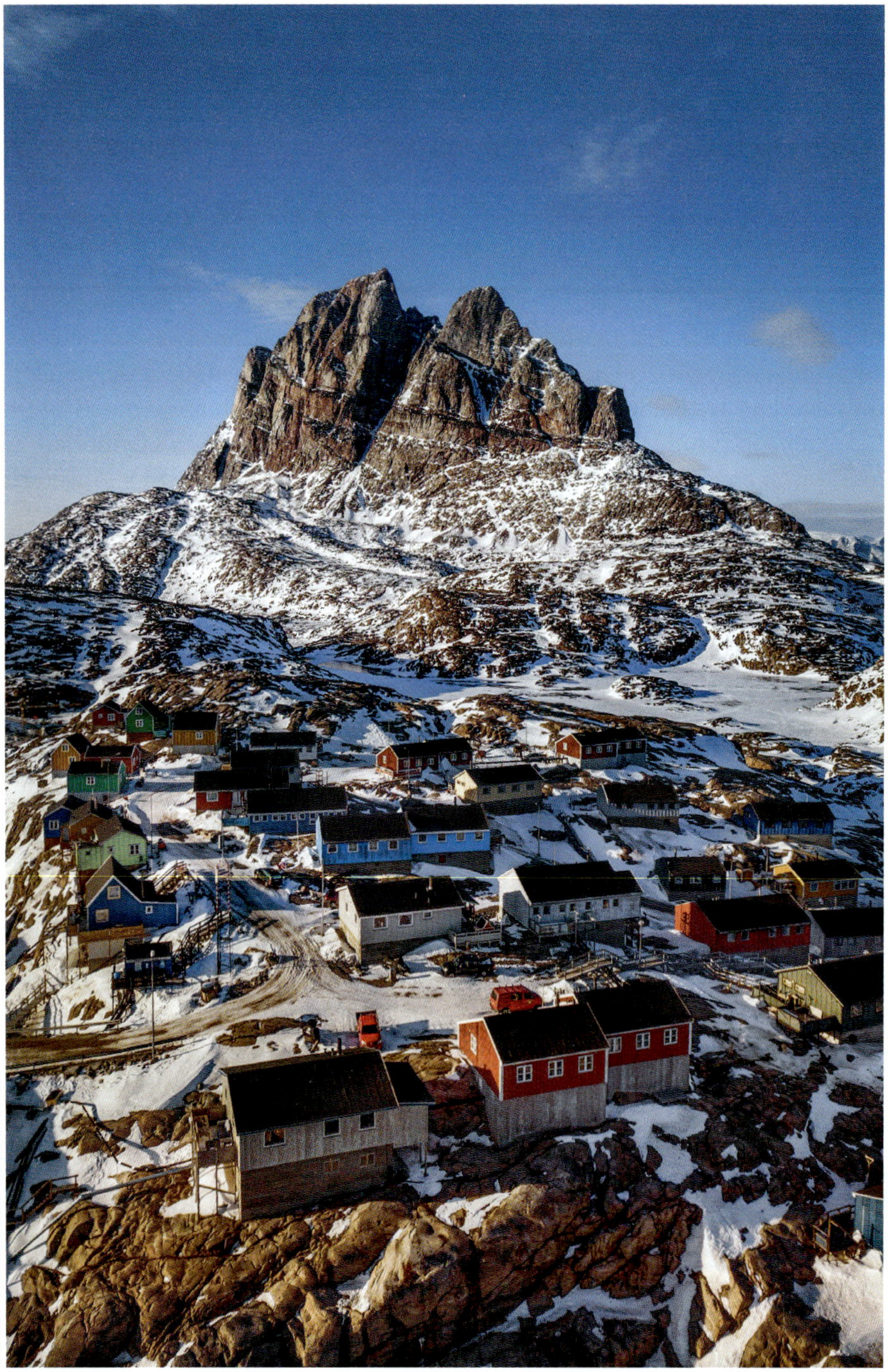

ANINGAAQ R. CARLSEN/VISIT GREENLAND, 2021

View of the town with the Uummannaq mountain in the background. Uummannaq means 'invigorating' or 'heart-shaped' and is named after the town landmark, the 1,170 m high mountain north of the town.

History of Uummannaq

In the 17th and 18th centuries, Europeans, primarily the Dutch, hunted for whales in the fjord system and engaged in tusk trade and other interaction with the Inuit population. This was considered problematic by the Royal Greenland Trading Department (KGH), but it was not until 1761 that a more permanent station was established for the Danes, and so the colony of Umanak was founded in 1763. Sealskin, narwhal and walrus tusk as well as polar bear skins were the main sources of income in the colony until the post-World War II modernisation.

From 1933 to World War II, marble was quarried in Mârmorilik (later Maarmorilik) relatively close to the settlement of Ukkusissat. Up to 300 workers, mainly from Denmark, lived in a barrack town by the open marble quarry. Besides modest interaction with the local Inuit, the marble quarry had limited impact on the district.

Until the modernisation after World War II, the colony of Umanak, with its many remote trading posts and settlements, served as mainly a hunting district and was relatively isolated. Still, the population gradually moved from the smaller settlements to the colony, which in 1938 had a population of 308, which grew to 449 by 1948, while 1,041 people lived in the district's remote trading posts and settlements.

During the post-war years, the district's revenue from seal hunting was in sharp decline, and the increasing fishing for Greenland shark was unable to compensate for the loss of revenue. The Danish administration's central reports, the Greenland Commission of 1948 (G-50) and the Greenland Committee of 1960 (G-60), which formed the basis of the modernisation, both assessed that the district did not have an adequate business base and recommended partial depopulation. Relatively many people were convinced to move to Qeqertarsuup Tunua (Disko Bay) or further south to join the cod fishing and fishing industry – however, only few were successful. Some went back, several of them to live as hunters across the district's settlements and smaller residences. Referring to G-50 and G-60, only modest investments in infrastructure were made in the district and the population was among the poorest in the country.

Marble quarrying at Maarmorilik was resumed from 1966 to 1973. From 1973, a lead and zinc mine, named the Black Angel after a distinctive marking on the mountainside, was started in the old marble quarry. The mine tunnel was located at approx. 600 m's height, and the ore was transported by cableway down to the crushing mill at the old marble quarry. The mine employed 200-360 people. Around 20 % of the workers where Greenlandic, many of whom came from the district. Subject to current rules and collective agreements, local Greenlandic workers received a lower salary than non-local workers, which, in 1977 led to the first real strike for equal pay in Greenland and where the National Confederation of Trade Unions of Greenland was one of the negotiating parties.

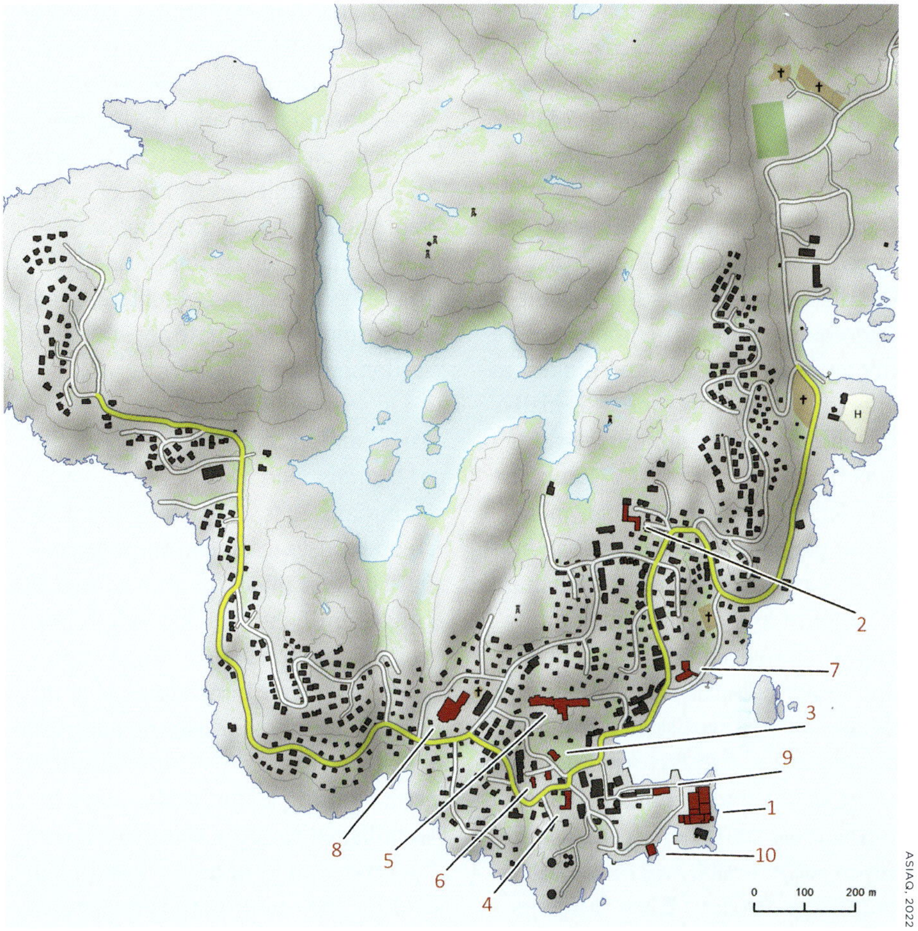

Uummannaq, street map. 1, 9 and 10) Fish factory. 2) Orphanage. 3) Uummannaq Church. 4) Municipal office. 5) Primary school. 6) Uummannaq Museum. 7) Health centre. 8) Sports centre.

Figure 47. Population trends in Uummannaq 1886-2021.

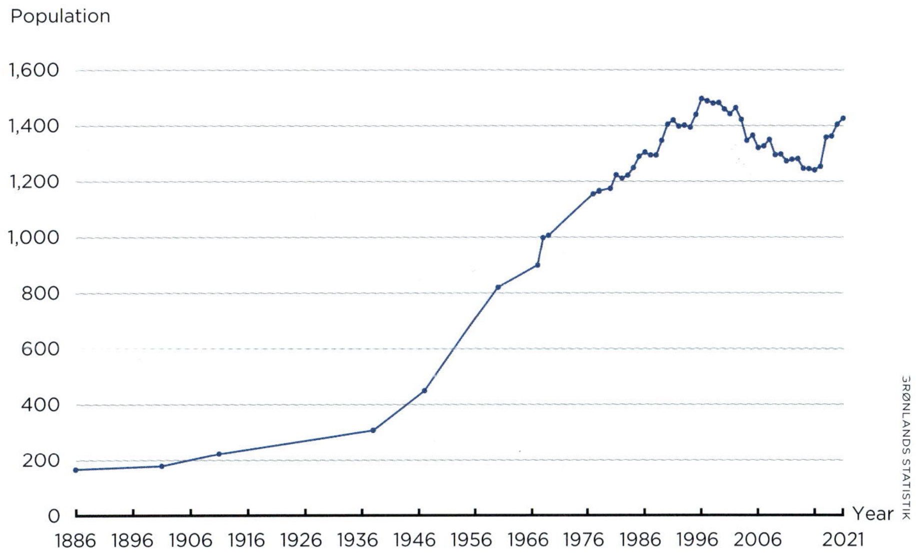

The Black Angel was decommissioned in 1990. There were plans to reopen the mine in 2010 and some preparations were made until 2015, but mining operations never resumed.

At the end of the 1970s, fishing and trading of Greenland halibut experienced a boom, which meant that the district's residents gradually went from being among the poorest to being among those with the highest average household income, especially in some of the settlements. Fishing of Greenland halibut makes Uummannaq one of the districts that, measured per capita, contribute the most to export income. Greenland halibut is caught from dinghies in summer and using dog sled or snowmobile and longline during ice formation, which is why the distance from fishing ground to fish factory must be as short as possible. People has therefore largely remained in the district's seven settlements, and in 2017, 981 out of the district's 2,236 residents lived in the settlements.

In 2017, the district's two northernmost settlements, Nuugaatsiaq and Illorsuit, were hit by a tsunami as a result of a major mountain landslide, and the two settlements were evacuated. The about 170 evacuated citizens were mainly relocated to Uummannaq or the other settlements in the district. In the summer of 2021, geological surveys identified a risk of more and larger mountain landslides in the fjord system, which led to considerations to evacuate several of the settlements in the district.

Development and geography of Uummannaq

Uummannaq is a very hilly town with steep roads, which can impede traffic. The vast majority of dwellings are single-family homes, a good proportion of which have running water, while the other households pick up water at tapping houses or have it delivered by tanker. For most dwellings with running water, drains have been established for grey water from the bathroom and kitchen, while in general for the town there is no sewer system for black water from toilets. Thus, most households use toilet buckets and collection of latrines, while some have a sludge tank.

The evacuation of Nuugaatsiaq and Illorsuit has brought significant construction activity in recent years, mainly in the town's northwest end, where the municipality plans future urban development.

In 1999, an airport was established at the settlement of Qaarsut on the Nuussuaq Peninsula, forming the south side of Uummannaq Fjord. From Qaarsut there is a helicopter link to Uummannaq and the district settlements. The intention was to improve the air service to Uummannaq that had until then been based on helicopter from Ilulissat, and Qaarsut was chosen due to more stable wind conditions and better space than on Uummannaq Island. It has since been debated whether the airport at Qaarsut has effectively improved or impaired air service between Uummannaq and Ilulissat.

Uummannaq receives supplies by ship from the Royal Arctic Line, allowing the town to export Greenland halibut every week from mid-May to mid-December. The first and the last ship may risk cancellation due to ice. In 2019, a restoration of the town's quayside facilities and an expansion of areas for the stores of the fish factories were initiated. In order to improve the conditions of dinghy fishermen, a breakwater is planned in what is known as Smørhavnen.

In addition to the necessary supply such as a shop, power plant, waterworks and telecommunications, Uummannaq has a hospital, a school that also teaches the children of the settle-

STUART MCINTYRE/BAM/RITZAU SCANPIX, 2015

The kindergarten in Uummannaq with the characteristic heart-shaped mountain in the background. The town is very hilly, making traffic challenging and the implementation and operation of any infrastructure

The port of Uummannaq. Of the town's working population, 36 % are engaged in hunting, fishing and fish processing, and in the district's settlements, it is 61 % of the population. Fishing is primarily for Greenland halibut, with dinghies during the open-water period and longlines from the sea ice in winter.

KÅRE HENDRIKSEN, 2020

ment in the graduating classes, and a boarding house for school children, as well as an orphanage that accepts children from all over the country.

Trade and industry in Uummannaq

With approximately 300 commercial hunters, most of whom have fishing as their primary activity, fishing and fish processing is the district's main industry and source of income. During the season, there are about 50 employees in the Royal Greenland factory, and somewhat fewer in the factories of Avannaa Seafoods and Sigguk Uummannaq ApS in Uummannaq. A similar number of people are employed in factories in the settlements. The Greenland Institute of Natural Resources in Nuuk estimates that fishing pressure in the district is above the sustainable level. A contributing cause is that some fishing vessels come from southern districts and fish on the Uummannaq quota.

Some jobs in Uummannaq can be found in administration and service and, to a lesser extent, building and construction. Growing tourism contributes ever more jobs both in summer and in winter, with dog sled tourism in particular attracting travellers. Local businesses offer accommodation for 40-50 people in lodging houses and flats. The town has a supermarket near the port as well as two convenience stores and a third under construction in the northwestern part of the town. Nightlife is centred around the bar called Parabolen.

Figure 48. Distribution of jobs by industry sector for people working in the town of Uummannaq, the settlements in Uummannaq district and Greenland as a whole, 2020.

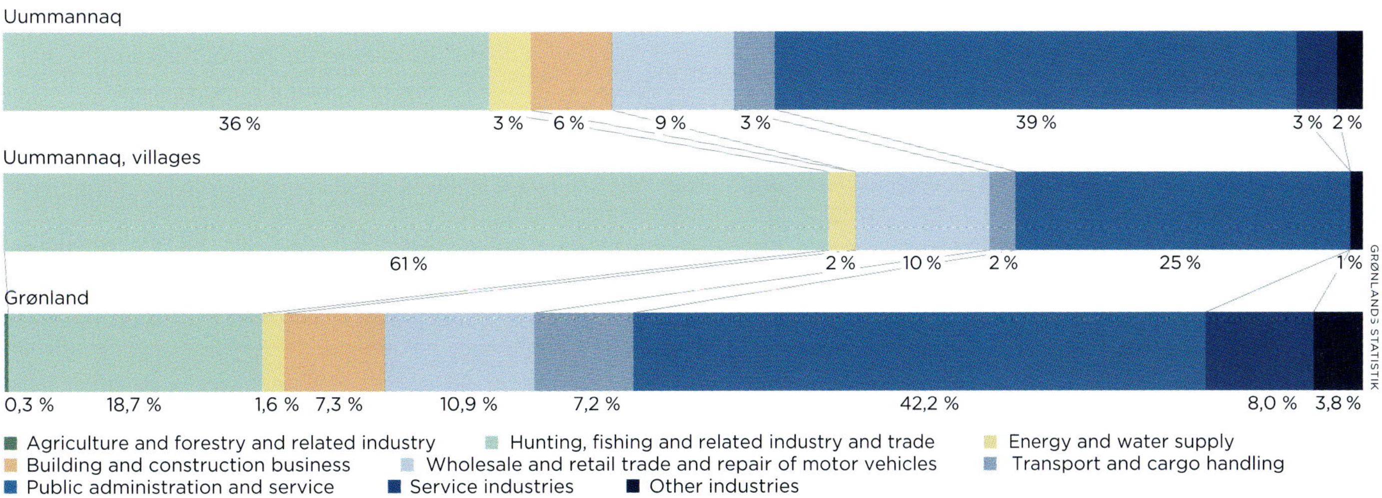

Upernavik

Upernavik means 'spring hunting ground', and the town is located on a relatively small island of the same name. In Upernavik district, the majority of the population lives in the settlements with a total of 1,655 inhabitants compared to 1,076 in the town. This means that 61% of the population lives in the district's nine settlements – and this is unusual for Greenland. It has a High Arctic maritime climate and thus relatively much precipitation.

History of Upernavik

The location of the colony of Upernavik can be attributed to the hunting ground of the Inuit at the place and the bay where sailing ships could lie safely at anchor sheltered from the prevailing winds. As one of the few Danish colonies, Upernavik was given a Greenlandic name.

The first attempt was made to establish Upernavik as a Danish colony in 1772, but due to the harsh weather conditions and the lack of boat traffic from the southerner colonies, the colonists left the area several times. In 1823, Upernavik became a remote trading post under Godhavn (Qeqertarsuaq) and in 1826 an independent colony. Upernavik was the northernmost official Danish presence and colony in Greenland until the Danish State assumed responsibility for Thule district in 1937. Until the inter-war period, the colony only included parts of the present-time Upernavik district; the area north of Holms Ø with, among other things, the settlement of Kullorsuaq being in 'no man's land'. In 1939, Upernavik had 240 inhabitants and grew to 367 in 1948, while the entire district had 1,459 inhabitants.

In the post-war years, the Danish administration estimated that the district – like Uummannaq – had no future business base and wanted most of the district's population to move to Qeqertarsuup Tunua (Disko Bay) and further south to participate in the cod fishery or work in the fishing industry. The Danish administration managed to convince a portion of the population to relocate – though without much success for the majority, which is why several moved back again. In parallel, the population was centralised in fewer settlements, from 24 settlements in 1938 to nine settlements today, and significant growth was seen in the town of Upernavik, where in 2003 the population peaked at 1,218.

Upernavik continued as a distinctive hunting district and for many years was one of Greenland's poorest. Even after the introduction of home rule in 1979, the central government discussed how to get portions of the population to move south. However, from the late 1980s the fishing of Greenland halibut was gradually increasing, and for a number of years, Upernavik has been one of the districts which, as measured per capita, has made the greatest contribution to the inshore fishing of Greenland halibut and export revenues. Greenland halibut fishing has increased the district's income base, and average household income has for periods been at the top of the list in some of Upernavik's settlements.

The majority of Greenland halibut fishing in the nearly 400 km long district takes place in summer by dinghy and in winter by longline from sea ice. It requires a relatively short distance to the fishing grounds and local trading posts, which is why there are fish factories in the town of Upernavik and seven of the district's settlements. Most hunters in the district live off a combination of fishing and hunting of seal, walrus, narwhal, polar bear and birds for their own household, as well as sales to the rest of the country that are still of vital importance.

Since the municipal amalgamation in 2009, when Upernavik had 1,157 inhabitants, there has been minor migration away from the town, partly because some administrative jobs moved to Ilulissat.

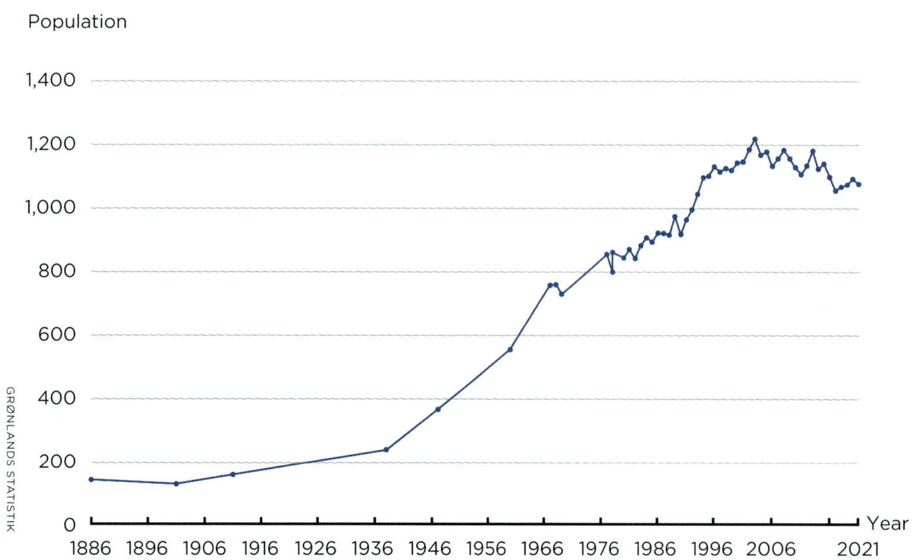

Figure 49. Population trends in Upernavik 1886-2021.

Development and geography of Upernavik

The town of Upernavik is located on a relatively small island facing the sea in a district characterised by an elongated archipelago with several very active glaciers.

The many single-family houses and semi-detached and terraced houses are located up a sloping mountainside, while the shop, factory and most public buildings are located at the port. The road network is limited and consists largely of steep dirt roads. Houses lie close to the roads without the possibility of establishing ditches, and this often causes problems with meltwater and grey water.

Despite the town's population numbers, no water supply has been established, and only the local hospital and old age home have running water. The households either have water delivered by tanker to the house's water tank, which is a relatively expensive option, or pick up water themselves at one of the town's tapping houses. The lack of water supply and modest consumption by citizens can be ascribed to the fact that the island has only a small lake for water supply that does not meet the real need and does not always provide good water quality. At the same time, the limited water resources imply that the Royal Greenland fish factory simply freezes the whole Greenland halibut, thereby failing to exploit potential income from fish processing. On the neighbouring island of Akia less than a kilometre to the southeast, large lakes are found, from which an adequate

Measured per capita, the approximately 2,700 people in the Upernavik district are among the largest contributors to Greenland's export revenue.

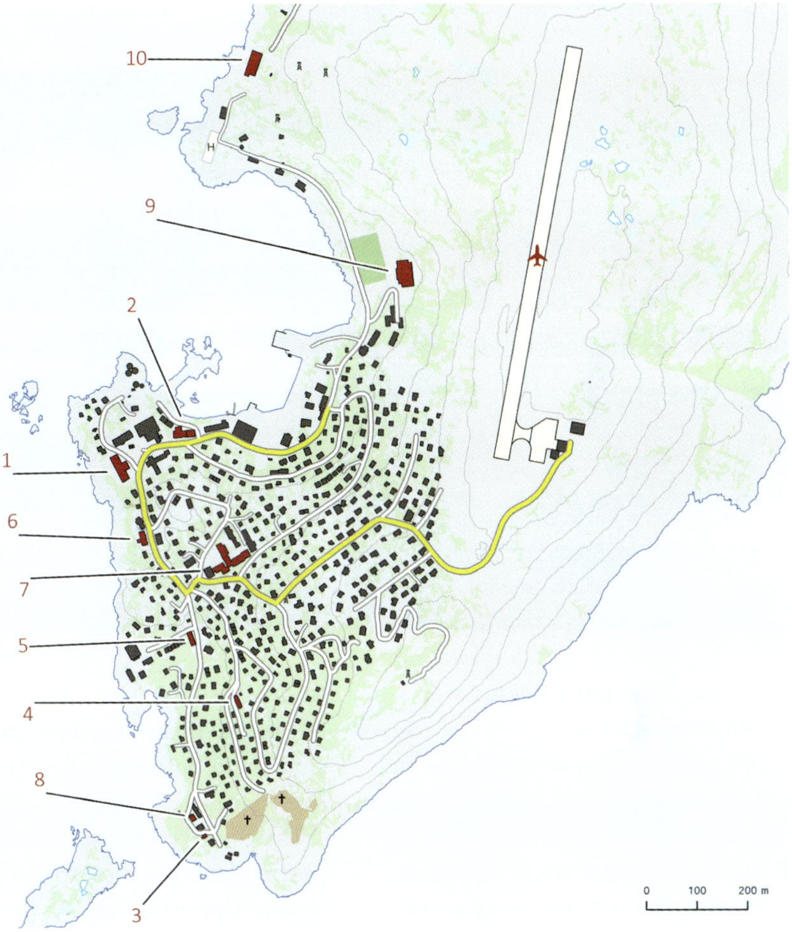

ASIAQ, 2022

Upernavik, street map. 1) Health centre. 2) and 10) Fish factory. 3) Old church. 4) Upernavik Church. 5) settlement hall. 6) Municipal office. 7) Primary school. 8) Museum. 9) Sports centre.

LINDA KASTRUP/RITZAU SCANPIX, 2015

MARTIN N JOHANSEN/BIOFOTO/RITZAU SCANPIX, 2014

In the Upernavik district, 61 % of the population lives in the nine settlements, where the main industry is Greenland halibut fishing, fish processing and hunting. In the cold winter months, when dog sledding and fishing from the sea ice are the norm, kamiks and polar bear skin trousers are the best attire.

water supply could potentially be established.

Upernavik has no sewer system and toilet buckets and collection of latrines are consistently used in the town, partly due to the sparse water resources.

There is limited storage space for boats and containers, and the shortage of space is a problem for private individuals and businesses.

Boat traffic to Upernavik is possible only from mid-May to the end of November. During his period, Royal Arctic Line's supply ship has weekly calls at Upernavik and the first and last ship could potentially be cancelled due to ice.

The country's highest-located airport is established on top of the island of Upernavik and in the winter months has a flight connection to Ilulissat a few times a week and in the summer somewhat more often.

The town is expanding to the north, which has made way for another fish factory established by Polar Seafood. To create opportunities for future business areas, reception station and residential areas, the municipality is extending the road. The purpose of the reception station is to decommission the open dump (waste landfill) and the outdated incinerator which creates odour and smoke nuisance in the southern and more historic part of the town.

Population, trade and industry of Upernavik

Out of the district's 1,769 working age residents, just under 400 have a commercial hunting licence and thus derive most of – or all – income from hunting and fishing. In addition, there are around 200 employees at the nine fish factories, making fishing of great importance.

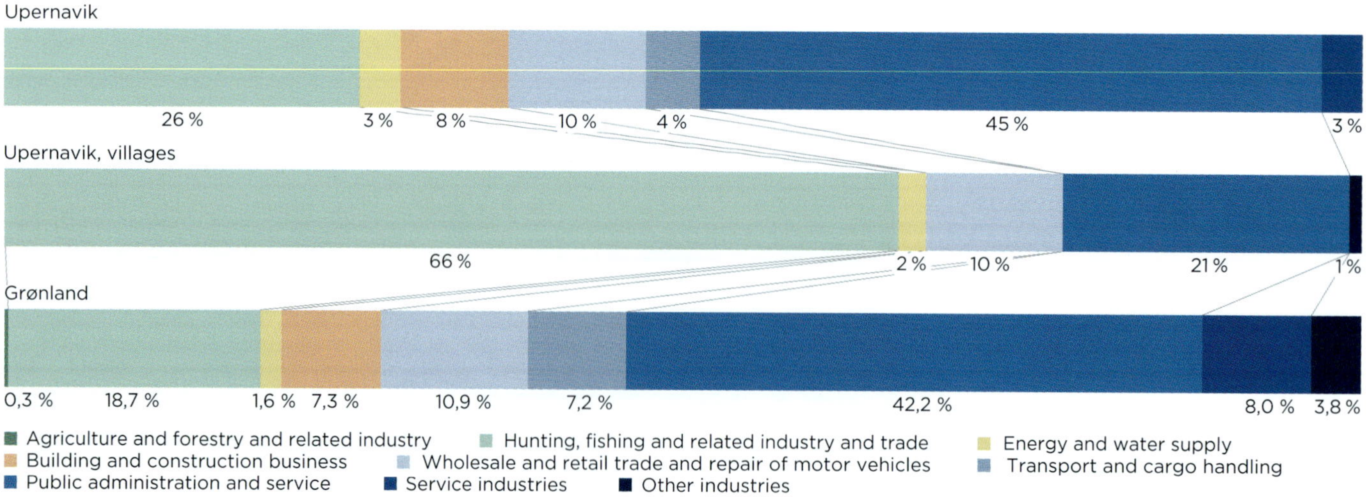

GRØNLANDS STATISTIK

Figure 50. Distribution of jobs by industry sector for people working in the town of Upernavik, the settlements in the Upernavik district and Greenland as a whole, 2020.

The school in Upernavik, which in the oldest grades also teaches pupils from most settlements, also contributes a number of jobs both at the school itself and the associated boarding house for school children. In addition, jobs in administration and service, as well as, to a lesser extent, building and construction. The town has a single supermarket (Pilersuisoq) near the port as well as two kiosks and a clothing shop.

Upernavik is considered to be an enterprising and well-functioning town, despite the challenged infrastructure. The town is experiencing a gradual increase in tourism both summer and winter and mainly with guests who want to experience 'original Greenland' with its many settlements and the life of hunters and fishermen. In summer, there are boat trips between the settlements and kayaking trips in the archipelago, and in winter, there is an increasing sledding tourism. Local businesses and private individuals offer accommodation for 40-50 people in a room, flat or an entire house.

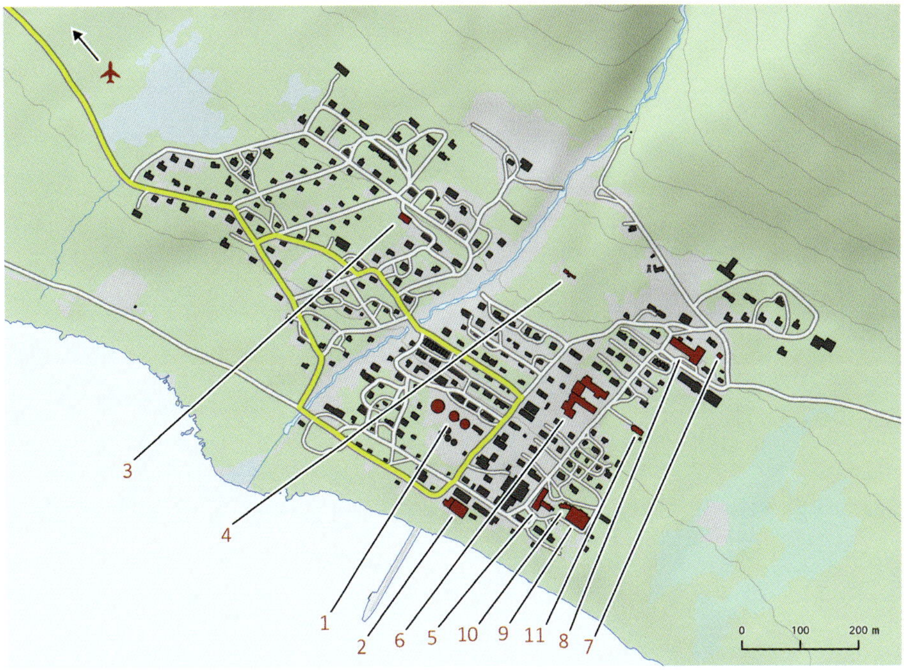

Qaanaaq, street map. 1) Water tanks. 2) Fish factory. 3) Lille Hotel. 4) Magnetic Observatory. 5) Municipal office. 6) Primary school. 7) Knud Rasmussens House. 8) Health centre. 9) Sports centre. 10) settlement hall. 11) Qaanaaq Church.

Qaanaaq

Qaanaaq is the current Greenlandic name for Thule and means 'caves by the beach'. Qaanaaq is Greenland's youngest and northernmost town with 619 inhabitants (2021). The town was established by the Danish State in 1953 after the relocation of the population from what was then Thule, which was called Uummannaq in Greenlandic. Thule had been named after the trading station established by Knud Rasmussen and Peter Freuchen in 1910 – a name inspired by the Antiquity designation of a happy island at the northernmost end of the world.

During the majority of the year when there is ice on the fjord, the Qaanaaq area has continental climate with extremely little precipitation – also known as High Arctic desert.

History of Qaanaaq

The population north of Qimusseriarsuaq (Melville Bay) is a later immigration from Canada than the other Inuit, and they define themselves as Inughuit, having their own language (dialect) and a distinct hunting culture. For example, Inughuit are the only ones in Greenland who lived in igloos on their hunting trips. Inughuit were formerly called Polar Eskimos by Europeans.

The population lived in isolation from the rest of Greenland with only sporadic contact with Inuit in the northern part of Upernavik district, and from the early 19th century, visits by single expeditions. Beginning in 1886, American Robert Peary conducted several expeditions in the area (see p. 110). This caused concern that the United States would lay claim to northernmost Greenland – and it became a contributing reason for The

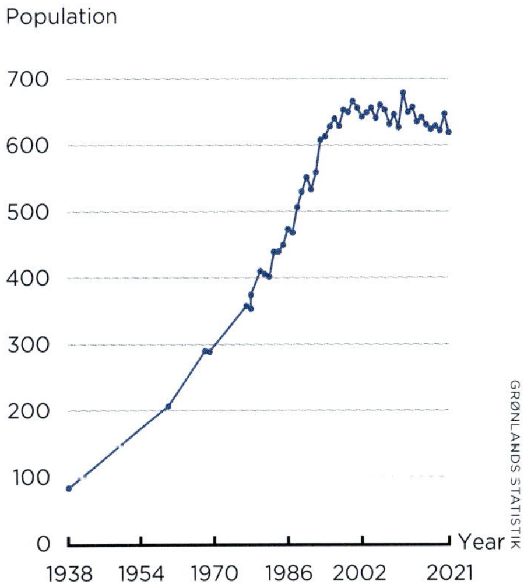

Figure 51. Population trends in Qaanaaq 1938-2021.

KIM INSUK/VISIT GREENLAND, 2019

Qaanaaq seen from south to north. The photo is taken from the newly constructed pier, which allows slightly easier access for ships to dock and manage cargo and passengers.

Literary Expedition by sled across Qimusseriarsuaq (Melville Bay) from 1903 to 1904. The Thule trading station at the Uummannaq settlement was established in 1910 to secure supplies for the population – and to ensure that the area became Danish. The trading station became the starting point for Knud Rasmussen's Thule expeditions, and Northwest Greenland north of Qimusseriarsuaq was named Thule in Danish.

Thule and the Thule expeditions had great importance in the inter-war period for Danish self-understanding, but despite several attempts, the running of the trading station was not taken over by the state until 1937, and the district was not given the same status as the old colonies; for example, Thule did not get status of a municipality but instead a Hunters' Council was established.

In 1951, the United States established Thule Air Base (Pituffik) on the great plain a few kilometres east of the Uummannaq settlement, and wanted to expand at the settlement in 1953. In May 1953, while it was still Arctic winter, Danish officials told the 116 Inughuit (27 families) to leave the area with few days' notice, and said that they would be given new houses at Qaanaaq. Therefore, 13 families travelled the 120 km north to Qaanaaq by dog sleds, while the other families travelled to other settlements in the district.

With the new Danish Constitutional Act of 5 June 1953, Greenland formally became an equal part of the Danish kingdom. It has later been discussed whether the reason for the hasty relocation was the desire not to clash with the Danish Constitutional Act.

Development and geography of Qaanaaq

On arrival in Qaanaaq, there was nothing. It was not until July, when it was possible to transport materials and workmen by ship, that a frantic establishment of the new Thule started, which was designated as the district's new centre.

The small replacement houses of between 14 and 24 m² built for the

Inughuit along the beach were not finished until September, when winter had set in – until then the families lived in tents. In the summer of 1954, the 'Danish town' with the necessary infrastructure, such as school, shop, telegraph station, etc., as well as dwellings for the Danish officials and their families were established along the main road perpendicular to the coast. The official residences in the Danish town were considerably larger and, as the first in Greenland, they were provided with district heating, while the replacement houses had coke stoves. Roughly speaking, the town has retained its structure with the community infrastructure gathered east of the river, while most of the development was seen west of the river.

Qaanaaq lies on a moraine slope with permafrost by a river in which water runs during the four summer months but is otherwise dry. In summer, two large water tanks are filled with river water for the winter, but this far from covers the need. Therefore, ice from icebergs is harvested on the sea ice with a dumper truck and wheel loader, and the ice is melted in a smelter. With climate change and thinner sea ice, this task becomes more dangerous and is at the same time extremely costly with a production price of around DKK 700 per cubic metre of water. Therefore, for the storage of river water, another water tank has been established, which was finished in the autumn of 2021.

The town has no sewer for black water, and the entire town, including the hospital, uses toilet buckets and collection of latrines. In the eastern part of the town, there are two parallel sewer lines for grey water. Both lines start at the hospital. Only the easternmost part, including the school, is connected and is today fully operational after a renovation in 2021. In the remaining part of town, the grey water is discharged to the ground, where in winter it creates ice formation at the foot of the houses.

Off the town is a reef that protects the coastline from swells, but also means that everything has to be barged in and out as there is no draught to a port. To smooth the barging, a pier was established to the reef in 2020.

The town is founded on permafrost, and this presents major challenges that are accelerating with climate change. Many of the houses, particularly west of the river, have suffered considerable subsidence damage, some even to such an extent that they are uninhabitable. This is a contributory cause of the town's major housing shortage, with many young people having to stay with their parents in small houses, even after they have had children themselves.

The sea ice, which is seen most of the year, obstructs shipping and navigation, so Qaanaaq and the district's three settlements receive only two annual supply ships in July and early September, respectively.

In 2001, a gravel runway was established about 7 km west of the town to ensure better connection to the rest of Greenland. Until then, the only connection was helicopter service to Pituffik (Thule Air Base) and aircraft from there, implying some limitations and requiring a permit by the US military. On the other hand, back then the airfare to the country's traffic hub Kangerlussuaq was consider-

Figure 52. Distribution of jobs by industry sector for people working in the town of Qaanaaq, the settlements in Qaanaaq district and Greenland as a whole, 2020.

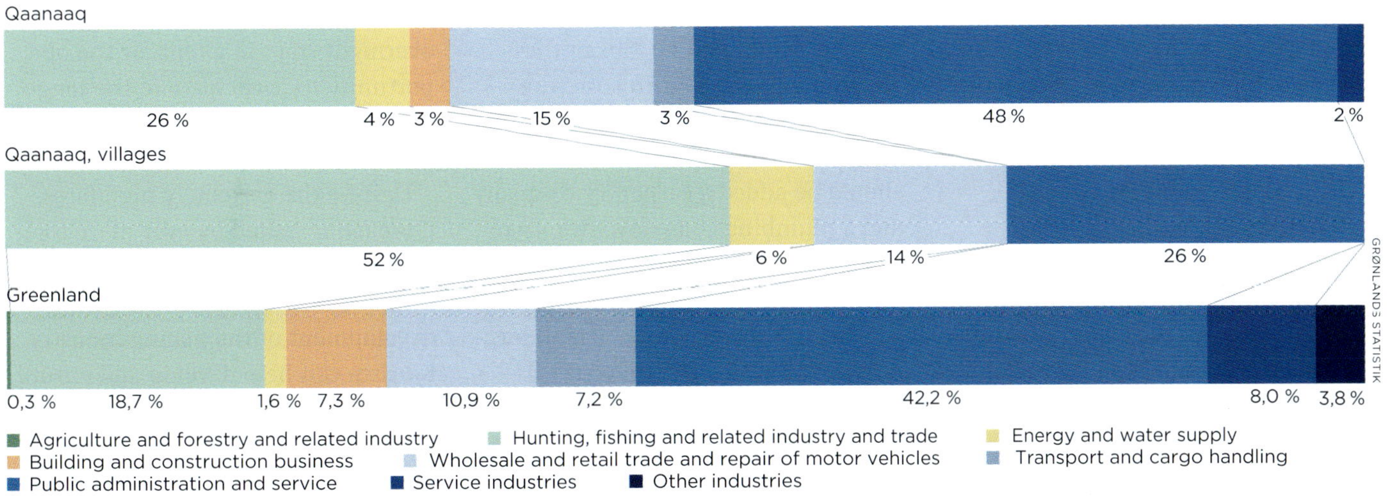

KÅRE HENDRIKSEN, 2015

On the way from Qaanaaq to Siorapaluk. In the Qaanaaq district, a considerable amount of transport between settlements is done by dog sled and the dog sled is used to transport the catch and in connection with Greenland halibut fishing.

ably cheaper than now, where, for example, a return flight to Nuuk costs around DKK 18,500 and requires an overnight stay each way. In summer, the road between the town and the airport is sometimes washed away during heavy thaw and rainstorms.

Business base of Qaanaaq

Since its establishment, the town has grown and its population peaked in 2011 at 678. The population has since decreased slightly, partly because some administrative jobs moved to Ilulissat following the municipal amalgamation in 2009.

The main industry – or the one that employed the most men – was until the beginning of the new millennium hunting of marine mammals such as seal, walrus, narwhal and beluga whale, as well as muskox, caribou, polar bears and birds. But conservation regulations have limited hunting, and the CITES regulation has reduced opportunities for the export of skins and tusks while the population has increased. It has reduced the income opportunities of the hunters. However, hunting remains an essential part of the business base for several families and contributes greatly to the town's food supply. The district has about 75 registered commercial hunters.

A small fish factory was established in Qaanaaq in the late 1980s, and since then there has been increasing longline fishing of Greenland halibut from the sea ice. This gives some hunters a monetary supplement. The factory has been expanded several times and is today operated in partnership between Royal Greenland and local company Inughuit Seafood. In winter, during the fishing season, there are about 12 jobs at the factory.

Furthermore, there are about 55 municipal jobs and 40 jobs in the government-owned infrastructure companies in the town and the three settlements of the district, to which should be added a carpentry company and a plumbing company with up to 15 employees during the season, as well as a small hotel and occasionally a kiosk and corresponding small businesses. Including the hunters, this means that there are about 230 jobs in the district with 474 (2021) people of working age.

Overall, Qaanaaq is the poorest district in Greenland with the lowest average household income, while average social benefits do not reflect the fact that such a large proportion of the population is outside the labour market, which can be attributed to hunting and its importance for subsistence economy.

Tourism

Qaanaaq district with the beautiful High Arctic scenery holds unique tourism development potentials that cater to a growing international segment. Summer allows for kayaking trips together with experienced hunters who still hunt narwhals from their kayaks, as well as longer sailing trips possibly combined with seal hunting and caribou hunting. The period of ice cover, which continues right through mid-July, offers long sled trips in the glow of the midnight sun. Several hunters offer sled hunting trips of up to several weeks with overnight stays in a tent, or the opportunity to participate in the longline fishing of Greenland halibut from the sea ice.

Despite the extremely high fares, Qaanaaq is visited by ever more tourists who want unique nature and cultural experiences. Combined with the development of the fishing industry, tourism can help develop and ensure a business base and livelihood for the Inughuit – the world's northernmost indigenous population group.

Nuna asiilasooq
Kalaallit nunagaarput
tamarmi qaqqartooq
Kangerluppassuit sinai
nunassaqqissisippaat,
tamaat sineriaa
qeqertat saangerpaat.

Land of great length,
That's what the Kalaallit have as a country,
among entire mountains,
The many fjords on the coast,
well suited as a country,
and the coast is speckled
with islands.

Nuannersoqaqaaq
angallavigigaanni
tamaat sineriak
Avannamut kujammullu
inunnik naapitsiffik
nunaqarfinnillu
uningavissalik.

It is so beautiful
when you journey
the entire coast.
Up north and down south,
there is a meeting place for people
and wherever you have land
there is a place to live.

Qaqqaasa saavini
kangerluillu paani
unipput inuii
Imartik pissaqarfigaat
atorluakkaminnik,
Kalaallimmi pigaat
soraajuerlutik.

At the feet of her mountains
and the mouths of her fjords
her inhabitants meet each other.
The sea is their domain
which they treat so well
It belongs to the Kalaallit
until the end of time!

Nuna asiilasooq, Greenland's unofficial national anthem, Jonathan Petersen, approx. 1920, unofficial translation.

AUTHORS OF THE INDIVIDUAL ARTICLES

LARGE BOXES AND THEMES ARE INDICATED IN ITALICS

NATURE AND LANDSCAPE

Chapter editor: Bo Elberling

Bo Elberling (CHAPTER INTRO), Andreas Westergaard-Nielsen and Jakob Abermann (CLIMATE), Dorthe Dahl-Jensen (THE ICE SHEET) Bo Elberling (*SCIENTIFIC EXPLORATION IN A HISTORICAL PERSPECTIVE*), Minik Rosing (GEOLOGY), Bo Elberling (*IKAITE COLUMNS AT IVITUTT*), Martin Ghisler (*GREENLANDIC GEMSTONES*), Bo Elberling (THE ICE-FREE LANDSCAPES), Elmer Topp-Jørgensen, Tom Christensen, Katrine Raundrup (FLORA AND FAUNA ON LAND), Elmer Topp- Jørgensen, Tom Christensen, Katrine Raundrup (*LEMMINGS AND RELEASE OF MUSKOXEN*), Mads Peter Heide-Jørgensen (*SEABIRDS*), Jens Fog Jensen (THE CULTURAL LANDSCAPES), Kirsten S. Christoffersen, Sille M. Myreng and Ole Geertz-Hansen (THE FRESH WATERS), Mads Peter Heide-Jørgensen (THE SEA AND THE FJORDS), Mads Peter Heide-Jørgensen (*LIFE IN THE OPEN WATERS – GREENLAND SHARK – THE HIGHLY DIVERSE BOTTOM FAUNA – SEALS IN THE GREENLANDIC WATERS – WHALE SPECIES AND WHALE HUNTING – THE POLAR BEAR*), Aart Kroon and Verner B. Ernstsen (COASTS), Bo Elberling and Kenneth Høegh (FARMING), Karen Nørregaard (*HORTICULTURE*), Henrik Meilby, Erik Dahl Kjær and Kenneth Høegh (*TREE PLANTING*), Elmer *Topp-Jørgensen, Tom Christensen and Katrine Raundrup* (BIODIVERSITY AND NATURE MANAGEMENT), Bo Elberling (*SYSTEMATIC ENVIRONMENTAL MONITORING*), Bo Elberling (*THE NATIONAL PARK IN NORTH AND EAST GREENLAND*), Bo Elberling (*SEASONAL POLLUTION IN NORTHEAST GREENLAND*).

HISTORY

Chapter editor: Inge Høst Seiding

Inge Høst Seiding (CHAPTER INTRO), Trap Denmark (*HISTORICAL OVERVIEW*), Mari Kleist (PALEO-INUIT), Mari Kleist and Pauline K. Knudsen (*THE STRONG TUNIT PEOPLE*), Christian Koch Madsen and Jette Arneborg (THE NORSE), Kirstine Møller, Asta Mønsted and Frederik Fuuja Larsen (THE INUIT CULTURE – PRE-COLONIAL PERIOD), Kirstine Møller, Asta Mønsted and Frederik Fuuja Larsen (*PRE-COLONIAL INUIT CULTURE— THE WOMAN — DUALITY AND BALANCE*), Bo Elberling (*MAPPING OF GREENLAND*), Kirstine Møller, Asta Mønsted and Frederik Fuuja Larsen (*THE MUMMIES FROM QILAKTISOQ (...)*), Marianne Jensen (THE COLONIAL PERIOD UNTIL THE WAR YEARS), Christian Koch Madsen and Jette Arneborg (*KUJATAA – FARMING ON THE BRINK OF THE ICE SHEET*), Knud Michelsen (*THE POLAR EXPLORER KNUD RASMUSSEN*), Jens Heinrich, Vivi Noahsen and Einar Lund Jensen (THE WAR YEARS AND SUBSEQUENT DECOLONISATION), Jens Heinrich, Vivi Noahsen and Einar Lund Jensen (HOME RULE (1979-2008)), Kåre Hendriksen (*POPULATION TRENDS FROM 1886 TO 2021*).

CULTURE

Chapter editor: Daniel Thorleifsen

Lisathe Møller (LANGUAGE), Arnaq Grove and Evi Kreutzmann (TRADITIONS AND TALES), Asta Mønsted (*INUIT WAY OF LIFE*), Arnaq Grove (*FOUR TALES ILLUSTRATED BY ARON OF KANGEQ – THE TALES OF ANAQORSUAQ AND QIVAAQIARSUK*), Naja Dyrendom Graugaard, Astrid Oberborbeck Andersen and Regine Møller (HUNTING CULTURE), Aviaja Lyberth Hauptmann and Natuk Lund Olsen (FOOD CULTURE), Daniel Thorleifsen and Bo Albrechtsen (MUSEUMS OF CULTURAL HISTORY AND HERITAGE),Bodil Karlshøj Poulsen (ASSOCIATION ACTIVITIES AND VOLUNTEERING), Flemming A. J. Nielsen (RELIGION AND RELIGIOUS COMMUNITIES), Thorkild Kjærgaard (*CHURCH BUILDING*) Inge Bisgaard (BUILDING CUSTOMS AND ARCHITECTURE), Nivi Christensen (VISUAL ARTS AND CRAFTS —*GREENLAND PAINTERS – HANS LYNGE AND JENS ROSING – TUPILAK*), Rosannguaq Rossen (FASHION AND DESIGN), Susanne Andreasen, Ruth Montgomery-Andersen and Arnajaraq Støvlbæk (THEATRE AND DANCE), Susanne Andreasen, Ruth Montgomery-Andersen, Arnajaraq Støvlbæk, Andreas Otte, Ununnquaq Petrussen, Poul Ringsted and Arnaq Grove (*DRUM DANCE AND DRUM SONG*), Susanne Andreasen, Ruth Montgomery-Andersen and Arnajaraq Støvlbæk (*MAKKA KLEIST*), Andreas Otte, Inunnguaq Petrussen and Poul Ringsted (MUSIC), Susanne Andreasen, Ruth-Montgomery-Andersen, Arnajaraq Støvlbæk, Andreas Otte, Inunnquaq Petrussen and Poul Ringsted (*KALATTUUT – GREENLANDIC STYLE*), Jette Rygaard and Susan Vanek (FILM), Arnaq Grove, Jette Rygaard and Susan Vanek (*THE WEDDING OF PALO*), Jette Rygaard and Aviaq Fleischer (LITERATURE), Jette Rygaard and Naja Paulsen (MEDIA).

SOCIETY AND BUSINESS

Chapter editor: Kuupik V. Kleist

Kuupik V. Kleist (INTRODUCTION), Kåre Hendriksen (*THE GREENLANDIC IN-*

SULAR COMMUNITY), Ulrik Pram Gad, Annemette Nyborg Lauritsen, Jon Rahbek-Clemmensen, Ella Loft, Mette Marie Stæhr Harder and Søren Thuesen (THE UNITY OF THE REALM AND THE DANISH STATE – *OTHER GOVERNMENT INSTITUTIONS – JOINT ARCTIC COMMAND*), Rasmus Leander Nielsen, Ulrik Pram Gad, Mitdlarak Lennert, Annemette Nyborg Lauritsen and Sophie Rud (SELF-GOVERNMENT), Ulrik Pram Gad and Sophie Rud (*PREMIERS OF NAALAKKERSUISUT*), Birger Poppel and Mariekathrine Poppel (POPULATION AND DEMOGRAPHY), Kåre Hendriksen (INFRASTRUCTURE), Tove Lading (HOUSING), Najaaraq Christiansen, Kåre Hendriksen, Birger Poppel (INDUSTRY AND LABOUR MARKET), Bo Elberling (*CONTINENTAL SHELF PROJECT*), Mitdlarak Lennert and Lars Demant-Poort (EDUCATION), Naja Paulsen and Jette Rygaard (*JOURNALISM PROGRAMME*), Christina Viskum Lytken Larsen, Ingelise Olesen (HEALTH AND CARE – PEQQINEQ – HEALTH AND BALANCE), Ulrik Pram Gad (PLANS).

MUNICIPALITIES AND TOWNS
Chapter editors: Bo Naamansen and Kåre Hendriksen

Kåre Hendriksen and Bo Naamansen (CHAPTER INTRO – *TOWN NAMES IN GREENLAND – EMPLOYMENT IN DISTRICTS AND TOWNS* – THE FIVE REGIONAL MUNICIPALITIES – SETTLEMENTS – SETTLEMENTS WITH AGRICULTURE AND SHEEP FARMING), Kåre Hendriksen and Najaaraq Demant-Poort (KOMMUNE KUJALLEQ), Bo Naamansen (*POPULATION TRENDS*), Kåre Hendriksen (NANORTALIK), Alibak Hard (QAQORTOQ), Kåre Hendriksen and Kenneth Høegh (NARSAQ), Bo Naamansen and Ole Guldager (NARSARSUAQ), Kåre Hendriksen and Najaaraq Demant-Poort (KOMMUNE QARGIK SERMERSOOQ), Kåre Hendriksen and Bo Naamansen (TASIILAQ), Kåre Hendriksen, Bo Naamansen and Kristian Hammeken (ITTOQQORTOORMIIT), Bo Naamansen (PAAMIUT – *NUUK*), Kåre Hendriksen and Najaaraq Demant-Poort (QEQQAA KOMMUNIA), Kåre Hendriksen and Bo Naamansen (MANIITSOQ), Kåre Hendriksen (SISIMIUT – *WORLD HERITAGE: AASIVISSUIT-NIPISAT*), Bo Naamansen (KANGERLUSSUAQ), Kåre Hendriksen and Najaaraq Demant-Poort (KOMMUNE QEQERTALIK), Kåre Hendriksen (KANGAATSIAQ – AASIAAT), Anne Mette Olsvig (QASIGIANNGUIT), Kåre Hendriksen (QEQERTARSUAQ), Kåre Hendriksen and Najaaraq Demant-Poort (AVANNAATA KOMMUNIA), Kåre Hendriksen and Karl Ejnar Madsen (ILULISSAT), Kåre Hendriksen (UUMMANNAQ – UPERNAVIK – QAANAAQ).

TRAP GREENLAND'S EDITORIAL TEAM

Niels Elers Koch
Editor-in-chief and Managing Director, Professor, dr.agro., dr.h.c.

Dorthe Gartmann Christiansen
Head of Secretariat, HR Manager

Erik Henz Kjeldsen
Head of Development

Rebecca Bailey
Volume and Text Editor, MA

Bo Elberling
Chapter editor of the chapter Nature and landscape, Professor, PhD in geology and D.Sc.

Mikkel Hytteballe Engell
Student Assistant

Vibeke Højgaard Frederiksen
Image Editor

Alice P. Halle
Senior Controller

Anders Tørring Hansen
Editorial Assistant

Anja Hørmann Hansen
Graphic Designer, production coordinator

Kåre Hendriksen
Chapter editor of the chapter Municipalities and towns, associate professor, PhD.

Kuupik V. Kleist
Chapter Editor of the chapter Society and business, former Premier, sociologist

Peter Lyth
Head of Communications, MSc (Econ. and Business Adm.)

Thea Martoft
Editorial Assistant

Lærke Termøhlen Mikkelsen
Digital Editor, MA

Bo Naamansen
Chapter editor of the chapter Municipalities and towns, director of Asiaq, MSc (Surveying, Planning and Land Management).

Inge Høst Seiding
Chapter Editor of the chapter History, Head of the Institute of Culture, Language & History, University of Greenland, PhD

Daniel Thorleifsen
Chapter editor of the chapter Culture, Director of the Greenland National Museum and Archives, MA

Emma Vestergaard
Student Assistant

TRAP DENMARK'S PROFESSIONAL COMMITTEE

Asbjørn Hellum (Chairman)
Former Director General of the Danish State Archives, The National Archives

Per Boje
Professor, D.Phil., University of Southern Denmark

Birte Christensen-Dalsgaard
Project Manager, lic.scient., DighumLab

Jens Morten Hansen
Former state geologist, Adjunct Professor, lic.scient., GEUS (Geological Survey of Denmark and Greenland)

Ove Hornby
Director, MA

Svend Larsen
Director, Royal Danish Library

Jørn Lund
Professor

Niels Lunde
Editor-in-chief, Dagbladet Børsen

Erling Friis Poulsen
Head of department, KL (Local Government Denmark)

SUBJECT EXPERTS, ETC.

Abermann, Jakob (b. 1981)
Dr. rer. nat. Assistant Professor at the Department of Geography and Regional Science, University of Graz, and project leader Asiaq, Greenland Survey.

Albrechtsen, Bo (b. 1968)
MA. Director of the Greenlandic House in Aalborg.

Andersen, Astrid Oberborbeck (b. 1979)
Ph.D. in Social Anthropology. Associate Professor at the Department of Culture and Learning, Aalborg University.

Andreasen, Susanne (b. 1978)
BA in Dramaturgy. Artistic Director for the National Theatre of Greenland.

Arneborg, Jette (b. 1952)
MA and Ph.D. in History and Medieval Archeology. Senior Researcher at the National Museum of Denmark.

Bisgaard, Inge (b. 1958)
MA in Architecture. Museum Curator at the Greenland National Museum & Archives.

Christensen, Nivi (b. 1987)
MA in Art History. Museum Director at Nuuk Art Museum.

Christensen, Tom (b. 1969)
MSc. in Biology. Chief Consultant the Department of Ecoscience, Aarhus University.

Christiansen, Najaaraq (b. 1982)
MSc. in Public Administration. Chief Consultant at Statistics Greenland.

Christoffersen, Kirsten Seestern (b. 1956)
Ph.D. in Biology. Professor at the Department of Biology, University of Copenhagen.

Dahl-Jensen, Dorthe (b. 1958)
Ph.D. and MSc. in Geophysics. Professor at the Niels Bohr Institute, University of Copenhagen.

Demant-Poort, Lars (b. 1976)
Ph.D. and MA in Education. Associate Professor at the University of Greenland.

Demant-Poort, Najaaraq (b. 1980)
MSc. in Political Science. Adjunct with Ph.D.-Course at the University of Greenland.

Elberling, Bo (b. 1968)
Ph.D. in Geology and dr.scient. in Geography. Professor at the Department of Geosciences and Natural Resource Management, University of Copenhagen.

Ernstsen, Verner Brandbyge (b. 1975)
Ph.D. in Marine Geology and MSc. in Physical Geography. Senior Researcher at GEUS (The Geological Survey of Denmark and Greenland).

Fleischer, Aviaq (b. 1985)
MA in Language, Literature and Media. Assistant Professor and Ph.D.-student at the Department of Language, Literature & Media, University of Greenland.

Gad, Ulrik Pram (b. 1973)
MSc. in Political Science and Ph.D. Senior Researcher at the Danish Institute for International Studies.

Geertz-Hansen, Ole (b. 1957)
Ph.D. in Freshwater Biology. Senior Scientist at the Greenland Institute for Natural Resources.

Ghisler, Martin (b. 1940)
Dr.phil. and Cand.scient. Emeritus in GEUS (The Geological Survey of Denmark and Greenland).

Graugaard, Naja Dyrendom (b. 1986)
Ph.D. in Culture and Learning. Postdoc at Department of Planning, Aalborg University.

Grove, Arnaq (b. 1951)
Ph.D. and MA in Danish and Communication. Associate Professor at the University of Greenland.

Guldager, Ole (b. 1968)
MSc. in Prehistoric Archeology. Curator at Narsaruaq Museum.

Hammeken, Kristian (b. 1978)
Cand.polyt. in Environmental Engineering and Bachelor in Engineering in Arctic Technology. Municipal Engineer at Sermersooq Municipality.

Hard, Alibak (b. 1971)
Service economist. Tourism consultant.

Harder, Mette Marie Stæhr (b. 1979)
Ph.D. in Social Sciences and MSc. in Political Science. Senior Lecturer at the Department of Political, Historical, Religious and Cultural Studies, Karlstad University, and Carlsberg postdoc from Department of Law, University of Copenhagen.

Hauptmann, Aviaja Lyberth (b. 1986)
Ph.D. and Microbiologist. Assistant Professor at the University of Greenland.

Heide-Jørgensen, Mads Peter (b. 1957)
Biologist. Professor, Dr.Scient., at the Greenland Institute for Natural Resources.

Heinrich, Jens (b. 1973)
Ph.D. and MA. Head of Greenland Representation in Copenhagen.

Hendriksen, Kåre (b. 1956)
Ph.D. and MA in Technical Environmental Management. Associate Professor in Sustainable Arctic Infrastructure at the Department of Planning, Aalborg University.

Høegh, Kenneth (b. 1966)
MSc. in Agriculture. Head of Greenland Representation in Washington, DC.

Jensen, Einar Lund (b. 1949)
Ph.D. in Arctic and Greenlandic Studies and MA in History and Greenlandic. Former Museum Curator at the National Museum of Denmark.

Jensen, Jens Fog (b. 1963)
Ph.D. in Prehistoric Archeology. Senior Researcher at the National Museum of Denmark.

Jensen, Manumina Lund (b. 1976)
MA. Ph.D.-student at the University of Greenland.

Jensen, Marianne (b. 1949)
MA in Cultural and Social History. Retired.

Kjær, Erik Dahl (b. 1964)
Dr.agro. and MSc. in Forest and Nature Management. Professor at the University of Copenhagen.

Kjærgaard, Thorkild (b. 1945)
Dr.phil. Historian. Taught until retirement at the University of Greenland.

Kleist, Kuupik V. (b. 1958)
Sociologist. President at ICC Greenland.

Kleist, Mari (b. 1974)
Ph.D. Archeologist. Associate Professor at the Department of Cultural and Social History, University of Greenland.

Knudsen, Pauline Kleinschmidt (b. 1960)
Archaeologist. Head of the Greenland Visitor Center.

Kreutzmann, Evi (b. 1979)
MA in Language, Literature and Media. School Advisor on History at National Education Agency.

Kroon, Aart (b. 1964)
Ph.D. in Physical Geography. Professor at the University of Copenhagen.

Lading, Tove (b. 1955)
Architect. Lecturer in Arctic Construction at DTU Construct (The Technical University of Denmark).

Larsen, Christina Viskum Lykten (b. 1980)
Ph.D. and MSc. in Sociology. Senior Researcher and Head of Health in Greenland, National Institute of Public Health, University of Southern Denmark, and Adjunct Senior Researcher at the Greenland Center for Health Research, Institute of Health and Nature, University of Greenland.

Larsen, Frederik Fuuja (b. 1967)
MA in Cultural and Social History. Museum Curator at the Greenland National Museum & Archives.

Lauritsen, Annemette Nyborg (b. 1965)
Ph.D. and MSc. in Social Sciences and Cultural Encounters. Head of Institute and Associate Professor, Ph.D., at the Institute of Social Science, Economics and Journalism, University of Greenland.

Lennert, Mitdlarak (b. 1988)
Ph.D. External Lecturer at the University of Greenland.

Loft, Ella Sigrid (b. 1994)
Stud.scient.pol. Intern at the Center for Arctic Security Studies, Royal Danish Defence College.

Madsen, Christian Koch (f. 1978)
Ph.D. Deputy Director at the Greenland National Museum & Archives.

Madsen, Karl Ejnar (b. 1978)
MSc. in Geography. Director at the Technical Department at Avannaata Municipality.

Markussen, Ulunnguaq (b. 1991)
MSc. in Social Sciences. Ph.D.-student at the University of Greenland.

Meilby, Henrik (b. 1963)
Ph.D. and MSc. in Forest and Nature Management. Professor at the University of Copenhagen.

Meldgaard, Morten (b. 1956)
Ph.D. and Dr.phil. Professor at the University of Greenland.

Michelsen, Knud (b. 1945)
MA. Senior Researcher at the Danish Arctic Institute.

Montgomery-Andersen, Ruth (b. 1957)
Doctor of Public Health, Midwife, Dancer, and Choreographer. Researcher and Project Manager for Greenland's first dance educational program.

Myreng, Sille Marie (b. 1989)
MSc. in Geography. Hydrologist at Asiaq, Greenland Survey.

Møller, Kirstine Eiby (b. 1987)
MA in Sustainable Heritage Management. Ph.D. at the Greenland National Museum & Archives.

Møller, Lisathe (b. 1973)
MA in Language, Literature and Media. Self-employed.

Møller, Regine (b. 1990)
MSc. in Social Sciences. Ph.D.-student at Aalborg University.

Mønsted, Asta (b. 1985)
Ph.D. in Prehistoric Archaeology, National Museum of Denmark.

Nielsen, Flemming A. J. (b. 1964)
Ph.D. and MTheol. Associate Professor and Head of the Department of Theology, University of Greenland.

Nielsen, Rasmus Leander (b. 1975)
MSc. in Political Science and Ph.D. Lecturer at the University of Greenland.

Noahsen, Vivi (b. 1988)
MA in Cultural and Social History. Head of Archives at the Greenland National Museum & Archives.

Nørregaard, Karen (b. 1942)
MPhil in Eskimology. Retired.

Naamansen, Bo (b. 1968)
MSc. in Surveying, Planning and Land Management. General Manager of Asiaq, Greenland Survey.

Olesen, Ingelise (b. 1961)
Midwife and MA in Health Promotion and Education. Research Coordinator at Health in Greenland, National Institute of Public Health, University of Southern Denmark, Institute of Health and Nature, University of Greenland.

Olsen, Natuk Lund (b. 1979)
Ph.D.-student. Head of Division at Government of Greenland.

Olsvig, Anne Mette (b. 1955)
Exam.pæd.fag., DPU, Head of Qasigiannguit Museum.

Otte, Andreas (b. 1983)
Ph.D. within Anthropology of Music. Communicator at the North Atlantic House, Odense.

Paulsen, Naja (b. 1959)
Journalist and MA in Rhetorics. Head of Department at the Department of Journalism, University of Greenland.

Petrussen, Inunnguaq (b. 1986)
MSc. in Public Administration. Political Advisor at the Parliament of Greenland.

Phil, Anton (b. 1980)
Curator at Greve Museum. President of the Nordic Flag Society.

Poppel, Birger (b. 1950)
MA in Economics. Project Chief Emeritus, Survey of Living Conditions in the Arctic (SLiCA) at Institute of Learning, University of Greenland.

Poppel, MarieKathrine (b. 1952)
Ph.D. and MA in Public Administration. Emerita at the University of Greenland.

Poulsen, Bodil Karlshøj (b. 1951)
Primary School Teacher and Master in Public Health. Retired.

Rahbek-Clemmensen, Jon (b. 1982)
Ph.D. and MSc. in Political Science. Associate Professor at the Center for Arctic Security Studies, Royal Danish Defence College.

Raundrup, Katrine (b. 1976)
Ph.D. in Biology. Researcher at the Greenland Institute for Natural Resources.

Ringsted, Poul Ole (b. 1937)
Primary School Teacher. Retired Head Teacher at Government of Greenland.

Rosing, Minik Thorleif (b. 1957)
Ph.D. in Geology. Professor at the Globe Institute, University of Copenhagen.

Rossen, Rosannguaq (b. 1982)
Ph.D. in Arctic Studies. Head of Department and Assistant Professor at the Department of Language, Literature and Media, University of Greenland.

Rud, Sophie (b. 1995)
MA-student at Aalborg University. Student Assistant at the Danish Institute for International Studies.

Rygaard, Jette (b. 1948)
Ph.D., mag.art. and cand.mag. in Literary Studies and Nordic Philology, Aalborg University. Associate Professor Emerita, University of Greenland.

Seiding, Inge (b. 1976)
Ph.D. Head of Institute and Associate Professor at the Institute of Culture, Language and History, University of Greenland

Støvlbæk, Arnajaraq (b. 1976)
BA from the Iceland Academy of the Arts. Self-employed.

Thorleifsen, Daniel (b. 1962)
MA. Director of the Greenland National Museum & Archives.

Topp-Jørgensen, Elmer (b. 1971)
MSc. in Biology. Special Consultant at the Department of Ecoscience, Aarhus University.

Thuesen, Søren T. (b. 1958)
Ph.D. and MA. Associate Professor at the Greenlandic and Arctic Studies Section, Department of Cross-Cultural and Regional Studies, University of Copenhagen.

Vanek, Susan B (b. 1987)
Master. Ph.D.-candidate at Binghamton University (State University of New York).

Westergaard-Nielsen, Andreas (b. 1981)
Ph.D. in Geography. Assistant Professor at the Department of Geosciences and Natural Resource Management, University of Copenhagen.

MAPS

Dam, Peder (b. 1976)
Ph.d. Museum curator, Museum Odense.

Kristensen, Rune Lyngbo (b. 1978)
GIS and Geodata specialist. Affiliated Asiaq, Greenland Survey.

FAKTATJEK

Andreasen, Jette (b. 1954)
Librarian.

TRANSLATION

Adhoc Translations

PLACE NAME INDEX

In case of wider mention of the locality, the reference is marked with bold.

T

U

V

W

Z

Ø

NAME INDEX

K

L

M

N

O

P

Q

R

S